Lecture Notes in Computer Science

Edited by G. Goos, Karlsruhe and J. Hartmanis, Ithaca
Series: I.F.I.P. TC7 Optimization Conferences

3

5th Conference on Optimization Techniques

Part I

Edited by R. Conti and A. Ruberti

Springer-Verlag
Berlin · Heidelberg · New York 1973

Prof. Dr. R. Conti
Istituto di Matematica
"Ulisse Dini"
Università di Firenze
Viale Morgagni 67/A
I-50134 Firenze/Italia

Prof. Dr. Antonio Ruberti
Istituto di Automatica
Università di Roma
Via Eudossiana 18
I-00184 Roma/Italia

AMS Subject Classifications (1970): 49-02, 49A40, 49B20, 49B35, 49B40, 49C05, 49C10, 49DXX, 65K05, 68A25, 68A45, 90-02, 90C05, 90C10, 90C20, 90C30, 90C50, 90C99, 90D05, 90D25, 90D99, 93-02, 93B05, 93B10, 93B20, 93B30, 93B35, 93B99, 93C20, 93E05, 93E20

ISBN 3-540-06583-0 Springer-Verlag Berlin · Heidelberg · New York
ISBN 0-387-06583-0 Springer-Verlag New York · Heidelberg · Berlin

Offsetprinting and bookbinding: Julius Beltz, Hemsbach/Bergstr.

PREFACE

These Proceedings are based on the papers presented at the 5th IFIP Conference on Optimization Techniques held in Rome, May 7-11, 1973. The Conference was sponsored by the IFIP Technical Committee on Optimization (TC-7) and by the Consiglio Nazionale delle Ricerche (Italian National Research Council).

The Conference was devoted to recent advances in optimization techniques and their application to modelling, identification and control of large systems. Major emphasis of the Conference was on the most recent application areas, including: Environmental Systems, Socio-economic Systems, Biological Systems.

An interesting feature of the Conference was the participation of specialists both in control theory and in the field of application of systems engineering.

The Proceedings are divided into two volumes. In the first are collected the papers in which the methodological aspects are emphasized; in the second those dealing with various application areas.

The International Program Committee of the Conference consisted of: R. Conti, A. Ruberti (Italy) Chairmen, Fe de Veubeke (Belgium), E. Goto (Japan), W. J. Karplus (USA), J. L. Lions (France), G. Marchuk (USSR), C. Olech (Poland), L. S. Pontryagin (USSR), E. Rofman (Argentina), J. Stoer (FRG), J. H. Westcott (UK).

Previously published optimization conferences:

Colloquium on Methods of Optimization. Held in Novosibirsk/USSR, June 1968.
(Lecture Notes in Mathematics, Vol. 112)

Symposium on Optimization. Held in Nice, June 1969.
(Lecture Notes in Mathematics, Vol. 132)

Computing Methods in Optimization Problems. Held in San Remo, September 1968.
(Lecture Notes in Operation Research and Mathematical Economics, Vol. 14)

TABLE OF CONTENTS

*paper not received

*paper not received

OPTIMAL CONTROL

STOCHASTIC CONTROL

MATHEMATICAL PROGRAMMING

NUMERICAL METHODS

*paper not received

Contents of Part II

(Lecture Notes in Computer Science, Vol. 4)

URBAN AND SOCIETY SYSTEMS

COMPUTER AND COMMUNICATION NETWORKS

---

* paper not received

BIOLOGICAL SYSTEMS

*paper not received

IDENTIFICATION OF SYSTEMS SUBJECT TO RANDOM STATE DISTURBANCE

by
A.V. Balakrishnan
Department of System Science
University of California, Los Angeles*

ABSTRACT

A theory of identification of a class of linear systems — lumped and distributed — in the presence of state or process noise. As a specific application, the problem of of identifying aircraft stability — control derivatives in turbulence is considered, and results obtained on actual flight data are included.

INTRODUCTION

The conscious use of mathematical models in system optimization has been growing rapidly in recent years. Perhaps the most spectacular example is the "world model" of Forrester [1]. If it is assumed that a model, before it can be used to predict the future, must first be verified on currently available data, then we are faced with the problem of "system identification" — of estimating unknown parameters in the model based on observed data. There is a large engineering literature (see [2]) on the subject of system identification and parameter estimation; much of it is however, in the nature of a collection of techniques with little or no precise mathematical framework. Often the authors make ad hoc simplifications for the avowed reason that the otherwise the mathematics involved is too complex. Whereas such a point of view may be harmless when theory plays a secondary role, as in the design of a physical system which can and is finally tested in the laboratory before the design is finalized, the situation is quite different in the identification problem where the end result is a set of numbers and we have really no means of being certain of their accuracy. Surely this would argue for extreme care in the mathematical formulation and analysis, and make the identification problem basically a mathematical and computational one.

In this paper we study the problem of identifying a linear system with time continuous description (rather than the discrete-time as in the bulk of the engineering literature), and furthermore taking into account unknown inputs (modelled as random state disturbance), usually omitted for reasons of mathematical complexity. Such a formulation turns out to be actually necessary in the practical problem of estimating aircraft parameters from flight data in turbulence, see [3].

More specifically, we can state the problem as follows: The observation $y(t;\omega)$, (in the usual notation, ω denotes the "sample point" corresponding to the random process) has the form:

$$y(t;\omega) = s(t;\omega) + n_1(t;\omega) \tag{1.1}$$

where $n_1(t;\omega)$ is the unavoidable measurement errors modelled as a white Gaussian process (as a reasonable idealization of a band-limited process of band-width large compared to that of the process $s(t;\omega)$). The system is assumed to be linear in its response to the (known) input $u(t)$, and to the unknown ("state") disturbance, assumed to be a physically realizable Gaussian process, and hence we can write:

$$s(t;\omega) = \int_o^t B(t-s)\ u(s)ds + \int_o^t F(t-s)\ n_2(s;\omega)ds \tag{1.2}$$

* Research supported in part under Grant No. 73-2492, Applied Mathematics Division, AFOSR U.S. Air Force.

where F(.) is determined from the spectral density of the disturbance process. We have omitted initial conditions, as we may, since we shall assume that the system is stable, that in fact:

$$||B(t)|| + ||F(t)|| = O(\exp -kt), \quad k > 0 \tag{1.3}$$

and allow for long observation time (only asymptotic properties of estimates are considered). We may also assume that the measurement noise in (1.1) is independent of the disturbance. The identification problem is that of estimating unknown parameters in the functions B(.), and F(.), from the observation $y(s,\omega)$, $0 \le s \le T$, and the given known input u(s), $0 \le s \le T$. Both $n_1(t,\omega)$ and $n_2(t,\omega)$ are white Gaussian with unit spectral density (matrix), and are independent of each other. It should be noted that instrument errors such as calibration error, bias error, time-delays etc., in the observation $y(t;\omega)$ are supposed to be known and corrected for.

Although seemingly simple, this formulation we hasten to point out, includes all known identification (and control) problems for linear systems. For example, the most commonly stated version:

$$\dot{x}(t) = A\, x(t) + B\, u(t) + F\, n_1(t;\omega) \tag{1.4}$$

$$y(t) = C\, x(t) + n_2(t;\omega) \tag{1.5}$$

is obviously of this form; in fact we readily see that:

$$F(t) = C\, e^{At}\, F$$

$$B(t) = C\, e^{At}\, B$$

and that (1.3) is satisfied if A is stable. The main simplification occurring in this example is of course that fact that the Laplace transforms of F(.) and B(.) are rational. In particular, (1.4) and (1.5) represent a "lumped parameter system," governed by ordinary differential equations. Our model includes also "distributed parameter" systems, governed by partial differential equations. For example, let G be a region in three-space dimensions with boundary Γ. Then we may consider a structure of the form (we omit the details):

$$\frac{\partial f}{\partial t} = \sum\sum a_{ij} \frac{\partial^2 f}{\partial x_i\, \partial x_j} + n(t,.,\omega) \tag{1.6}$$

$$f(t,.) \text{ on } \Gamma = u(t,.)$$

$$y(t) = C\, f(t,.) + n_1(t,\omega) \tag{1.7}$$

where C is a linear transformation with finite dimensional range. "Solving" (1.6), we see that (1.7) can again be expressed in the terms (1.1), (1.2). What is important to note is that the formulation (1.1), (1.2) requires only the specification (within unknown parameters) of the functions B(.) and F(.). Yet another example is one in which the state space is finite dimensional but the state noise does not have a rational spectrum. A practical instance of this is the problem of identifying aircraft parameters in turbulence characterized by the Von Karman spectrum. In other words, the Laplace transform of B(.) is rational, but:

$$\left| \int_o^\infty e^{2\pi i f t}\, F(t)dt \right|^2 = \frac{k\, f^4}{(a+f^2)^{17/6}}$$

Although we shall not go into it here, we wish to note that it is also possible to formulate stochastic control problems in terms of (1.1), (1.2). Thus we may seek to minimize

$$\int_o^T E[Qs(t;\omega),\ s(t;\omega)]\, dt$$

subject to constraints on the control $u(t)$, which is required to be adapted to the observation process $y(t,\omega)$. See [4] for more on this.

Estimation Theory

Let θ denote the vector of unknown parameters that we need to identify. We shall assume that there is a true value θ_o (in other words, no "model" error), and that it lies in a known interval, $I(\theta)$. Let $\hat{\theta}_T$ denote an estimate of θ_o based on $y(s)$, $0 \le s \le T$. We shall be interested only in estimates with the following two properties:

(i) asymptotically unbiased: $E(\hat{\theta}_T) \to \theta_o$

(ii) consistent: $\hat{\theta}_T$ converges in a suitable sense to θ_o as $T \to$ infinity

We shall show that it is possible to find an estimate with these properties by invoking the classical method of "maximum likelihood."

For this we can proceed along one of two [essentially equivalent, as we shall show] points of view concerning the noise process $n_1(t;\omega)$. In the first (the Ito or Wiener process point of view), we write (1.1) in the integrated version:

$$Y(t;\omega) = \int_o^t s(\sigma;\omega)d\sigma + W_1(t;\omega) \tag{2.1}$$

where $W_1(t;\omega)$ is a Wiener process on $C(0,T)$, the Banach space of continuous functions on $[0,T]$. Then for each fixed θ (assuming or "given" θ) the process $Y(t;\omega)$, $0 \le t \le T$ with sample functions in $C[0,T]$, induces a measure thereon which is absolutely continuous with respect to the Wiener measure (induced by $W_1(t,\omega)$). Moreover we can then calculate the corresponding Radon-Nikodym derivative, which is then our "likelihood" functional, denoted $p(Y;\theta;T)$, and the maximum likelihood estimate $\hat{\theta}_T$ is the value of θ which yields the maximum in I_θ (or, actually in a subinterval). The calculation of the derivative $p(Y;\theta;T)$ can be accomplished by one of two ways. The first method is to use the Krein factorization technique (see his monograph [5]). Thus let

$$m(t;\theta) = \int_o^t B(t-s)\, u(s)ds \tag{2.2}$$

and let

$$R(\theta;t;s) = E[(s(t;\omega) - m(t))(s(s;\omega) - m(s))^*] \tag{2.3}$$

E denoting expected value. Consider the operator

$$Rf=g; \quad g(t) = \int_o^T R(t;s)\, f(s)ds \tag{2.4}$$

mapping $L_2(0,T)$ into itself. Then R is non-negative definite and trace-class. Let I denote the identity operator. Then, following the procedure invented by Krein [5], we can "factorize" $(I+R)^{-1}$ as:

$$(I+R)^{-1} = (I-\mathscr{L}^*)\,(I-\mathscr{L}) \tag{2.5}$$

where $\mathscr{L}$ is a uniquely determined Volterra operator, such that furthermore, $(\mathscr{L}+\mathscr{L}^*)$ is trace-class (nuclear). The probabilistic interpretation is that the process:

$$Z(t;\omega) = \tilde{Y}(t;\omega) - \int_o^t \hat{\tilde{y}}(s;\omega)ds \tag{2.6}$$

where

$$\tilde{Y}(t;\omega) = Y(t;\omega) - \int_o^t m(s;\theta)ds \tag{2.7}$$

$$\hat{y}(t;\omega) = \int_o^t L(t;s)\ dY(s;\omega), \tag{2.8}$$

$L(t;s)$ being the kernel corresponding to $\mathscr{L}$ (and of course depends on θ), is a Wiener process. Moreover from this fundamental result, we can readily deduce that (cf [8]):

$$p(Y;\theta;T) = \exp\left(-\frac{1}{2}\right)\left\{\int_o^T ||\hat{y}(t;\omega) + m(t)||^2 dt - 2\int_o^T [m(t) + \hat{y}(t;\omega),\ dY(t;\omega)]\right\} \tag{2.9}$$

where the integral:

$$\int_o^T [\hat{\tilde{y}}(t;\omega),\ dY(t;\omega)] \tag{2.10}$$

is to be calculated as an Ito integral. This is fine, except that there is a fundamental practical difficulty in calculating (2.10) in that the observation is never actually such that $W_1(t;\omega)$ is a Wiener process. More accurately, one should model $n_1(t;\omega)$ as a band-limited process, of band large compared with that of $s(t;\omega)$. On the other hand, if we do this, then we cannot calculate the likelihood functional. Hence we may proceed in the following approximate way: we assume in theory that $W_1(t;\omega)$ which is band-limited, but of large bandwidth, so that, following [6], and exploiting the fact that $(\mathscr{L} + \mathscr{L}^*)$ is trace-class, we have the approximations:

$$\hat{\tilde{y}}(t;\omega) \doteq \int_o^t L(t;s)\left(y(s;\omega) - m(s;\theta)\right)ds$$

$$\int_o^T [m(t,\theta),\ dY(t;\omega)] \doteq \int_o^T [m(t;\theta),\ y(t;\omega)]dt$$

and the crucial approximation [see [6]]:

$$\int_o^T [\hat{\tilde{y}}(t;\omega),\ dY(t;\omega)] = \int_o^T [\hat{\tilde{y}}(t;\omega),\ y(t;\omega)]dt - \frac{1}{2}\mathrm{Tr.}\ (\mathscr{L}+\mathscr{L}^*) \tag{2.11}$$

Using these, we can finally write:

$$p(Y;\theta;T) = \exp\left(-\frac{1}{2}\right)\int^T ||\hat{\tilde{y}}(t;\omega) + m(t;\theta)||^2 dt - 2\int_o^T\left[\left(m(t;\theta) + \hat{\tilde{y}}(t;\omega)\right),\ y(t;\omega)\right]dt + \mathrm{Tr.}\ (\mathscr{L} + \mathscr{L}^*) \tag{2.12}$$

We shall next (briefly) show that we obtain the same expression for the likelihood functional, using a slightly different point of view, which may be somewhat closer to reality. Thus we take the view that $n_1(t;\omega)$ is of bandwidth sufficiently large that we can assume it to be "white noise"-$n_1(t;\omega)$ for each ω is an element of $L_2(0,T)$ with Gauss measure defined on the (field of) cylinder sets. Such a measure cannot be countably additive on the Borel sets. Denoting the measure by p_G, we have that ("weak distribution")

$$\int \exp i[n_1(.\omega),h] \quad dp_G = \exp - ||h||^2$$

for every h in $L_2(0,T)$. Our basic sample point ω can now be taken to be in the associated product Hilbert space H to accommodate $n_1(t,\omega)$ and $n_2(t;\omega)$, with Gauss measure thereon. Then we can rewrite (1.1), (1.2) as:

$$y(t;\omega) = m(t;\theta) + s(t;\omega) + G\, n(t;\omega) \tag{2.13}$$

$$s(t;\omega) = \int_0^t F(t-s)D\, n(s;\omega)ds \tag{2.14}$$

where

$$n(t;\omega) = \begin{bmatrix} n(t;\omega) \\ n(t;\omega) \end{bmatrix}$$

$$GG^* = I; \quad DG^* = 0; \; DD^* = I \text{ (Identity matrix in appropriate dimensions)}$$

Then with $\mathscr{L}$ defined as before, we can readily see that

$$\Big(y(t;\omega) - m(t;\theta)\Big) - \int_0^t L(t;s)\Big(y(s;\omega) - m(s;\theta)\Big)\, ds = z(t;\omega) \tag{2.15}$$

again defines white noise:

$$E\left(\exp i\left(\int_0^T [z(t;\ ),\ h(t)]\ dt\right)\right) = \exp - ||h||^2$$

for h in H. Allowing for finitely additive measures, we can again show that the measure induced by $y(t;\omega)$ is absolutely continuous with respect to the Gauss measure induced by $n_1(t;\omega)$, and that the derivative can be expressed:

$$\left(\exp\left(-\frac{1}{2}\right)\left\{\int_0^T [z(t;\omega),\ z(t;\omega)]dt - \int_0^T [y(t;\omega),\ y(t;\omega)]dt + \text{Tr.}\ (\mathscr{L} + \mathscr{L}^*)\right\}\right. \tag{2.16}$$

which upon simplification is readily seen to yield the same expression as (2.12). The main feature of (2.16) is that it is the formal direct analogue of what is obtained in the case when the observation consists of a finite number of points. (See [7]).

Next we shall show how to obtain $\mathscr{L}$ by exploiting the state space representation, again in the generality of non-rational transforms of B(.) and F(.)

STATE SPACE REPRESENTATION

When the Laplace transforms of B(.), and F(.) are rational we can obtain the (finite-dimensional) state representation (as (1.4), (1.5)) for (1.1), (1.2) by a well-known procedure. Moreover, the determination of $\mathscr{L}$ using this representation turns out to nothing more than recursive filtering, with the corresponding likelihood functional expressions as given in [8]. To be specific, we shall now assume that:

u(t) is p-by-1

B(.) is n-by-p

F(.) is n-by-m

Then the dimension of $n(t;\omega)$ in (2.13), (2.14) is (n+m-by-1), D is m-by-n+m, G is n-by-n+m. We shall work with the form (2.13), (2.14). Let $\mathcal{H}$ denote the Hilbert space $L_2(0,\infty)^n$ of n-by-one functions. In addition to (1.3), we also assume that:

$$||\dot{B}(t)|| + ||\dot{F}(t)|| = 0(\exp -kt),\ k > 0 \tag{3.1}$$

dot representing derivatives. This is probably stronger than we need. Then we have:

<u>Theorem</u>

Under conditions (1.3) and (3.1), there exist linear transformations B, F, mapping E_p and E_m into $\mathcal{H}$ such that

$$x(t;\omega) = \int_o^t T(t-s)\left(B\ u(s) + FD\ n(s;\omega)\right) ds \tag{3.2}$$

being the generalized solution of :

$$\dot{x}(t;\omega) = A\ x(t;\omega) + B\ u(t) + FD\ n(t;\omega) \tag{3.3}$$

where A is the operator with:

domain of A = $[f\varepsilon\mathcal{H}$, f absolutely continuous and derivative $f'\varepsilon\ \mathcal{H}]$,

$Af = f'$

and T(t) is the semigroup generated by A. Moreover (1.2), or equivalently (2.14), can be expressed:

$$y(t;\omega) = C\ x(t;\omega) + G\ n(t;\omega) \tag{3.4}$$

where C is a linear bounded transformation mapping $\mathcal{H}$ into E_n. Here B can be taken as:

$$B\ u \sim (-L) \sum_o^\infty \dot{B}(s+nL)u \quad , \quad u\varepsilon E_p$$

$$F\ w \sim (-L) \sum_o^\infty \dot{F}(s+nL)w \quad , \quad w\varepsilon E_m$$

where L is a fixed positive number, and then C is defined by:

$$Ch = \frac{1}{L} \int_o^L h(s)ds,\ h\varepsilon\ \mathcal{H}$$

<u>Proof</u>

Although the genesis of this representation was by a long and tedious route, it can be verified quite quickly. First of all, the conditions (1.3) and (3.1) yield the convergence in the norm of $\mathcal{H}$ of both series:

$$\sum_o^\infty \dot{B}(s+nL)u \quad \text{and} \quad \sum_o^\infty \dot{F}(s+nL)w$$

Next, it is immediately verified that

$$CT(t)B = B(t)u,\quad t \geq 0$$

$$CT(t)F\ w = F(t)w,\ t \geq 0$$

The interested reader will find it useful to rewrite (3.3) as a nonhomogeneous first order partial differential equation. Note that the state space we have obtained need have no relation to the original state space, if any. The state space we have obtained is reduced, and the controllable part coincides with the controllable part of the original state space. See [9]. The controllable part is finite dimensional if and only if the Laplace transforms or B(.) and F(.) are rational.

Using this representation, we shall next provide a (constructive, "real-time") characterization of $\mathscr{L}$. In fact, formally, the final results can be written exactly as in the case where the state space is finite dimensional. The proofs however will be quite different. First let us fix θ, and let

$$\tilde{y}(t;\omega) = y(t;\omega) - m(t;\theta); \quad \tilde{x}(t;\omega) = x(t;\omega) - \int_0^t T(t-s)Bu(s)ds$$

Note first of all that

$$\tilde{x}(t;\omega) = \int_0^t T(t-s)\ FD\ n(s;\omega)ds \tag{3.5}$$

and hence

$$E(||\tilde{x}(t;\omega)||^2) < \infty \quad 0 \le t \le T$$

Let $\eta(t;\omega)$ denote the mapping:

$$\eta(t;\omega) = \tilde{y}(s;\omega), \qquad 0 \le s \le t \le T$$

mapping ω into $L_2(0,t)^n$. Then we can calculate the conditional expectation

$$E\left(\tilde{x}(t;\omega)|\eta(t;\omega)\right) = \hat{\tilde{x}}(t;\omega)$$

or, equivalently, the minimizing (linear bounded) operator

$$\text{Min.}\ E\left(||\tilde{x}(t;\omega) - L\ \eta(t;\omega)||^2\right)$$

over the class of linear bounded transformations L mapping $L_2(0,t)$ into $\mathscr{H}$. From (10) it is known that the minimizing operator is given by

$$L_o(t) = E\left(\tilde{x}(t;\omega)\eta(t;\omega)^*\right)\left(E\left(\eta(t;\omega)\eta(t;\omega)^*\right)\right)^{-1}$$

Let

$$R_x(t;s) = E\left(\tilde{x}(t;\omega)\tilde{x}(s;\omega)^*\right)$$

Then we can see that:

$$L_o(t) = R_{12}(t)\left(I + R_{22}(t)\right)^{-1} \tag{3.6}$$

where

$$R_{12}(t)f = h; \qquad h = \int_0^t R_x(t;s)\ C^*\ f(s)ds$$

$$R_{22}(t)f = g; \qquad g(s) = \int_0^t C\ R_x(s;\sigma)\ C^*\ f(\sigma)d\sigma,\ 0 < s < t.$$

Note furthermore that

$$R_x(t;s) = T(t-s)\ R_x(s;s) \qquad \text{for}\ \ s < t \tag{3.7}$$

We can draw all the results we need from these. First let us define:

$$\tilde{y}(t;\omega) - C\,\hat{\hat{x}}(t;\omega) = z_o(t;\omega) \tag{3.8}$$

and let

$$\left(I + R_{22}(t)\right)^{-1} = \left(I - G(t)\right) \tag{3.9}$$

so that in particular we have:

$$R_{22}(t) - R_{22}(t)\,G(t) = G(t) = R_{22}(t) - G(t)\,R_{22}(t) \tag{3.10}$$

Using (3.9) in (3.8) we have:

$$z_o(t;\omega) = \tilde{y}(t;\omega) - \int_o^t R_{22}(t;s)\,\tilde{y}(s;\omega)ds + \int_o^t R_{22}(t;s)ds \int_o^t G(t;s;\sigma)\tilde{y}(\sigma;\omega)d\sigma$$

where we denote the kernel corresponding to G(t) by $G(t;s;\sigma)$, and exploiting (3.10) we have:

$$z_o(t;\omega) = \tilde{y}(t;\omega) - \int_o^t G(t;t;s)\,\tilde{y}(s;\omega)ds$$

and comparing with the Krein factorization (Appendix II in [8], for example), we obtain that

$$\int_o^t G(t;t;s)\,\tilde{y}(s;\omega)ds = C\,\tilde{x}(t;\omega) = \int_o^t L(t;s)\,\tilde{y}(s;\omega)ds$$

with $L(t;s)$ given by (2.8), thus characterizing $\mathscr{L}$, and hence also:

$$z_o(t;\omega) = z(t;\omega)$$

Next let us express $\hat{\hat{x}}(t;\omega)$ in terms of $z(s;\omega)$, $0 \le s \le t$. For this, let us note that we may regard $\mathscr{L}$ as mapping $L_2(0,t)$ into itself (being a Volterra operator), and interpreting the adjoint ($\mathscr{L}_t^*$) correspondingly, we have:

$$(I - \mathscr{L}_t^*)\;(I - \mathscr{L}) = \left(I + R_{22}(t)\right)^{-1}$$

and using this we can write:

$$\hat{\hat{x}}(t;\omega) = R_x(t;s)\;C^*\left(z(s;\omega) - \int_s^t G(\sigma;s;\sigma)\;z(\sigma;\omega)d\sigma\right)ds \tag{3.11}$$

where we have used the fact that $G(\sigma;\sigma;s)^* = G(\sigma;s;\sigma)$. We can rewrite this as:

$$\hat{\hat{x}}(t;\omega) = \int_o^t R_x(t;s)\;C^* z(s;\omega)ds - \int_o^t\left(\int_o^\sigma R_x(t;s)\;C^*\;G(\sigma;s;\sigma)ds\right)z(\sigma;\omega)d\sigma \tag{3.12}$$

From:

$$\hat{\hat{x}}(t;\omega) = \int_o^t R_x(t;s)C^*\left(y(s;\omega) - \int_o^t G(t;s;\sigma)\;y(\sigma;\omega)d\sigma\right)ds$$

it follows that:

$$E\left(\hat{\tilde{x}}(t;\omega)\ \tilde{x}(t;\omega)^*\right) = \int_0^t R_x(t;s)C^*CR_x(s;t)ds - $$

$$- \int_0^t R_x(t;s)C^*G(t;s;\sigma)ds \int_0^t CR_x(\sigma;t)d\sigma \tag{3.13}$$

Now using the right side of (3.10) we have:

$$CR_x(s;\sigma)C^* - \int_0^t G(t;s;\tau)\ CR_x(\tau;\sigma)C^*d\tau = G(t;s;\sigma)$$

and hence:

$$G(t;s;t) = CR_x(s;t)C^* - \int_0^t G(t;s;\tau)CR_x(\tau;t)C^*d\tau$$

Substituting this we have:

$$\int_0^t R_x(t;s)C^*\ G(t;s;t)ds = \int_0^t R_x(t;s)C^*\ CR_x(s;t)C^*ds$$

$$- \int_0^t R_x(t;s) \int_0^t C^*G(t;s;\tau)CR(\tau;t)C^*d\tau\ ds$$

$$= E\left(\hat{\tilde{x}}(t;\omega)\ \tilde{x}(t;\omega)\right)C^* \tag{3.14}$$

using (3.13). But letting

$$P(t) = E\left(||\tilde{x}(t;\omega) - \hat{\tilde{x}}(t;\ )||^2\right)$$

$$= E\left(\tilde{x}(t;\omega)\tilde{x}(t;\omega)^*\right) - E\left(\hat{\tilde{x}}(t;\omega)\tilde{x}(t;\omega)^*\right)$$

we have from (3.12), using (3.14) and (3.7) that

$$\hat{\tilde{x}}(t;\omega) = \int_0^t T(t-s)\ P(s)C^*\ z(s;\omega)ds \tag{3.15}$$

which is the expression in terms of $z(.;\omega)$ we sought. It should be noted that the line of reasoning for this proof is quite different from the usual (e.g. [8]) in that no Martingale theory is invoked. Now (3.15) in turn yields that

$$E\left(\hat{\tilde{x}}(t;\omega)\ \hat{\tilde{x}}(t;\omega)^*\right) = \int_0^t T(t-s)P(s)C^*CP(s)T(t-s)^*ds$$

and hence we have that:

$$P(t) = \int_0^t T(t-s)F\ F^*\ T(t-s)^*ds - \int_0^t T(t-s)P(s)C^*\ CP(s)T(t-s)^*ds$$

and hence for x,y in the domain of A^* we have:

$$[\dot{P}(t)x,y] = [P(t)A^*x,y] + [P(t)x,\ A^*y] + [FF^*x,y] - [P(t)C^*CP(t)x,y] \tag{3.16}$$

with $P(0) = 0$. Also from:

$$C\hat{\hat{x}}(t;\omega) = \int_0^t CT(t-s)P(s)C^* \left(\tilde{y}(s;\omega) - C\hat{\hat{x}}(s;\omega)\right) ds$$

we have that, denoting by K the operator:

$$Kf = g; \quad \int_0^t CT(t-s)P(s)C^* f(s)ds = g(t), \quad 0 \le t \le T$$

mapping $L_2(0,T)^n$ into itself,

$$C\hat{\hat{x}}(.;\omega) = (I - K)^{-1} K\, y(.;\omega) \tag{3.17}$$

where K is a Volterra operator of Hilbert-Schmidt type. The operator P(s) is clearly non-negative definite and nuclear. From (3.17) we have that

$$\mathscr{L} = (I - K)^{-1} K$$

and hence

$$\text{Trace } (\mathscr{L} + \mathscr{L}^*) = \text{Tr. } (K + K^*) = 2 \int_0^T CP(s)C^* ds \tag{3.18}$$

Finally, we can also see that $\hat{\hat{x}}(t;\omega)$ is the generalized solution of

$$\dot{\hat{\hat{x}}}(t;\omega) - A\, \hat{\hat{x}}(t;\omega) = P(t)C^* \left(y(t;\omega) - C\hat{\hat{x}}(t;\omega)\right) \tag{3.19}$$

being the recursive form sought. Finally we note that we can write:

$$z(t;\omega) = y(t;\omega) - C\, \hat{x}(t;\omega) \tag{3.20}$$

where

$$\hat{x}(t;\omega) = \hat{\hat{x}}(t;\omega) + \int_0^t T(t-s)\, B\, u(s)ds$$

and $\hat{x}(t;\omega)$ is readily verified to be the generalized solution of

$$\dot{\hat{x}}(t;\omega) - A\, \hat{x}(t;\omega) = P(t)C^* \left(y(t;\omega) - C\hat{x}(t;\omega)\right) + Bu(t) \tag{3.21}$$

We have thus obtained the quantities in (2.16) as solutions of differential equations, solvable in real time. Thus (2.16) can be written, using (3.20), (3.21), and (3.7):

$$p(Y;\theta;T) = \exp - \frac{1}{2} \left\{ \int_0^T [C\hat{x}(t;\omega), C\hat{x}(t;\omega)]dt - 2 \int_0^T [y(t;\omega), C\hat{x}(t;\omega)]dt + 2 \text{ Tr. } \int_0^T CP(t)C^* dt \right\} \tag{3.22}$$

COMPUTATION OF THE ESTIMATE

Let us next examine the computation of the estimate of the unknown parameter vector. First it is necessary to study some properties of the functional defined by (2.9), or, in the form in which we shall actually use it, (3.22). We shall use $p(y;\theta;T)$ to indicate the functional generally, and $p(Y;\theta;T)$ to indicate the value when we use the actual observation, in which case it will be called the "likelihood functional." Let Ω denote the sample space $L_2(0,T)^n$, and p_G the Gauss measure thereon. When we have

$$\int_\Omega p(y;\theta;T)\, dp_G = 1$$

so that using

$$q(y;\theta;T) = \text{Log } p(y;\theta;T)$$

(so that $q(Y;\theta;T)$ is the log-likelihood functional), we have

$$\int_\Omega \left(\nabla_\theta q(y;\theta;T)\right) p(y;\theta;T)\, dp_G = 0$$

(since the dependence on θ is sufficiently smooth). In particular then:

$$E\left(\nabla_\theta\, q(Y;\theta_o;T)\right) = 0$$

Similarly (analogous to the familiar classical arguments):

$$\int_\Omega \frac{\partial^2}{\partial\theta_j\,\partial\theta_i} q(y;\theta;T)\, p(y;\theta;T)\, dp_G$$

$$= - \int_\Omega \left(\frac{\partial}{\partial\theta_i} q(y;\theta;T)\right)\left(\frac{\partial}{\partial\theta_j} q(y;\theta;T)\right) p(y;\theta;T) dp_G$$

where $\{\theta_i\}$ denote the components of θ. In particular then:

$$E\left(\frac{\partial^2}{\partial\theta_j\partial\theta_i} q(Y;\theta_o;T)\right) = - E\left(\left(\frac{\partial}{\partial\theta_i} q(Y;\theta_o;T)\right)\left(\frac{\partial}{\partial\theta_j} q(Y;\theta_o;T)\right)\right) \tag{4.1}$$

These relations are important to us, because our technique consists in miximizing the log likelihood functional in the known interval in which θ_o lies; or, more accurately, we seek a root of the gradient of the log likelihood functional:

$$\nabla_\theta\, q(Y;\theta;T) = 0 \tag{4.2}$$

using in particular the Newton-Raphson technique, but avoiding the use of second derivatives, thanks basically to (4.1). First we note that, from (3.22), we have:

$$\frac{\partial}{\partial\theta_i} q(\theta;Y;T) = (-1)\left\{ \int_o^T \left[\left(y(t;\omega) - C\hat{x}(t;\omega),\ \frac{\partial}{\partial\theta_i} C\hat{x}(t;\omega)\right)\right] dt + \text{Tr.} \int_o^T \frac{\partial}{\partial\theta_i} CP(t)C^* dt \right\} \tag{4.3}$$

Let $M(\theta;T)$ denote the matrix with components m_{ij} defined by:

$$m_{ij}(\theta;T) = \int_o^T \left[\frac{\partial}{\partial\theta_i} C\hat{x}(t;\theta),\ \frac{\partial}{\partial\theta_j} C\hat{x}(t;\theta)\right] dt \tag{4.4}$$

Then our basic "identifiability condition" is that

$$M(\theta_o) = \underset{T\to\infty}{\text{Limit}}\ (1/T)\ E\left(M(\theta_o;T)\right) \tag{4.5}$$

be positive definite; to check this, in practice, since θ_o is unknown, we must verify that this is satisfied for every θ in the interval in which θ_o lies. (The positive definiteness of $M(\theta_o)$ assures it in a sufficiently small neighborhood of θ_o). Our basic algorithm then is:

$$\theta_{n+1} = \theta_n + \left(M(\theta_n;T)/T\right)^{-1}\left(\nabla_\theta\, q(Y;\theta_n;T)/T\right) \tag{4.6}$$

The asserted positive definiteness also assures the positive definiteness of $M(\theta_n;T)/T$ for all large enough T in a sufficiently small neighborhood of θ_o. For these aspects see [8]. Let us also note here that

$$E\left(m_{ij}(\theta_o;T)\right) = E\left(\left(\frac{\partial}{\partial\theta_i} q(Y;\theta_o;T)\right)\left(\frac{\partial}{\partial\theta_j} q(Y;\theta_o;T)\right)\right) \tag{4.7}$$

which is most easily verified by using the Ito-integral version of the likelihood functional, (2.9). Assuming that

$$\lim_{T\to\infty} \frac{1}{T}\int_o^T [u(t),\, u(t+s)]dt$$

exists and is continuous in s, we can proceed to evaluate $M(\theta)$ as in [8]. Reference may also be made to [8] for the asymptotic (ergodic) properties of unbiasedness and consistency, for the case where the system is finite dimensional (Laplace transforms of F(.) and B(.) are rational). The general case considered here can be treated in similar fashion, although space does not permit inclusion here.

EXPERIENCE ON ACTUAL DATA

We turn now to indicate briefly the results obtained using our computational technique on actual (as opposed to simulated) data. The dynamic system considered arises from the longitudinal mode perturbation equations for an aircraft in wind-gust (turbulence) (Rediess-Iliff-Taylor, see [3]). This is a system where only a "rational" approximation of the spectrum of turbulence is used, so that the total system is finite dimensional. Leaving the many essential details to the comprehensive work of Iliff [11], the state space formulation of the problem is as follows: (see also [3]):

$$\dot{x}(t) = A\,x(t) + B\,u(t) + F\,n_2(t)$$

$$V(t) = C\,x(t) + D\,u(t) + G\,n_1(t)$$

where $n_1(.)$, and $n_2(.)$ are independent white Gaussian, and the matrices in the equations have the form:

$$A = \begin{bmatrix} Z_1 & 0 & 1 & Z_1 \\ 0 & 0 & 1 & 0 \\ M_1 & 0 & M_3 & M_1 \\ 0 & 0 & 0 & -\frac{\bar{v}}{1000} \end{bmatrix}$$

$$B = \begin{bmatrix} Z_4 \\ 0 \\ M_4 \\ 0 \end{bmatrix} \qquad F = \sigma \begin{bmatrix} 0 \\ 0 \\ 0 \\ \frac{1}{20\bar{v}} \end{bmatrix}$$

$$C = \begin{bmatrix} 0 & 0 & 1 & 0 \\ 0 & 1 & 0 & 0 \\ \dfrac{10M_1 - \bar{v}Z_1}{g} & 0 & \dfrac{10M_3}{g} & \dfrac{10M_1 - \bar{v}Z_1}{g} \\ k_1 & 0 & \dfrac{32k_1}{\bar{v}} & k_1 \end{bmatrix}$$

$$D = \begin{bmatrix} 0 \\ 0 \\ \dfrac{10M_4 - \bar{v}Z_4}{g} \\ 0 \end{bmatrix}$$

g = acceleration due to gravity

G = diag. [.0005, .0001, .01, .00001]

The lettered entries are unknown, except for $\bar{v}$, which is known (=1670). Note that the turbulence power is an unknown parameter.

Figure 1 shows the complete time history of the observation v(t) (four components) subdivided into various regions for later identification, as well as the input time-history. Estimates were computed over the various subregions each by three methods:

Method I: Neglecting the measurement noise on the angle of attack measurement (v_4) and following the corresponding maximal likelihood technique developed in [3]. This is reasonable for this particular example at high turbulence levels.

Method II: This is the method developed herein, see also [7].

Method III: This was a "check" method, in which the turbulence was ignored completely in the model.

The results are summarized in Figure 2. Sample means and variances of the estimates obtained over the different data-regions are shown, along with the wind-tunnel values as well estimates obtained on other turbulence-free (smooth air) flights. It can be seen that Method II yields the most consistent estimates. It also turns out that Method II is the least in computational time — the estimates converging in fewer iterations. It can also be seen that ignoring the turbulence leads to the worst results. For more discussion see [11]. The remaining figures indicate the nature of the "fit" obtained using the estimated coefficients to the observed data. Figure 3 shows the close agreement provided by Method II. Figures 4 and 5 indicate how much worse the agreement is on the same stretch of data if the turbulence is <u>not</u> accounted for.

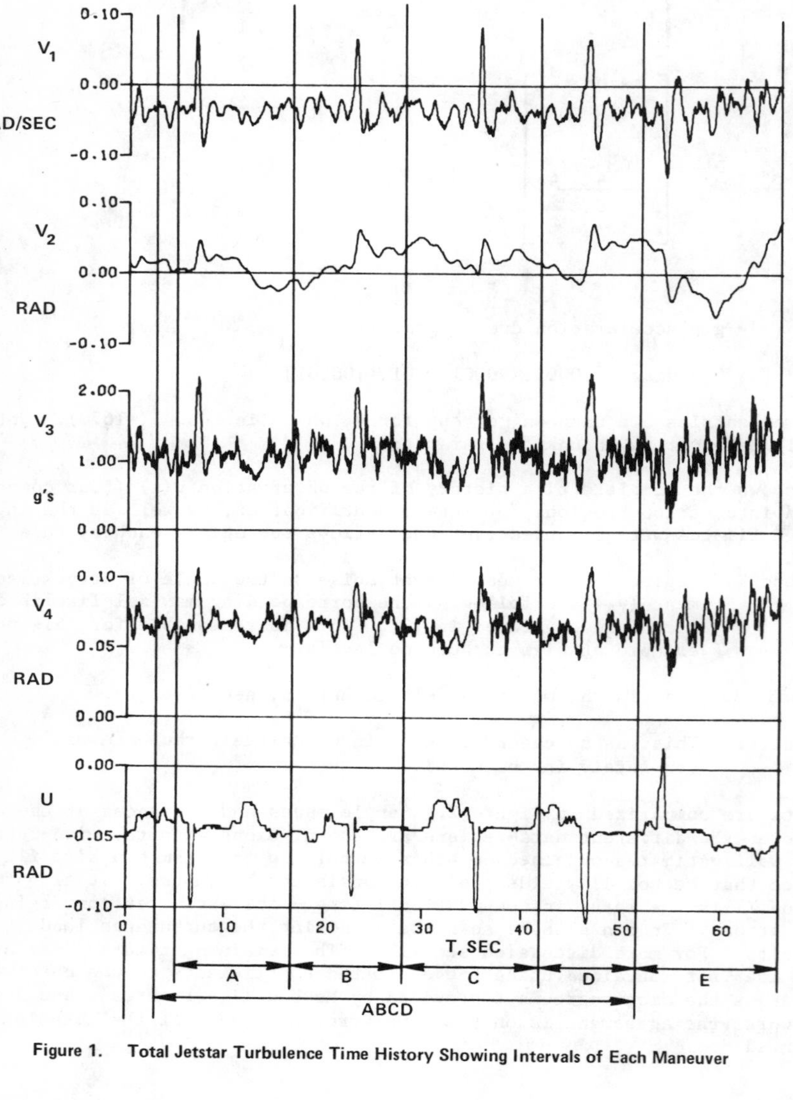

Figure 1. Total Jetstar Turbulence Time History Showing Intervals of Each Maneuver

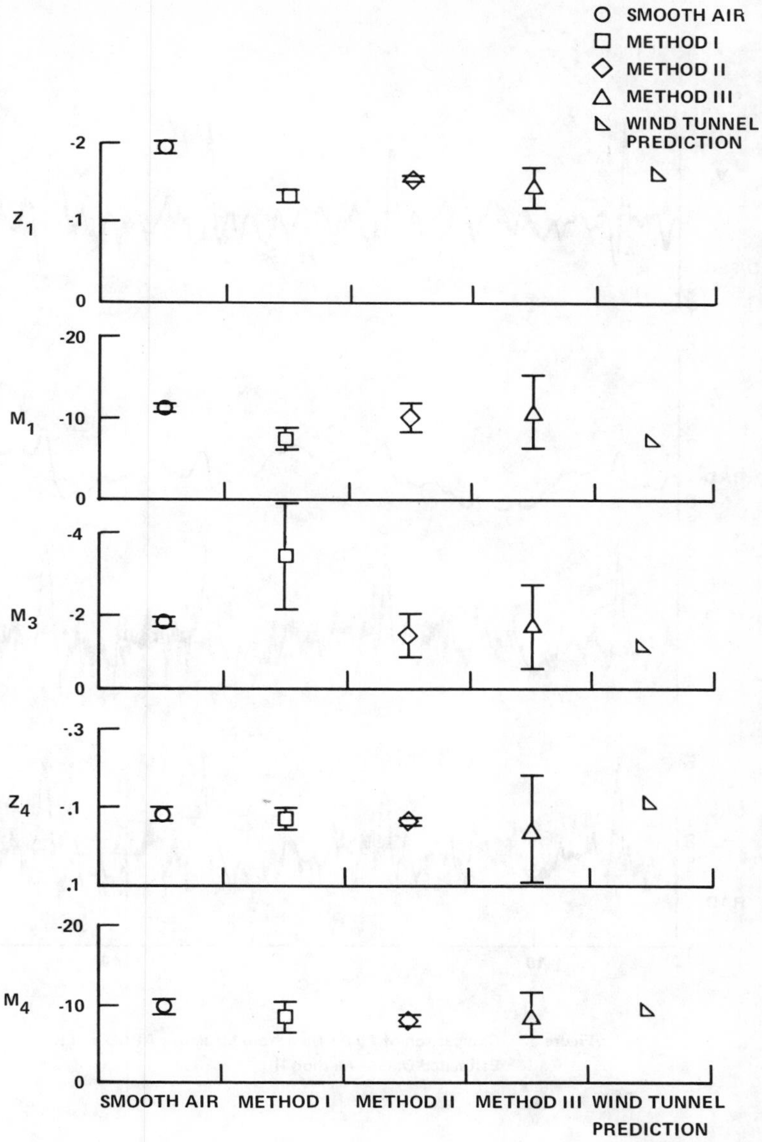

Figure 2. Means and Standard Deviations for Five Methods of Estimating Coefficients

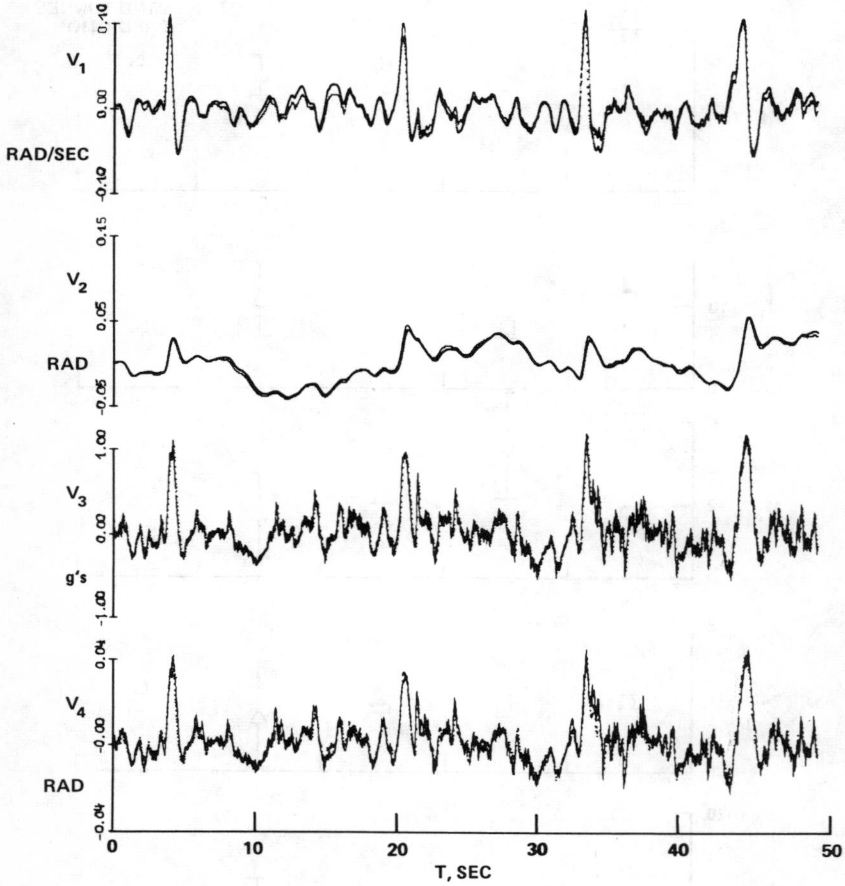

Figure 3. **Comparison of Flight Data from Maneuver ABCD and the Estimated Data – Method II**

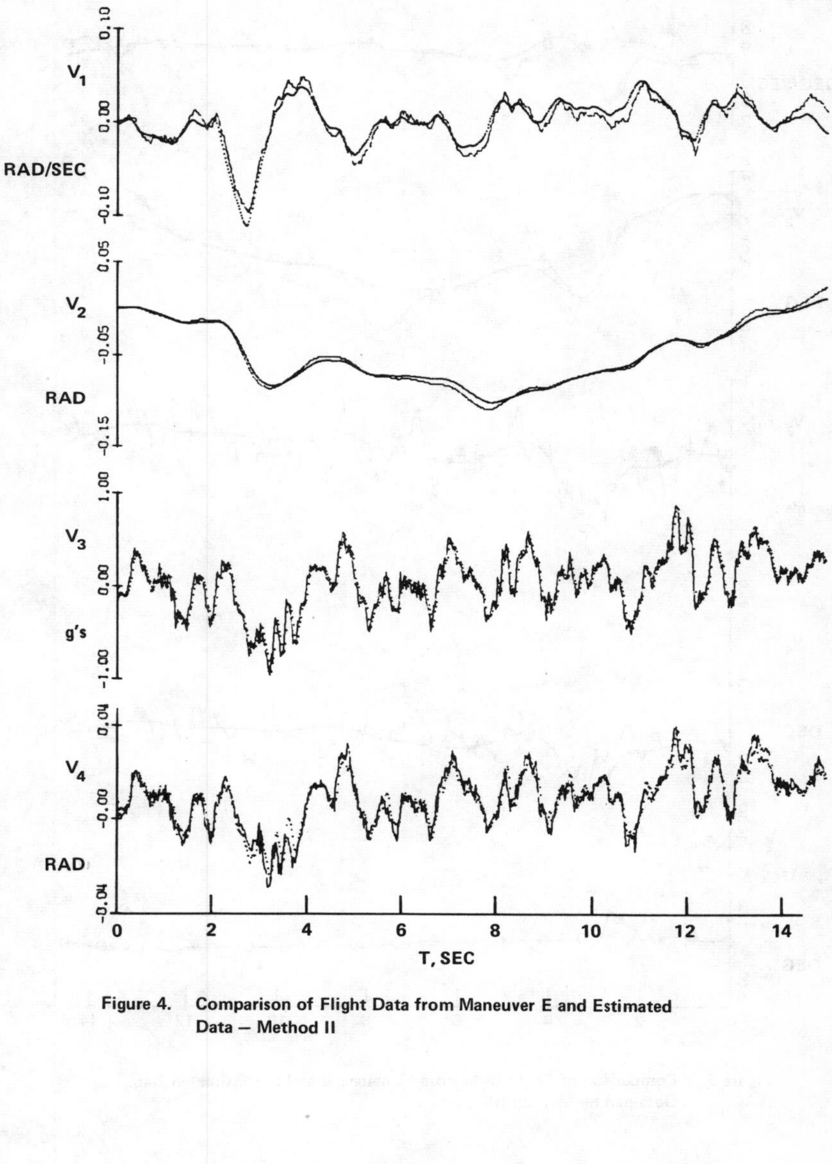

Figure 4. **Comparison of Flight Data from Maneuver E and Estimated Data – Method II**

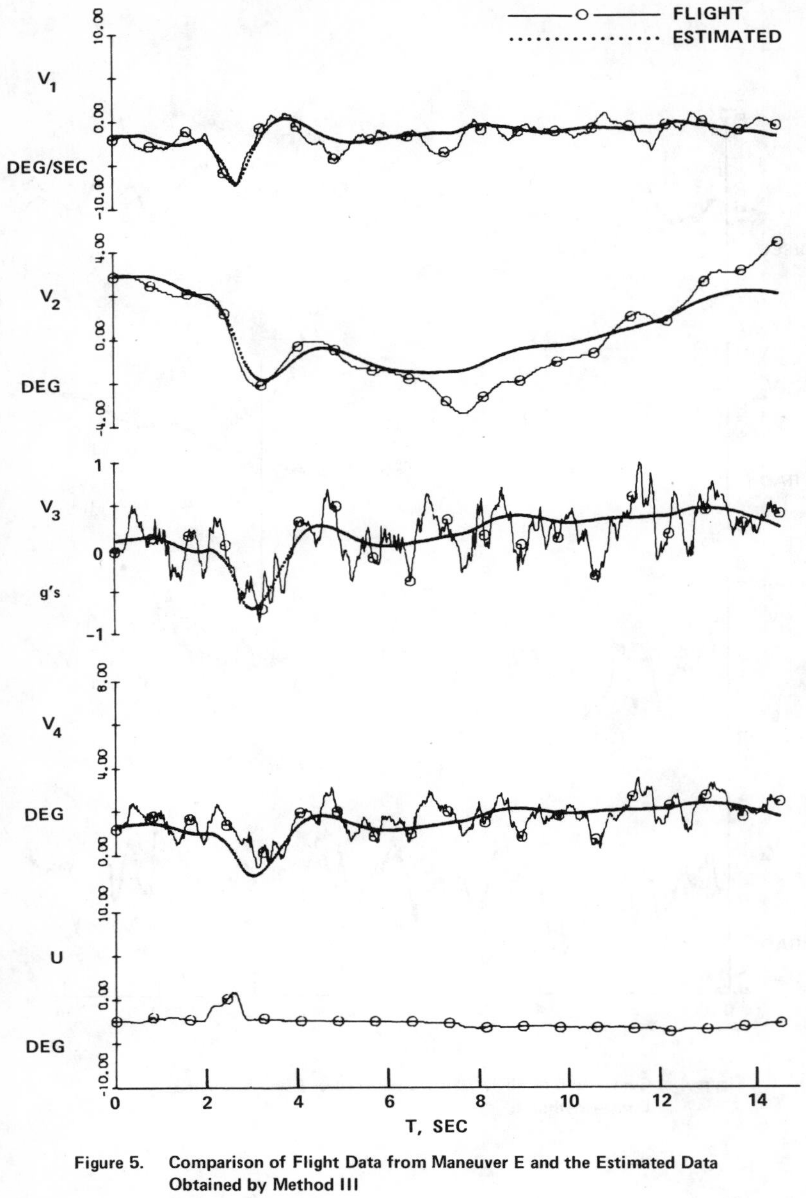

Figure 5. Comparison of Flight Data from Maneuver E and the Estimated Data Obtained by Method III

REFERENCES

1. Forrester, J.W.: "World Dynamics," Wright-Allen Press, 1971.

2. Proceedings of the Second IFAC Symposium on Identification and Process Parameter Estimation, Prague, Czechoslovakia, June 1970.

3. Balakrishnan, A.V.: Identification and Adaptive Control: An Application to Flight Control Systems, Journal of Optimization Theory and Applications, March 1972.

4. Balakrishnan, A.V.: Identification and Adaptive Control of Non-dynamic Systems, Proceedings of the Third IFAC Symposium on Identification and Process Parameter Estimation, June 1973.

5. Gohberg, I., and Krein, M.G.: Volterra Operators, AMS Translations, 1970.

6. Balakrishnan, A.V.: On the Approximation of Ito Integrals using Band-limited Processes (to be published, SIAM Journal on Control, February 1974).

7. Balakrishnan, A.V.: Modelling and Identification Theory: A Flight Control Application, in "Theory and Applications of Variable Structure Systems," Academic Press, 1972.

8. Balakrishnan, A.V.: Stochastic Differential Systems, Vol. 84, Lecture Notes on Economics and Mathematical Systems, Springer-Verlag, 1973.

9. Balakrishnan, A.V.: "System Theory and Stochastic Optimization," in "Network and Signal Theory," Peter Peregrinns Ltd., 1972.

10. Balakrishnan, A.V.: Introduction to Optimization Theory in a Hilbert Space, Lecture Notes on Operations Research and Mathematical Systems, Vol. 42, Springer Verlag, 1971.

11. Iliff, K.W.: Identification and Stochastic Control, An Application to Flight Control in Turbulence, UCLA Ph.D Dissertation, 1973.

ADAPTIVE COMPARTIMENTAL STRUCTURES IN BIOLOGY AND SOCIETY

R.R. Mohler and W.D. Smith

INTRODUCTION

Many dynamical processes, which are relevant to man, his biochemistry and the society of which he is a part, are modelled conveniently as adaptive control systems with appropriate compartmental structures. It is assumed here that the systems of concern may be described or accurately approximated by precise mathematical relationships. Such equations are derived generally from established physical laws, intuition, or experimental data. If this is not the case, it may be possible to use a linguistic approach such as suggested by Zadeh for the development of a parallel modeling and identification theory.

Even with the assumption that the system may be quantified conventionally, however, it is not possible to say that the necessary theory of modelling and identification has been developed. In more cases than not, linear system models, despite their popularity in the literature, fail because of their rigid structure. Still, as is shown here, certain results from linear system theory may be utilized sometimes to derive effective models of complex adaptive processes. This convenience is due to the linear-in-state structure of the bilinear model which is suggested in this paper. Some bilinear system theory and its application to dynamic modeling has been developed and much of this work is summarized by Mohler. The proceedings of the first U.S. - Italy Seminar on Variable Structure Systems, which was an outgrowth of common bilinear system research, further establishes the base for modeling and identification as they relate to adaptive processes in general |See Mohler and Ruberti|.

Biological and societal processes, like those generally found throughout nature, are distributed as well as adaptive. Still such processes are usually ameanable to discretization according to compartments. For example, biological compartments may be formed for anatomical convenience, and societal compartments according to geographical and governmental convenience. Suppose a single substance X is distributed among n compartments between which it diffuses naturally at a rate described by conservation equations of the form

---

R.R. Mohler is with the Department of Electrical and Computer Engineering Oregon State University, Corvallis,Oregon 97331.USA.(Presently visiting University of Rome)

W.D. Smith is with the University of New Mexico Medical School, Albuquerque, New Mexico 87106. USA.

The research reported here is supported by NSF Grant GK 33249

$$\frac{dx_i}{dt} = \sum_{j=1}^{n}{}' \phi_{ij} - \sum_{k=1}^{n}{}' \phi_{ki} + \phi_{ia} - \phi_{ai} + p_i - d_i \ , \tag{1}$$

$i = 1,\dots,n,$

where x_i is the concentration of substance X in the ith compartment, ϕ_{ij} is the flux of X from the jth to the ith compartment, ϕ_{ia} is the flux of X from the environment of uniform concentration x_a to the ith compartment, p_i is the rate of production and d_i is the rate of destruction of X in the ith compartment. The primed summation denotes deletion of the ith term. Compartmental volumes are assumed constant.

The flux to the ith compartment from the jth compartment is described by

$$\phi_{ij} = \rho_{ij} x_j \quad , \tag{2}$$

where i, j=1,...,n,a and ρ_{ij} is an exchange parameter which may be constant or may be a multiplicative control. Additive control may be synthesized through the net production of substance $(p_i - d_i)$ which is zero for a conversative system. (For brevity, any non manipulative terms have been neglected). It is readily seen from Eqs. (1) and (2) that a collection of terms leads to a bilinear system of the following form:

$$\frac{dx}{dt} = Ax + \sum_{k=1}^{m} B_k u_k x + Cu \quad , \tag{3}$$

where $x \varepsilon R^n$ is the state vector; $u \varepsilon R^m$ is the control vector; A, B_k and C are n×n, n×n and n×m matrices respectively. Again, x is composed of compartmental quantities of substance X; additive control may arise from net production, and multiplicative control from those exchange parameters which are manipulated. While, additive and multiplicative controls are independent variables for the type of process considered here, in general they might not be. Therefore, it is assumed here that zeros appear in the appropriate positions of B_k and C to maintain independent additive and multiplicative control variables where necessary. Those exchange parameters which can not be manipulated result in constant coefficients of the A matrix. Here, it is assumed that compartmental capacities do not change significantly. While the process may include transport of more than one substance, no generality is lost in utilizing the above form of equations. In the human body such processes are controlled by what is called homeostasis.

BIOLOGICAL PROCESSES

A single cell may represent a convenient compartment for modeling a more complicated organism. Nutrients are generated for cell growth and cell division by biochemical processes within the cell itself and by transfer across the cell membrane. Enzymes effect control by manipulation of membrane permeability and by triggering the formation of proteins and nucleic acids and their precursors.

In addition to the natural diffusion from compartments of high concentration to those of low concentration, there exists an active transport from low concentration to high concentration with the necessary energy supplied by cellular metabolism. Such active transport as derived by Hill and Kedem is described by a nonlinear function (division of summations) of compartmental densities. The process may be approximated, however, by a bilinear equation |See Mohler|.

The development considered here has assumed that the process involves mass transport. Again the concept is more general and in physiology alone includes numerous processes such as body fluid and electrolyte balance |See Alvi|, thermal regulation |Milsum|, kinetics of material ingested or excreted from the body |Snyder|, control of carbon dioxide in the lungs |Grodins| as well as kinetics of metabolities in cell suspensions |Sheppard and Householder|

Tracer experimentation

For biological processes it is usually impossible to measure the necessary quantities directly. Frequently, it is possible, however, to estimate compartmental system parameters by observing the behavior of a tracer, such as a radioisotope or a dye, which has been injected or ingested into the system |Sheppard and Householder|. If it is assumed that the tracer in every compartment is distributed uniformly through substance X, that labeled and unlabeled substances behave identically, and that the tracer does not appreciably affect the compartmental system behavior, then the tracer system with specific compartmental activity $a_i(t)$ is described by

$$\frac{d(x_i a_i)}{dt} = \sum_{k=1}^{n}{}' \phi_{ik}(t)\, a_k - \sum_{q=1}^{n}{}' \phi_{qi}(t)\, a_i + \phi_{ia}(t) a_{ia}(t)$$

$$- \phi_{ai}(t)\, a_i - d_i(t)\, a_i + f_i(t) \ , \tag{4}$$

$i = 1,\dots,n$,

where $a_i(t)$ is specific activity of tracer in the ith compartment, $a_{ia}(t)$ is the specific activity of substance in influx $\phi_{ia}(t)$, and $f_i(t)$ is the influx of tracer which is inserted directly into compartment i. Normally, tracer experiments are conducted with the compartmental process in "equilibrium" (+), and it is further assumed that the process is closed, $\phi_{ai} = \phi_{ia} = 0$, and conservative, $p_i = d_i = 0$, $i=1,\dots,n$.

Let tracer be inserted directly into p compartments with its behavior measured in q compartments. Then, with the above assumptions

$$\frac{da}{dt} = S\, a + X_d^{-1} P f \tag{5}$$

and

$$w = Q\, a \quad , \tag{6}$$

where $a \varepsilon R^n$ is the specific activity vector; X_d is the diagonal matrix of elements $x_i,\dots,x_n$; P is an input matrix with unity elements in each row corresponding to a compartment which is accessible for tracer insertion and with zero elements elsewhere; f is the p-vector of inserted total activity fluxes, and therefore $X_d^{-1} P f$ is the vector of inserted specific activity fluxes; w is the output q-vector of observed compartmental specific activities. Also, $S = [s_{ij}]$ with

$$s_{ij} = \phi_{ij}/x_i \ , \qquad j = 1,\dots,n,\ i \neq j$$

and

$$s_{ii} = -\frac{1}{x_i} \sum_{k=1}^{n} \phi_{ik} \ .$$

(+) I.e., with $\dot{x}_i = \dot{\phi}_{ij} = \dot{\phi}_{ai} = \dot{\phi}_{ia} = \dot{p}_i = \dot{d}_i = 0$; $i,j = 1,\dots,n, i \neq j$, but possible net flux between compartments and with environment.

If the tracer enters the system by natural flux routes from $\acute{p}$ separate external sources of labeled stubstance, each with its own tracer specific activity a_i, $i = 1,\ldots,\acute{p}$, then

$$\frac{da}{dt} = S\,a + X_d^{-1}\,\Phi_{ad}\,\acute{P}\,\acute{a} \tag{7}$$

and

$$w = Q\,a \;. \tag{8}$$

Here Φ_{ad} is a diagonal matrix of elements $\phi_{ia} \cdots \phi_{na}$, $\acute{a}$ is a $\acute{p}$-dimensional tracer source activity vector, and $\acute{P}$ is the $n\times\acute{p}$ output matrix with a single unity in each row, i, for which $a_{ia} \neq 0$.

In practice, the subject might be placed in a radioactive bath, and its absorption of tracer monitored in a so-called "soak up" experiment. In "wash out" experiments, the subject is saturated to a given specific activity of tracer, and wash out is observed in a tracer free environment. In either case, the presence of all compartments can not be detected unless the tracer system, Eqs. (7) and (8), is completely controllable and completely observable.

Obviously, the tracer system is linear, and may be analyzed by classical linear system theory. For example, the conventional rank tests for controllability and observability may be applied to either tracer system to see that sufficient accessibility is available to identify the minimum realization of the system. For the time variant case, rank tests may still be applied to two convenient time-variant test matrices to show that the model is a minimal realization |See D'Angelo|. Standard procedures may then be used to synthesize the minimal realization from the input-output observations. These, inturn, dictate the necessary tracer experiments.

Adaptation and Control

Cells, perhaps the most basic of biological compartments, multiply in a manner similar to that of populations of biological species |Mohler|. This process of cellular fission also may be likened to that of nuclear fission and the generation of radioactive particles |Mohler and Shen|. Cells may multiply in a controlled manner to synthesize an organ or an organism in the evolution of life which must adapt to its environment. Just as the cellular processes are regulated in the body by homeostasis, so are the combinations of these cells working together to perform integrated specific tasks in the functioning of a living organism or human being.

In the regulation of body temperature, for example, heat (additive control) is produced internally by basal metabolism and heat is lost by means of respiration, excretion, evaporation from skin and radiation from skin. Transfer of heat between compartments (such as skin, muscle and core of various internal organs) is regulated by means of vasomotor control of blood vessels which effectively alters conductivity of heat transfer as a bilinear control mechanism. The control policy in this case is established by the hypothalamus in the brain based on temperature feedback information.

It is convenient to examine a two-compartment model of thermoregulation since its behavior can be studied in a state plane. Skin and core are the compartments selected since temperatures in these compartments are most meaningfult to system behavior and they are monitored by the central nervous system for regulation of temperature. |See Mohler|

Water Balance

Water, comprising on the average about sixty percent of the body weight, must be carefully regulated as part of the overall body process. About two-thirds of body wa ter is located within the cells, and the remainder is in the so-called extracellular compartment. The extracellular compartment or intravasculer compartment includes plasma, the interstitual compartment in which tissue cells are bathed and the transcellular compartment. In some cases there is a distinct anatomical boundary, such as wall membrane, but for the interstitual compartment such boundaries are not so definitely defined. The transcellular compartment, which includes digestive secretions, is separated from plasma by a continuous layer of epithelial cells and capillary membrane.

Tracer experiments are used to obtain the distribution of water throughout these compartments. The intracellular compartment can not be so isolated, and must be accounted for by taking differences.

Drinking, eating, skin absorption and respiratory absorption are all mechanisms by which water enters the body's extracellular compartment. Oxidative water, a metabolic end product, varies with diet and metabolic rate, and becomes an influx to the intracellular compartment. Effluxes from the extracellular compartment to the environment include urine, fecal water, skin evaporation and respiration.

Water balance is regulated by homeostasis through pressure forces and membrane permeabilities. Feedback control again is established through the hypothalamus. An amount of water equal to two-thirds of the blood volume is exchanged between the blood and the extravascular fluid each minute.

The kidney with its thousands of filtration passages called nephrons is the key component in body water balance. It processes blood plasma through its nephrons to produce urine. The water loss in each nephron is regulated by the action of the antidiuretic hormone (ADH) on the tubule and the collecting duct permeability |Pitts|. Glomerular filtration rate (GFR), which is collected at the arterial end of the nephron, and the renal plasma flow (RPF) may be regulated to control water loss. RPF is a function of arteriolar resistance in the kidney which is controlled by constrictor muscles, again parametric control. While details of the process are quite complicated, water excretion in urine may be approximated by a bilinear system with efflux of form

$$\phi_u = u_k \ (c_g + b_g w_b) \quad , \tag{9}$$

where control u_k is a function of permeability (determined by ADH), arteriolar resistance and osmotic pressure driving water from renal tubules; w_b, body water, is a sum of compartmental amounts of water which in turn may be state variables. C_g , b_g const.

Though to a much lesser degree, water balance is regulated by parametric control through the skin and through the lungs. Vasomotor control of blood flow and hormone regulated permeability control water losses. Vasomotor constriction reduces hydrostatic pressure and thus reduces water transfer to the skin, and also by dropping skin temperature reduces avaporation. In the lungs the net loss is regulated parametrically by ventilation rate and vasomotor control of respiratory blood vessels. Normally, the skin and the lungs play a much lesser role in controlling water balance of mammals than do the kidneys. But under adverse conditions they can be significant.

Water loss through skin is extremely complicated and still not very well understood. It is transported in three stages: exchange between the blood and the dermal connective tissue, exchange between the dermis and the epidermis, and finally exchange between the epidermis and the surroundings. The exchange takes place by filtration and diffusion. The force which induces capillary filtrarion is described by the Starling Hypothesis. This hypothesis states that the net force driving water out of a capillary is equal to the hydrostatic pressure difference minus the osmotic pressure difference with both differences taken between the blood and tissue |Ruch and

Patton, p. 625|. This results in a net outward force in the capillaries at the arteriolar end and a net inward force at the venous end. Water diffusion across capillary walls is regulated by an effective permeability, and it further diffuses through layers of skin fibers and through duct openings. Then, water loss between skin and surrounding air is usually assumed to be proportional to the difference between saturated vapor pressure at skin temperature and the vapor pressure of the air |Chew, p. 88|.

Consequently, it can be shown that net water loss through the skin is approximately described during equilibrium operation by

$$\phi_{sa} = \rho_s (w_d - w_e) = k_s (w_e p_{se} - r p_{sa}) ,$$

where ρ_s is effective permeability, w_d is dermis water concentration, w_e is epidermis water concentration, p_{se} is saturated vapor pressure of skin, p_{sa} is saturated vapor pressure of air, and r is relative humidity. Again, the role of parametric adaptive control is apparent. This skin water loss may then be further related to more basic mammalian control functions as follows:

$$\phi_{sa} = k_s \left[\frac{\dfrac{\rho_s(t)\gamma w_b(t) + k_s r p_{sa}}{\rho_s(t)} - r p_{sa}}{k_{s2} - \dfrac{k_{s3}}{\left(\dfrac{1}{w_b(t)} + k_{s1}\right) u_s(t)} + k_s} \right] , \tag{10}$$

where $w_b(t)$ is amount of body water, $u_s(t)$ is portion of blood flowing to skin, and k_{si} $(i = 1,2,3)$ are constants. Here, a bilinear model could be used as an approximation, again more accurate than a linear model. A similarly complex relationship may be derived for parametric control of water loss through the lungs. In the lungs, it can be shown that ventilation rate is a parametric control variable. Some animals can considerably alter their body water by control of breathing.

For convenience, tracer experuments were conducted on a wild house mouse in a controlled environment in order to obtain a simple model for its water balance. A simplified single compartment which is used merely to demonstrate the procedure is shown in fig. 1. Obviously, a more meaningful model would include more compartments such as for intracellular water, extracellular water and storage water.

Equilibrium acclimation graphs averaged over several animals of drinking influx, food-water influx, urine efflux and net exchange by skin and lungs are provided by fig. 2 |Haines|. The solid curve in fig. 2 (a) connects experimental values of drinking water influxes given at various levels of water deprivation. The dashed curve connects the calculated values of water influx from food. It is found that this influx for the unconstrained mouse is nearly proportional to body mass. Urinary and fecal daily water losses are charted by dash-dot curves on fig. 2 (b). The solid curve joins experimental data for net evaporative water effluxes at different drinking levels. As drinking water is reduced, it shows that urinary-fecal and evaporative losses are decreased. Fig. 2 (b) further breaks down evaporative losses through skin (short dashes) and through evaporation (long dashes).

A typical long-term tracer wash out record for an equilibrium (ad libitum) animal is shown in fig. 3. After tritiated water injection, periodic samples of body water were analyzed for radioactivity. Note that the radioactive decay of tritium (half life of 12.5 years) has negligible effect on the experiment. Obviously, the de cline can be closely fitted by a single negative exponential over the time interval shown. From the single compartmental equation for body water change

$$\frac{dw_b}{dt} = \phi_{in} - \phi_{out} \tag{11}$$

and the specific tracer activity change

$$\frac{da}{dt} = -\frac{\phi_{in}}{w_b} a \quad , \tag{12}$$

it is seen that in equilibrium the negative slope of the tracer experiment graph, fig. 3, in magnitude is inversely proportional to the quantity of body water.

If the specific activity plotted in fig. 3, were plotted from separate experimental data for evaporate and plasma specific activities, their graphs would have the same slope as shown in fig. 3, but each would have its own level. This can be explained by a two compartment model of: (1) internal water in equilibrium with blood plasma, and (2) extravascular fluid in skin and lungs affected by evaporative exchange.

SOCIETAL PROCESSES

There have been numerous special journal issues, books, conferences and comprehensive reports devoted to such major areas of society as urban dynamics, public systems, health care, transportation and economics. Economics, with Keynes as a pioneer researcher, probably has received most of the concentration. Again an adaptive structure, such as that of the bilinear system, is necessary to satisfy well-developed intuitive concepts which in many cases have been substantiated experimentally. As noted by Runyan, gross simplifications to a very complicated system, in many instances, has hindered rather than helped interdisciplinary work in economics. The adaptive bilinear structure offers a significant improvement in this respect, and parametric control by economic investment is a good example of its use.

For population transfer models, arrival rates of different classes are equal to appropriate population class terms multiplied by corresponding migration attractiveness multipliers. Again, to yield a bilinear model.

In socioeconomics, adaptive control may be synthesized through such quantities as legislation, land zoning, taxes and capital investment.

Transportation Systems

Transportation represents one particularly interesting socioeconomic process. It may be desired to develop a model in order to study the transportation needs of a nation, state or region. A significant part of this study might include an analysis of transportation effectiveness to regulate socioeconomic development. It is apparent from historical developments of railroad lines, freeways, air service and shipping that industrial development and socioeconomic development in general follow the transportation networks.

A systems overview of the transportation process is shown by fig. 4. This block diagram shows the major interactions and feedback loops of demography, economics, transportation planning and transportation network dynamics.

The simplicity of this functional diagram is somewhat deceiving, since each block represents a major system in its own right,numerous interconnections and feedback loops. Fortunately, this study need only concern itself with transportation related socioeconomic and environmental processes. While land-use and pollution are not shown as separate subsystems in fig. 4, they are inherently part of the transportation planning/controller and the transportation dynamics subsystems. Land-use planning plays a most significant role in the generation of transportation policies which are shown as inputs to the network dynamics.

The transportation "plant" dynamics, which includes people and commodity multimodal transportation, is composed of three major components, viz., trip generation, distribution and assignment (including modal choice).

The trip generation portion establishes a relationship between trip end-volume and such socioeconomic factors as land use, demography and economics. Demographic information utilized here includes total population, school enrollment and number of household units in a zone of interest. Useful economic data includes total employment, average income, automobile ownership and retail sales. Hosford in a Missouri study showed that only a few basic factors are necessary to estimate trip generation with an effective trip attraction factor entering the process as a parametric control.

The trip-distribution model assigns trips from the generation model to competing destination zones or compartments. Three widely used adaptive models are: (1) the Fratar model, (2) the intervening opportunities model, and (3) the gravity model |Heanne and Pyers|.

The Fratar model expands existing highway travel patterns by a growth factor for each origin and destination zone. The future trips are then loaded on to transportation networks according to some assignment technique based on travel time, convenience, etc. The intervening opportunities model is an exponential probability model based on the simple relation that a trip will terminate in a region or compartment is equal to the probability that this region contains an acceptable destination, times the probability that an acceptable destination closer to the origin of the trip has not been found. The gravity model, perhaps the most popular trip distribution model, is analogous to Newton's gravitational law of masses. Consequently, the traffic between zones or compartments is proportional to the product of the compartmental populations and inversely proportional to the square of the distance between the compartments of interest. Frequently, the gravity model is further adapted to a particular system by means of some parametric multiplier or attraction factor.

The traffic-assignment model allocates trip interchanges to a specific transportation system. This includes modal selection as well as network assignment. Most traffic assignment models are based on a desired quality of minimum path length. But more realistic models should include convenience, comfort, cost, safety, etc. Greenshields derived a mathematical expression for such a quality of traffic flow.

Transportation models of this kind may be used to formulate development plans for a statewide system which must be integrated with similar national models. Like the national model which is an expansion of the Northeast Corridor study, statewide studies must include well formed goals, diverse socioeconomic systems, numerous transportation modes and geometrical nodes and different concepts of quality of transportation as they relate to an overall quality of life. Five state-wide transportation studies of a superficial nature are compared by Hazen.

REFERENCES

Alvi, Z.M., "Predictive Aspects of Monitored Medical Data", Ph. D. dissertation, School of Engineering, University of California, Los Angeles, 1968.

Chew, R.M., "Water Metabolism of Mammals", Physiological Mammalogy, vol. II (Eds. W.V.Mayer and R.G.VanGelder), Academic Press, New York, 1965.

D'Angelo, H., Linear Time-Varying Systems: Analysis and Synthesis, Allyn and Bacon, Boston, 1970.

Greenshields, B.D., "The Measurement of Highway Traffic Performance", Traffic Engineer, 39, 26-30, 1969.

Grodins, F.S., Control Theory and Biological Systems, Columbia Press, New York,1963.

Haines, H., Personal Communication, University of Oklahoma.

Hazen, P.I., "A Comparative Analysis of Statewide Transportation Studies", Highway Research Record, 401, 39-54, 1972.

Heanne, K.E. and Pyers, C.E., "A Compartative Evaluation of Trip Distribution Procedures", Highway Research Record, 114, 20-50, 1965.

Hill, T.L. and Kedem, O., "Studies in Irreversible Thermodynamics III. Models for Steady-state and Active Transport across Membranes". J. Theoret. Biology 10, 399-441, 1966.

Hosford, J.E., "Development of a Statewide Traffic Model for the State of Missouri", Final Report of Project No. 2794-P, Midwest Research Institute, Kansas City, Missouri, 1966.

Keynes, J.M., General Theory of Employment, Interest and Money, Harcourt, Brace and Co., New York, 1935.

Mohler, R.R., Bilinear Control Processes with Applications to Engineering, Ecology and Medicine, Academic Press, New York, to be published.

Mohler, R.R. and Ruberti, A., Eds., Theory and Application of Variable Structure Systems, Academic Press, New York, 1972.

Mohler, R.R. and Shen, C.N., Optimal Control of Nuclear Reactors, Academic Press, New York, 1970.

Milsum, J.H., Biological Control Systems Analysis, McGraw Hill, New York, 1966.

Pitts, R.F., Physiology of the Kidney and Body Fluids, Year Book Medical Publishers, Chicago, 1963.

Ruch, T.C. and Patton, H.D., Physiology and Biophysics 19th edition, W.B. Saunders, Philadelphia, 1965.

Runyan, H.M., "Cybernetics of Economic Systems", IEEE Transactions Systems, Man and Cybernetics, SMC-1, 8-17, 1971.

Snyder, W.S., et al., "Urinary Excretion of Tritium following Exposure of Man to HTO-a two Exponential Model", Phys. Med. Biology 13, 547-559, 1968.

Sheppard, C.W. and Householder, A.S., "The Mathematical Basis of the Interpretation of Tracer Experiments in Closed Steady-state Systems", J. Applied Physics, 22, 510-520, 1951.

Zadeh, L.A., "Outline of a New Approach to the Analysis of Complex Systems and Decision Processes", IEEE Transactions Systems, Man and Cybernetics, SMC-3, 28-44, 1973.

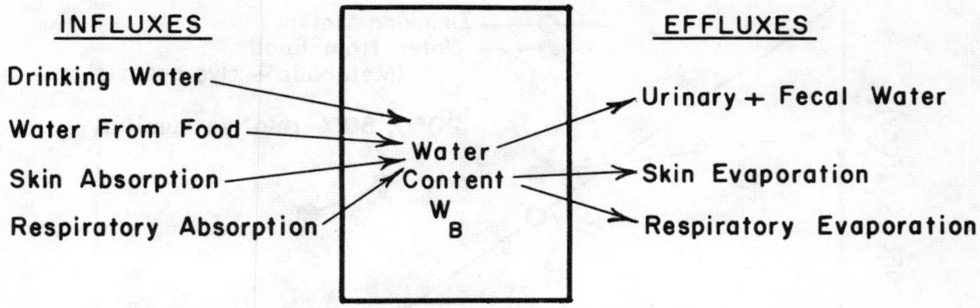

Fig. 1. A Simplified Water Exchange Model for the Experimental Wild House Mouse.

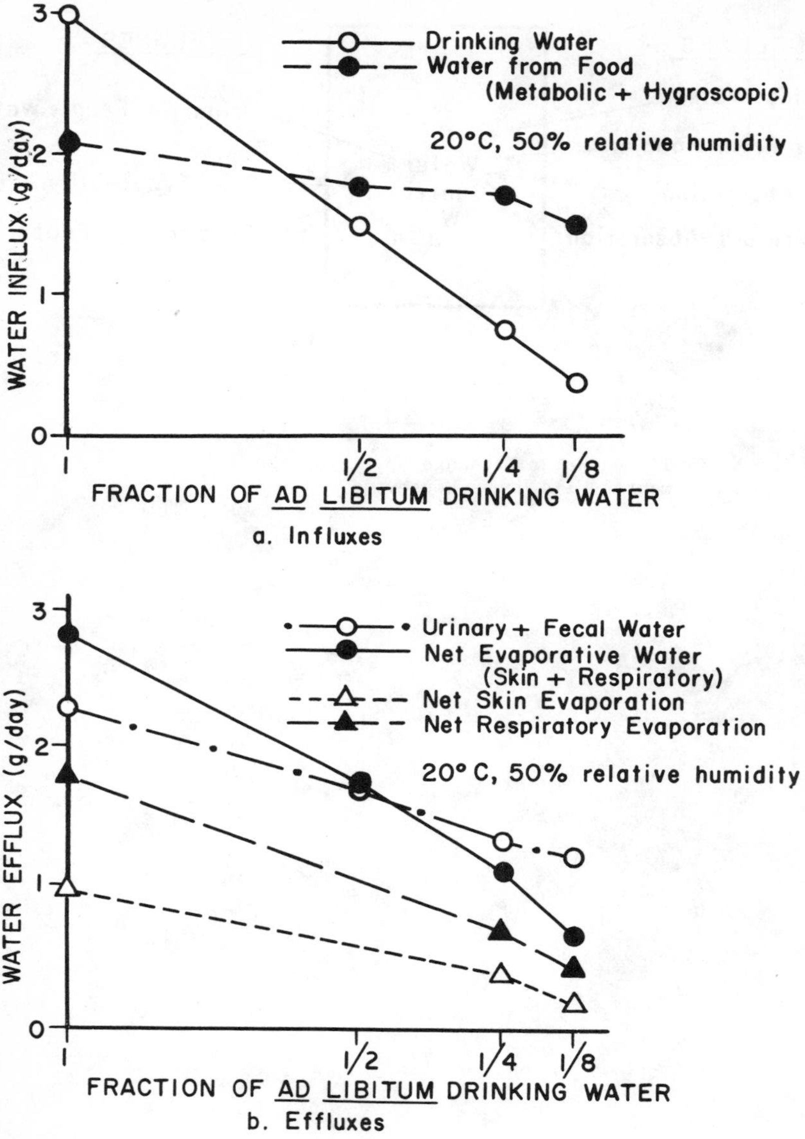

Fig. 2. "Equilibrium" Water Fluxes Versus Water Deprivation in the Wild House Mouse.

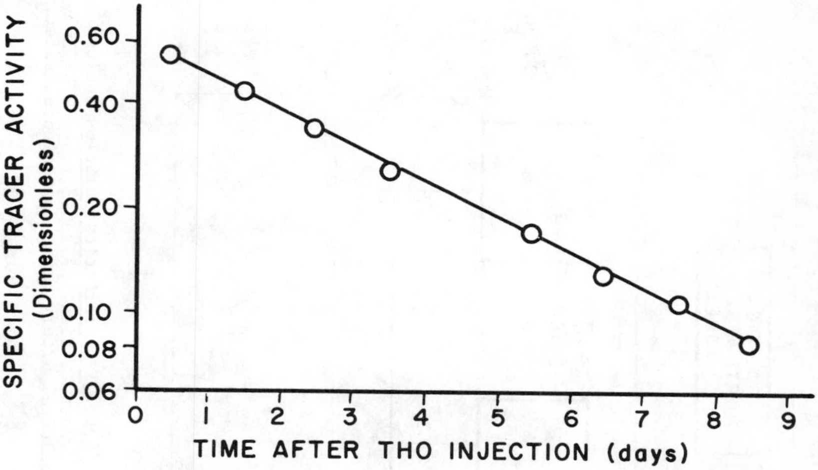

Fig. 3. Specific Activity Versus Time After THO Injection in an Ad Libitum Wild House Mouse (Semilog Grid).

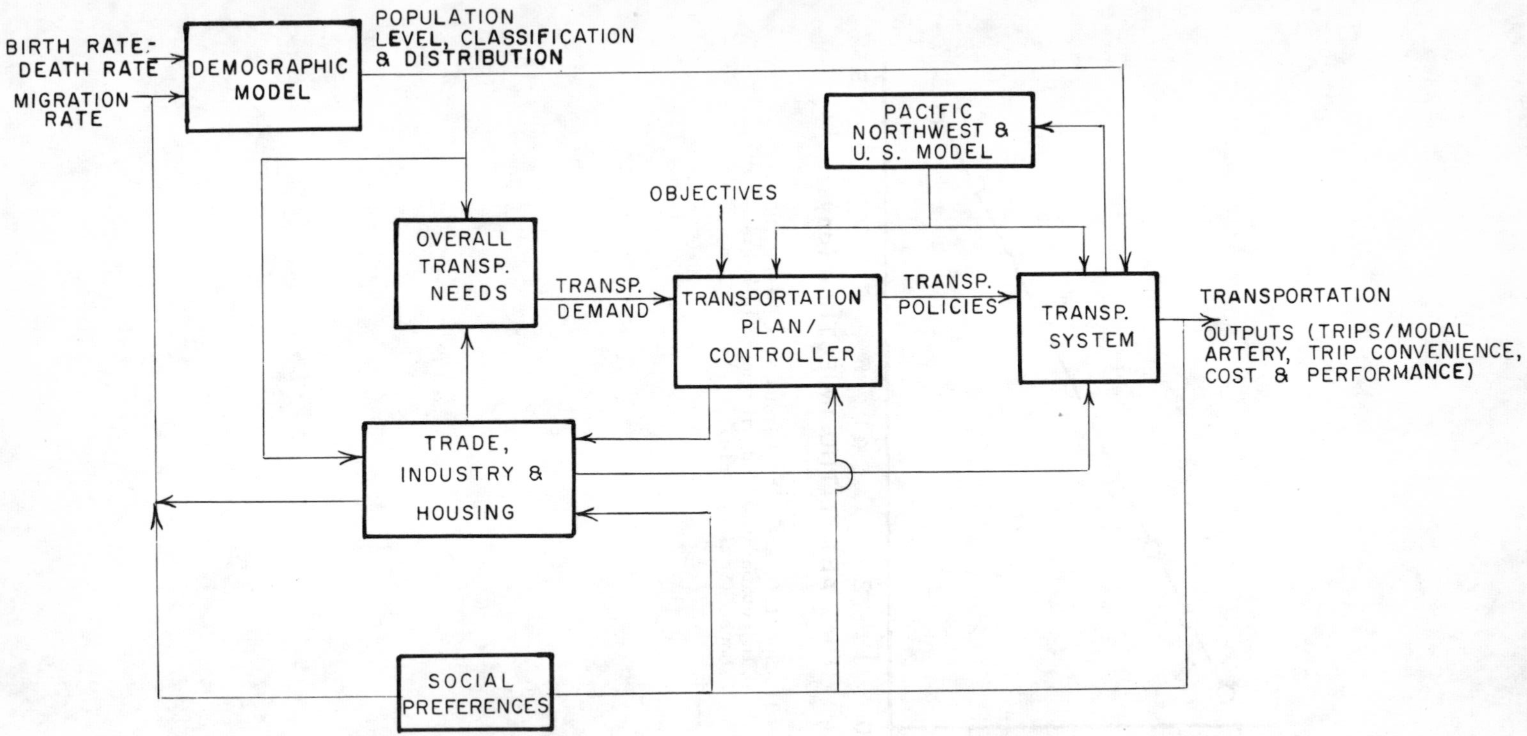

Note: Arrows generally represent multivariable outputs. E.g., Transportation policies include components of land use, legislation, taxation, investment in transportation for rapid transit, highways, airports and seaports. Transportation Outputs include trips/modal artery, trip convenience, safety and cost, pollution (CO, NO_x, HC), etc.

Fig. 4 Interconnection levels of Dynamic Socioeconomic & Transportation Systems

ON THE OPTIMAL SIZE OF SYSTEM MODEL

Mohamed Z. Dajani
University of Louvain
Louvain-la-Neuve, Belgium

ABSTRACT

The problem of finding the best size for the system model is considered. By explicitly synthesizing the complexity cost of a proposed model, one is able to transform this problem into a constrained integer programming problem.

1. INTRODUCTION

Finding the "best" model is a dilemma that has occupied the attention of an ever increasing number of investigators. Since complicating the model may increase its accuracy, and the "best" model is the optimal compromise between the conflicting demands of better accuracy and simpler models [1], [2], [3].

In this brief paper, attention is focused on the problem of finding the best dimension of system model that incorporates, newly, quantitatively the optimization of model complexity, or inversely its accuracy. Note that model complexity is not singularly produced by the increase of the model's dimension but is also affected by the choice of a realization method (inter connections, and number of operators).

For a physical process with measurable input-output pairs it is desired to find a best model subject to some general quantitative performance criterion

$$J_G = J_{Estimator} + J_{Control} + J_{Complexity} - J_{Accuracy} \quad . \qquad (1)^{*}$$

In order to proceed one has to start from a rough model and improve upon it as time advances. Let

$$\dot{\underline{x}} = \underline{f}(\underline{x},\underline{u};\underline{\alpha},\underline{t}) \quad ; \quad \underline{x} : (s\times 1) \quad ; \quad \underline{u} : (m\times 1) \qquad (2)$$

be the initial system state model of s-order with known initial condi-

* Definition of symbols are given in the list of symbols.

tions $\underline{x}(0)$, and s being a changable variable. Find the optimal value of $s = \hat{s}$, and that of $u = \hat{u}$ such that J_G is minimized. The $+$ and $-$ operators in (1) are symbolic and their exact nature is determined as per the application, e.g.,

$$+\{\cdot\} \triangleq e^{+\{\cdot\}} \quad , \quad -\{\cdot\} = e^{-\{\cdot\}} \, . \tag{3}$$

2. ANALYSIS

Defining the complexity of the model cost, $J_{complexity}$ as the integration of submodels subcosts, one needs to define the following <u>per unit</u> (p.u.) set of subcosts :

C_{state} : p.u. cost of deploying a single model (estimator) state ;

$C_{control}$: p.u. cost of mapping a single state (estimate) into a single closed loop control ;

$C_{control\ energy}$: p.u. cost of input energy per control channel ;

C_{gain} : p.u. cost for solving one Riccati equation to evaluate the controller gain (estimator gain) ;

$C_{interconnection}$: p.u. cost of interconnecting two different states with a unit gain.;

Assuming an arbitrary number of interconnections, g. Therefore,

$$\begin{aligned} J_{complexity} &= 2s \times C_{state} + (m\times s) \times C_{control} + m \times C_{control\ energy} \\ &\quad + s(s-1) \times C_{gain} + 2g \times C_{interconnection} \end{aligned} \tag{4}$$

is the explicit expression of $J_{complexity}$ as a function of the variables s and a chosen set of p.u. costs and intercomplexity index g. At this point, it is worth noting that $J_{accuracy}$ will be chosen in accordance with prevailing forms in the literature, i.e., the weighted sum of the positive difference between the process output and the model output, e.g.,

$$J_{accuracy} \triangleq \begin{cases} \sum_i a_i (y_i - y_{m_i})^2 \\ \sum_i b_i |y_i - y_{m_i}| \\ \vdots \\ \text{etc} \end{cases} \quad ; \text{ for the same input sequence .} \tag{5}$$

Although it is not explicitly evident, $J_{accuracy}$ is an implicit function of the model order, s.

The remaining two performance indices $J_{control} + J_{estimation}$ are of the familiar quadratic-weighted-integral type.

The most natural approach at this point seems to apply the Pontryagin maximum principle. However, its direct application will reveal no result and indicates its inability to handle optimization for non-fixed configuration system. A generalization of the maximum principle to this type of problems is certainly a useful extension.

Ideally one wishes to optimize J_G by simultaneously optimizing the modeling cost, $J_M \triangleq J_{complexity} + J_{accuracy}$, and the stochastic control cost, $J_{s.c.} \triangleq J_{estimation} + J_{control}$. It is true that both group of costs are interdependent, however simultaneous optimization of J_M and $J_{s.c.}$ is analytically formidable. In the present treatment a suboptimal procedure is adopted in which, first J_M is optimized, and then second $J_{s.c.}$ is optimized using the results of the first step.

In this paper only the optimization process of J_M is discussed since that of $J_{s.c.}$ is widely treated in literature, e.g. [4]. In the next section an algorithm (essentially an integer programming algorithm) is presented whose output is the optimal value of $s \triangleq \hat{s}$.

3. THE PROBLEM DEFINITION

Find the minimum of $J_{complexity}$ such that :

$$1 \leq s \leq S^{*} \text{ , (inequality constraint) ,} \tag{6}$$

where S^{*} the integer representing the model order at which case $[\Phi^T\Phi]$ is a singular matrix. Φ is a characteristic matrix of the system whose entries are the delayed input and output of the same system. For details see Åström and Eykhoff [1].

As typical of integer programming problems the solution for the optimal model size $\hat{s}$ is not trivial and requires to meet certain conditions most of which are justified in our present formulation [5].

4. CONCLUSION

It is hoped that this presentation will stimulate deeper interest in the engineering problem of finding the most feasible system model by explicitly including a complexity cost that reflects the model realization effort and its versatility. The model identification problem is transformed into a constrained integer programming problem. Natural extension of this methodology is the generalization of the maximum principle or the calculus of variations to systems with not fixed configuration.

5. BIBLIOGRAPHY

[1] Åström, K.J., and Eykhoff, P. System identification - A survey. Automatica, Vol. 7, pp. 123-162, 1971.

[2] Woodside, C.M. Estimation of the order of linear systems. Automatica, Vol. 7, pp. 727-733, 1971.

[3] Sage, A.P., and Melsa, J.L. System identification. Academic Press, New York, 1972.

[4] Special issue on linear-quadratic-Gaussian problem. IEEE Trans. on Automatic Control, Vol. 16, Dec. 1971.

[5] Graves, L.G., and Wolfe, P. Recent advances in Mathematical programming, McGraw-Hill, Inc., New York, 1963.

6. LIST OF SYMBOLS

$\underline{x}$: (s×1) model state vector
$\underline{u}$: (m×1) system control vector
$\underline{\alpha}$: (q×r) model random parameter matrix.
s : the model size (dimension of state vector, $\underline{x}$)
a_i, b_i : time-varying weighting coefficients

Acknowledgement.

The author wishes to thank the Faculty of Applied Science and in particular his colleagues at the Unité de Mécanique.Appliquée of the University of Louvain for their continuous support.

INFORMATION-THEORETIC METHODS FOR MODELLING AND ANALYSING LARGE SYSTEMS

Raymond E. Rink
Department of Electrical
and Computer Engineering
Oregon State University
Corvallis, Oregon

Introduction

The pairwise, causal interactions between components and between variables in a complex system may be viewed as processes of communication. That is, for the values of one variable to affect the values of another, there must be a direct path (channel) which conveys information to the latter about the values of the former. This point of view can provide some useful methods for modelling and analysing large systems, as will be outlined in the following sections.

Transmission Model of a System

The first step in modelling a real system involves choosing a suitable set of state variables (if the model is to be Markovian) and identifying the direct interactions between them.

The statement that "the value of variable x at time t has a causal effect on the value of variable y at time t + 1" implies that x(t) and y(t+1) are not statistically independent, over the ensemble of all possible time series generated by the system. This hypothesis can be tested by measuring the transmission (mutual information) between x(t) and y(t+1), given some data. If the system is stationary and ergodic, this transmission, denoted* $T(x,y'|y)$, can be measured from a single time-series of data. It is easily shown that $T(x,y'|y)$ is zero if and only if x(t) and y(t+1) are statistically independent. This idea was apparently first discussed by Watanabe [1], and was called "interdependence analysis" by him. Conant [2] later showed how the measured transmissions may be used to group the system variables into strongly connected subsystems.

The transmission is defined as

$$T(x,y'|y) = H(y'|y) - H(y'|x,y)$$

where the H's are entropy functions. If x and y are discrete variables (or quantized continuous variables) with L possible levels each, the entropies can be written as

$$H(y'|y) = -\sum_{i=1}^{L} \sum_{j=1}^{L} P_{y',y}(i,j)\log P_{y'|y}(i|j)$$

and $$H(y'|x,y) = -\sum_i \sum_j \sum_k P_{y',y,x}(i,j,k)\log P_{y'|y,x}(i|j,k),$$ where $P_{y'|y}(i|j)$ is

---

*y' = y(t+1) is, in most dynamic systems, most strongly dependent on y(t) itself. This dependence is eliminated by conditioning the transmission on y, leaving a more sensitive measure of interaction.

the probability that y' will take on its ith value, given that y had its jth value, etc. These probabilities are, of course, not known without prior knowledge and must be estimated from frequency data. If N values of each of the variables are available, then the estimated transmission is

$$\hat{T}(x,y'|y) = -\sum_{i=1}^{L}\sum_{j=1}^{L} \frac{N_{y',y}(i,j)}{N} \log \frac{N_{y',y}(i,j)}{N}$$
$$+ \sum_i\sum_j\sum_k \frac{N_{y',y,x}(i,j,k)}{N} \log \frac{N_{y'|y,x}(i|j,k)}{N} .$$

As N becomes very large, $\hat{T}$ will approach T if the data are statistically well-behaved. In practice, $N \gg L^3$ usually implies that $\hat{T} \simeq T$.

One weakness of the method is that the converse to the statement of the previous paragraph does not hold, i.e. it is not true that statistical dependence necessarily implies a direct, causal effect of x(t) on y(t+1). It may be that a third variable influences both x(t) and y(t+1) in such a way that the apparent transmission $T(x,y'|y)$ is spurious. There is, however, the following simple procedure which can be used, provided that enough data are available, to identify those interactions that are direct and causal.

Suppose that there are n state variables $\{x_1,x_2,\ldots\ldots x_n\}$ in a minimal representation of the system, and that the causal actions upon x_1 of all other (n-1) variables are to be determined. The first step is to measure all the pairwise transmissions and to find the i_1 for which $\hat{T}(x_{i_1},x_1'|x_1)$ takes its maximum value. This identifies a variable x_{i_1}, which is tentatively assumed to have a causal action on x_1. If $\hat{T}(x_{i_1},x_1'|x_1)$ is, in fact, spurious, then

$$H(x_1'|\{x_i|i\neq i_1\}) = H(x_1'|\{x_i\}).$$

In other words, the addition of x_{i_1} to the set of conditioning variables does not produce a further decrease in the conditional entropy of x_1 if x_{i_1} is not directly connected to x_1. Whereas, if x_{i_1} is directly connected to x_1 and there is noise anywhere in the system, the value of x_{i_1} at time t is not uniquely determined by the values of the other (n-1) variables at time t, and the conditional entropy of x_1 at time t+1 will be further reduced by the addition of x_{i_1} to the conditioning set. This will have to be checked at the end of the procedure.

Assuming, then, x_{i_1} to be causal, the second step is to remeasure the transmissions between the other (n-2) variables and x_1 conditioned on x_{i_1}, to find the i_2 which maximizes $\hat{T}(x_{i_2},x_1'|x_1,x_{i_1})$.

If this maximum is not zero, it identifies a variable x_{i_2} which is tentatively taken to have a causal action on x_1 that is not mediated by x_{i_1}. Again, the conditional entropies of x_1 with and without x_{i_2} in the conditioning set will have to be checked at the end of the procedure, in order to verify this assumption.

The third step is to remeasure the transmissions between the remaining (n-3) variables and x_1 conditioned on both x_{i_1} and x_{i_2}, and to identify the x_{i_3} which has the largest, etc. The process continues for s_1 steps with $s_1 \leq (n-1)$, until the remaining conditional transmissions are all zero. Of course, any transmission that is found to be zero at any step need not be remeasured, for that variable is then known to be noncausal for x_1.

At the end of the procedure it is necessary to check whether the net transmissions

$$\begin{aligned}\hat{T}(x_{i_j}, x_1' \mid x_1, x_{i_1}, x_{i_2}, \ldots . x_{i_{j-1}}, x_{i_{j+1}}, \ldots . . x_{i_{s_1}}) \\ = \hat{H}(x_1' \mid x_1, x_{i_1}, x_{i_2}, \ldots . . x_{i_{j-1}}, x_{i_{j+1}}, \ldots . . x_{i_{s_1}}) \\ - \hat{H}(x_1' \mid x_1, x_{i_1}, x_{i_2} \ldots \ldots \ldots , x_{i_{s_1}})\end{aligned}$$

are all nonzero for $j=1,2,\ldots . s_1$, thus confirming that the s_1 identified variables are indeed causal for x_1. If the net transmission is zero for any j, then that x_{i_j} is not causal and may be deleted from the subset, which, in turn, requires the net transmissions to be re-evaluated for the remaining (s_1-1) variables. If one of these is zero, that variable must also be deleted, etc. The final result of the process is a set of (s_1-r_1) variables with nonzero net transmissions to x_1. If this process is repeated for each of the state variables, the overall result is the transmission model of the system. It can be shown diagrammatically with nodes and arrows, where the arrows are labeled with the "strength" of the net transmission.

Obviously, $s = \max\{s_i\}$ cannot be very large if this procedure is to be effective, given a finite amount of data and computing time. Conditional probabilities which come from an (s+1)-dimensional distribution with L discrete levels along each dimension are encountered in the course of the computation, and good frequency estimates of these probabilities would require $N \gg L^{s+1}$ values for each of the state variables. The number of computer operations is similarly large.

This does not prevent the method from being used for large-scale systems, however, provided that the interconnections are relatively sparse. The state-dimension n could be several hundred, and, if the maximum number of interconnections s which impinge on any state variable were not more than, say, ten, the method could be used with L = 4 or L = 8. Many large-scale systems are of this type, where variables interact directly only with neighboring variables, as in a diffusion process. Spatially distributed socio-economic and ecological systems in general would have this property.

Functional Form of the Dependencies

The transmission modelling approach described in the last section has some superficial similarity to correlation methods, especially to the partial-coherence-function models of linear systems. Those methods are really much more restricted, however, since they assume a linear form of relationship between variables, whereas the transmission model refers to statistical dependence in general, without prior assumption of any particular functional form.

Of course, the functional forms must still be specified, if the model is to be used for extensive analysis. The huge number of conditional transition probabilities that comprise a purely stochastic model of a large system may well be unmanageable for simulation or optimization studies.

One approach to the needed simplification is to assume that the state variables are governed continuously by a dynamic system of equations, each having a rate-multiplier structure of the type used by Forrester, e.g.

$$\overset{\circ}{x}_1 = x_1[(a_1+n_1)\prod_{j=1}^{s_1-r_1} f_{1j}(x_{i_j}) - (b_1+m_1)\prod_{j=1}^{s_1-r_1} g_{1j}(x_{i_j})]. \tag{1}$$

Here the rate of accumulation of x_1 is homogeneous of first degree in x_1, as is the case in ecological and socio-economic systems where birth and death rates for species; growth and decay rates for commodities and capital, etc., all are proportional to the existing level of the variable.

The factors (a_1+n_1) and (b_1+m_1) include the "normal" exponential growth and decay parameters a_1 and b_1, i.e. the average exponents that would be found in the absence of feedback from other variables. They also include the "noise" fluctuations $n_1(t)$ and $m_1(t)$ that may occur in the exponent due to randomness in the generation or degeneration process for the species or commodity.

The factors $f_{1j}(x_{i_j})$ or $g_{1j}(x_{i_j})$ (normally only one of them would vary from unity for a particular x_{i_j}) represent the modulation in a parameter value that is caused by feedback from another variable in the system. In cases where the system is a cybernetic one, these multipliers can perhaps be interpreted as endogenous control variables whose values are being optimized in some sense. Thus, the general problem of identification of f_{1j} or g_{1j} would involve finding an appropriate performance index and then synthesizing f_{1j} or g_{1j} as a function of x_{i_j}, i.e. the dual optimization problem. We will consider here a special case, but one which is frequently plausible; namely the "bang-bang" control law

$$f_{1j}(x_{i_j}) = 1 + \frac{d_{1j}}{2}\,\mathrm{Sgn}(x_{i_j} - \bar{x}_{i_j}). \tag{2}$$

Here $\bar{x}_{i_j}$ is the "normal" value of x_{i_j}, and $(1\pm d_{1j}/2)$ is the modifying factor that is applied to the normal growth parameter a_1, depending on whether $x_{i_j} > \overline{x_{i_j}}$ is favorable or unfavorable to the growth of x_1. The only parameter to be determined is d_{1j}, the "depth of modulation," which will be seen to be related to the transmission. We may remark in passing that, if (Πf_{i_j}) is considered to be a composite control variable, the dynamic system is of the <u>bilinear</u> type, a class whose excellent controllability and performance qualities have been analyzed elsewhere [3,4,5].

Consider the difference equation

$$y(t+1) = y(t) + y(t)[a+n(t)]\ [1 + \frac{d}{2}\,\mathrm{Sgn}(x(t) - \bar{x})]. \tag{3}$$

This is equivalent to the equation

$$z(t+1) \equiv \frac{y(t+1)}{y(t)} - 1 = a[1 + \frac{n(t)}{a}]\ [1 + \frac{d}{2}\ \mathrm{Sgn}(x(t) - \bar{x})]. \qquad (4)$$

This transformation would correspond, in an actual identification process, to taking the data $\{y(t)\}$, forming the ratios of successive values, and subtracting unity.

The precise relationship between $T(x,z')$ and d depends on the statistics of $x(t)$ and $n(t)$. If the data $\{x\}$ come from a stationary process, the "normal" value $\bar{x}$ could be taken as that value which is exceeded by one-half the data. This is equivalent to assuming that $\mathrm{Sgn}(x(t)-\bar{x})$ is binary with probabilities (1/2, 1/2), regardless of the distribution of $x(t)$ itself. Also, in many cases of large scale systems (e.g. socio-economic) the fluctuations $n(t)$ represent the aggregate decision noise of a collection of individuals, hence will tend to be normally distributed.

Under these particular assumptions, a Monte Carlo simulation was used to determine the transmission $\hat{T}(x,z')$ for various values of c and d, where c/2 is the standard deviation of $n(t)/a$. Values of x and n were generated by random number generators and z' computed. The simulated "data" $\{x\}$ and $\{z\}$ were then scaled and sorted into L=16 quantum intervals and the entropies calculated from the frequency estimates of probability.

The resulting transmission values are plotted in Figure 1, as functions of d/2 with c/2 as parameter. Clearly, the transmission value alone is not enough to determine d if c is also unknown, as would usually be the case with real data.

The residual entropy $H(z'|x)$ provides the entra information that is needed, and it is plotted in Figure 2. If the measured entropy (e.g. $\hat{H}(z'|x) = 1.85$ nats) is drawn as a horizontal line on Figure 2. the intersections of that line with the parametric family provide several pairs of values (c/2, d/2) which, when transferred to Figure 1, can be fitted to a curve as shown. The intersection of this curve with the horizontal line corresponding to the measured value of transmission (e.g. $\hat{T} = .563$) gives the values of c/2 and d/2 (e.g. $c/2 \simeq 0.25$ and $d/2 = 0.435$).

For the multivariable case where, for example, the interactions of $(s_1 - r_1)$ variables $\{x_{i_j}\}$ with x_1 are to be modelled by equations of the form of (1) and (2), the data can be transformed as above to yield $\{z_1\}$. If the standard deviations of m_1 and n_1 are known, the corresponding curves of Figure 1 can be used directly to find the value of $d_{1j}/2$ for each measured net transmission $\hat{T}(x_{i_j}, x_1')$.

If, on the other hand, the standard deviation of n_1, say, is unknown, but it is known that $m_1 = 0$ and that all interactions are of the type f_{1j} rather than g_{1j}, then Figures 1 and 2 can be used together to estimate the $d_{1j}/2$ values. (This case may occur, for example, in socio-economic systems where birth rates and capital generation rates result from individual decisions and have, in the aggregate, both causal and random fluctuations, whereas death rates and depreciation rates do not.) The procedure involves simply taking the interactions one by one, treating the other

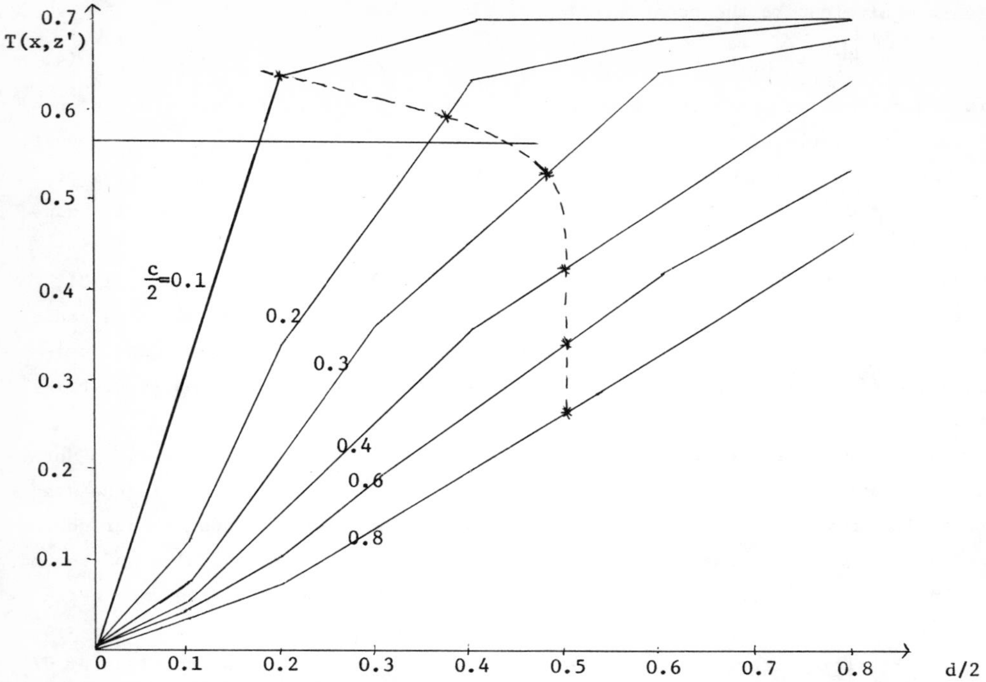

figure 1. Transmission as a function of d/2, with c/2 as parameter.

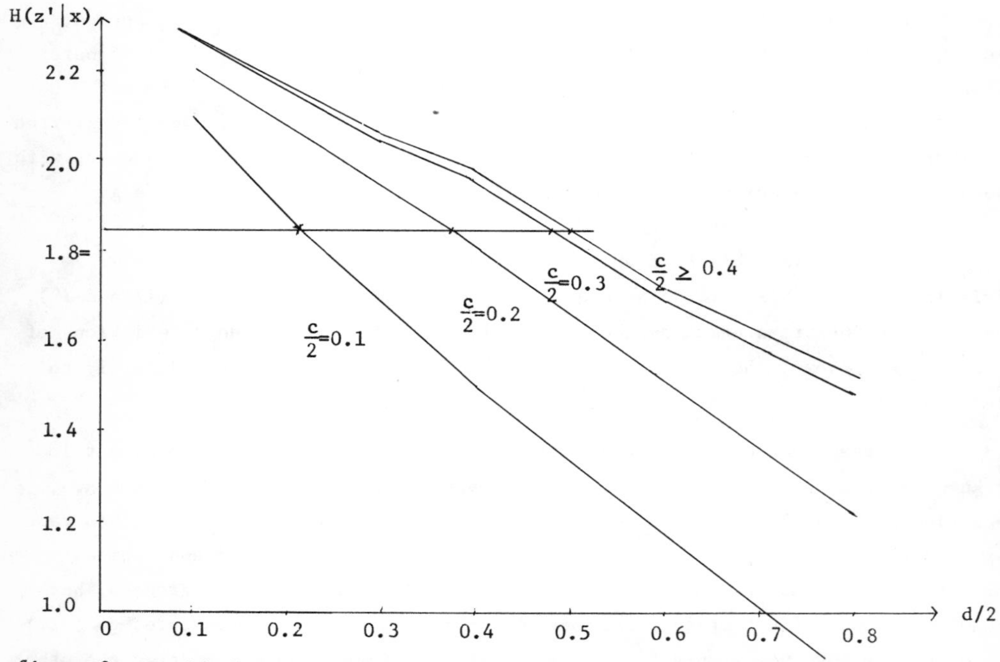

figure 2. Residual Entropy as a function of d/2, with c/2 as parameter

factors as a resultant, normally-distributed "noise" factor

$$(a+n_1) \prod_{j \neq k} f_{1j}(x_{i_j}) \equiv (a+\tilde{n}_{1k})$$

with residual entropy

$$\hat{H}(z_1' | x_{i_k}) = \hat{H}(z_1') - \hat{T}(x_{i_k}, z_1).$$

Here $\tilde{n}_{1k}$ has a complicated expansion in terms of n_1 and the functions

$$\frac{d_{1j}}{2} \operatorname{Sgn}(x_{i_j} - \bar{x}_{i_j}), \; j \neq k,$$

and the only justification for treating it as having a normal distribution lies in the tendency toward normality for functions of many stochastic variables, i.e. the law of large numbers.

In the more general case where both types of interactions f_{1j} and g_{1j} are known to exist and the standard deviations of n_i and m_i are unknown, more general procedures than those developed here will be needed.

Information and Performance

Quite apart from the parametric type of model discussed in the last section, there are some direct inferences that can be made about the potential performance of a large system, based on the transmission diagram. This is possible because there is, in any feedback control system, an intimate relationship between the amount of information available to the controller and the error performance of the system. In fact, it can be shown [6] that if the error entropy is taken as the measure of performance for a linear system, the reduction in error entropy achieved by feedback control is bounded by the information available to the controller.

Each state variable in a large system may be imagined to have a local controller which is receiving information about the values of certain other variables over separate communication channels, corresponding to those paths in the transmission diagram which impinge on the state variable in question. These channels represent a certain composite channel capacity ("information-gathering capacity" would be a more descriptive term) that is available to the local controller.

This composite local information capacity is not, however, simply the sum of the impinging transmission rates. The latter must be smaller than capacity if the channel reliability is to be adequate. Although Shannon's Coding Theorem shows that error-free transmission can be achieved at rates approaching the channel capacity, this can only be achieved at the expense of increasingly long time-delays for encoding and decoding the data, and arbitrarily long delays cannot be tolerated in real-time control. On the other hand, the total transmission rate will not be unnecessarily small, either, since control effectiveness increases with the amount of reliable data that is used. Thus, there is some optimal transmission between zero and capacity.

It has been shown [7] that, if the plant and controller are linear and certain other conditions are met, the optimal feedback transmission rate R satisfied the equation

$$2R = nE(R),$$

where n is the number of state variables being transmitted and E(R) is the channel reliability function. This equation always has a unique solution between zero and capacity. Conversely, if the transmission rate and n are known, the channel capacity can be found as C(R,n).

Thus, the composite local capacity C_1 available for the control of state variable x_1 can be estimated by calculating the total net transmission rate

$$R_1 = \frac{1}{T} \sum_{j=1}^{s_1-r_1} \hat{T}(x_{i_j}, x_1' | x_{i_1}, \ldots\ldots x_{i_{j-1}}, x_{i_{j+1}} \ldots\ldots x_{i_{s_1-r_1}})$$

where T is the sampling (coding) time interval, and then calculating

$$C_1 = C(R_1, s_1 - r_1).$$

The total information capacity in the system is then

$$C_t = \sum_{i=1}^{n} C_i.$$

Consider now the potential performance of the system in the extreme case where the total information-gathering capacity C_t is centralized, i.e. where all "local control" of state variables is replaced by a centralized controller which has the objective of optimizing a single performance index for the system as a whole.

If that performance index is the mean square state error and if the inherent "plant" mechanisms of the system are linear (not an unreasonable assumption for many socio-economic processes like population, capital, etc., as mentioned previously), then the optimum form of control law is linear state feedback and the separation theorem applies. In this case it can be shown [7,8] that the total mean square state error (m.s.e.) is bounded by

$$E[||x(t)-x_{optimum}||^2] \leq N^2(T) \sum_{i=2}^{\infty} \lambda_i(T) \frac{F(n,C_t,T)}{1-\gamma(T)\cdot F(n,C_t,T)} + e^2(T) \qquad (5)$$

where the summation gives the m.s.e. due to estimation error, i.e. limited information, and $e^2(T)$ is the m.s.e. resulting from the controller's limitations alone. $N^2(T)$ is the mean square state disturbance that would be caused by the exogenous noise acting on the uncontrolled system over the sample interval T and $\gamma(T)$ is the square of the largest eigenvalue of the state transition matrix of the uncontrolled system. The $\lambda_i(T)$ are the squares of the largest eigenvalues of matrices Φ_i which govern the estimation-induced state error dynamics according to the expression

$$\Delta x(t) = \sum_{i=2}^{\infty} \Phi_i [x(t-iT) - \bar{x}(t-iT)],$$

where $\bar{x}(t-iT)$ is the real-time controller's estimate of the state vector of time (t-iT). The function F is defined by

$$F(n,C_t,T) = (2A+k) \exp [-2R_t^* C_t T/n]$$

where A is an algebraic function of n, C_t, and T; k is a parameter between zero and one which measures the relative entropy power of the exogenous noise source compared to that of a white Gaussian source, and R_t* is the solution of the equation $2R_t^* = nE(R_t^*)$.

Several important conclusions can be drawn from the bound (5). For given state dimension n and controller gain matrix, the m.s.e. is a function of only C_t, the total information capacity, and T, the sampling-coding time. There exists the possibility of _information instability_ if $1 - \gamma(T)\cdot F(n,C_t,T)$ is not greater than zero. In particular, if

$$\lim_{T\to 0} [2A(n,C_t,T) + k] > 1,$$

then T must be greater than zero for the system to be informationally stable [8]. This result differs from the case of ordinary sampled-data control, where only $e^2(T)$ is present in the m.s.e. and the optimum sampling interval T is always zero.

Thus, for given n,C_t, and optimum central controller, there exists a $T_{opt} \geq 0$ which minimizes the bound (5). The relationship between these minimum values and C_t provides the link between the potential performance of a large-scale system and its transmission diagram.

REFERENCES

1. Watanabe, S., _IRE Transactions on Information Theory_, PGIT-4, 84 (1954).

2. Conant, R., "Detecting Subsystems of a Complex System," _IEEE Transactions on Systems, Man, and Cybernetics_, pp. 550-553, (1972).

3. Rink, R. E. and R. R. Mohler, "Completely Controllable Bilinear Systems," _SIAM Journal on Control_, Vol. 6, 477-487 (1968).

4. Mohler, R. R. and R. E. Rink, "Control with a Multiplicative Mode," _Journal of Basic Engineering_, Vol. _91_, 201-206 (1968).

5. Mohler, R. R., _Bilinear Control Processes, with Application to Engineering, Ecology, and Medicine_, Academic Press, N.Y. (1973).

6. Wiedemann, H. L., "Entropy Analysis of Feedback Control Systems," in _Advances in Control Systems_, C. T. Leondes, Ed., Academic Press, N.Y., Vol. 7 (1969).

7. Rink, R. E., "Optimal Utilization of Fixed-Capacity Channels in Feedback Control," _Automatica_, Vol. _9_, 251-255 (March 1973).

8. Rink, R. E., "Coded-Data Control of Linear Systems," (to be published).

A NEW CRITERION FOR MODELING SYSTEMS

By Lawrence W. Taylor, Jr.

NASA Langley Research Center
Hampton, Virginia

The problem of modeling systems continues to receive considerable attention because of its importance and because of the difficulties involved. A wealth of information has accumulated in the technical literature on the subject of systems identification and parameter estimation; references 1 through 5 are offered as summaries. The problem that has received most attention is one in which the form of the system dynamics is known, input and noisy output data are available, and only the values of the unknown model parameters are sought which optimize a likelihood function or the mean-square response error. It would seem with the variety of estimates, the algorithms, the error bounds, and the convergence proofs, and the numerical examples that exist for these systems identification problems, that any difficulties in obtaining accurate estimates of the unknown parameter values should be a thing of the past. Unfortunately, difficulties continue to confront the analyst. Perhaps the most important reason for difficulties in modeling systems is that we do not know the form of the system dynamics. Consequently, the analyst must try a number of candidate forms and obtain estimates for each. He is then confronted with the problem of choosing one of them with little or no basis on which to base his selection. It is tempting to use a model with many parameters since it will fit the measured response error best. Unfortunately, it is often the case that a simpler model would be better for predicting the system's response because the fewer number of unknown parameters could be estimated with greater accuracy. It is this problem of the analyst that this paper addresses and offers a criterion which can be used to select the best candidate model. Specifically, a numerical example will be used to illustrate the notions expressed in references 4, 6, and 7.

SYSTEMS IDENTIFICATION WITH MODEL FORMAT KNOWN

The systems identification problem of determining the parameters of a linear, constant-coefficient model will be considered for several types of estimates. It will be shown that maximum-likelihood estimates can be identical to those which minimize the mean-square response error. The subject algorithm, therefore, can be used to obtain a variety of similar estimates. Attention is also given to the calculation of the gradient that is involved in the algorithm and to the Cramer-Rao bound which indicates the variance of the estimates.

Problem Statement

The problem considered is that of determining the values of certain model parameters which are best with regard to a particular criterion, if the input and noisy measurements of the response of a linear, constant-coefficient system are given. The system to be identified is defined by the following equations:

$$\dot{x} = Ax + Bu \qquad y = Fx + Fu + b \qquad z = y + n$$

where x, u, y, b, n, z are the state, control, calculated response, bias, noise, and measured response, respectively. The unknown parameters will form a vector c. The matrices A, B, F, and G and the vectors b and $x(0)$ are functions of c.

Minimum Response Error Estimate

One criterion that is often used in systems identification is the mean-square difference between the measured response and that given by the model. A cost function which is proportional to the mean-square error can be written as

$$J = \sum_{i=1}^{N} (z_i - y_i)^T D_1 (z_i - y_i)$$

where D_1 is a weighting matrix and i is a time index. The summation approximates a time integral. The estimate of c is then

$$\hat{c} = \underset{c}{\text{ARG MIN}}(J)$$

where ARG MIN means that vector c which minimizes the cost function H.

Linearize the calculated response y with respect to the unknown parameter vector c

$$y_i = y_{i_0} + \nabla_c y_i (c - c_0)$$

where y_{i_0} is the nominal response calculated by using c_0, $\nabla_c y_i$ is the gradient of y with respect to c, and c_0 is the nominal c vector.

Substituting y_i into the expression for J and solving for the value of c which minimizes J yields

$$\hat{c} = c_0 + \left[\sum_{i=1}^{N} (\nabla_c y_i)^T D_1 \nabla_c y_i\right]^{-1} \left[\sum_{i=1}^{N} (\nabla_c y_i)^T D_1 (z_i - y_{i_0})\right]$$

If this relationship is applied iteratively to update the calculated nominal response and its gradient with respect to the unknown parameter vector, the minimum-response error estimate $\hat{c}$ will result. The method has been called quasi-linearization and modified Newton-Raphson. The latter seems more appropriate since the Newton-Raphson (ref. 1) method would give

$$c_{k+1} = c_k + \left[\nabla_c^2 J_k\right]^{-1} \left[\nabla_c J_k\right]$$

where

$$\nabla_c J_k = -2 \sum_{i=1}^{N} (\nabla_c y_{i_k})^T D_1 (z_i - y_{i_k})$$

$$\nabla_c^2 J_k = 2 \sum_{i=1}^{N} (\nabla_c y_{i_k})^T D_1 \nabla_c y_{i_k} - 2 \sum_{i=1}^{N} (\nabla_c^2 y_{i_k})^T D_1 (z_i - y_{i_k})$$

The second term of $\nabla_c^2 J_k$ diminishes as the response error $(z_i - y_{i_k})$ diminishes. The modified Newton-Raphson method is identical to quasi-linearization if the term is neglected.

Maximum Conditional Likelihood Estimate

Another criterion that is often used is to select c to maximize the likelihood of the measured response when c is given

$$\hat{c} = \underset{c}{\text{ARG MAX}}\{P(z|c)\}$$

where

$$P(z|c) = \frac{1}{(2\pi)^{N(MZ)/2}|M_1|^{N/2}} \exp\left[-\frac{1}{2}\sum_{i=1}^{N}(z_i - y_i)^T M_1^{-1}(z_i - y_i)\right]$$

The maximum conditional likelihood estimate of the unknown parameters will be the set of values of c which maximize $P(z|c)$, if it is recognized that y is a function of c. If it is noted that the maximization of $P(z|c)$ occurs for the value of c which minimizes the exponent, the maximum conditional likelihood estimate is the same as that which minimizes the mean-square response error provided the weighting matrix D_1 equals M_1^{-1}.

Maximum Unconditional Likelihood (Bayesian) Estimate

The unconditional probability density function of z can be expressed as

$$P(z) = P(z|c)P(c)$$

The probability of c relates to the a priori information available for c before use is made of the measurements z.

The unconditional probability of z is then

$$P(z) = \frac{1}{(2\pi)^{[(MC)+N(MZ)]/2}|M_1|^{N/2}|M_2|^{1/2}} \exp\left[-\frac{1}{2}\sum_{i=1}^{N}(z_i - y_i)^T M_1^{-1}(z_i - y_i) - \frac{1}{2}(c - c_0)^T M_2^{-1}(c - c_0)\right]$$

Again, the expression is maximized by minimizing the exponent. Setting the gradient with respect to c equal to zero and solving yields

$$\hat{c}_{k+1} = \hat{c}_k + \left[\sum_{i=1}^{N}(\nabla_c y_{i_k})^T M_1^{-1}\nabla_c y_{i_k} + M_2^{-1}\right]^{-1}\left[\sum_{i=1}^{N}(\nabla_c^T y_{i_k})^T M_1^{-1}(z_i - y_{i_k}) - M_2^{-1}(\hat{c}_k - c_0)\right]$$

An identical estimate would have resulted if a weighted sum of mean-square response error and the mean-square difference of c and its a priori value are minimized, provided the weighting matrices used equaled the appropriate inverse error covariance matrices. Consequently, the same algorithm can be used for both estimates

$$\hat{c} = \text{ARG}\min_c\left[\sum_{i=1}^{N}(z_i - y_i)^T M_1^{-1}(z_i - y_i) + (c - c_0)^T M_2^{-1}(c - c_0)\right]$$

Variance of the Estimates

A very important aspect of systems identification is the quality of the estimates. Since the estimates themselves can be considered to be random numbers, it is useful to consider their variance, or specifically their error covariance matrix. The Cramer-Rao bound (ref. 1) provides an estimate of the error covariance matrix.

$$E\left\{(\hat{c} - c_{true})(\hat{c} - c_{true})^T\right\} = \left[\sum_{i=1}^{N} (\nabla_c y_i)^T M_1^{-1} \nabla_c y_i + M_2^{-1}\right]^{-1}$$

If the a priori information is not used, M_2^{-1} will be the null matrix.

Importance of Testing Results

The importance of testing a model once estimates of the unknown parameters have been made, cannot be overemphasized. The test can be to predict response measurements not included as part of the data used in obtaining the model parameters. If the fit error, J, is greater for this test than for a similar test using a model with fewer unknown parameters, it would indicate that the data base is insufficient for the more complicated model. The testing should be repeated and the number of model parameters reduced until the fit error for the test is greater for a simpler model. The testing procedure should be used even if the model format is "known" since the data base can easily lack the quantity and the quality for estimating all of the model parameters with sufficient accuracy.

Reference 4 provides an example of a model which failed a test of predicting the system's response. Figure 1 shows how the fit error varied for both the data used to determine the model parameters and that of the test, versus the number of model parameters. The lower curve shows that as additional unknown parameters are included, the fit error decreases. When the models were tested by predicting independent response measurements, however, the upper curve shows the models having more unknown parameters performed more poorly than simpler models. Fortunately, more data were available but had not been used in the interest of reducing the computational effort. When four times the data were used to determine the parameters of the same models, the results shown in figure 2 were more successful. The fit error for the predicted response tests continued to decrease as the number of unknown model parameters increased, thereby validating the more complicated models.

Unless such tests are performed, the analyst is not assured of the validity of his model. Unfortunately, reserving part of the data for testing is costly. A means of making such tests without requiring independent data is discussed in a later section of the paper.

SYSTEM IDENTIFICATION WITH MODEL FORMAT UNKNOWN

One never knows in an absolute sense what the model format is of any physical system because it is impossible to obtain complete isolation. Even if the model format was known with certainty, it is necessary to test one's results against a simpler model, a point made in an earlier section. Regardless of the cause, the requirement is the same: a number of candidate models must be compared on a basis which reflects the performance of each in achieving the model's intended use. For the purpose of this discussion, it will be assumed that the model's intended use is to predict the response of a system and that a meaningful measure of the model's performance is a weighted mean-square error.

A Criterion for Comparing Candidate Models

Let us continue the discussion of the problem stated earlier using the same notation. The weighted mean-square response error which was minimized by the minimum response error estimate was

$$J = \sum_{i=1}^{N} (z_i - y_i)^T D_1 (z_i - y_i)$$

Let us denote the weighted mean-square response error which corresponds to testing the model's performance in predicting the system's response as

$$J^1 = \sum_{i=1}^{N} (z^1{}_i - y_i)^T D_1 (z^1{}_i - y_i)$$

where z^1 is measured response data that is not part of z which is used to determine the model parameters. For convenience, we consider the input, u, to be identical in both cases.

The criterion suggested for comparing candidate models is the expected value of J^1. If it is possible to express the expected value, $E\{J^1\}$, in terms not involving actual data for z^1, then a considerable saving in data can be made and improved estimates will result from being able to use all available data for estimating.

Let us examine first the expected value of the fit error with respect to the data used to determine estimates of the unknown parameters.

We can express the response error as

$$z - y = y_{true} + n - y \approx y_{true} + n - y_{true} - \nabla_c y(\hat{c} - c_{true})$$

$$= n - \nabla_c y(\hat{c} - c_{true})$$

assuming the response can be linearized with respect to the model parameters over the range in which they are in error.

The expected value of the fit error, $E\{J_1\}$, becomes

$$E\{J\} = E\left\{\sum_{i=1}^{N} (z_i - y_i)^T D_1 (z_i - y_i)\right\}$$

$$= E\left\{\sum_{i=1}^{N} \left[n - (\nabla_c y_i)(\hat{c} - c_{true})\right]^T D_1 \left[n - (\nabla_c y_i)(\hat{c} - c_{true})\right]\right\}$$

Expanding we get

$$E\{J\} = E\left\{\sum_{i=1}^{N} n_i{}^T D_1 n_i\right\} - 2E\left\{\sum_{i=1}^{N} n_i{}^T D_1 (\nabla_c y_i)(\hat{c} - c_{true})\right\}$$

$$+ E\left\{\sum_{i=1}^{N} (\hat{c}^T - c^T_{true})(\nabla_c y_i)^T D_1 (\nabla_c y_i)(\hat{c} - c_{true})\right\}$$

If a maximum likelihood estimate is used, or if the minimum mean-square response error estimate is used with a weighting equal to the measurement error covariance matrix, then we can write

$$D_1 = M^{-1}$$

$$\hat{c} = c_{true} + \left[\sum_{i=1}^{N} (\nabla_c y_i)^T M^{-1} (\nabla_c y_i)\right]^{-1} \left[\sum_{i=1}^{N} (\nabla_c y_i)^T M^{-1} n_i\right]$$

Again, linearization is assumed and it is noted that

$$z_i - y_{true_i} = n_i$$

After substituting we get

$$E\{J\} = E\left\{\Sigma\, n_i^T M^{-1} n_i\right\} - 2E\left\{P^T Q^{-1} P\right\} + E\left\{P^T Q^{-1} Q Q^{-1} P\right\}$$

$$= E\left\{\Sigma\, n_i^T M^{-1} n_i\right\} - E\left\{P^T Q^{-1} P\right\}$$

where

$$P = \sum_{i=1}^{N} (\nabla_c y_i)^T M^{-1} n_i$$

$$Q = \sum_{i=1}^{N} \nabla_c y_i \, M^{-1} \nabla_c y_i$$

Next, let us examine expected fit error $E\{J\}$ of a model used to predict response measurements, z^1, which are independent of the data z, used to determine the estimates of the model.

We can again express the expected fit error as

$$E\left\{J^1\right\} = E\left\{\sum_{i=1}^{N} n^1{}_i^T D_1 n^1{}_i\right\} - 2E\left\{\sum_{i=1}^{N} n^1{}_i^T D_1 \nabla_c y_i (\hat{c} - c_{true})\right\}$$

$$+ E\left\{\sum_{i=1}^{N} \left(\hat{c}^T - c_{true}^T\right)(\nabla_c y_i)^T D_1 \nabla_c y_i (\hat{c} - c_{true})\right\}$$

Note that the only difference between the above expression and that obtained earlier for $E\{J\}$ is that the noise vector is n^1 instead of n. The same expression can be used for $\hat{c}$ as before since it is the estimate of c based on the data, z, that is desired.

$$\hat{c} = c_{true} + \left[\sum_{i=1}^{N} (\nabla_c y_i)^T M^{-1} \nabla_c y_i\right]^{-1} \left[\sum_{i=1}^{N} (\nabla_c y_i)^T M^{-1} n_i\right]$$

Substituting the above expression for $\hat{c}$, and M^{-1} for D_1 we get

$$E\left\{J^1\right\} = E\left\{\sum_{i=1}^{N} n^1{}_i{}^T M^{-1} n^1{}_i\right\} - 2E\left\{\sum_{i=1}^{N} n^1{}_i{}^T M^{-1} \nabla_c y_i Q^{-1} \sum_{i=1}^{N} (\nabla_c y_i)^T M^{-1} n_i\right\}$$

$$+ E\left\{P^T Q^{-1} P\right\}$$

where P and Q are defined as before. Since the noise vector, n^1 and n, are uncorrelated, that is

$$E\left\{n_i n^1{}_i{}^T\right\} = 0 \quad \text{for all } i \neq j$$

then the second term is zero.

Since the noise vectors n and n^1 have the same covariance matrix, M, we can write

$$E\left\{\sum_{i=1}^{N} n^1{}_i{}^T M^{-1} n^1{}_i\right\} = \text{TRACE}\left[E\left\{\sum_{i=1}^{N} n^1{}_i{}^T M^{-1} n^1{}_i\right\}\right]$$

$$= \text{TRACE}\left[E\left\{\sum_{i=1}^{N} n^1{}_i n^1{}_i{}^T\right\} M^{-1}\right]$$

$$= \text{TRACE}\left[NI\right] = N \cdot MZ$$

where N is the number of time samples and MZ is the number of measurement quantities. Also

$$E\left\{\sum_{i=1}^{N} n_i{}^T M^{-1} n_i\right\} = N \cdot MZ$$

We can now express $E\{J^1\}$ in terms of $E\{J\}$ as

$$E\{J^1\} = E\{J\} + 2E\{P^T Q^{-1} P\}$$

Examining the second term

$$E\{P^T Q^{-1} P\} = E\left\{\sum_{i=1}^{N} n_i{}^T M^{-1} \nabla_c y_i Q^{-1} \sum_{i=1}^{N} (\nabla_c y_i)^T M^{-1} n_i\right\}$$

After taking the trace of the scalar and reordering the vectors we get

$$E\{P^TQ^{-1}P\} = \text{TRACE}\left[\sum_{i=1}^{N} \nabla_c y_i Q^{-1} \sum_{j=1}^{N} (\nabla_c y_j)^T M^{-1}\right]$$

Because the noise is uncorrelated at unlike times, the term simplifies to

$$E\{P^TQ^{-1}P\} = \text{TRACE}\left[E\left\langle \sum_{i=1}^{N}\sum_{j=1}^{N} n_j n_i^T M^{-1} \nabla_c y_i Q^{-1} (\nabla_c y_j)^T M^{-1}\right\rangle\right]$$

Finally, we have that the expected fit error for the case of testing the model's prediction of the system's response as

$$E\{J^1\} = J + 2\,\text{TRACE}\left[\sum_{i=1}^{N} \nabla_c y_i Q^{-1} \sum_{j=1}^{N} (\nabla_c y_j)^T M^{-1}\right]$$

Since it is available, the actual fit error, J, is used instead of its expected value. The intent of the new criterion, $E\{J^1\}$, is that it be used instead of J in determining the level of model complexity that is best.

An Application of the New Modeling Criterion

The control input and response time histories of the numerical example used to demonstrate the use of the new modeling criterion are shown in figure 3. The dynamics resemble that of the lateral-directional modes of an airplane and are described in detail in reference 5. The level of noise that has been added to the calculated time histories is also indicated in figure 3. The sampling rate was 10 samples per second. The unknown parameters in the A, B, and b matrices can number as many as 10, 9, and 4, respectively. Only 16 parameters have values other than zero.

Five sets of calculated responses to which noise was added, were analyzed using the algorithm of reference 5. The analysis was repeated, allowing a different number of parameters to be determined by the algorithm. The fit error, J, in each case was averaged for the five sets of responses, and plotted as the lower curve of figure 4 as a function of the number of unknown parameters allowed to vary. As the number of unknown parameters was increased, terms were always added and never substituted. Because of this it can be argued that the fit error, J, should be monotonically decreasing as the number of unknown parameters increases. It can be seen in figure 4, however, that a point is reached beyond which J increases. The reason for the increase is because there is a point reached where convergence is a problem because of the essentially redundant unknown parameters. Although some of the five cases for 23 unknown parameters showed a decrease in J, others did not. The analyst might be tempted to settle for the model having 19 unknown parameters since the calculated response best fits the "measured response." It would be a mistake to do so; one that is often made.

In addressing the question "Which of the candidate models best predict system response?" a test was applied. The results of the test were then compared to the modeling criterion in the following way. For each set of unknown parameters determined, the corresponding calculated response was compared to the true response to which independent noise was added. The resulting fit errors were again averaged and plotted in figure 4 as the upper curve. The test performed simulates the practice of reserving actual response data for model testing purposes only. As a result of these tests it can be seen that a model with only 10 unknown parameters is best in terms of predicting system response. This is six parameters fewer than the "true model." That is to say, there was less error in setting six parameters to zero than the error resulting from trying to determine more than 10 unknown parameters. The number of parameters that best predict system response will increase if additional data are used or if the noise is decreased.

The purpose of the new modeling criterion is to eliminate the need of independent response measurements comparing the model candidates, thereby allowing all of the response data to be used in determining the unknown parameters. The new criterion was calculated as part of the system's identification algorithm and the results are shown in figure 5. The results of testing the candidate models given in the previous figure are included in figure 5 for comparison. The minimum in the curve for the new criterion occurs at 10 unknown parameters as was the case for the previous test results. This indicates that the new criterion can be used instead of testing the model candidates against independent response data.

The Problem of Too Many Candidate Models

Although it is a great help to have a criterion for comparing candidate models, there remains a problem of an excessive number of candidate models. A simple calculation illustrates the enormous number of candidate models that result from the combinations of unknown parameters that can be used for the simple example used to illustrate the modeling criterion. The total number of possible candidate models exceeds 8 million. Even though it is possible to greatly reduce the amount of calculation effort by neglecting changes in the gradient of the response with respect to the unknown parameters, $\nabla_c y$, solving a set of simultaneous algebraic equations would still be required for each model. Consequently, the calculations involved for 8 million candidate models becomes an economic consideration. It is estimated that testing all of the 8 million candidate models would require about one hour on a CDC-6600 computer. As the maximum number of unknown parameters increases, the number of possible candidate models rapidly becomes astronomical.

In practice, the analyst has enough understanding of the dynamics of the system being modeled to know to some degree which terms are primary and which are less important. It would be valuable, however, if one did not have to rely on the analyst's judgment. The problem of searching for the best candidate model, therefore, remains a formidable problem worthy of attention.

REFERENCES

1. Balakrishnan, A. V.: Communication Theory. McGraw-Hill Book Company, C. 1968. IFAC Symposium 1967 on the Problems of Identification in Automatic Control Systems (Prague, Czechoslovakia), June 12-17, 1967.

2. Cuenod, M., and Sage, A. P.: Comparison of Some Methods Used for Process Identification. IFAC Symposium 1967 on the Problems of Identification in Automatic Control Systems (Prague, Czechoslovakia), June 12-17, 1967.

3. Eykhoff, P.: Process Parameter and State Estimation. IFAC Symposium 1967 on the Problems of Identification in Automatic Control Systems (Prague, Czechoslovakia), June 12-17, 1967.

4. Taylor, Lawrence W., Jr.: Nonlinear Time-Domain Models of Human Controllers. Hawaii International Conference on System Sciences (Honolulu, Hawaii), January 29-31, 1968.

5. Taylor, Lawrence W., Jr., and Iliff, Kenneth W.: Systems Identification Using a Modified Newton-Raphson Method - A Fortran Program. NASA TN D-6734, May 1972.

6. Taylor, Lawrence W., Jr.: How Complex Should a Model Be? 1970 Joint Automatic Control Conference (Atlanta, Georgia), June 24-26, 1970.

7. Akaike, H.: Statistical Predictor Identification. Ann. Inst. Statist. Math., Vol. 22, 1970.

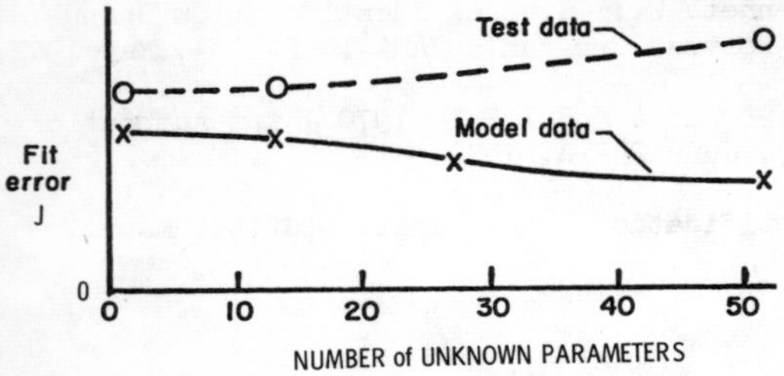

Figure 1. An example of insufficient data/parameters. Pilot tracking case. T = 60 seconds.

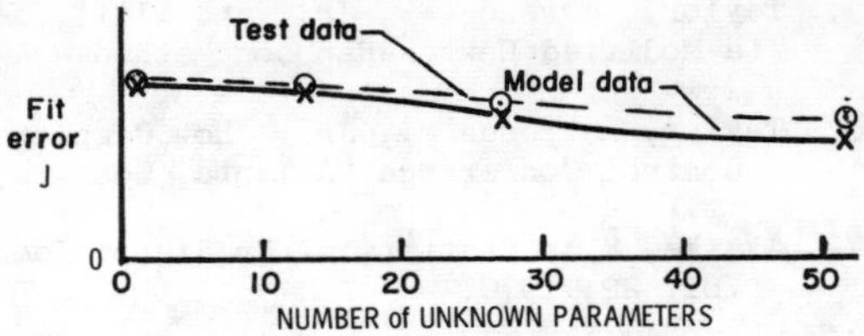

Figure 2. An example of sufficient data/parameters. Pilot tracking case. T = 240 seconds.

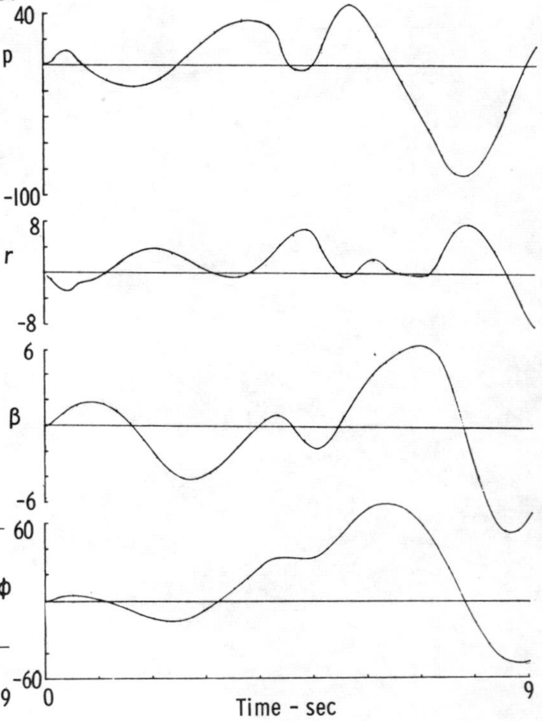

Figure 3. Response histories for example problem. Lateral-directional case.

Units in Degrees, Degrees per Sec

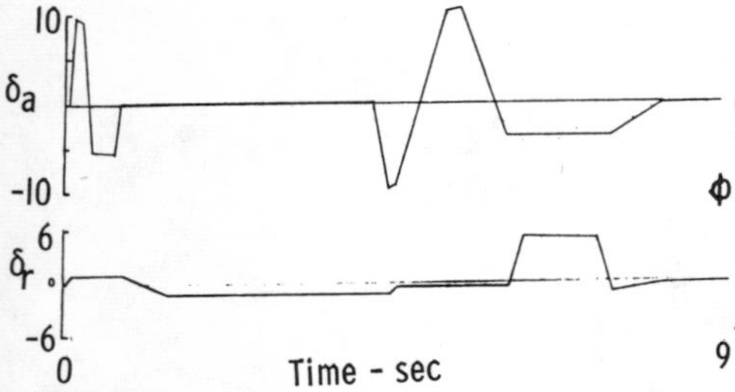

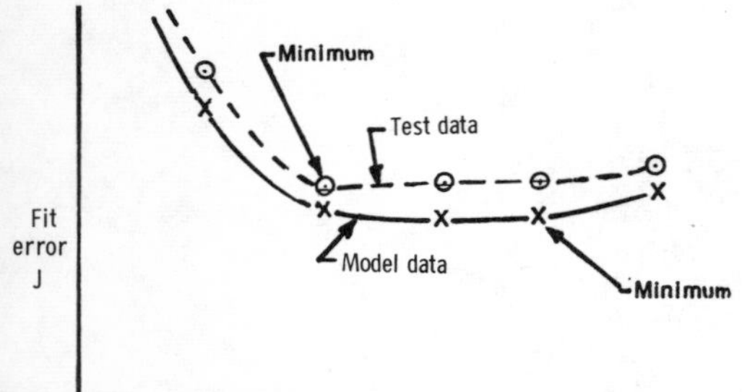

Figure 4. Comparison of fit error using model and test data. Lateral-directional case.

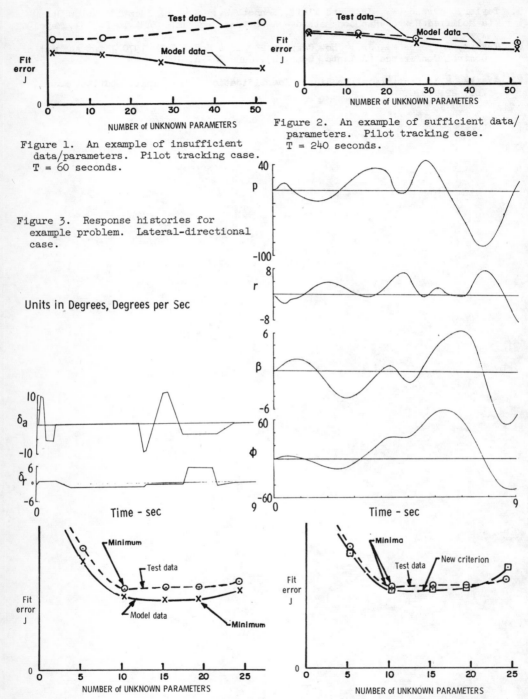

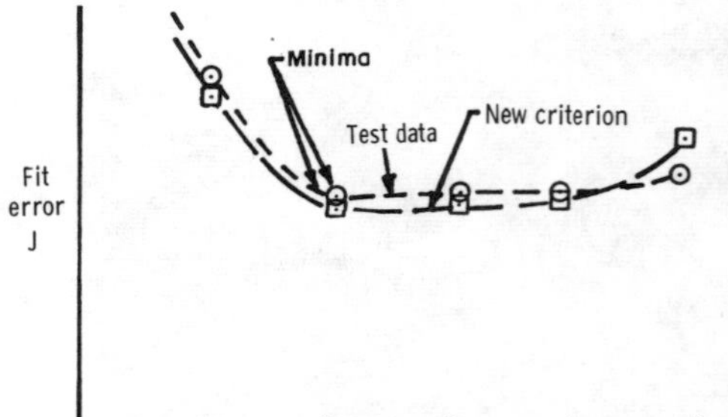

Figure 5. Comparison of fit error using test data and new criterion. Lateral-directional case.

STOCHASTIC EXTENSION AND FUNCTIONAL RESTRICTIONS OF ILL-POSED ESTIMATION PROBLEMS

Edoardo Mosca

Facoltà di Ingegneria
Università di Firenze

Abstract - In Theorem 1 it is shown that under mild conditions the minimum-variance smoothed-estimation of a Gaussian process $y(t)$ with covariance $R_y(t,\tau)$ from noise-corrupted measurements $z(t)=y(t)+n(t)$, $t\in T$, with $n(t)$ Gaussian with covariance $R_n(t,\tau)$, is equivalent to a purely deterministic optimization problem, namely, finding $y\in H(R_y)$ that minimizes the functional

$$L(y|z) = \| y \|^2_{R_n} - 2\,\mathrm{Re}\left[\langle z, y\rangle_{R_n}\right] + \| y \|^2_{R_y} .$$

Here $\langle\cdot,\cdot\rangle_R$ denotes inner-product of the reproducing kernel Hilbert space (RKHS) $\mathbb{H}(R)$. Theorems 2 and 3 deal with the more general situation where y is a set of linear measurements on an unknown function $w(\alpha)$, $\alpha\in A$.

The above stochastic-deterministic equivalence provides valuable insight and important consequences for modelling. An example shows how this equivalence allows one to make use of Kalman-Bucy filtering in a purely deterministic problem of smoothing.

1. Introduction

The present paper evolves from some ideas of the author (Mosca 1970 a, b, 1971 a, b, 1972 a, b, c), and is very close in spirit with some recent works by Kimeldorf and Wahba (1970, 1971), and Parzen (1971). It deals with a problem that often occurs in experimental work, for instance, in indirect-sensing experiments. Roughly speaking, the problem consists of reconstructing an unknown function or signal w from a set of linear measurements on w and from the knowledge of the class to which the signal belongs. Specifically, it is assumed that:

(a1). The unknown w is an element of a complex Hilbert space $\mathcal{H}$ of functions mapping their domain A into the p-fold Cartesian product C^p of the complex numbers C. The inner product of $\mathcal{H}$ is denoted by $\langle\cdot,\cdot\rangle_{\mathcal{H}}$, and the corresponding norm by $\|\cdot\|_{\mathcal{H}}$. In most applications, either $\mathcal{H}=L_2(A)$, where A is a Lebesgue-measurable subset of the r-fold Cartesian product R^r of the real line R, or $\mathcal{H}=l_2(A)$, where A denotes a subset of the set of all integers.

(a2). A <u>linear measurement</u> on an element $w\in\mathcal{H}$ equals the value of a continuous linear functional

$$\langle w, x\rangle_{\mathcal{H}}$$

where, by the Riesz representation theorem (Naylor and Sell, 1971), $x\in\mathcal{H}$ is the representator of the functional.

Under the above assumptions, the problem can be stated as follows.

Problem 1. Given the observations

$$z(t) = y(t) + n(t) \quad , \; t \in T,$$

equal to the sum of an additive noise n(t) and a set of linear measurements on $w \in \mathcal{H}$

$$y(t) = \langle w, x(\cdot,t) \rangle_{\mathcal{H}} \quad , \tag{1}$$

with $\{x(\cdot,t), t \in T\}$ a known family of functions in $\mathcal{H}$,

$$x(\cdot,t) \in \mathcal{H} \quad , \; t \in T,$$

find a suitable estimate of the unknown $w \in \mathcal{H}$. ▨
With reference to Problem 1 some further assumptions are needed:

(a3). The observation set T is a Lebesque-measurable subset of the q-fold Cartesian product R^q of the real line R.

(a4). The additive noise n is a sample function from a zero-mean stochastic process with covariance kernel

$$R_n(t,\tau) \triangleq E\left[n(t)\, n^*(\tau)\right] < \infty \quad , \quad t,\tau \in T.$$

Example 1. Let u be the input to an n-dimensional linear system with continuous coefficients

$$\begin{aligned} \dot{\underline{x}}(t) &= \underline{A}(t)\,\underline{x}(t) + \underline{b}(t)\, u(t) \\ y(t) &= \underline{c}(t)\,\underline{x}(t) \end{aligned}$$

starting from the initial state $\underline{x}(t_o) = \underline{x}_o$. By observing the system output y over T= $[t_0, t_1]$, it is desired to reconstruct the initial state $\underline{x}_o$ and the system input over the same interval T. The variation of constants formula (Brockett, 1970) gives immediately

$$\begin{aligned} y(t) &= \underline{c}(t)\,\underline{\phi}(t,t_o)\,\underline{x}_o + \int_{t_o}^{t_1} h(t,\alpha)\, u(\alpha)\, d\alpha \\ &= \langle \underline{x}_o, \underline{\phi}'(t,t_o)\,\underline{c}'(t) \rangle_{E^n} + \langle u, h(t,\cdot) \rangle_{L_2[t_o,t_1]} \quad , \; t \in [t_o,t_1], \end{aligned}$$

where: $\underline{\phi}(t,\alpha)$ is the state-transition matrix for $\dot{\underline{x}}(t) = \underline{A}(t)\,\underline{x}(t)$;

$$h(t,\alpha) \triangleq \underline{c}(t)\,\underline{\phi}(t,\alpha)\,\underline{b}(\alpha)\; 1(t-\alpha)$$

is the system impulse-response; 1(t) the unit step function; E^n is the Euclidean n-space; and $L_2[t_o, t_1]$ is the space of all real-valued square-integrable functions on $[t_o, t_1]$. If one defines $\mathcal{H}$ as the direct sum (Halmos, 1957)

$$\mathcal{H} = E^n \oplus L_2[t_o,t_1]$$

and

$$x(\cdot,t) \triangleq \{\underline{\phi}'(t,t_o)\,\underline{c}'(t),\; h(t,\cdot)\} \in \mathcal{H} \quad ,$$

the observed output y can be expressed in terms of the unknown

$$w(\cdot) \triangleq \{\underline{x}_o, u(\cdot)\} \in \mathcal{H}$$

as in (5). ▨

If $\mathcal{H}$ is infinite-dimensional and T a continuous observation set, under fairly general conditions it has been shown (Mosca, 1970 a, b, 1971 a), that no linear unbiased estimate of $w \in \mathcal{H}$ exists. This means that there is no $w \in \mathcal{H}$ such that $\langle w, x(\cdot,t)\rangle_{\mathcal{H}} = \hat{y}(t) \quad , \quad t \in T,$ where y(t) is the minimum-variance linear estimate of the value y(t) taken on by y at $t \in T$. Following Hadamard's terminology (Hadamard, 1952), Problem 1 is referred to as an ill-posed estimation problem.

The chief goal of this paper is to compare the two main approaches that have been devised to deal with the class of ill-posed problems described above. The first approach (Franklin, 1970, Turchin, Kozlov, and Malkevich, 1971), called hereafter the stochastic-extension approach, consists of reinterpreting the unknown w as a sample function from a random process with zero-mean and covariance kernel R_w defined on A x A. Thus, known methods of statistical inference can be used to get a minimum-variance linear estimate of $w(\alpha)$, $\alpha \in A$. The second approach (Mosca, 1971 b, 1972 a) called hereafter the functional-restriction approach, consists of constraining the unknown w to a linear subset of $\mathcal{H}$ of well-behaved functions χ on A, completely characterized by an Hermitian nonnegative-definite kernel K_χ on A x A. In functional analysis this linear set of functions is denoted by $H(K_\chi)$, and called the reproducing kernel Hilbert space (RKHS) with reproducing kernel (RK) K_χ.

2. Stochastic Extensions

For the sake of simplicity, hereafter $\mathcal{H}$ is assumed to be equal to $L_2(A)$. This makes the discussion somewhat less abstract, and the notation less cumbersome. However, it should be clear that the conclusions that will be obtained are valid for more general Hilbert spaces $\mathcal{H}$.

Let R_w be the covariance kernel of the zero mean process w

$$R_w(\alpha,\alpha_1) \triangleq E\left[w(\alpha)\, w^*(\alpha_1)\right] < \infty \quad , \quad \alpha,\alpha_1 \in A.$$

The stochastic nature of w induces a randomness on y. In fact, because of (1), y turns out to be a zero-mean stochastic process with covariance kernel

$$\begin{aligned} R_y(t,\tau) &\triangleq E\left[y(t)\, y^*(\tau)\right] \\ &= \int_A\int_A x(\alpha_1,\tau)\, R_w(\alpha,\alpha_1)\, x^*(\alpha,t)\, d\alpha\, d\alpha_1 \,, \quad t,\tau \in T. \end{aligned} \tag{2}$$

Eq. (2) can be rigorously justified by using Fubini-Tonelli theorem (Royden, 1968) in conjunction with some auxiliary assumption on R_w, for instance, the finiteness of its trace

$$\mathrm{Tr}\, R_w \triangleq \int_A R_w(\alpha,\alpha)\, d\alpha < \infty .$$

Assuming the processes w and n uncorrelated

$$E[w(\alpha)n^*(t)] = 0\,, \qquad \alpha \in A,\ t \in T,$$

the crosscorrelation between w and z is simply

$$\rho(\alpha,t) \triangleq E[w(\alpha)z^*(t)] = E\left[w(\alpha)\int_A w^*(\alpha_1)x(\alpha_1,t)d\alpha_1\right]$$
$$= \int_A R_w(\alpha,\alpha_1)\,x(\alpha_1,t)\,d\alpha_1 .$$

Using a formula (Parzen, 1967) that gives the minimum-variance linear estimate in terms of a RKHS inner-product[1], in the present instance one has

$$\hat{w}(\alpha) \triangleq E[w(\alpha)\,|\,z(t), t\in T] = \langle z, \rho^*(\alpha,\cdot)\rangle_{R_y+R_n}$$
$$= \langle z(t), \int_A R_w(\alpha_1,\alpha)\,x^*(\alpha_1,t)\,d\alpha_1\rangle_{R_y+R_n} .$$

It is of some interest to note that the image of $\hat{w}$ under the transformation (1) mapping functions on A into functions on T coincides with the minimum-variance linear estimate $\hat{y}$ of y. In fact,

$$\int_A \hat{w}(\alpha)x^*(\alpha,t)d\alpha = \langle z(\tau), \int_A\int_A x(\alpha,t)R_w(\alpha_1,\alpha)x^*(\alpha_1,\tau)\,d\alpha\,d\alpha_1\rangle_{R_y+R_n}$$
$$= \langle z, R_y(\cdot,t)\rangle_{R_y+R_n} = E[y(t)\,|\,z(\tau),\tau\in T] \triangleq \hat{y}(t). \tag{3}$$

From now on, w and n are assumed to be Gaussian. Most of the results that follow, however, can also be reinterpreted for non-Gaussian processes under the usual provision of translating strict-sense concepts into their corresponding wide-sense versions. The following theorem, whose proof can be found in the Appendix, is the key result that allows one to relate the stochastic-extension approach and the functional-restriction approach.

Theorem 1. Let the two probability measures induced on the space of sample functions on T by the kernels R_y+R_n and R_n be equivalent. Then the minimum-variance estimate $\hat{y}(t)$, $t\in T$, of y(t) based on z is with probability one the element in $H(R_y)$ minimizing the functional

$$F(y|z) = \ell(y|z) + \|y\|^2_{R_y} \tag{4}$$

over all $y \in H(R_y)$ iff the two probability measures corresponding to R_y+R_n and R_n are strongly equivalent in the Hájek's sense. ▨

---

[1] Throughout this paper the following notational convention is adopted. Given an Hermitian nonnegative-definite kernel P, H(P) will denote the unique corresponding RKHS, $\langle\cdot,\cdot\rangle_P$ the inner product of H(P), and $\|f\|_P = +\sqrt{\langle f,f\rangle_P}$ the norm of $f \in H(P)$.

In (4) the functional

$$\ell(y|z) \triangleq \|y\|^2_{R_n} - 2\,\mathrm{Re}\left[\langle z, y \rangle_{R_n}\right] \tag{5}$$

equals minus the natural log of the probability density functional, Radon-Nikodym derivative or likelihood ratio, of the probability measure induced on the space of sample functions by the process

$$\{z(t) = y(t) + n(t),\ t \in T,\ y \text{ fixed in } H(R_y)\}$$

with respect to the probability measure corresponding to the process $\{z(t)=n(t),\ t \in T\}$. Notice also that Theorem 1 gives a necessary and sufficient condition for y to be in $H(R_y)$. This condition, namely strong equivalence, if[2]

$$R_n(t,\tau) = \delta(t-\tau)\ ,$$

i.e., n is white, is simply

$$Tr\,R_y \triangleq \int_T R_y(t,t)\,dt < \infty\ .$$

3. Functional Restrictions

Let $\mathcal{H}$ be again an arbitrary functional Hilbert space, and Γ be the linear transformation mapping $\mathcal{H}$ into functions y on T according to (1). Let $\mathcal{R}[\Gamma]$ be the range of Γ. Next, let $\mathcal{X} \subset \mathcal{H}$ be the space of all minimum-norm solutions χ of (1), when y runs over $\mathcal{R}[\Gamma]$. $\mathcal{X}$ can also be equivalently defined as the closed subspace of $\mathcal{H}$ spanned by the t-indexed family of functions $\{x(\cdot, t),\ t \in T\}$. Then, the restriction Ψ of Γ to $\mathcal{X}$:

$$[\Psi\chi](t) \triangleq \langle \chi, x(\cdot,t) \rangle_{\mathcal{H}} = y(t)$$

has the properties listed in the following

__Lemma__ (Mosca 1970a, b, 1971 a, b, 1972a)

i) The range of Ψ coincides with RKHS $H(K_y)$ with RK given by

$$K_y(t,\tau) = \langle x(\cdot,\tau), x(\cdot,t) \rangle_{\mathcal{H}}\ ,\quad t,\tau \in T. \tag{6}$$

ii) The transformation $\Psi : \mathcal{X} \to H(K_y)$ establishes a __congruence__ or isometric isomorphism between $\mathcal{X}$ and $H(K_y)$, i.e.

$$\begin{matrix} \chi_1, \chi_2 \in \mathcal{X} \\ \Psi\chi_1 = y_1\ ,\ \Psi\chi_2 = y_2 \end{matrix} \quad \Longrightarrow \quad \langle \chi_1, \chi_2 \rangle_{\mathcal{H}} = \langle y_1, y_2 \rangle_{K_y}\ .$$

---

[2] If n is white, the discussion can be made mathematically rigorous by using the notion of a generalized process (Gelfand and Vilenkin, 1964). On a more qualitative basis, one can think of $H(\delta)$ as the pseudo-RKHS $L_2(T)$.

iii) The inverse transformation Ψ^{-1}: $H(K_y) \to \mathcal{X}$ of Ψ is given by

$$[\Psi^{-1}y](\alpha) = \langle y, x^*(\alpha,\cdot)\rangle_{K_y} , \qquad \alpha \in A, \tag{7}$$

where the right-hand side of this equation defines the preimage of y under Ψ in the norm-topology of $\mathfrak{H}$.

Properties i)-iii) have naturally direct relation with the problem of the present paper if $\mathfrak{H}$ is identified with $L_2(A)$. With these premises, we are now ready to describe the functional -restriction approach through the steps that follow.

1. Choose an Hermitian nonnegative-definite kernel K_χ in such a way that $H(K_\chi) \subset \mathcal{X}$. Thus, as is proved in the Appendix, for any $\chi \in H(K_\chi)$

$$y(t) = \langle \chi, x_r(\cdot,t)\rangle_{K_\chi} \tag{8}$$

where

$$x_r(\alpha,t) \triangleq \int_A x(\alpha_1,t)\, K_\chi^*(\alpha_1,\alpha)\, d\alpha_1 . \tag{9}$$

From the Lemma it follows immediately that the transformation defined by (8) establishes a congruence between $H(K_\chi)$ and RKHS $H(K_y)$ with RK

$$K_y(t,\tau) = \int_A\int_A x(\alpha_1,\tau)\, K_\chi(\alpha,\alpha_1)\, x^*(\alpha,t)\, d\alpha\, d\alpha_1 , \qquad t,\tau \in T. \tag{10}$$

Moreover, if y and χ are related as in (8),

$$\chi(\alpha) = \langle y, x_r^*(\alpha,\cdot)\rangle_{K_y} , \qquad \alpha \in A. \tag{11}$$

2. Among all y on the hypersphere

$$\| y \|_{K_y} = \| \chi \|_{K_\chi} = \rho$$

find the one, call it $\check{y}_\rho$, minimizing the functional

$$\ell(y|z)$$

defined by (5).

3. Increase the radius ρ until large variations $\delta\rho$ produce small variations of

$$|z(t) - \check{y}_\rho(t)| , \quad t \in T,$$

Denote the function y obtained in this way by $\check{y}$.

4. The preimage of $\check{y}$ under the transformation (8) is the <u>functional-restricttion estimate</u> $\check{\chi}$ of χ.

$$\check{\chi}(\alpha) = \langle \check{y}, x_r^*(\alpha,\cdot)\rangle_{K_y} , \qquad \alpha \in A. \ \blacksquare \tag{12}$$

Notice, that, by introducing the Lagrange multiplier λ, steps 2 and 3 are equivalent to minimizing the functional

$$L_\lambda(y|z) \triangleq \ell(y|z) + \lambda \| y \|^2_{K_y} , \tag{13}$$

with suitable λ , among all $y \in H(K_y)$. To this end, it has already been shown (Mosca, 1972a) that an iterative algorithm of the steepest-descent type can be used to compute $\check{y}$ and $\check{\chi}$ iff the two probability measures corresponding to $K_y + R_n$ and R_n are strongly equivalent. Finally, it is to be noticed, that, regardless of the interpretation that led us to (13), the problem of minimizing the functional (13) is the classical problem of smoothing in applied mathematics.

4. Equivalence between Stochastic Extensions and Functional Restrictions

Comparing (13) with (4), and taking into account the basic role played in both cases by the condition of strong equivalence, one can conclude that if this condition is fulfilled and

$$K_y = \lambda R_y \qquad \text{so that} \qquad \|y\|^2_{R_y} = \lambda \|y\|^2_{K_y} ,$$

then the stochastic-extension (minimum-variance) estimate $\hat{y}$ equals the functional-restriction estimate $\check{y}$. In practice, the comparison between the two approaches is even simpler for in the stochastic-extension approach R_y and R_n are specified up to multiplicative positive constants. This means that in practical instances $\hat{y}$ has to be found by minimizing

$$F_\lambda(y|z) \triangleq \ell(y|z) + \lambda \|y\|^2_{R_y} \tag{14}$$

instead of (4). As in (13), here λ must be estimated adaptively from the actual received data $z(t)$, $t \in T$. $L_\lambda(y \mid z)$ and $F_\lambda(y \mid z)$ are now of the same form, and therefore the next theorem follows.

Theorem 2 (Equivalence between $\check{y}$ and $\hat{y}$). Let K_y be the RK used in the functional-restriction estimation of y, and R_y the covariance kernel used in the stochastic-extension estimation of y. Let

$$K_y = R_y,$$

and the two probability measures corresponding to $R_y + R_n$ strongly equivalent. Then the functional-restriction estimate of y equals with probability one the stochastic-extension estimate of y

$$\check{y}(t) = \hat{y}(t) , \quad t \in T . \quad ▨$$

Next theorem concerns the estimation of χ.

Theorem 3 (Equivalence between $\check{\chi}$ and $\hat{\chi}$). Let K_χ be the RK used in the functional-restriction estimation of χ, and R_w be the covariance kernel used in the stochastic-extension estimation of w. Let

$$K_\chi = R_w \tag{15}$$

be such that

$$H(K_\chi) \subset \mathscr{X} . \tag{16}$$

Finally, let the two probability measures corresponding to $R_y + R_n$ and R_n be strongly equivalent. Then, the functional-restriction estimates equal with probability one the corresponding stochastic-extension estimates, namely,

$$\check{y}(t) = \hat{y}(t)\ ,\quad t \in T,$$
$$\check{\chi}(\alpha) = \hat{\chi}(\alpha)\ ,\quad \alpha \in A. \tag{17}$$

<u>Proof.</u> The first of (17) follows as a corollary of Theorem 1. The second of (17) can be proved by noting that (8), being a congruence, is 1-1 between $H(K_\chi)$ and $H(K_y)$, and that, being $\hat{\chi}$ in $H(K_\chi)$,

$$\langle \hat{\chi}, x_r(\cdot,t)\rangle_{K_\chi} = \int_A \hat{\chi}(\alpha)\, x^*(\alpha,t)\, d\alpha = \hat{y}(t)\ ,\quad t \in T,$$

where the last equality was proved in (3).

5. Conclusions.

The equivalence that has been proved to exist between functional restrictions and stochastic extensions in estimation, provides valuable insight and important consequences for modelling. To those more inclined to accept deterministic rather than stochastic formulations, the above results tell that the covariance kernel R_w used in the estimation of w can be chosen on the grounds of the analytic properties of the RKHS $H(R_w)$. The opposite is evidently also true. More importantly, one can make direct use of efficient algorithms developed within the stochastic framework to find functional-restriction estimates.

<u>Example 2 (Smoothing).</u> Let $z(t)$, $t \in T = [t_0, t_1]$, be an observed process. The problem is to find the function y on T with absolutely continuous first derivative and with

$$y(t_0) = \dot{y}(t_0) = 0$$

minimizing the functional

$$L(y|z) = \int_T [y^2(t) - 2z(t)y(t)]\,dt + \int_T [\ddot{y}(t)]^2 dt.$$

This is a completely deterministic problem of smoothing, the functional to be minimized being a compromise between the smoothness of the estimate and fidelity to the data. However, in order to make use of known algorithms, like Kalman-Bucy filtering, developed and usually applied in a purely stochastic context, we will reformulate the problem in a convenient way.

First, note that the first integral at the right-hand side of the last equation can be interpreted as the probability density functional $\ell(y\,|\,z)$ corresponding to the process

$$z(t) = y(t) + n(t)\ ,\quad t \in T,$$

with y a given deterministic signal and n a zero-mean white Gaussian process with covariance

$$R_n(t,\tau) = \delta(t-\tau)\ ,\quad t,\tau \in T.$$

Second, consider the set $\mathcal{Y}$ of all signals y generated as outputs on T of the finite-dimensional linear system

$$\mathcal{S}:\begin{cases}\ddot{x}(t) = u(t)\\ y(t) = x(t)\\ x(t_0) = \dot{x}(t_0) = 0\end{cases} \qquad \text{or} \qquad \begin{cases}\dot{\underline{x}}(t) = \begin{pmatrix}0 & 1\\ 0 & 0\end{pmatrix}\underline{x}(t) + \begin{pmatrix}0\\ 1\end{pmatrix}u(t)\\ y(t) = (1,0)\,\underline{x}(t)\\ \underline{x}(t_0) = 0\end{cases}$$

driven by inputs $u \in L_2(T)$:

$$\mathcal{Y} = \{y : T \to R \mid y = \mathcal{S}u,\ u \in L_2(T)\}.$$

$\mathcal{Y}$ can also be seen to be the class of all functions on T with absolutely continuous first derivative, with second derivative square-integrable on T, and such that $y(t_0) = \dot{y}(t_0) = 0$.

Therefore the **desired** y that solves the smoothing problem must be an element of $\mathcal{Y}$. Each $y \in \mathcal{Y}$ has the representation

$$y(t) = \int_T u(\alpha)\, h(t-\alpha)\, d\alpha$$

where $h(\alpha) \triangleq \alpha \cdot 1(\alpha)$ is the impulse response of $\mathcal{S}$. Therefore, according to part i) of the Lemma, $\mathcal{Y}$ is the RKHS with RK

$$K_y(t,\tau) = \int_T h(\tau-\alpha)\, h(t-\alpha)\, d\alpha$$
$$= (t\wedge\tau)\left[t\tau - \frac{1}{2}(t+\tau)(t\wedge\tau) + \frac{1}{3}(t\wedge\tau)^2\right], \quad t,\tau \in T,$$

where $\wedge$ denotes minimum. Moreover, according to part ii) of the Lemma, there must be a closed subspace $\mathcal{X}$ of $L_2(T)$ congruent with $H(K_y)$, i.e., for each $y \in H(K_y)$ there is a unique $u \in \mathcal{X}$ such that

$$\| y \|^2_{K_y} = \| u \|^2_{L_2(T)}$$

But, being $\ddot{y}(t) = u(t)$, it must also be

$$\| u \|^2_{L_2(T)} = \| \ddot{y} \|^2_{L_2(T)}$$

Whence the conclusion is

$$\| y \|^2_{K_y} = \int_T [\ddot{y}(t)]^2 dt.$$

The initial problem is therefore equivalent to finding the element $y \in H(K_y)$ that minimizes the functional

$$L(y|z) = \ell(y|z) + \| y \|^2_{K_y}.$$

Being the condition of strong equivalence $\mathrm{Tr}K_y < \infty$ fulfilled, according to Theorem 2, this can be looked at as a stochastic-extension estimation problem. For this problem we already know [compare (2) and (10)] that the finite-dimensional linear system above driven by white noise generates at its output a process with covariance K_y. Therefore, the smoothed estimate we are looking for can be computed by first Kalman-Bucy filtering the observations $z(t)$, $t \in T$, as in the figure below, and then smoothing the output of the Kalman-Bucy filter (Kailath and Frost, 1968, Rhodes, 1971).

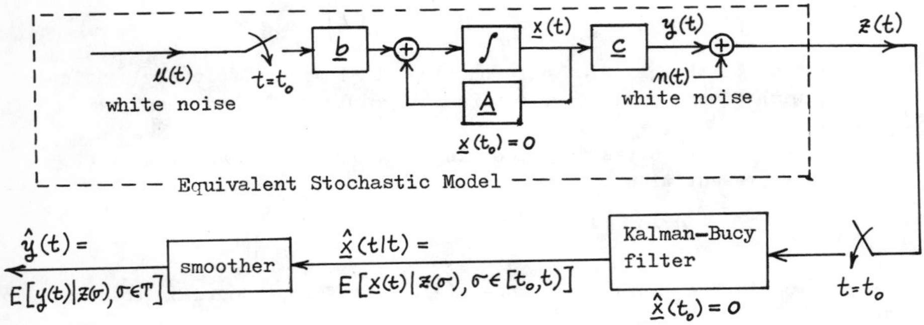

APPENDIX

Proof of Theorem 1. The assumed equivalence between the two probability measures induced by $R_y + R_n$ and R_n can be expressed, as Kalliampur and Oodaira (1963) showed, by the fact that R_y can be represented by the expansion

$$R_y(t,\tau) = \sum_{j=1}^{\infty} \lambda_j \varphi_j(t) \varphi_j^*(\tau) \quad , \quad t,\tau \in T$$

where

$$\sum_{j=1}^{\infty} \lambda_j^2 < \infty ,$$

$\{\varphi_j\}$ is an orthonormal sequence of functions in $H(R_n)$, and the functional series converges pointwise to $R_y(t,\tau)$ on $T \times T$. Now let $\{\varphi_j\} \cup \{\theta_\gamma\}$ be a complete orthonormal sequence in $H(R_n)$. Thus

$$R_n(t,\tau) = \sum_{j=1}^{\infty} \varphi_j(t) \varphi_j^*(\tau) + \sum_{\gamma \in \Gamma} \theta_\gamma(t) \theta_\gamma^*(\tau) \quad , \quad t,\tau \in T.$$

Therefore, using the two previous expansions,

$$[R_y + R_n](t,\tau) = \sum_{j=1}^{\infty} (1+\lambda_j) \varphi_j(t)\varphi_j^*(\tau) + \sum_{\gamma \in \Gamma} \theta_\gamma(t)\theta_\gamma^*(\tau)$$

From this expansion, one sees that $\{\sqrt{1+\lambda_j}\, \phi_j\} \cup \{\theta_\gamma\}$ is a complete orthonormal sequence in $H(R_y + R_n)$. Making use of (3), we get

$$\hat{y}(t) = \left\langle z, \sum_{j=1}^{\infty} \lambda_j \varphi_j(\cdot) \varphi_j^*(t) \right\rangle_{R_y+R_n} = \sum_{j=1}^{\infty} \frac{\lambda_j}{\sqrt{1+\lambda_j}} \left\langle z, (1+\lambda_j)^{1/2} \varphi_j(\cdot) \right\rangle_{R_y+R_n} \varphi_j(t)$$

$$= \sum_{j=1}^{\infty} \frac{\lambda_j}{\sqrt{1+\lambda_j}} z_j \varphi_j(t) ,$$

where the second equality holds by continuity of the inner product, and $z_j \triangleq \langle z, (1+\lambda_j)^{1/2} \varphi_j(\cdot) \rangle_{R_y+R_n}$ are independent, identically distributed, Gaussian random variables with zero-mean and unit variance.

We are now ready to show that

$\hat{y} \in H(R_y)$ with pr. 1 $\Longleftrightarrow$ $R_y + R_n$ and R_n are strongly equivalent.

In fact,

$$E \| \hat{y} \|^2_{R_y} = E\left[\sum_{j=1}^{\infty} \frac{\lambda_j}{1+\lambda_j} |z_j|^2 \right] = \sum_{j=1}^{\infty} \frac{\lambda_j}{1+\lambda_j} .$$

Therefore,

$$\hat{y} \in H(R_y) \text{ with pr. } 1 \Longleftrightarrow \sum_{j=1}^{\infty} \frac{\lambda_j}{1+\lambda_j} < \infty \Longleftrightarrow \sum_{j=1}^{\infty} \lambda_j < \infty .$$

In the above line, the first equivalence is a consequence of Kolmogorov 0-1 Law (Loève, 1963), whereas the second equivalence can be proved by using the positivity of the λ_j's. The condition $\sum_{j=1}^{\infty} \lambda_j < \infty$ has been called by Hájek (1962) strong equivalence between the probability measures corresponding to $R_y + R_n$ and R_n.

The fact that under the strong equivalence condition $\hat{y}$ minimizes (4) follows easily. ▨

<u>Proof of (8).</u> If $\chi \in H(K_\chi)$, by the reproducing property of $H(K_\chi)$, one has

$$\chi(\alpha) = \langle \chi, K_\chi(\cdot, \alpha) \rangle_{K_\chi} , \quad \alpha \in A .$$

Hence

$$\int_A \chi(\alpha_1)\, x^*(\alpha_1, t)\, d\alpha_1 = \int_A \langle \chi, K_\chi(\cdot, \alpha_1) \rangle_{K_\chi} x^*(\alpha_1, t)\, d\alpha_1$$

$$= \left\langle \chi(\alpha), \int_A x(\alpha_1, t)\, K_\chi^*(\alpha_1, \alpha)\, d\alpha_1 \right\rangle_{K_\chi} . \quad ▨$$

REFERENCES

BROCKETT, R.W., 1970, Finite Dimensional Linear Systems (New York: Wiley)

FRANKLIN, J.N., 1970, J. Math. Analysis Applic., 31, 682-716.

GELFAND, I.M., and VILENKIN, N.Ya., 1964, Generalized Functions, Vol. 4 (New York: Academic);

HADAMARD, J., 1952, Lectures on Canchy's Problem in Linear Partial Differential Equations (New York: Dover)

HAJEK, J., 1962, Czech Math. J., 87, 404-444.

HALMOS, P.R., 1957, Introduction to Hilbert Space (New York: Chelsea Publishing Co.)

KAILATH, T. and FROST, P., 1968, I.E.E.E. Trans. Automatic Control, 13, 655-660.

KALLIAMPUR, G. and OODAIRA, H., 1963, Time Series Analysis, M. Rosenblatt, Ed. (New York: Wiley)

KIMELDORF, G.S. and WAHBA, G., 1970, The Annals of Math. Statistics, 1970, 41, 245-502; 1971, J. of Math. Analysis Applic., 1, 82-95.

LOÈVE, M., 1963, Probability Theory, (Princeton, N.J.: Van Nostrand)

MOSCA, E., 1970 a, Preprints 2nd Prague IFAC Symp. Identification and Process Parameter Estimation, Paper 1.1, June (Prague: Academia); 1970 b, Functional Analytic Foundations of System Engineering, Eng. Summer Conf. Notes, The University of Michigan, Ann Arbor, July; 1971 a, I.E.E.E. Trans. Inf. Theory, 17, 686-696; 1971 b, Proc. I.E.E.E. Decision and Control Conf., December; 1972 a, I.E.E.E. Trans. Automatic Control, 17, 459-465; 1972 b, Int. J. Systems Science, 357-374; 1972 c, I.E.E.E. Trans. Inf. Theory, 18, 481-487.

NAYLOR, A.W. and SELL, G.R., 1971, Linear Operator Theory in Engineering and Science (New York: Holt, Rinehart, and Winston)

PARZEN, E., 1967, Time Series Analysis Papers (San Francisco: Holden-Day); 1971, Proc. Fifth Princeton Conf. Inf. Sciences and Systems, March.

RHODES, I.B., 1971, I.E.E.E. Trans. Automatic Control, 16, 688-706

ROYDEN, H.L., 1968, Real Analysis (New York: Macmillan)

TURCHIN, V.F., KOZLOV, V.P., and MALKEVICH, M.S., 1971, Soviet Physics Uspekhi (English translation), 13, 681-703.

REGRESSION OPERATOR IN INFINITE DIMENSIONAL VECTOR SPACES AND ITS APPLICATION TO SOME IDENTIFICATION PROBLEMS

Andrzej Szymanski
Institut of Mathematics
University of Lodz
POLAND

Introduction

The paper is an attempt to transfer the notion of regression and its properties known for numerical random variables into the theory of Banach valued random variables and to apply that theory to some identification problems.

Assuming that the trajectories of stochastic processes realized in the system are elements of a separable Banach space, we may regard them as Banach valued random variables and the problem of identification can be transfered into the domain of the infinite dimentional probability theory.

On the basis of the generalized Radon-Nikodym theorem given by Rieffel for the Banach valued random variables, the definition of the conditional expectation of a random variable with respect to any σ field has been formulated in the paper and its existence and uniqueness have been proved. The Kolmogoroff theorem on the form of the conditional expectation has also been transfered. That makes it possible to formulate the definition of the regression operator for the mentioned random variables.

The problem of identification can be reduced to the problem of finding the regression operator on the basis of the realizations of stochastic processes in the input and output of the system.

Part II of the paper refers to some properties of the regression operator in a separable Hilbert space. On the basis of the given here approximation theorem an approximative algorithm has been found for linear operators and one example given.

1. The Regression Operator in a Banach Space

1.1. Basic concepts and definitions.

Let us denote by $(\Omega, \mathfrak{F}, p)$ a probability space, where $\mathfrak{F}$ is some σ-field of subsets of the set Ω and p is normed complete measure defined on $\mathfrak{F}$. Let X be a separable Banach space over the field of real numbers.

DEFINITION 1.1. By the X-valued random variable ξ we mean a strongly measurable (with respect to $\mathfrak{F}$) mapping of the space Ω into the space X.

DEFINITION 1.2. An element of a Banach space X defined as follows

$$E\xi = (B) \int_{\Omega} \xi(\omega)\, p(d\omega)$$

is called the mean value or the expectation $E\xi$ of the X-valued random variable. The integral is meant in the sense of Bochner.

We assume more ever that all the random variables considered in this paper are Bochner integrable.

1.2. Conditional expectation of an X-valued random variable with respect to a σ-field.

DEFINITION 1.3. An $\mathfrak{A}$-measurable random variable $E(\eta|\mathfrak{A})$ which satisfies, for every $A \in \mathfrak{A}$, the relationship

$$(B) \int_A E(\eta|\mathfrak{A})(\omega)\, p(d\omega) = (B) \int_A \eta(\omega)\, p(d\omega)$$

is called the conditional expectation of X-valued random variable η with respect to a σ-field $\mathfrak{A} \subset \mathfrak{F}$.

The existence and uniqueness of the conditional expectation for the random variables considered here follows from the generalization of the classical Radon-Nikodym theorem for an X-valued measures, published by N. Rieffel ([5], p. 466)

Analogously as for real random variables we have the following:

PROPERTY 1.1. If $\zeta = \alpha\xi + \beta\eta$, then for any σ-field $\mathfrak{A} \subset \mathfrak{F}$ the equality

$$E(\zeta|\mathcal{A}) = \alpha E(\xi|\mathcal{A}) + \beta E(\eta|\mathcal{A})$$

holds.

PROPERTY 1.2. If a random variable ζ is measurable with respect to a σ-field $\mathcal{A}$, then

$$E(\zeta|\mathcal{A}) = \zeta$$

1.3. Conditional expectation with respect to a random variable and concept of regression operator.

Let ξ and η be measurable mappings of the set Ω into separable Banach spaces X_1 and X_2 respectively. Let $\mathcal{B}_i$ (i=1,2) denote the σ-field of borelian subsets of the space X_i. With the use of the mapping ξ we transpose the measure from the σ-field $\mathcal{B}_1$ into the σ-field $\mathcal{B}_1$, accepting for every set $B\in\mathcal{B}_1$

$$p_\xi(B) = p(\xi^{-1}B)$$

Let us define the σ-field $\xi^{-1}\mathcal{B}_1$ as follows

$$\xi^{-1}\mathcal{B}_1 = \left\{ A\subset\Omega : \bigvee_{B\in\mathcal{B}_1} (A=\xi^{-1}B) \right\}$$

It follows from the above remarks that $\xi^{-1}\mathcal{B}_1 \subset \mathcal{L}$.

DEFINITION 1.4. The conditional expectation of the random variable η with respect to the σ-field $\xi^{-1}\mathcal{B}_1$ is called the conditional expectation of the random variable η with respect to the random variable ξ and denoted by $E(\eta|\xi)$. We have

$$E(\eta|\xi) \overset{df}{=} E(\eta|\xi^{-1}\mathcal{B}_1)$$

For the conditional expectation $E(\eta|\xi)$ the following theorem analogous to that of Kolmogoroff is true

THEOREM 1.1. If ξ is X_1-valued random variable and η is a X_2-valued random variable, then there exists a borelian mapping $\Upsilon: X_1 \to X_2$ i.e. a measurable mapping with respect to the σ-field $\mathcal{B}_1$ and such that $\Upsilon^{-1}B\in\mathcal{B}_1$ for $B\in\mathcal{B}_2$, which satisfies the relationship

$$E(\eta|\xi)(\omega) = \Upsilon[\xi(\omega)]$$

for almost all $\omega \in \Omega$, Ψ being determining uniquely precisely to equivalence (in the induced measure p_ξ).

The Kolmogoroff theorem for X-valued random variables may be proved by use of the Radon-Nikodym-Rieffel theorem analogously as in the classical case ([6]).

DEFINITION 1.5. A borelian mapping M: $X_1 \to X_2$ satisfying the relationship

$$M[\xi(\omega)] = E(\eta|\xi)(\omega) \qquad \text{for almost all } \omega \in \Omega$$

is called t h e r e g r e s s i o n o p e r a t o r of the X_2-valued random variable η with respect to the X_1-valued random variable ξ

The existence and uniqueness of the regression operator follows from the generalized Kolmogoroff theorem.

1.4. Identification theorem.

It is easy to verify that, for mentioned in this paper random variables, we have following

PROPERTY 1.3. If Banach valued random variables ξ and η are independent in the sense of [4], than

$$E(\eta|\xi) = E\eta$$

holds with probability 1.

THEOREM 1.2. (Identification theorem) Let the random variables ξ and ζ be independent and η denoted as

$$\eta = U\xi + \zeta$$

where U is borelian mapping $X_1 \to X_2$.

The regression operator of η with respect to ξ i d e n t i f i e s the borelian mapping U iff

$$E\zeta = \Theta$$

Proof. N e c e s s a r i t y. Let us consider the random variable $\zeta = \eta - U\xi$. Taking its expectation and using the assumption $U\xi = E(\eta|\xi)$ we have

$$E\zeta = E\eta - E(U\xi) = E\eta - E[E(\eta|\xi)] = \Theta$$

S u f i c c i e n c y. Considering the conditional expectation of η with respect to ξ we have

$$E(\eta|\xi)=E(\eta-U\xi+U\xi|\xi)=E(\eta-U\xi|\xi)+E(U\xi|\xi)=E(\zeta|\xi)+U\xi$$

By the virtue of the property 1.3

$$E(\zeta|\xi)=E\zeta = \Theta$$

holds. Hence

$$E(\eta|\xi)=U\xi$$

holds with probability 1, which was to be proved.

1.5. Interpretation of identification theorem in domain of controll theory.

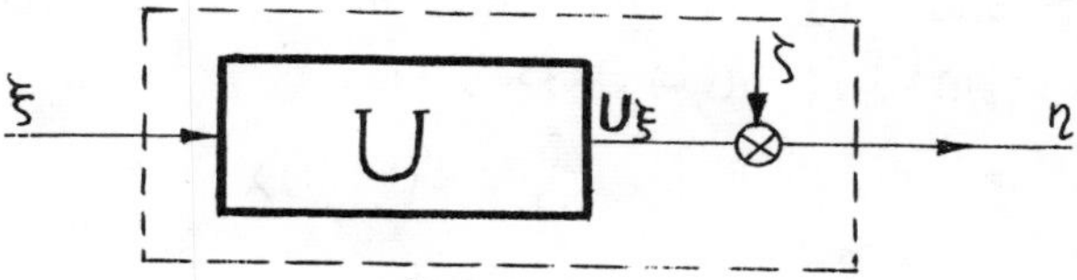

Fig. 1

Let us consider the system described by the borelian operator U. We may observe the stochastic processes ξ and η in the input and output of this system.

Assuming that the trajectories of stochastic processes ξ, η and ζ realized in the system are elements of a separable Banach space we may regard them as Banach valued random variables and the problem of identification can be transfered into the domain of the infinite dimentional probability theory.

The identification theorem says that in the case shown in Fig. 1 if ξ and ζ are independent and $E\zeta = 0$ we can search the regression operator η with respect to ξ to find the borelian operator U.

Then due to the identification theorem the problem of identification can be reduced to the problem of finding the regression operator on the basis of the observed realizations of stochastic processes in the input and output of system.

The next chapter deals with the searching of the regression operator in the case when the space X is a Hilbert space.

2. Properties of the Regresssion Operator in a Hilbert Space

2.1. Basic remarks.

Let H_i (i=1,2) denote separable Hilbert Spaces. We assume that all the random variables appearing in this chapter satisfy the condition

$$\int_{\Omega} \| \xi(\omega) \|^2 p(d\omega) < \infty$$

It is easy to verify that for the above considered random variables following Lemma holds.

LEMMA 2.1. If $\mathfrak{M} \subset \mathfrak{F}$ and a H_2-valued random variable is $\mathfrak{M}$-measurable by condition $\int_{\Omega} \| E(\eta | \mathfrak{M})(\omega) \|^2 p(d\omega) < \infty$, then

$$E[\langle \xi, \eta \rangle | \mathfrak{M}] = \langle \xi, E(\eta | \mathfrak{M}) \rangle$$

2.2. The relationship between the regression operator in Hilbert space and the least square method.

Let ξ be a H_1-valued random variable and η a H_2-valued random variable. We denote the scalar product in the space H_2 as $\langle \cdot, \cdot \rangle$.

Denote by $\mathcal{B}_i$ the σ-field of borelian subsets of the spaces H_i (i=1,2). Consider the class $\mathfrak{U}$ of operators acting from H_1 into H_2 with the following properties:

i) U are borelian operators i.e. measurable in the sense

$$U^{-1}(B) \in \mathcal{B}_1 \qquad \text{for } B \in \mathcal{B}_2$$

ii)
$$\int_{\Omega} \| U[\xi(\omega)] \|^2 p(d\omega) < \infty$$

For the regression operator in Hilbert space the following theorem is true

THEOREM 2.1. The functional $E\|\eta - U\xi\|^2$ attains its minimum in the class $\mathfrak{U}$ by the condition $\int_{H_1} \| M(x) \|^2 p_{\xi}(dx) < \infty$ iff

$$(U\xi)(\omega) = E(\eta | \xi)(\omega)$$

holds with the probability 1 i.e. iff U is the regression operator M.

P r o o f. By the properties of scalar product in Hilbert space and by the assumed condition $\int_{\Omega}\|E(\eta|\xi)(\omega)\|^2 p(d\omega)<\infty$ we obtain

$$E\|\eta-U\xi\|^2=E\|\eta-E(\eta|\xi)\|^2+E\|E(\eta|\xi)-U\xi\|^2+2E\left[\langle\eta-E(\eta|\xi),\ E(\eta|\xi)-U\xi\rangle\right]$$

Considering the last term we have

$$E\left[\langle\eta-E(\eta|\xi),\ E(\eta|\xi)-U\xi\rangle\right]=E\left\{E\left[\langle\eta-E(\eta|\xi),E(\eta|\xi)-U\xi\rangle|\xi\right]\right\}$$

By the lemma 2.1 we obtain

$$E\left[\langle\eta-E(\eta|\xi),\ E(\eta|\xi)-U\xi\rangle|\xi\right]=\left\langle E\left\{\left[\eta-E(\eta|\xi)\right]|\xi\right\},\ E(\eta|\xi)-U\xi\right\rangle$$

It follows from the properties 1.1 and 1.2 that

$$E\left\{\left[\eta-E(\eta|\xi)\right]|\xi\right\}=E(\eta|\xi)-E(\eta|\xi)=\Theta$$

Thus we have

$$E\left[\langle\eta-E(\eta|\xi),\ E(\eta|\xi)-U\xi\rangle|\xi\right]=\langle\Theta,\ E(\eta|\xi)-U\xi\rangle=0$$

The last term in estimated sum equals zero and the remaining ones are non negative, thus the expression to be estimated attains minimum iff the equality

$$(U\xi)(\omega)=E(\eta|\xi)(\omega)$$

holds with the probability 1, which was to be proved.

2.3. Approximation theorem.

Let a H_2-valued random variable ζ be independent from H_1-valued random variable ξ. Let us assume that the expectation of ζ is equal zero. Let a random variable η be expressed as

$$\eta=A\xi+\zeta$$

where A is a linear operator. According to the identification theorem the operator A is the regression operator of η with respect to the random variable ξ. By the virtue of the theorem 2.1 the operator A, as the regression operator minimizes the functional $E\|\eta-A\xi\|^2$ in the class $\mathcal{U}$. The minimum of this functional we denote as δ^2. We have

$$\delta^2=E\|\eta-A\xi\|^2$$

Let $\{e_i^{(j)}\}$ i=1,2,..., j=1,2 denote orthonormal bases in spaces H_j (j=1,2). Then we may show random variables ξ and η uniquely in the form

$$\xi=\sum_{i=1}^{\infty}\xi_i e_i^{(1)} \qquad \text{and} \qquad \eta=\sum_{i=1}^{\infty}\eta_i e_i^{(2)}$$

Let us determine the random variables $\xi^{(N)}$ and $\eta^{(N)}$ as

$$\xi^{(N)}=\sum_{i=1}^{N}\xi_i e_i^{(1)} \qquad \text{and} \qquad \eta^{(N)}=\sum_{i=1}^{N}\eta_i e_i^{(2)}$$

By the above notations the following theorem holds.

THEOREM 2.2 (Main theorem) If the linear regression operator A is bounded and if operators A_N exist and satisfy conditions

$$E\,\|\eta^{(N)}-A_N\,\xi^{(N)}\|^2=\min \qquad \text{for } N=1,2,\ldots$$

where the minimum is taken over all the finite dimensional operators A_N with matrices of order N, then

$$\lim_{N\to\infty} E\,\|\eta-A_N\xi\|^2=\delta^2$$

where

$$\delta^2=E\,\|\eta-A\xi\|^2$$

P r o o f. Let us denote

$$\delta_N^2=E\,\|\eta-A_N\xi\|^2$$

By the theorem 2.1 we have

$$\delta^2\leqslant\delta_N^2$$

We may show the expression $A\xi$ uniquely in the form

$$A\xi=\sum_{j=1}^{\infty}\sum_{i=1}^{\infty}a_{ji}\,\xi_i e_j^{(2)}$$

and analogously

$$A_N\xi=\sum_{j=1}^{N}\sum_{i=1}^{N}a_{ji}^{(N)}\xi_i e_j^{(2)}$$

Thus we may express the errors δ^2 and δ_N^2 in the following way

$$\delta^2=E\,\|\eta-A\xi\|^2=E\Big[\sum_{j=1}^{\infty}(\eta_j-\sum_{i=1}^{\infty}a_{ji}\,\xi_i)^2\Big]$$

and

$$\delta_N^2=E\,\|\eta-A_N\xi\|^2=E\Big[\sum_{j=1}^{\infty}(\eta_j-\sum_{i=1}^{\infty}a_{ji}^{(N)}\xi_i)^2\Big]$$

Taking into account the fact that $a_{ji}^{(N)}=0$ for i=N+1, N+2, ..., we can write

$$\delta_N^2=E\Big[\sum_{j=1}^{N}(\eta_j-\sum_{i=1}^{N}a_{ji}^{(N)}\,\xi_i)^2\Big]+E\Big[\sum_{j=N+1}^{\infty}\eta_j^2\Big]$$

The operator A_N minimizes the functional δ_N^2, since it is easy to see that

$$\delta_N^2 \leq E\left[\sum_{j=1}^{N} (\eta_j - a_{ji}\xi_i)^2\right] + E\left[\sum_{j=N+1}^{\infty} \eta_j^2\right]$$

From the last expression we can obtain

$$\delta_N^2 \leq \delta^2 + \varkappa_N$$

where

$$\varkappa_N = E\left[\sum_{j=1}^{N} \left(\sum_{i=N+1}^{\infty} a_{ji}\xi_i\right)^2\right] - E\left[\sum_{j=N+1}^{\infty} \left(\eta_j - \sum_{i=1}^{\infty} a_{ji}\xi_i\right)^2\right] + E\left[\sum_{j=N+1}^{\infty} \eta_j^2\right]$$

It is easy to see that $\left(\eta_j - \sum_{i=1}^{\infty} a_{ji}\xi_i\right)^2 \to 0$ and $\sum_{j=N+1}^{\infty} \eta_j^2 \to 0$ because the respective series are convergent.

Let us consider the first term of the expression of the $\varkappa_N$. We have

$$\sum_{j=1}^{N} \left(\sum_{i=N+1}^{\infty} a_{ji}\xi_i\right)^2 = \sum_{j=1}^{N} (A\xi - A\xi^{(N)})_j^2 \leq \|A\xi - A\xi^{(N)}\|^2$$

but

$$\|A\xi - A\xi^{(N)}\|^2 \leq \|A\|^2 \|\xi - \xi^{(N)}\|^2$$

Thus

$$E\left[\sum_{j=1}^{N} \left(\sum_{i=N+1}^{\infty} a_{ji}\xi_i\right)^2\right] \leq \|A\|^2 \, E\|\xi - \xi^{(N)}\|^2$$

From the above considerations and from the fact that the operator A is bounded follows that

$$\varkappa_N \to 0$$

summarizing the obtained results we have

$$\delta^2 \leq \delta_N^2 \leq \delta^2 + \varkappa_N$$

where

$$\varkappa_N \to 0$$

Thus $\delta_N^2 \to \delta^2$, which was to be proved.

We may call the above proved theorem the approximation theorem because it gives the way of approximation of the linear bounded regression operator A by means of the finite dimensional operators A_N, when all the conditions are satisfied. This approximation is understood in the sense of the minimization of square mean error.

The approximation theorem enables in a general case of a linear bounded regression operator the derivation of the approximative algorithm. This makes the statistical estimation of the operators A_N and the use of computer for calculation possible.

2.3. Approximative algorithm.

Let us assumpt that $E\xi = 0$ and $E\eta = 0$. On the contrary we may consider the random variables $\xi' = \xi - E\xi$ and $\eta' = \eta - E\eta$.

Let us denote by $\Lambda_{\xi^{(N)}}$ and $\Lambda_{\xi^{(N)}\eta^{(N)}}$ the covariance matrices defined in the following way

$$\Lambda_{\xi^{(N)}} = \left[\operatorname{cov}(\xi_i, \xi_j)\right]_{i,j \leq N}$$

and

$$\Lambda_{\xi^{(N)}\eta^{(N)}} = \left[\operatorname{cov}(\eta_j, \xi_i)\right]_{i,j \leq N}$$

THEOREM 2.3. If $\det \Lambda_{\xi^{(N)}} \neq 0$ then the operators A_N determined by the theorem 2.2 can be expressed in the following form

$$A_N = \Lambda_{\xi^{(N)}\eta^{(N)}} \cdot \Lambda_{\xi^{(N)}}^{-1}$$

P r o o f. Let us consider the expression $E\|\eta^{(N)} - A_N\xi^{(N)}\|^2$. From the properties of scalar product and by the above notations we may obtain

$$E\|\eta^{(N)} - A_N\xi^{(N)}\|^2 = E\|\eta^{(N)}\|^2 + \sum_{i=1}^{N}\sum_{k=1}^{N}\operatorname{cov}(\xi_i, \xi_k)\sum_{j=1}^{N} a_{ji}^{(N)} a_{jk}^{(N)} - 2\sum_{j=1}^{N}\sum_{i=1}^{N} a_{ji}^{(N)}\operatorname{cov}(\eta_j, \xi_i)$$

We have the following equation after the differentiation of the above expression

$$\sum_{k=1}^{N} a_{j_0,k}^{(N)}\operatorname{cov}(\xi_k, \xi_{i_0}) = \operatorname{cov}(\eta_{j_0}, \xi_{i_0})$$

for $i_0, j_0 = 1, 2, \ldots, N$

Using the notation of $\Lambda_{\xi^{(N)}}$ and $\Lambda_{\xi^{(N)}\eta^{(N)}}$ we can write

$$A_N \cdot \Lambda_{\xi^{(N)}} = \Lambda_{\xi^{(N)}\eta^{(N)}}$$

Because the matrix $\Lambda_{\xi^{(N)}}$ by the assumption is non singular we have

$$A_N = \Lambda_{\xi^{(N)}\eta^{(N)}} \cdot \Lambda_{\xi^{(N)}}^{-1}$$

This ends the proof.

From the theorem 2.3 the approximative algorithm follows. Determination of the elements of the covariance matrices on the basis of the observed realizations needs the statistical approach, which will considered in the next chapter.

3. Statistical approach to the approximation of linear bounded regression operator

3.1. Statistical approximation theorem.

Let us denote by $x_k^{(l)}$ - k-th realization of the l-th component of ξ, analogously by $y_k^{(l)}$ - of the random variable η. We can determine the estimator of covariance $\mathrm{cov}(\xi_i, \xi_j)$ by the formula

$$\widehat{\mathrm{cov}}(\xi_i, \xi_j) = \frac{1}{n-1} \sum_{k=1}^{n} x_k^{(i)} x_k^{(j)}$$

The above estimator converges stochastically to the covariance $\mathrm{cov}(\xi_i, \xi_j)$.

If we denote the matrices $\hat{\Lambda}^{(n)}_{\xi^{(N)}}$ and $\hat{\Lambda}^{(n)}_{\xi^{(N)}\eta^{(N)}}$ by the formulas

$$\hat{\Lambda}^{(n)}_{\xi^{(N)}} = \left[\frac{1}{n-1} \sum_{i=1}^{n} x_k^{(i)} x_k^{(j)} \right]_{i,j \leq N}$$

and

$$\hat{\Lambda}^{(n)}_{\xi^{(N)}\eta^{(N)}} = \left[\frac{1}{n-1} \sum_{i=1}^{n} x_k^{(i)} y_k^{(j)} \right]_{i,j \leq N}$$

all the elements of those matrices will converge stochastically to the elements of the matrices $\Lambda_{\xi^{(N)}}$ and $\Lambda_{\xi^{(N)}\eta^{(N)}}$.

Let us denote

$$\hat{A}_N^{(n)} = \hat{\Lambda}^{(n)}_{\xi^{(N)}} \cdot \hat{\Lambda}^{(n)}_{\xi^{(N)}\eta^{(N)}}$$

We raise the problem of convergence of the sequence of the funcionals $E\|\eta - \hat{A}_N^{(n)}\xi\|^2$ to the value δ^2, when $N \to \infty$ and $n \to \infty$.

By this notions the following theorem is true.

THEOREM 3.1. If the H_1-valued random variable ξ and the H_2-valued random variable η are independent and if $\det \hat{A}_N^{(n)} \neq 0$, and $E\xi^4 < \infty$, then

$$\lim E\|\eta - \hat{A}_N^{(n)}\xi\|^2 = \delta_A^2$$

where

$$\delta_A^2 = E\|\eta - A\xi\|^2$$

P r o o f. By the theorem 2.1, by the Hölder inequality and using the properties of scalar product we have

$$\delta_A^2 \leq E\|\eta - \hat{A}_N^{(n)}\xi\|^2 \leq \delta_N^2 + E\|(A_N - \hat{A}_N^{(n)})\|^2 + 2\sqrt{E\|\eta - A_N\xi\|^2 \cdot E\|(A_N - \hat{A}_N^{(n)})\xi\|^2}$$

Let us consider now the term $E\|(A_N-A_N^{(n)})\xi\|^2$. By the Hölder's inequality we have

$$E\|(A_N-\hat{A}_N^{(n)})\|^2 \leq \sqrt{E\|A_N-A_N^{(n)}\|^4 . E\|\xi\|^4}$$

but

$$E\|A_N-\hat{A}_N^{(n)}\|^4 = \sum_{i=1}^{N}\sum_{j=1}^{N}\sum_{k=1}^{N}\sum_{l=1}^{N} E\left\{\left[a_{ij}^{(N)}-(aN^{(n)})_{ij}\right]^2 \left[a_{kl}^{(N)}-(a_N^{(n)})_{kl}{}^2\right]\right\}$$

Using again the Hölder's inequality and by the theorem 6.31 ([2]) we have

$$\lim_{n\to\infty} E\left[a_{ij}^{(N)}-(\hat{a}_N^{(n)})_{ij}\right]^4=0$$

Since

$$E\|(A-\hat{A}_N^{(n)})\xi\|^2 \to 0$$

Summarizing the obtained results we have

$$\delta_A^2 \leq E\|\eta-\hat{A}_N^{(n)}\xi\|^2 \leq \delta_N^2 + \varkappa_n$$

where

$$\varkappa_n \to 0 \quad \text{if} \quad n\to\infty$$

Thus by the above results and by the theorem 2.2 we obtain

$$\lim E\|\eta - A_N^{(n)}\xi\|^2 = \delta_A^2$$

which was to be proved.

3.2. Example.

Let us consider an example of application of the statistical approximation theorem to the identification of the dynamical system described by the differential equation

$$\frac{d^2y}{dt^2} + a_1\frac{dy}{dt} + a_2y=x$$

Let us assume that the input $x(t)$ is a trajectory of a stochastic process $\xi(\omega,t)$ and that the function $x(t)$ is square integrable on the interval $[0,T]$. Then x is an element of Hilbert space L_2. By the same assumptions concerning all the others processes ξ and η we can apply the statistical identification theorem to the identification of the mentioned system.

All the observed functions must be shown in the following way

$$z^{(1)}(t)=\sum_{k=1}^{\infty} a_k^{(1)}\varphi_k(t)$$

where φ_k are any orthonormal functions choosed for the considered case respectively.

The observed signal must be sampled and the coefficients $a_k^{(l)}$ for given realization l are determined by the condition

$$\sum_{i=1}^{m}\left[z^{(l)}(t_i)-\sum_{k=1}^{N}a_k^{(l)}\varphi_k(t_i)\right]^2=\min$$

The coefficients are determined in such a way that they minimize the above square mean error

Having all the coefficients of the observed realizations we can make computations of the matrices $\hat{\Lambda}_{\xi(N)}$, $\hat{\Lambda}_{\xi(N)\ \eta(N)}$ and $\hat{A}_N^{(n)}$. Determining of the number N giving a good approximation is the problem, which is not solved in this paper. After the computation of the estimator $\hat{A}_N^{(n)}$ we may estimate the output signal as $\hat{A}_N^{(n)}$. Obviously, the matrix $\hat{A}_N^{(n)}$ depends on the choosing of the orthonormal system.

Conclusions

The main purpose of this paper was the transfer of the notion of regression into the theory of Banach valued random variables. This approach enables the formulation of identification problem in the domain of the infinite dimensional probability theory.

In the special case, when the random variables are Hilbert valued we state that the regression operator minimizes the square mean error. This makes the proof of the approximation theorem possible in the case of the linear bounded regression operator. Given further the statistical approximation theorem enables practically the identification of linear systems.

The problem unsolved in this paper is as well the approximation of the nonlinear regression operator.

It seems to the author that the general formulation of identification problem presented in his paper should help with solving also these problems.

R e f e r e n c e s

[1] Dunford N., Schwartz J. - Linear Operators, Part I, London 1958

[2] Goldberger A.S. - Econometric Theory, John Wiley and Sons. Inc., New York - London - Sydney 1964

[3] Hille E. -Functional Analysis and Semi-groups, New York 1948

[4] Mourier E. - Les element aleatoires dans un espace de Banach, Annals de L'Institut Henri Poincare, 13(1953), pp. 161-244

[5] Rieffel N. - The Radon-Nikodym theorem for Bochner Integral, Transactions of the American Mathematical Society, 131 (1968), pp. 466-487

[6] Renýi A. -Probability theory, Academia Kiado, Budapest 1972

AN APPROACH TO IDENTIFICATION AND OPTIMIZATION IN QUALITY CONTROL

W. Runggaldier
Facoltà di Statistica
Università
Padova (Italy)

G. Romanin Jacur
Laboratorio di Elettronica
Industriale del CNR
Padova (Italy)

1. INTRODUCTION

In what follows a procedure of simultaneous identification and control is presented, applied to the problem of quality control of a production line where items are classified as either good or defective. There two different subproblems may be considered :

1) Checking the process performance, which is actually an identification problem.
2) Designing a sampling procedure, where the defective items sampled may be replaced by non-defective ones. The sampling procedure is to meet some given requirements, generally keeping the average outgoing ratio of defectives below a desired level, at the minimum sampling cost. This is actually a problem of optimal control.

1.1 The model. Taking ideas from dual control, we suggest a unified procedure for the simultaneous identification and control, as shown in fig. 1.

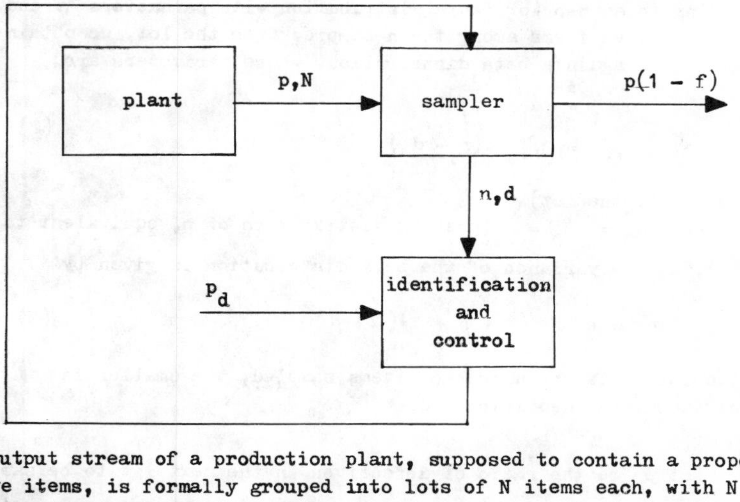

Fig. 1

The output stream of a production plant, supposed to contain a proportion p of defective items, is formally grouped into lots of N items each, with N a suitable number. A sampler, operating at frequency f, inspects n = Nf items from each lot, and replaces the d defectives found by nondefectives. The average outgoing ratio of defectives thus becomes $p(1 - f)$. In fact, the average number of defectives still in the lot after sampling is given by

$$E_p(d_{tot} - d) = p \cdot N - p \cdot f \cdot N = pN(1 - f) \tag{1}$$

(d_{tot} : total number of defectives in the lot).

The numbers n and d are used both to get better and better estimates of p, (identification), and to update the sampling frequency f (control). The cycle repeats for each lot.

1.2 The objective. Designing the procedure in such a way that it not only identifies the value of p, but also makes us aware of any change of p, at the same time guaranteeing that the average outgoing ratio of defectives does not exceed a desired level p_d, all this at minimum total cost.

We assume p changes in only one direction, namely it increases, i.e. the production process can only deteriorate.

2. IDENTIFICATION AND CONTROL IN CASE OF CONSTANT PARAMETER

For expository purposes the procedure is first presented assuming p does not change with time.

2.1 The identification. Following the ideas of Bayesian statistics, we consider p distributed according to an a-priori beta distribution with parameters α and β. Having observed d defective items among the n sampled from the lot, we obtain an a-posteriori distribution, again a beta distribution, whose parameters are

$$\begin{aligned} \alpha_{i+1} &= \alpha_i + d_i \\ \beta_{i+1} &= \beta_i + (n_i - d_i) \end{aligned} \tag{2}$$

(The index i refers to the lot).

Initially we take $\alpha_0 = \beta_0 = 1$ (uniform distribution of p, equivalent to absence of information). As the variance of the beta distribution is given by

$$\sigma^2 = \alpha \cdot \beta / (\alpha + \beta + 1)(\alpha + \beta)^2 \tag{3}$$

we see that the larger is the number of items sampled, the smaller is σ, implying an incrasingly accurate identification.

2.2 The control. Let p_1 be the ratio of defectives in the next lot to be sampled. Sampling with frequency f, the average outgoing ratio of defectives becomes (see(1)) $p_1(1 - f)$. Being p constant, $E(p_1) = p$, to meet the objective (1.2), the sampling frequency f must satisfy the condition

$$p(1 - f) \lesssim p_d \tag{4}$$

along with minimizing the cost, which here is the sampling cost supposed to be a linear function of f, i.e. $A \cdot f$ (A: cost of sampling N items).

The only information we have about the value of p is its beta distribution updated so far. Denoting by m and σ its mean value and standard deviation respectively, by the Chebyshev inequality the probability of p exceeding $\hat{p} = m + K\sigma$ is at most $1/K^2$. We therefore may suppose p not greater than $\hat{p}$ and solve for f by substituting in (4) $\hat{p}$ for p, thereby satisfying our objective except for a probability with known upper bound.

For what follows a good value for K was found to be K = 3. Then f becomes

$$f = 1 - p_d/\hat{p} \qquad (5)$$
$$(\hat{p} = m + 3\sigma)$$

Initially f is set equal to $1 - p_d$ equivalent to $\hat{p} = 1$.

Fig. 2 shows a flow chart of the procedure described so far (The dotted cases are not to be considered for the present case).

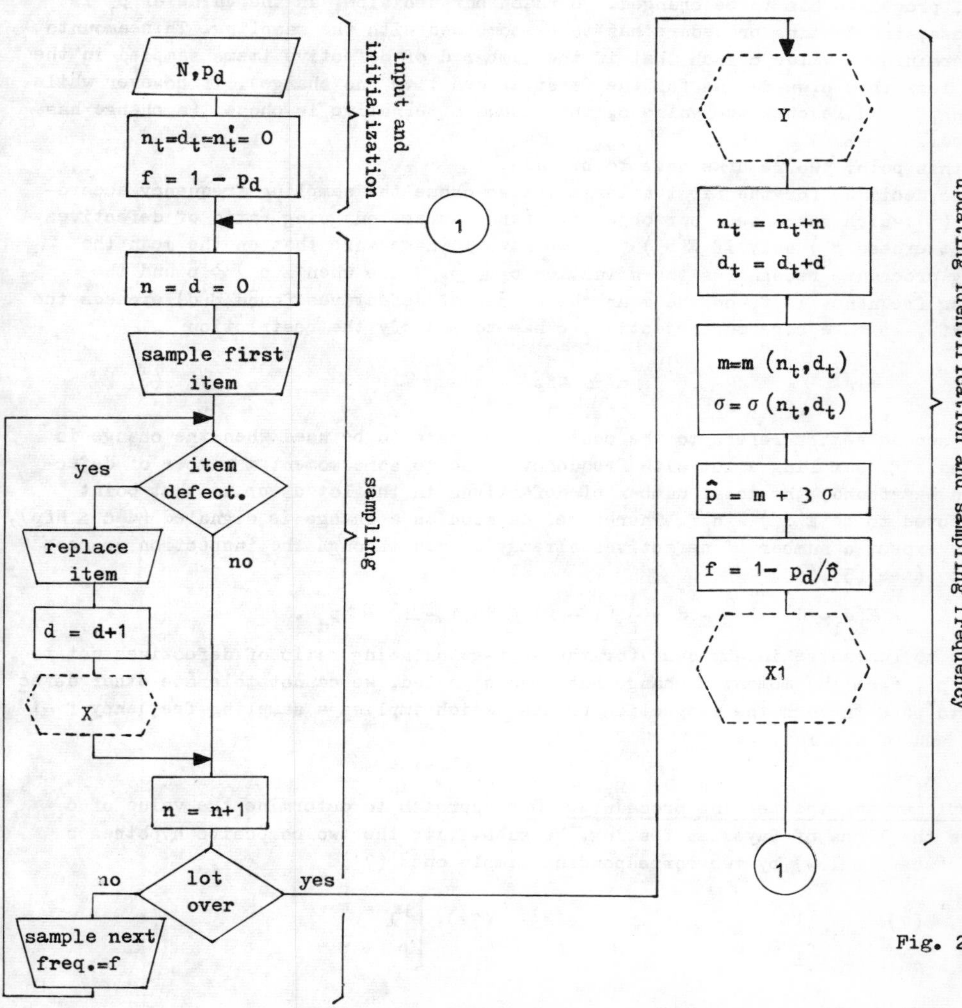

Fig. 2

3. THE GENERAL CASE OF THE PARAMETER VARYING WITH TIME

Relaxing the condition introduced in the previous section 2., the ratio of defectives p in the production is now allowed to vary with time. Following our objective we then have to decide from lot to lot whether p has changed. For this purpose a testing procedure is derived.

3.1 Defining a testing procedure.

Being at the beginning of a new lot we denote again by p_1 its ratio of defectives. For a p that does not change, the beta distribution updated so far may be considered to hold.

Then, (see Sec. 2.2), except for a small probability, we have $p < \hat{p}$. Also $E(p_1) = p$. Therefore, if $p_1 < \hat{p}$ we decide for no change in the lot. In this case for the identification and control we proceed as for p constant. If however $p_1 \geq \hat{p}$, we decide a change has occurred. In this latter case a new identification is started and the control procedure has to be changed. To reach our decision, as the value of p_1 is not known, the testing procedure has to be combined with the sampling. This amounts to determining a value c such that if the number d of defective items sampled in the lot is less than c, we decide for the first alternative (no change). If however while sampling, d reaches the value c, the second alternative is chosen (a change has occurred).

At this point two remarks have to be made.

Since, deciding for the first alternative, we chose the sampling frequency according to (5), which guarantees our objective (the average outgoing ratio of defectives does not exceed p_d) only if $E(p_1) < \hat{p}$, we have to make sure that on the mean the testing procedure recognizes the situation $p_1 \geq \hat{p}$. Since when $E(p_1) \geq \hat{p}$ and the sampling frequency is f, on the mean the number of defectives found E(d) exceeds the value $Nf\hat{p}$, to make sure our objective, c has to satisfy the restriction

$$c \leq Nf\hat{p} \tag{6}$$

The second remark refers to the control procedure to be used when the change is signaled. If, sampling a lot with frequency f, up to some moment a number of defectives d was found, the total number of defectives in the lot d_1 up to that point is expected to be $E(d_1) = d/f$. Therefore, as soon as a change is signaled ($d=c \leq Nf\hat{p}$), we must expect a number of defectives already passed through the inspection process given by (see (5),(6))

$$E(d_1 - d) = \frac{d}{f} - d = \frac{d}{f}(1 - f) \leq N\hat{p}(1 - f) = Np_d ,$$

so that to guarantee in any case for the average outgoing ratio of defectives not to exceed p_d, from the moment a change has been signaled, we cannot tolerate other defectives to pass through the inspection process, which implies a sampling frequency f = 1 to the end of the lot.

3.2 Implementing the testing procedure.

Our approach to determine the value of c follows the lines of Bayesian testing. We substitute the two composite hypotheses of 3.1 (see 7 below) by two corresponding simple ones (7')

$$(7) \quad \begin{cases} p_1 < \hat{p} \\ p_1 \geq \hat{p} \end{cases} \qquad\qquad (7') \quad \begin{cases} p_1 = m \\ p_1 = \hat{p} \end{cases}$$

(m is as before the mean value of the beta distribution updated so far).

The probability for a type - I error (to decide for the second alternative while the first one holds) is then given by (8). Similarly for a type-II error (to decide for the first alternative while the second one holds) the probability is expressed by (9).

$$P_I = \sum_{d=c}^{n} \binom{n}{d} m^d (1-m)^{n-d} = 1 - \sum_{d=0}^{c-1} \binom{n}{d} m^d (1-m)^{n-d} \tag{8}$$

$$P_{II} = \sum_{d=0}^{c-1} \binom{n}{d} \hat{p}^d (1-\hat{p})^{n-d} \tag{9}$$

where n is the value of Nf rounded to the nearest integer, i.e. the number of items sampled from the whole lot. Introducing cost factors C_I and C_{II} for the type-I respectively type-II errors and assuming equal a-priori probability for the two alternative hypotheses (7'), the cost related to the testing procedure is given by

$$C = \frac{1}{2} (C_I P_I + C_{II} P_{II}) \tag{10}$$

We want the value of c to minimize this cost. Supposing for a moment such a value of c exists and is unique, we find it by examining how C varies for two successive values of c (c and c + 1). This amounts to calculating the difference

$$C(c+1)-C(c) = \frac{1}{2}\binom{n}{c} \left[C_{II}\hat{p}^c(1-\hat{p})^{n-c} - C_I m^c (1-m)^{n-c} \right] \tag{11}$$

and examining the sign of the expression within brackets, i.e.

$$C_{II}\hat{p}^c(1-\hat{p})^{n-c} - C_I m^c (1-m)^{n-c} \tag{12}$$

Increasing the value of c by one, the first term of (12) is multiplied by $\hat{p}/(1-\hat{p})$ and the second one by $m/(1-m)$, where obviously $\hat{p}/(1-\hat{p}) > m/(1-m)$.

As a result, (12) and consequently (11) is a strictly increasing function of c, implying the minimum of C, if it exists, to be unique. The corresponding minimizing value of c is the one at which (12) changes sign (from negative to positive). For reasonable choices of C_I and C_{II} this changing of sign, if any, occurs in the range $c > 0$, as (12) is negative for $c = 0$. Otherwise by the choice of C_I and C_{II} a sampling one by one would have been more convenient. Since (see(6)) the value of c is not allowed to exceed $Nf\hat{p}$, if (12) does not change its sign in the range from 0 to $Nf\hat{p}$, we set $c = Nf\hat{p}$.

Instead of determining at the beginning of each cycle the corresponding value of c and use it for the test, we may proceed in a recursive manner thereby greatly simplifying the testing procedure. We start by calculating (12) for c = 0. Then, for each defective item found during the sampling, we calculate (12) recursively for the next value of c. As soon as either (12) changes sign or the number d of defectives sampled reaches the value $Nf\hat{p}$, we decide for the second alternative (change) continuing the sampling with f=1 to the end of the lot and starting a new identification. If however by the end of the lot none of the two situations has occurred, we decide for the first alternative (no change) and proceed with identification and control as for p constant. The whole procedure with the details can be seen from the flow chart of Fig. 2 inserting in the dotted cases X and X1 the flow segments of Fig. 3 (the dotted cases Y, Y1, Z are not to be considered here).

Sampling the first lot, the flow segment X is left out.

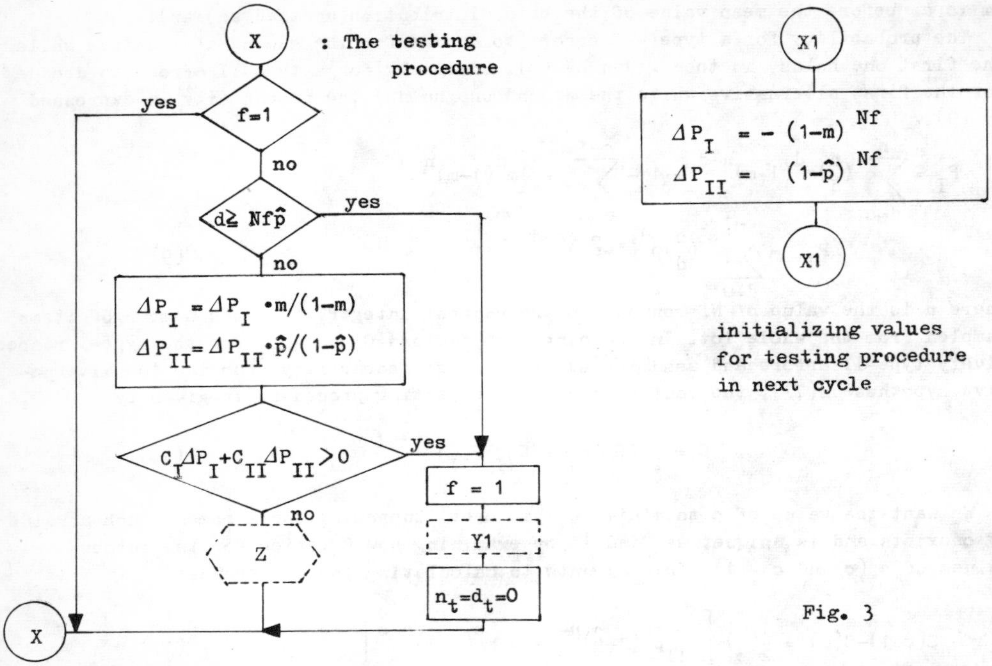

Fig. 3

<u>3.3 Cost factors and double testing.</u> If we make the wrong decision that a change has occurred, the n_t items sampled since the present identification has started and from which we derived the relative information, are no longer considered, as a new identification begins. Therefore, to reach the same stage of information as before, n_t more items have to be sampled. This implies the choice of $C_I = A/N \cdot n_t$ (remember from 2.2, A is the cost of sampling N items). C_{II} is chosen as the cost of having a lot (N items) whose outgoing ratio of defectives on the mean exceeds the value p_d. At the moment a type-I error occurs, n_t might be large, implying a large value of C_I and consequently of the cost related to the test.

In order to possibly reduce this cost and also to improve the reliability of the test, the procedure described in the previous section is extended to a double testing procedure including a second test to be performed in the lot that immediately follows the one where a change has been signaled, with the aim of making a check on the decision taken in the previous lot.

The modified procedure is as follows. If, during the sampling of a lot, the test leads us to decide that a change has occurred, we go on sampling with f=1 to the end of the lot and, while a new identification is started, the old one is continued. In the following lot, sampled with frequency f corresponding to the new identification, along with the normal test the second test, based upon the old identification, is performed.

If this second test confirms the previous decision, we definitely abandon the old identification and go on with the new one. If however the second test rejects the previous decision, only the old identification is continued and the sampling frequency for the following lot chosen according to it.

With the modifications just seen, the cost factor C_I for the normal test can be reduced to the value A, as in case of a wrong decision taken in a lot, previous its rejection in the following, we have to sample about N items.

The second test follows the lines of the normal test determining c' such that for

$d \geq c'$ by the end of the lot, the change previously signaled is confirmed, otherwise rejected. The value of c' however is not obtained this time by minimizing a cost, but by setting it equal to $Nf\hat{p}'$ where $\hat{p}' = m' + 3\sigma'$ (m', σ' referring to the old beta distribution). This choice of c' is motivated by the following two considerations. First, the lot is sampled at higher frequency, implying more accurate decisions. Secondly, the cost factors related to the second test are difficult to determine being partly due to chance.

The whole procedure as described so far can be seen in more details from the previous flow charts (Fig. 2, Fig. 3) inserting the flow segments Y and Y1 of Fig. 4 in the corresponding dotted cases. (The dotted case Z is not yet to be considered).

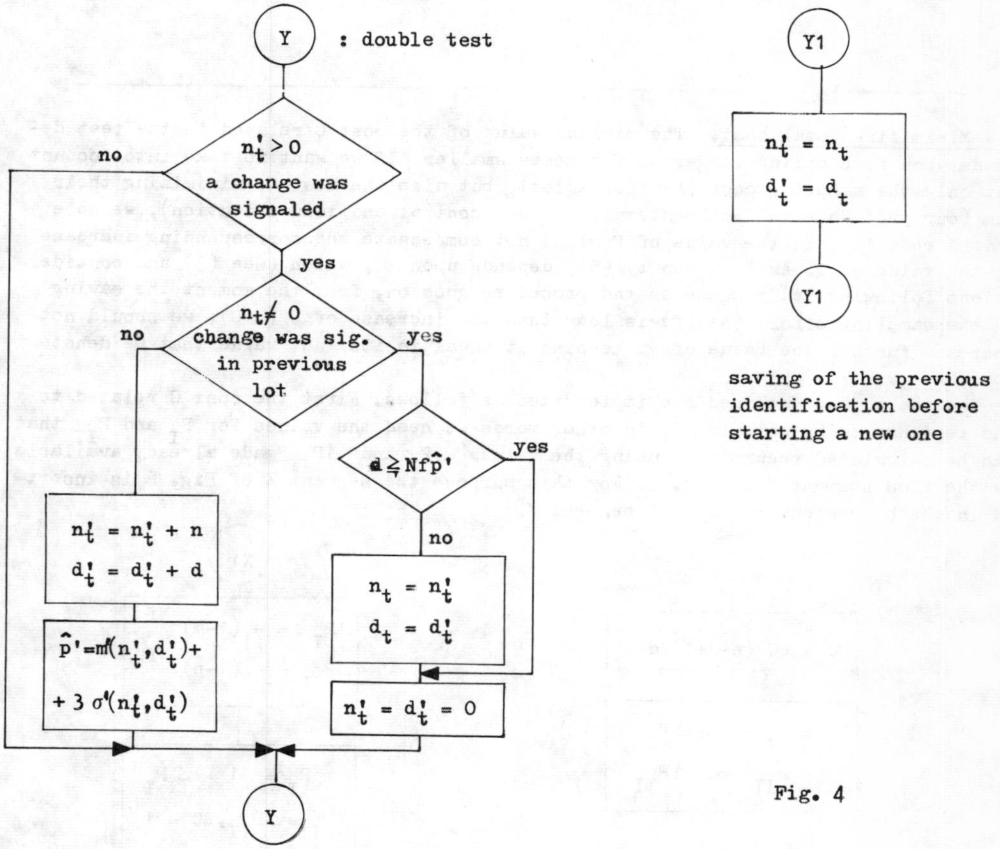

Fig. 4

In Fig. 5 a diagram is shown, where on the horizontal axis each segment represents a lot and where on the vertical axis we plot the corresponding sampling frequency.

The * indicates the point where a change signal occurs.

In the lot following the one in which the double test is performed, the sampling frequency is represented by the upper dotted line in case the change has been confirmed, otherwise by the lower one. The values of f used for this diagram are derived from the results of an actual simulation run with $N = 400$, $p = 0.1$, $p_d = 0.06$.

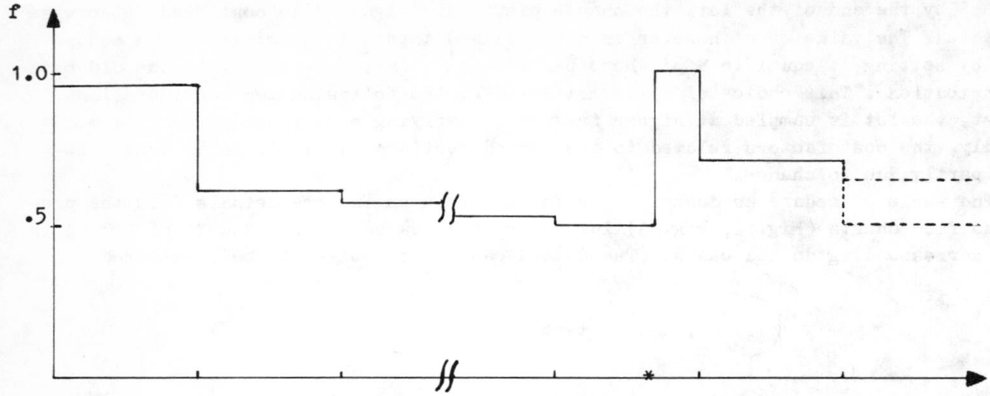

3.4 Minimizing total cost. The minimum value of the cost C related to the test depends upon f, becoming larger as f becomes smaller. If we want to take into account not only the sampling cost (A•f for a lot), but also the cost C, minimizing their sum (our procedure in fact contemplates both control and identification), we note that a reduction in the value of f might not compensate the corresponding increase in the value of C. As f, given by (5), depends upon σ, which (see (3) and considerations following it) reduces as the procedure goes on, from the moment the saving in the sampling effort ($A\cdot\Delta f$) is less than the increase of C (ΔC), we should not decrease further the value of σ keeping it equal to its last value that we denote by σ_{min}.

The ideas just outlined are implemented as follows. First the cost C related to the test has to be calculated, in other words we need the values for P_I and P_{II} that can be calculated recursively using the values ΔP_I and ΔP_{II} made already available by the flow segment X of Fig. 3. For this purpose the segment Z of Fig. 6 is inserted in the corresponding case of segment X.

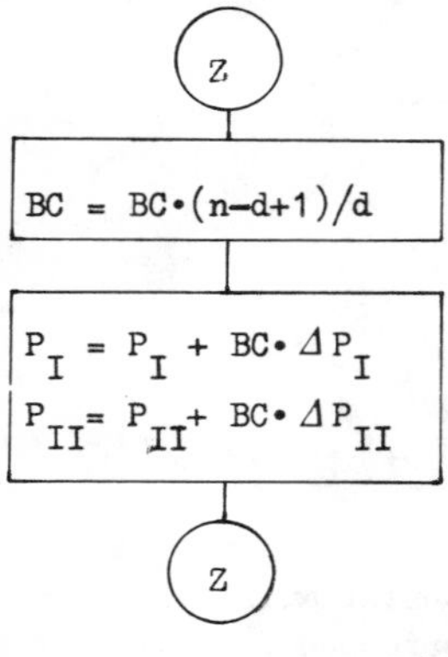

Fig. 6

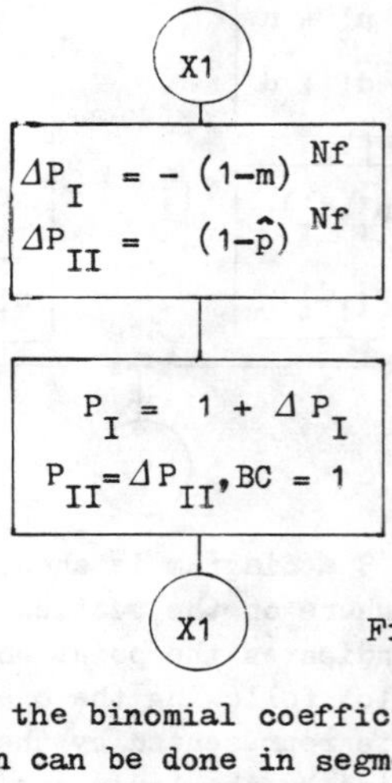

Fig. 7

The symbol BC in the segment Z of Fig. 6 stands for the binomial coefficient $\binom{Nf}{d}$. The values BC, P_I and P_{II} have to be initialized, which can be done in segment X1 of the flow modified as in Fig. 7.

If in section X of the flow the test is fullfilled (d = c), we have already the values of P_I and P_{II} we need. If not, at the end of the lot we will go through the steps of section X again simulating a ficticious sampling, until the test is fullfilled. Entering section Y we then calculate the value of C.(This last part is not reported in the flow as it is not difficult to imagine how it performs). Next we proceed to the comparison of ΔC (difference between the value of C just calculated and the one of the previous lot) with $A \cdot \Delta f$ (Δf is the difference between the sampl ing frequency used in the previous lot and the present one).

The comparison is made except if a change has been signaled in the previous lot. If ΔC exceeds $A \cdot \Delta f$, σ will be kept constant ($\sigma = \sigma_{min}$) for the lots that follow. The comparison is not performed, if previously σ had already been set equal to a σ_{min}, the same if we are at the end of the first two lots.

In case a change was signaled that has been confirmed, a new identification begins and the whole procedure starts anew. If the change has not been confirmed, we proceed as described above excluding from consideration the lot, where the double test is performed.

Both costs considered above depend not only upon f, they depend also upon N, considered here an input chosen in a suitable way. A suitable N means it should neither be too large, to allow a convenient updating both of the identification and the control, nor too small, to prevent an excessive randomness in the ratio of defec tives from lot to lot.

We found a reasonable choice for N the number 500.

4. CONCLUSION

Comparing the procedure described here with the standard procedures used in quality control,we foundthat we are better off.The above procedure in fact is an adaptive one,while those are not.There is however a cost to be paid for adaptivity,namely the computing we need for each cycle.

The way we approached the identification of the possibly time-varying parameter p is not the only possible one,as also a regression analysis could serve the scope.If however the variation of p shows a rather irregular pattern,a regression analysis could not be sufficiently adequate and to obtain better adaptivity,a procedure as the one described would be preferable.

Looking for possible generalizations of the above method,we remark that for each lot,the quality control problem as described here may also be visualized as a particular decision process over a Markov chain,where transition probabilities are only partially known and may also vary with time.We therefore think the method may be extended to such more general problems.

As the restriction,that the expected ratio of defectives in the final production should not exceed the value p_d,involves the unknown parameter p,one could also think of approaching similarly stochastic programming problems with restrictions involving unknown and possibly time-varying parameters.

Acknowledgement:This work has been supported by the Consiglio Nazionale delle Ricerche,Rome (Italy).

IDENTIFICATION DE DOMAINES

Jean CEA

I. M. S. P.

PARC VALROSE

06034 - NICE CEDEX

1 - INTRODUCTION, EXEMPLES :

Dans certains problèmes, d'identification ou de contrôle, l'élément à identifier ou à contrôler est un ouvert de $\mathbb{R}^n$ ou, ce qui revient au même, sa frontière.

Soit $\mathcal{O}$ une famille d'ouverts de $\mathbb{R}^n$; à l'ouvert $\Omega \in \mathcal{O}$ on associe un problème noté $J(\Omega)$; on a donc une fonction

$$J : \mathcal{O} \longrightarrow \mathbb{R}$$
$$\Omega \longrightarrow J(\Omega)$$

les problèmes qui se posent sont alors les suivants :

PROBLEME 1 :

$$\inf_{\Omega \in \mathcal{O}} J(\Omega)$$

PROBLEME 2 :

déterminer $\Omega_{opt} \in \mathcal{O}$ tel que

$$J(\Omega_{opt}) \leq J(\Omega)$$
$$\forall \Omega \in \mathcal{O}$$

En général, la définition de J utilise une "fonction d'état" et une "observation".

EQUATION D'ETAT :

$$\Omega \in \mathcal{O} \longrightarrow y_\Omega \in V$$

où y_Ω est "la" solution dans un espace V de l'équation

$$E(\Omega, y) = o$$

En général, $\mathbf{V}$ sera un espace de fonctions ou de distributions définies sur Ω , et $E(\Omega,y) = 0$ sera une équation aux dérivées partielles.

Si H désigne un nouvel espace et C_Ω un opérateur de V dans H, $C_\Omega : \mathbf{V} \longrightarrow H$, on observera $C_\Omega \; y_\Omega$, le résultat de <u>l'observation</u> sera désigné par h_g ; on aura donc la <u>fonction coût</u> :

$$J(\Omega) = \frac{1}{2} || C_\Omega \; y_\Omega - h_g ||^2_H$$

EXEMPLE 1 :

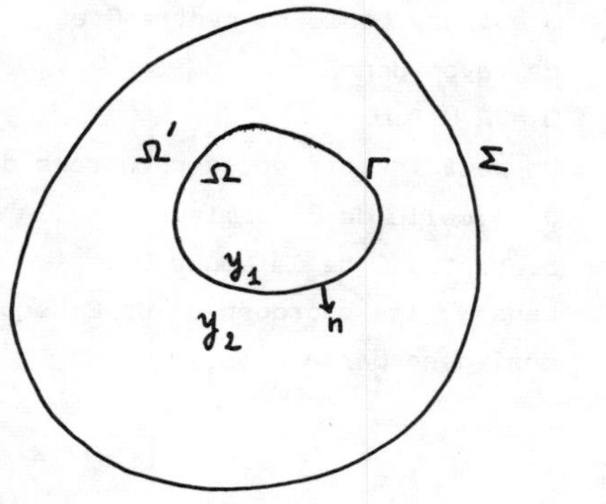

$D = \Omega \cup \Gamma \cup \Omega$

Dans cet exemple, l'équation d'état est une équation de transmission : k désigne un nombre positif donné, et f un élément donné dans $L^2(D)$:

$$(1.1)\begin{cases} -k \; \Delta y_1 + y_1 = f\big|_\Omega & (\Omega) \\ -\Delta y_2 + y_2 = f\big|_\Omega & (\Omega') \\ \left.\begin{array}{l} y_1 = y_2 \\ k \dfrac{\partial y_1}{\partial n} = \dfrac{\partial y_2}{\partial n} \end{array}\right] & (\Gamma) \\ y_2 = 0 & (\Sigma) \end{cases}$$

L'opérateur Δ est le laplacien ; à partir de Ω (et des données fixes, k, f) on a donc défini les fonctions y_1 et y_2 ou mieux la fonction y_Ω dont les restrictions à Ω et à Ω' sont y_1 et y_2.

Si $K \subset D$, on peut alors considérer les fonctions coûts suivantes

$$(1.2) \qquad J(\Omega) = \frac{1}{2} || y_\Omega - h_g ||^2_{L^2(K)}$$

ou

$$(1.3) \qquad J(\Omega) = \frac{1}{2} || y_\Omega - h_g ||^2_{H^1(K)}$$

Si $K \subset \Sigma$, on peut considérer une autre fonction coût

$$(1.4) \qquad J(\Omega) = \frac{1}{2}||y_\Omega - h_g||^2_{L^2(K)}$$

un problème qui entre dans le cadre de cet exemple est le suivant : trouver la forme optimale Ω_{opt} d'un diélectrique de constante k telle qu'un champ électrique soit minimum dans un domaine donné K.

EXEMPLE 2.

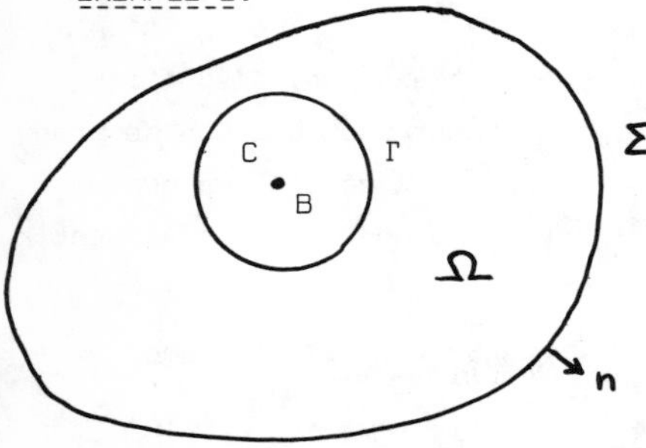

B est une boule de centre C et de rayon donné r

$D = \Omega \cup \Gamma \cup B$

On peut identifier la recherche de Ω à celle de B ou mieux à celle du centre C de la boule B :

Dans $\mathbb{R}^2$ les coordonnées de C seront désignées par $a = (a_1, a_2)$

Ainsi, dans ce cas la "vraie" variable sera a !

A partir de a on définit y_a solution de :

$$(1.5) \qquad \begin{cases} -\Delta y + y = f\big|_\Omega & (\Omega) \\ \dfrac{\partial y}{\partial n} = 0 & (\Sigma) \\ y = 0 & (\Gamma) \end{cases}$$

On peut maintenant définir des fonctions coûts comme dans l'exemple 1 ; Dans cet exemple B pourrait représenter, par exemple, la section droite d'un tube par lequel on apporte ou on évacue de la chaleur.

De façon plus générale, on pourrait faire dépendre Γ de $a \in \mathbb{R}^n$; On se ramène alors à une minimisation d'une fonction J définie sur un sous ensemble de $\mathbb{R}^n$ (dans l'exemple 2, c'est l'ensemble où peut se trouver le centre de la boule B).

Notons qu'au point de vue numérique pour chaque évaluation de J il faudra discrétiser (5) et résoudre le système approché.
Ce genre de problèmes a été étudié par KOENIG et ZOLESIO (thèse de 3ième cycle, Université de NICE).

EXEMPLE 3 :

Il s'agit de la famille des problèmes à frontière variable ; ce genre de problèmes est étudié, avec les méthodes présentées dans cette communication, par A. DERVIEUX et B. PALMERIO.

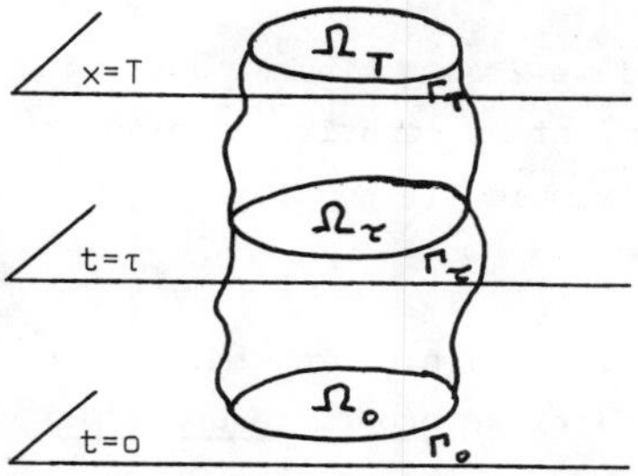

$Q = \bigcup_{o < \tau < T} \Omega_\tau$

Il s'agit de déterminer y et Γ_t, $o < t < T$ de façon que les relations (6) et (7) aient lieu

$$(1.6)\quad \begin{cases} \dfrac{\partial y}{\partial t} - \Delta y = f & (Q) \\ \dfrac{\partial y}{\partial n} = h & (\Gamma_t) \\ y(.,t=o) = y_o(.) & (\Omega_o) \end{cases}$$

$$(1.7)\quad y = g \qquad (\Gamma_t)$$

les éléments f, g, h, y sont donnés.

On peut transformer ce problème en un problème de minimisation : à tout Q on associe la solution y_Q de (1.6) et on pose

$$J(Q) = \frac{1}{2} \int_o^T ||y_Q - g||^2_{L^2(\Gamma_t)} dt$$

Déterminer y et Q tels que (1.6) (1.7) soient vérifiées revient à trouver Q_{opt} par lequel $J(Q_{opt}) = o$

ENSEMBLE $\mathcal{U}$:

Dans ce qui suit nous nous intéressons à des ouverts $\Omega \subset D$ où D est donné, $\bar{D}$ étant un compact de $\mathbb{R}^n$. Si χ_Ω désigne la fonction caractéristique de Ω , on désigne par $\mathcal{U}$ l'ensemble des u, u = $\chi_\Omega, \Omega \in \mathcal{O}$

TOPOLOGIE sur Ω ou sur $\mathcal{U}$:

On désigne par $\Omega \vartriangle \Pi$ la différence symétrique entre Ω et Π ;

$x \in \Omega \vartriangle \Pi \Longleftrightarrow \quad x \in \Omega \cup \Pi,\ x \notin \Omega \cap \Pi$

On dira que $\Pi_n \longrightarrow \Omega$ lorsque $\text{mes}(\Omega \vartriangle \Pi_n) \longrightarrow 0$ lorsque $u \longrightarrow +\infty$.

Cela revient à prendre sur $\mathcal{U}$ une topologie du type $L^p(D)$, $1 \leq p < +\infty$.

2 - EXISTENCE D'UNE SOLUTION

En général, l'existence d'une solution du problème 2 aura lieu si par exemple $\mathcal{U}$ est compact et J continu. Il est donc important de connaître des ensembles d'ouverts qui soient compacts pour, par exemple, la topologie $L^2(D)$:

En voici un exemple :

Soit Ω_o un ouvert non vide contenu dans D ; soit $\Pi(\theta,h,r)$ la famille des ouverts Ω de D qui contiennent Ω_o et qui vérifient la <u>propriété uniforme</u> de cône suivante : $\forall\ \Omega \in \Pi(\theta,h,r)$, $\forall\ x \in \partial\Omega$, il existe un cône C_x de sommet x, d'angle 2θ , de côté h tel que $\forall\ y \in \Omega \cap B(x,r)$ on ait $y + C_x \subset \Omega$; $B(x,r)$ désigne la boule de centre x et de rayon r

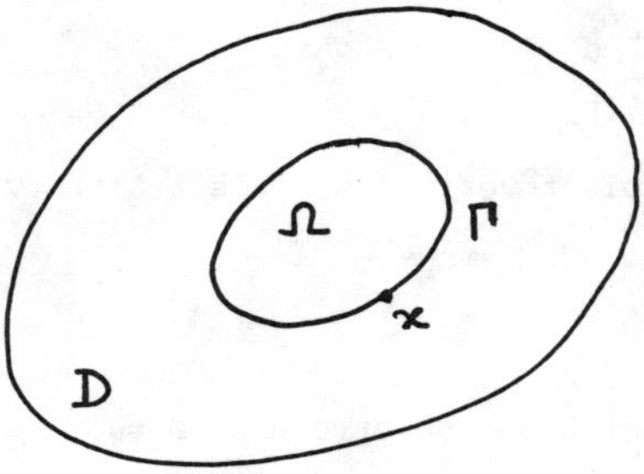

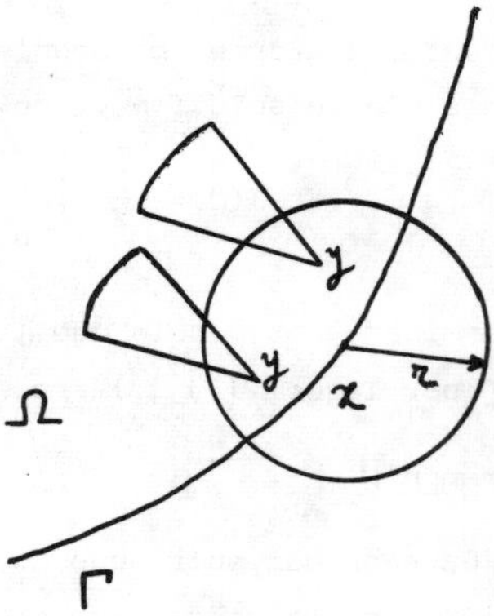

D. CHENAIS a démontré le

THEOREME : $\Pi(\theta,h,r)$ un compact dans $L^2(D)$.
Notons qu'on peut aussi caractériser $\Pi(\theta,h,r)$ par une condition uniforme de Lipschitz relative à la frontière des $\Omega \in \Pi(\theta,h,r)$.

3 - DEVELOPPEMENT DE J ; CONDITION NECESSAIRE D'OPTIMALITE.

Soient Ω et $\Omega + \delta\Omega$ deux éléments de $\mathcal{\Omega}$, u et $u + \delta u$ les fonctions caractéristiques associées ; on s'intéresse aux fonctions J qui admettent un développement du type :

$$J(u + \delta u) = J(u) + T_1(u, \delta u) + T_2(u,\delta u) \tag{3.1}$$

où $T_1(u,\delta u)$ (resp. $T_2(u,\delta u)$ est un infiniment petit d'ordre 1 (resp d'ordre supérieur à 1, par exemple d'ordre 2)

On suppose que

$$T_1(u,\delta u) = S(u, u + \delta u) - S(u,u) \tag{3.2}$$

où $v \longrightarrow S(u,v)$ est linéaire et continue ; si par exemple

$\mathcal{U} \subset L^p(D)$ $1 \leq p < +\infty$, si p' vérifie $\frac{1}{p} + \frac{1}{p'} = 1$, alors $S(u,v)$ s'écrira

$$\left\{\begin{array}{l} S(u,v) = \displaystyle\int_D G_u(x)\, v(x)\, dx \\ G_u \in L^{p'}(D) \end{array}\right. \tag{3.3}$$

un cas intéressant est celui où $p = 1$; $G_u \in L^\infty(D)$.

En termes d'ouverts, on peut écrire ces relations sous la forme

$$J(\Omega + \delta\Omega) = J(\Omega) + T_1(\Omega,\delta\Omega) + T_2(\Omega,\delta\Omega) \tag{3.1'}$$

$$T_1(\Omega,\delta\Omega) = S(\Omega,\Omega + \delta\Omega) - S(\Omega,\Omega) \tag{3.2'}$$

où $\Pi \longrightarrow S(\Omega,\Pi)$ est <u>additive</u> et <u>continue</u>

$$(3.3)' \quad \begin{cases} S(\Omega,\Pi) = \displaystyle\int_{\Pi} G_\Omega(x)\,dx \\ G_\Omega \in L^{p'}(D) \end{cases}$$

Notons que si $\delta\Omega^+ \subset \complement\,\Omega$, $\delta\Omega^- \subset \Omega$, $\Omega + \delta\Omega = (\Omega \cup \delta\Omega^+) \setminus \delta\Omega^-$

on a

$$(3.4) \qquad T_1(\Omega,\delta\Omega) = \int_{\delta\Omega^+} G_\Omega(x)\,dx - \int_{\delta\Omega^-} G_\Omega(x)\,dx$$

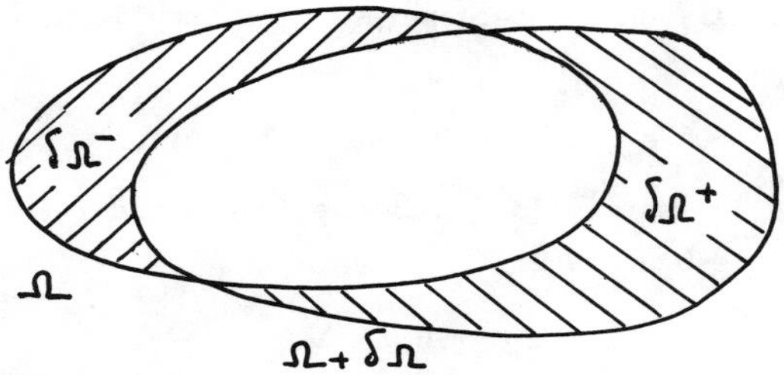

EXEMPLE 1 :

Dans le cadre de l'exemple 1 du n° 1, en choisissant J défini par (1.2), introduisons l'état "adjoint" $p_\Omega = (p_1,p_2)$ par :

$$(3.5) \quad \begin{cases} -k\Delta p_1 + p_2 = (y - y_d)\big|_\Omega & (\Omega) \\ -\Delta p_2 + p_2 = (y - y_d)\big|_\Omega & (\Omega') \\ p_1 = p_2 & (\Gamma) \\ k\dfrac{\partial p_1}{\partial n} = \dfrac{\partial p_2}{\partial n} & \\ p_2 = o & (\Sigma) \end{cases}$$

alors, on démontre que

$$(3.6) \qquad G_\Omega = (k - 1)\ \nabla y_\Omega \cdot \nabla p_\Omega$$

avec $\nabla v(x) = \text{grad. } v(x)$ cf. J. CEA, A. GIOAN, J. MICHEL.

POINT CRITIQUE :

$\Omega \in \mathcal{O}$ est un point critique de J, si il existe $\Omega > o$ tel que $T_1(\Omega, \delta\Omega) \geq o$
$\forall \quad \delta\Omega$ tel que $\Omega + \delta\Omega \in \mathcal{O}$, mes $(|\delta\Omega|) \leq r$; $|\delta\Omega| = (\Omega + \delta\Omega) \Delta \Omega$
Notons que dans un cadre hilbertien, lorsqu'il n'y a pas de contrainte, la condition analogue est grad $J(u) = o$.

En tenant compte de (3.1)', (3.2)', (3.3)' et (3.4) on démontre la

PROPOSITION 3.1 :

Ω est un point critique de $\mathcal{O}$ si et seulement si

$$(3.7) \begin{cases} G_\Omega(x) \leq o & \forall \ x \in \Omega \\ G_\Omega(x) \geq o & \forall \ x \notin \Omega \end{cases}$$

Naturellement on a la

PROPOSITION 3.2 :

Si Ω est optimal alors Ω est critique. Ainsi les relations (3.7) constituent une une condition nécessaire pour que l'ouvert Ω soit une solution du problème 2.

4 - APPROXIMATION D'UN POINT CRITIQUE.

4.1 . La méthode des approximations successives :

Elle est basée sur les conditions (3.7) d'optimalité : soit Ω , définissons $T(\Omega)$ par

$$x \in T(\Omega) \Longleftrightarrow G_\Omega(x) < o$$

si l'ensemble où G_Ω s'annulle est de mesure nulle, (3.7) s'écrit

$$(3.7)' \qquad \Omega = T(\Omega)$$

On a donc l'algorithme suivant :

$$\begin{cases} \Omega_o \text{ donné} \\ \Omega_{n+1} = T(\Omega_n) \quad n = o,1,\ldots \end{cases}$$

Naturellement, il faudra modifier cet algorithme si $\Omega \in \mathcal{O}$ n'entraîne pas $T(\Omega) \in \mathcal{O}$.

Cet algorithme a donné d'excellents résultats numériques, la convergence ayant en général lieu après une ou deux itérations, toutefois, la convergence théorique de cet algorithme reste encore à établir.

4.2. Les méthodes du type gradient :

CAS PARTICULIER 1 :

(Etudié par BENDALI - DJAADANE - Thèse de 3ième cycle - Université d'Alger)

L'ouvert Ω est défini comme l'image d'un ouvert fixe B de $\mathbb{R}^n$ par une application $t \longrightarrow x = X(a,t)$ qui dépend d'un paramètre $a \in K \subset \mathbb{R}^p$.

K est un ensemble compact ; Ainsi la "vraie" variable n'est pas Ω mais a et J est donc une fonction de a :

Pour déterminer grad J(a), on commence par calculer

$$T_1(\Omega,\delta\Omega) = \int_{\Omega+\delta\Omega} G_\Omega(x)\, du - \int_\Omega G_\Omega(x)\, du,$$ où en réalité

G est indicé par a ; en fait cette relation s'écrit

$$T_1(a,\delta a) = \int_B G_a(X(a+\delta a,t))\, J(a+\delta a,t)\, dt - \int_B G_a(X(a,t))\, J(a,t)\, dt$$

où J représente le Jacobien du changement de variable. A partir de là, en développant G_a et J, l'infiniment petit étant δa, on obtient la différentielle de J par rapport à δa et donc le gradient de J en a. On peut alors utiliser n'importe quelle méthode du type gradient.

CAS PARTICULIER 2 : (étudié par KOENIG et ZOLESIO)

La famille $\mathcal{O}$ d'ouverts de $\mathbb{R}^n$ est définie comme étant l'image d'une partie $\mathcal{A}$ d'un espace de Banach A par une application u vérifiant les propriétés suivantes :
- u est injective
- $u \in \mathcal{C}^0(\mathcal{A},\mathcal{O}) \cap \mathcal{C}^1(\mathcal{A},\mathcal{C}'(D))$

La "vraie" variable $a \in \mathcal{A}$ peut alors être un élément de $\mathbb{R}^n$ (exemple 2) ou un élément d'un espace fonctionnel (problème "à frontière libre")

On sait aussi dans ce cas expliciter le gradient.

CAS PARTICULIER 3 :

Dans $\mathbb{R}^2$ pour simplifier, Ω est repéré à l'aide d'une fonction

$\varphi : [0,1] \longrightarrow [0,1]t,\ x \in \Omega \Longleftrightarrow t \in]0,1[,\ x \in \varphi(t)$

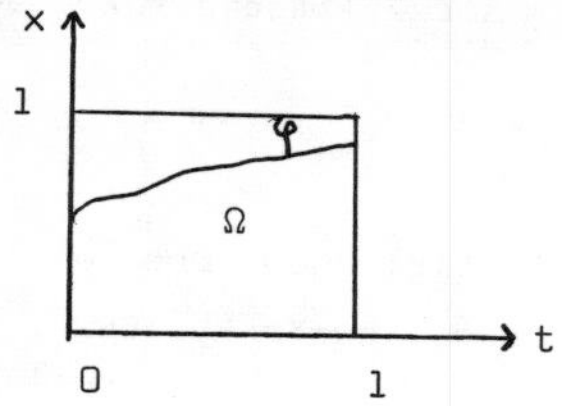

La "vraie" variable est ici φ. Dervieux et Palmerio obtiennent, en passant par les relations intermédiaires (3.1), (3.2), (3.3), une relation du type :

$$J(\varphi + h\psi) = J(\varphi) + h \int_0^1 G_\varphi(t).\ \psi(t)\,dt + \ldots$$

Cette fois h est l'infiniment petit ; on a négligé les termes d'ordre supérieur à 1 ;

Cette relation permet de savoir dans quelle "direction" il faut se déplacer pour déterminer J ; de plus si on avait $t \longrightarrow \varphi(a,t)$ $t \in [0,1]$, $a \in K \subset \mathbb{R}^p$, on pourrait obtenir facilement le gradient de J par rapport à a.

CAS PARTICULIER 4 :

On suppose que $\bar{D} = \bigcup_{i \in I} \bar{w}_i$; en fait les w_i sont des "éléments finis" : ouverts de D, deux à deux disjoints. Ω sera repéré par un ensemble d'indices I_Ω : $\bar{\Omega} = \bigcup_{i \in I_\Omega} \bar{w}_i$; En reprenant les principes de la méthode classique du gradient, et en les modifiant pour les appliquer au problème actuel, J.CEA, A. GIOAN et J. MICHEL proposent un algorithme convergent. Les résultats numériques sont bons mais inférieurs toutefois à ceux de la méthode des approximations successives.

B I B L I O G R A P H I E

Nous donnons une bibliographie très sommaire, limitée aux cas particuliers étudiés ;

BERGMAN S., SCHIFFER M.
Kernel functions and elliptic differential equations in mathematical physics, Academic Press, New-York, 1953

CEA J., GIOAN A., MICHEL J.
Quelques résultats sur l'identification de domaines. A paraître dans "CALCOLO" 1973.

CHENAIS D., Un résultat d'existence dans un problème d'identification de domaine, note aux C.R. Acad. Sci. PARIS, t 276, (12 Fév. 1973).

DERVIEUX. A. PALMERIO. B. Identification de frontière dans le cas d'un problème de Dirichlet, note aux C.R. Acad Sci. PARIS, t.275 (20 Novembre 1972).

HADAMARD. J. Mémoire sur le problème d'analyse relatif à l'équilibre des plaques élastiques encastrées, oeuvres de Jacques Hadamard, C.N.R.S., PARIS, 1968.

KOENIG. M. ZOLESIO, J.P. Sur la location d'un domaine de forme donnée, thèse 3ième cycle, I.M.S.P., Université de NICE, 13 Mars 1973.

THE MODELLING OF EDIBLE OIL FAT MIXTURES

J.O. Gray - J.A. Ainsley
Dept. of Electrical Engineering
U.M.I.S.T.
Sackville Street, Manchester
U.K.

Introduction

This paper is concerned with a practical application of straightforward modelling and optimisation methods to an industrial process which promises to yield very attractive economic returns. The process consists of the large scale batch production of edible oil fat mixtures such as margarine and shortening products.

The batch production of edible oil fat mixtures requires the mixing of quantities of raw material to meet both economic and taste criteria. The latter is the more difficult to quantify but it is found that the variation with temperature of the ratio of the liquid to solid phase of the mixture has an important bearing on its taste and tactile properties and hence on the acceptability of the product. A measure of this ratio is obtained by dilatometery measurements at specific temperatures. The dilatometery temperature profile of a compound oil fat mixture is, however, a highly non-linear function of the behaviour of the individual constituents and much skill and experience is required by the analyst in the determination of constituent fractional proportions to ensure that the final compound meets the required dilatation temperature specification. Accurate computer models of this non-linear behaviour for a range of compounds would be a valuable aid to the designer of edible oil fat mixtures both in speeding the design process and reducing wastage due to error.

This paper deals initially with the derivation from experimental data of regression models to describe the dilatation temperature profiles of binary and tertiary interactive oil fat mixtures. These models are subsequently incorporated in a computer design program to calculate mixture constituent ratios which will yield a best fit to any specified mixture dilatation temperature profile. Economic restraints are incorporated by setting an upper bound to the fractional proportion of any of the constituent components. Such bounds will vary with the seasonal changes in raw material costs and are at the discretion of the designer.

Dilatation and the oil correction factor

If a quantity of solid fat is heated, the volume of fat during heating can be represented by the curve ABCD in Figure 1. Between point A and point B the fat is completely solid and between C and D it is in a purely liquid condition. The dilatation of a fat at any temperature is defined as the expansion during isothermal melting and is represented in Figure 1 by the difference in respective volumes between the partially melted fat at point P and the subcooled liquid fat at point R. The dilatation is usually denoted by the symbol D_T where the subscript refers to the temperature at which the dilatation is measured. Dilatation figures are usually related to 25 grams of fat with the volume being expressed in mm^3.

The measurement of dilatation depends on a knowledge of the subcooled liquid line CQ which is normally difficult to determine experimentally due to the inherent instability of the subcooling process. CQ must therefore be obtained by extrapolation of the easily measured liquid line CD. The coefficient of the oil thermal expansion has been shown[(1)] to be of the form

$$E = \lambda + \gamma T$$

The constants λ and γ have been given a range of values by various experimenters[(1)] and, depending on the value used, the resulting dilatation figure can lie within a spread of 36 dilatation units at 20°C. This spread narrows as the degree of extrapolation decreases and, as it is impractical to determine coefficients λ and γ for every possible combination of fat and oils, some compromise figure must be determined and generally adopted in the measurements. A degree of uncertainty is thus present in any quoted dilatation figure and this will be reflected in the expected accuracy of any analytical model derived.

The modelling of binary mixtures

The data from six binary oil mixtures were available for modelling. The mixtures were as follows:

1. PALM OIL + LARD
2. HARDENED FISH OIL + LARD
3. HARDENED FISH OIL + PALM OIL
4. COCO OIL + PALM OIL
5. HARDENED FISH OIL + COCO OIL
6. LARD + COCO OIL

The dilatations for each of these mixtures was measured at various compositions and the following temperatures: 20°C, 30°C and 37°C.

A typical set of experimental results is shown in Figure 2 which demonstrates the general non-linear form of the mixing process. To improve the symmetry of the raw data a dimensionless dilatation d was adopted where

$$d = \frac{f(X_\alpha)}{AX_\alpha + BX_\beta} = g(X_\alpha)$$

Here X_α, X_β are the constituent concentrations,

A,B are the respective pure component dilatations

$f(X_A)$ is the raw data

and $g(X_\alpha)$ is a least squares regression model obtain from the raw data.

Polynomial forms of up to fourth order were determined using a regression analysis computer program to obtain the optimum $g(X_\alpha)$ function and, for each polynomial form, both the total and residual variances were calculated with the significance of the model being examined by means of the F test method[2]. A typical regression result is shown in TABLE 1 where the F test percentage figure indicates the significance of the model. The standard error is given in terms of dilatation units when the model is applied to the original data. The 95% confidence limits of the polynomial coefficients are also listed.

TABLE 1

A typical regression result given by the F test method. Binary mixture.

Hardened fish oil lard mixture

$$d_{20} = 1.0 - 0.65X_F + 0.57X_F^2 + 0.13X_F^3 - 0.06X_F^4$$

F-test 99% Standard error ± 7 units

Coefficient	Limit
1.0	± 0.004
-0.65	± 0.01
0.57	± 0.01
0.13	± 0.01
-0.06	± 0.01

The predicted error for the best d_{20} models of mixtures 2, 3, 4 and 5 lay within a band of approximately 40 dilatation units at an F test significance level of 99%. The most ill fitting model was that of mixture 1 where the predicted error band was 150 dilatation units. At 30°C the error band was small for all mixtures due to a more accurate oil correction figure and the smaller absolute values of the dilatations.

Modelling of tertiary mixtures

Four mixtures of oils consisting of three interactive components were modelled. These were:

1. HARDENED FISH OIL + PALM OIL + COCO OIL
2. HARDENED FISH OIL + LARD + PALM OIL
3. HARDENED FISH OIL + LARD + COCO OIL
4. LARD + PALM OIL + COCO OIL

The measured dilatations of these mixtures were available at various concentrations and the following temperatures: 20°C, 30°C and 37°C. In this case, as shown in Figure 3, there are two possible ways of proceeding to formulate the model and a range of polynomial regression models were postulated which had the two general forms

$$D_T = K_0 + K_1a + K_2b + K_3c + K_4a^\sigma + K_5b^\sigma + K_6c^\sigma$$

$$D_T = K_0 + K_1a + K_2b + K_3c + K_4\delta^\sigma{}_{ac} + K_5\delta^\sigma{}_{bc} + K_6\delta^\sigma{}_{ab}$$

where a, b, c = AX, BX, CX

$$\delta_{ac}, \delta_{bc}, \delta_{ab} = d_{ac}\ |A + C|,\ d_{bc}\ |B + C|,\ d_{ab}\ |A + B|$$

$$\sigma = 1, 1.5, 2.0, 2.5 \ldots$$

The predicted D_{20} figures for for best models of mixtures 1, 2 and 3 were within a band of 50 dilatation units at an F test significance level of 99%. The corresponding D_{30} and D_{37} figures for these mixtures were 40 and 20 dilatation units respectively. The best model for mixture 4 produced corresponding error bands of 80, 90 and 94 dilatation units. A typical regression result is shown in TABLE 2. Here a variable is declared redundant if the chance that it is redundant is in excess of 50%.

TABLE 2

Typical regression result given by F test
Tertiary Mixture

Hardened fish oil/palm oil/coco oil mixture

$$\begin{aligned} D_{20} = {} & 1178 - 3.3FX_F - 0.96PX_P - 1.5CX_C \\ & + 0.19d_{FP}\ FX_F + PX_P + 1.65d_{FC}\ FX_F + CX_C \\ & + 0.32D_{CP}\ CX_C + PX_P + 0.001\ FX_F{}^2 \\ & - 0.002\ PX_P{}^2 - 0.0004\ CX_C{}^2 \end{aligned}$$

F-test 99.9 Standard error ± 22 dilatation units

Coefficient	Limit	Coefficient	Limit
1178	± 4	0.002	± 0.00017
-3.3	± 0.035	0.0004	± 0.00018
-0.96	± 0.049	0.19	± 0.039
-1.5	± 0.048	1.65	± 0.037
0.001	± 0.00003	0.32	± 0.059

The term - 0.0004 is redundant

Nuclear magnetic resonance spectroscopy

Nuclear magnetic resonance spectroscopy methods can be employed to give a direct reading of liquid in a solid/liquid sample and thus, in this case, it is possible to determine the solid fat index at any temperature and hence the dilatation figure. Due to errors in the oil correction calculation, more accurate models will only result if a more absolute measurement criterion such as the solid fat index is adopted. Some preliminary measurements have been taken using this method and a comparison with similar dilatometery measurements is given in Figure 4. Future experimental data will be based entirely on NMRS measurements.

Synthesising optimum three component oil mixtures

A computer program based on the simplex hill climbing routine and incorporating the best regression models of tertiary oil mixtures was devised to calculate the composition of a three component oil fat mixture which meets a required dilatation temperature specification. The program which is written in Fortran IV requires 8K of computer store. There are eight steps in its execution.

(1) The desired mixture dilatation values are entered.
(2) The appropriate oil mixture is chosen.
(3) Pure component dilatation values are entered.
(4) The maximum desired value of each constituent in the mixture is chosen. This value will be determined by economic or other criteria at the disposal of the analyst. The stoichiometric mixture relationship is assumed to be satisfied.
(5) Starting point of search is chosen.

Data entry is now complete and the following steps are automatically undertaken by the program:

(6) By altering each constituent by ±0.5% six points circling the chosen starting point Po are obtained as shown in Figure 5.
(7) The objective function e is obtained at each of the seven points

$$e = (E_{20} - S_{20})^2 + (E_{30} - S_{30})^2 + (E_{37} - S_{37})^2$$

where E_{20}, E_{30}, E_{37} are the mixture dilatations at 20°C, 30°C and 37°C calculated from the regression models and S_{20}, S_{30}, S_{37} are the specified mixture dilatations.

(8) The smallest objective function is determined and this point used as an initial point in a new search. The search is terminated if the initial point has the smallest value of objective function. If the search meets a boundary then the optimum solution is that point which crosses the boundary. As shown in Figure 5 the search near the apex of the triangle is essentially constrained to two directions.

No facility exists in the simple hill climbing program used to determine whether the minimum obtained is local or global. Instead a map of the objective function is printed out which can be visually scanned by the operator to obtain the global minimum search area.

Typical computer results are shown in TABLE 3 for four oil mixtures where a constant arbitrary dilatation profile is specified. This profile was appropriate for the first two oil mixtures and a good design resulted.

The D_{37} figure specified for mixture 3 was, however, unrealistically low as LARD PALM COCO mixtures are generally characterised by D_{37} values in the range 50 → 120 dilatation units. The attempt by the hill climbing routine to meet this figure had a subsequent deleterious effect on the D_{30} figure obtained. A similar situation occurs with the HFO.LARD.PALM.D_{30} mixture.

Discussion of results

The regression methods used in the derivation of the mixture models were found to be generally successful in dealing with the inherent non-linearities of component interaction and, once dilatation estimate errors are eliminated, better models should be obtained.

Possible improvements in the optimisation program include the incorporation of a sensitivity routine to inform the operator as to which component variation has the greatest effect on the dilatation figures and the insertion of a command interceptor. The latter routine would allow the operator to re-enter the design program at any point, change a parameter value and exit directly to ascertain the effect of his action on the dilatation contour. This would enable a fast iterative design to be completed with a minimum of data entry.

As the entire program only occupies some 8k words of computer store it is suitable for use on a small, inexpensive, dedicated computer which could also be used for on-line control of the mixing system from the tank farm. Alternatively, the program is suitable for use on a time shared terminal connected via a land link to a remote central processor. The economics of the latter approach are now being investigated.

A computer program of this type will completely remove the tedium of design calculation and ensure a specified dilatation temperature profile for the final product batch. The associated reduction in product waste due to design error will, of course, result in significantly lower production costing over a wide range of product mixtures.

Conclusions

The oil correction calculation gives rise to a large error in dilatation readings, particularly at low temperatures, and this error is reflected in the derived regression models. A more absolute measurement technique than normal dilatometery methods is thus necessary if accurate models are required. The optimisation program developed for synthesising oil mixtures is simple in concept, easy to use and gives good results provided a realistic dilatation contour is specified.

References

(1) Boekenoogen, H.A. Analysis and Characterisation of Oil Fats and Fat Products, 1966 Vol. 1. (Chichester: John Wiley - Interscience Publications).

(2) Perry, J.H. Chemical Engineers Handbook, 1963, 4th Edn. (New York: McGraw-Hill Book Co.).

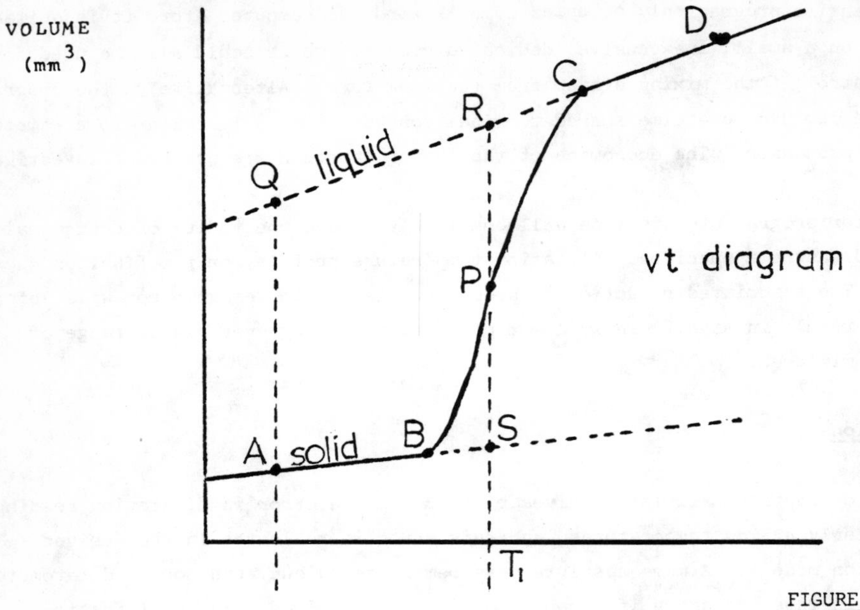

FIGURE 1

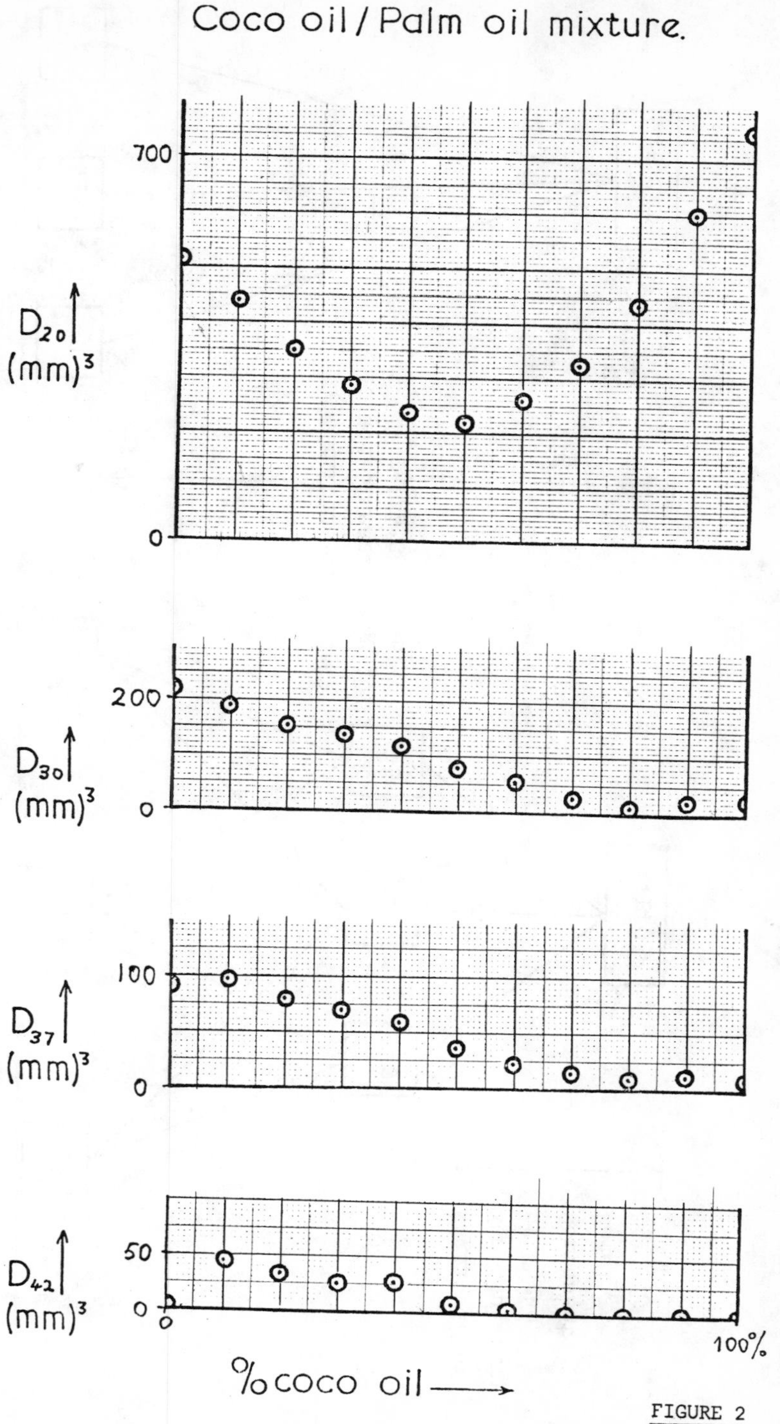
Coco oil / Palm oil mixture.
700
0
D_{20}
$(mm)^3$
200
0
D_{30}
$(mm)^3$
100
0
D_{37}
$(mm)^3$
50
0
0
D_{42}
$(mm)^3$
100%
% coco oil

FIGURE 2

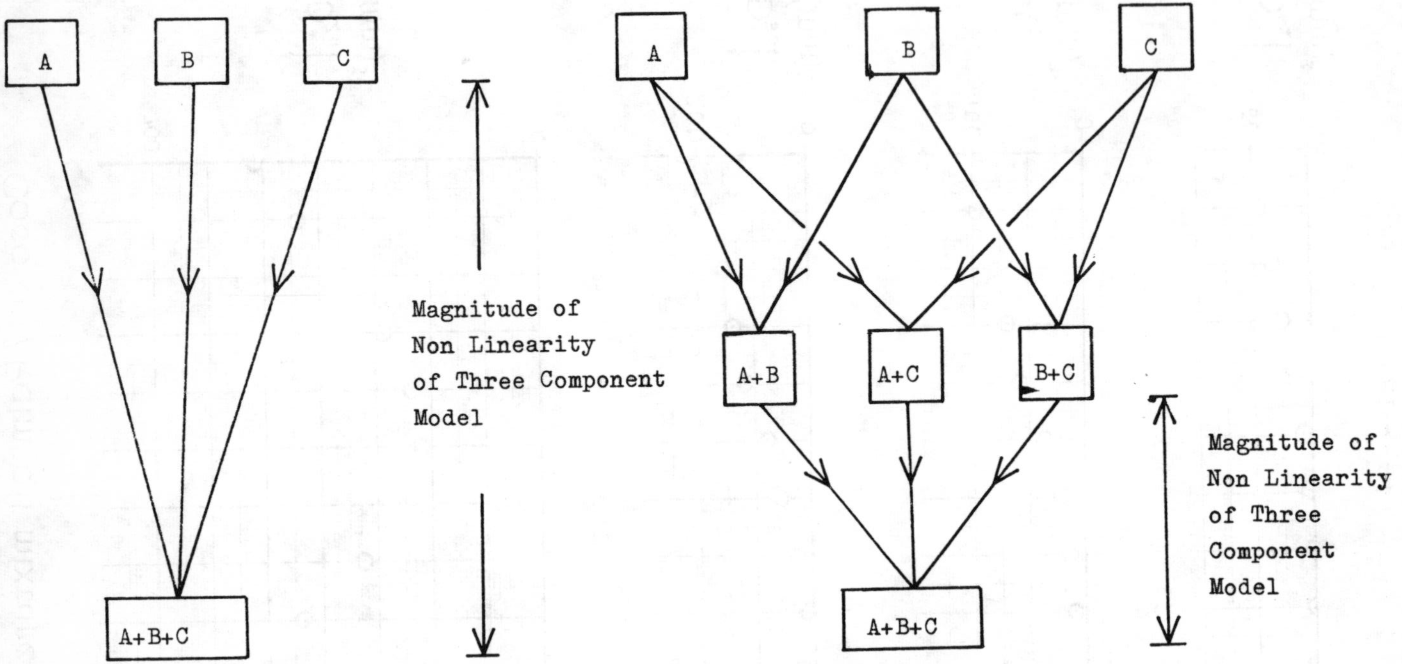
A
B
C
A+B+C
Magnitude of
Non Linearity
of Three Component
Model
A
B
C
A+B
A+C
B+C
A+B+C
Magnitude of
Non Linearity
of Three
Component
Model

FIGURE 3

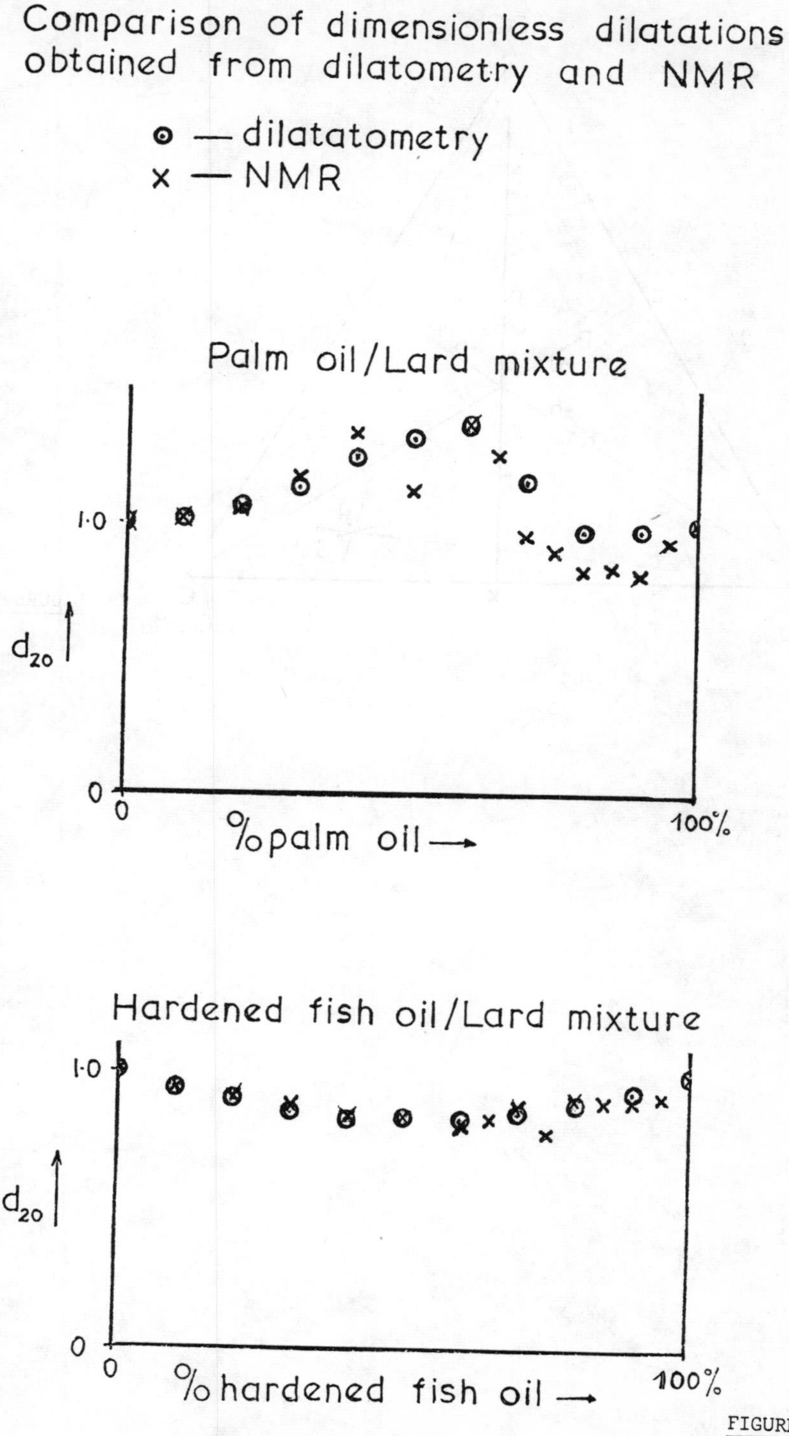
Comparison of dimensionless dilatations obtained from dilatometry and NMR
⊙ — dilatatometry
× — NMR
Palm oil/Lard mixture
1·0
d20
0
0
%palm oil →
100%
Hardened fish oil/Lard mixture
1·0
d20
0
0
%hardened fish oil →
100%

FIGURE 4

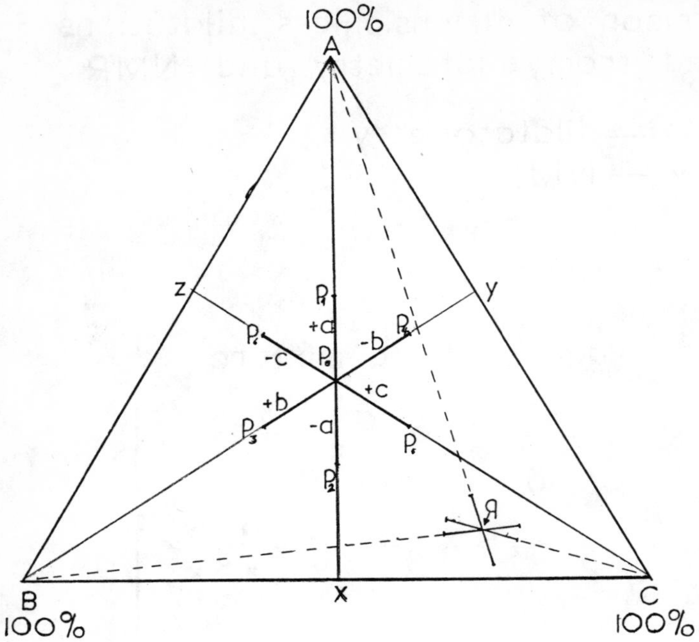

FIGURE 5

ARBITARY SPECIFIED MIXTURE DILATATION D20 = 550; D30 = 200; D37 = 10

MIXTURE		HFO	PALM	COCO	HFO	LARD	COCO	LARD	PALM	COCO	HFO	LARD	PALM
Assumed	D20	685	535	748	658	600	748	790	625	748	680	695	525
Component	D30	203	251	24	203	172	24	306	225	24	193	266	225
Dilatations	D37	8	129	10	8	81	10	169	91	10	0	141	91
Maximum													
Desired	B_1 =	0.97	H		0.97	H		1.0	L		0.95	H	
Fractional	B_2 =	0.5	P		1.0	L		1.0	P		1.0	L	
Composition	B_3 =	0.2	C		0.2	C		0.2	C		1.0	P	
Optimum		0.8303	H		0.7470	H		0.4523	L		0.8104	H	
Mixtures		0.1211	P		0.1852	L		0.4147	P		0.0090	L	
		0.0485	C		0.0688	C		0.1330	C		0.1806	P	
Calculated	D20	552			565			567			586		
Dilatation	D30	196			207			166			151		
	D37	9			12			76			32		
Errors	D20	2			15			17			36		
(Dilatation)	D30	-4			7			-34			-49		
(Units)	D37	-1			2			66			22		

TABLE 3

FREE BOUNDARY PROBLEMS AND IMPULSE CONTROL

J.L. LIONS

University of PARIS VI

and

I.R.I.A., Rocquencourt

INTRODUCTION

Free boundary problems classicaly arise in Physics, one of the simplest examples being the melting of ice. We do not intend here, by no means, to review this family of problems, which are very numerous, and have been studied by a large number of different methods. We only want to give brief indications on the possible use of the so-called variational inequalities (V.I) in these problems, the interest being there to connect the free boundary problems and the optimization theory ; this is the subject of Section 1 below.

We consider next impulse control problems, as introduced in A. Bensoussan and the Author [3] 1).

One can show that the optimal cost function is characterized by the solution of a new free boundary problem. Mathematical analogies with problems solved in Section 1 lead us to introduce a new tool, which (somewhat) extends the V.I : the quasi-variational inequalities (Q.V.I), as introduced in [2] by A. Bensoussan, M. Goursat and the Author for the stationary case and in [3]2) by A. Bensoussan and the Author for the evolution case. This is the object of Section 2.

Complete reports on the many problems which arise along these lines will be given in [3]3), 4), 5) [9]. Other problems directly related to this note are studied in [11].

The results briefly presented in Section 2 extend the results obtained by various authors in the so-called s - S policy. Let us quote here the works of Scarf [12], Veinott [14]. (Cf. also the bibliographies of these works).

The methods introduced in this short note show the strong connections which exist between free boundary problems in Physics, in problems of management and in techniques of optimization.

I. FREE BOUNDARY PROBLEMS AND V.I.

1.1 An example.

Let $\mathcal{O}$ be a bounded open set of $\mathbb{R}^n$ (1), with smooth boundary Γ. Let A be a partial differential operator given in $\mathcal{O}$ by

$$A\varphi = -\sum_{i,j=1}^{n} \frac{\partial}{\partial x_i}\left(a_{ij}(x)\frac{\partial \varphi}{\partial x_j}\right) + \sum_{i=1}^{n} a_i(x)\frac{\partial \varphi}{\partial x_i} + a_o(x)\varphi \tag{1.1}$$

where a_{ij}, a_i, $a_o \in L^\infty(\mathcal{O})$. We suppose that A is elliptic in the following sense : if $\varphi, \psi \in H^1(\mathcal{O})$ (2) and if we define

$$a(\varphi,\psi) = \sum_{i,j=1}^{n} \int_{\mathcal{O}} a_{ij}(x)\frac{\partial \varphi}{\partial x_j}\frac{\partial \psi}{\partial x_i}\,dx + \sum_{i=1}^{n}\int_{\mathcal{O}} a_i \frac{\partial \varphi}{\partial x_i}\psi\,dx + \int_{\mathcal{O}} a_o\,\varphi\psi\,dx \tag{1.2}$$

then

$$a(\varphi,\varphi) \geqslant \alpha\,\|\varphi\|^2_{H^1(\mathcal{O})}\quad,\ \alpha > 0. \tag{1.3}$$

The problem we want to consider is the following : we look for a function u in $\mathcal{O}$ such that

$$Au - f \leqslant 0 \text{ in } \mathcal{O}, \tag{1.4}$$

$$u - \psi \leqslant 0 \text{ in } \mathcal{O} \tag{1.5}$$

$$(Au-f)(u-\psi) = 0 \text{ in } \mathcal{O} \tag{1.6}$$

$$\frac{\partial u}{\partial \nu} = 0 \quad (3) \tag{1.7}$$

where f, ψ are given in $\mathcal{O}$, and where $\frac{\partial}{\partial \nu}$ denotes the conormal derivative associated with A.

---

(1) In the physical applications, $n = 1,2,3$. In the problems of impulse control that we study in Section 2, the dimension n can be very large.

(2) $H^1(\mathcal{O}) = \{\varphi \mid \varphi, \frac{\partial \varphi}{\partial x_1}, \ldots, \frac{\partial \varphi}{\partial x_n} \in L^2(\mathcal{O})\}$, φ being real valued ;

$$\|\varphi\|^2_{H^1(\mathcal{O})} = \sum_{i=1}^{n}\int_{\mathcal{O}}\left(\frac{\partial \varphi}{\partial x_i}\right)^2 dx + \int_{\mathcal{O}}\varphi^2\,dx.$$

(3) Actually this boundary condition should be dropped at boundary points (if any) where $u = \psi$.

Of course $\mathcal{O}$ is divided in two regions where respectively $Au = f$ and $u = \psi$, with (1.4) and (1.5) being satisfied everywhere, and these regions being not given ; the interface between these two regions is a free surface.

We now transform the problem (1.4)...(1.7) into a V.I. One can check that it is equivalent to finding u satisfying

$$(1.8) \qquad \left| \begin{array}{l} a(u,v-u) \geq (f,v-u) \qquad \forall\, v \leq \psi \ , \\ u \leq \psi . \end{array} \right.$$

The problem (1.8) is what is called a variational inequality (V.I) following [13] and [10]; it follows from these works that under the hypothesis (1.3), the problem (1.8) admits a unique solution.

1.2 Various remarks.

Remark 1.1.

A large number of problems in Physics and in Mechanics lead to problems of this type, in stationary or non stationary cases ; in the latter case, one has to use the tool of V.I. of evolution, introduced in [10]. Examples and variants arising in elasto-plasticity, in non-newtonian fluids etc.. are studied in the book [6].

Remark 1.2.

In the problem alluded to in Remark 1.1 , one obtains a V.I on the "physical" unknown function u . A very important step forward to the theory is due to C. Baiocchi [1] who showed that classical free boundary problems arising in infiltration theory can be reduced - on a new " non physical" unknown function, say $\tilde{u}$ - to a V.I. on $\tilde{u}$. This idea, properly used, permits the solution of a number of other free boundary problems ; cf. [4][5][7].

Remark 1.3.

For the problems of V.I. numerical methods are known and numerically implemented ; a report is given in the book of Glowinski, Trémolières and the Author [8].

II. IMPULSE CONTROL PROBLEMS.

2.1. Formulation of the problem.

We give here a very formal presentation ; for precise statements we refer to [3], 1) and 3). We explain the problem on an example. We assume that we have, at time t, an amount $x = \{x_1, \ldots, x_n\}$ of goods ([1]) . We suppose that we know the cumulative demand $D(t,s)$ between t and $s > t$; $D(t,s)$ can be deterministic or stochastic. A policy v_{xt} is a double sequence :

$$v_{xt} = \left\{ \theta^1_{xt} , \xi^1_{xt} ; \theta^2_{xt} , \xi^2_{xt} ; \ldots \right\} \tag{2.1}$$

where $t \leqslant \theta^1_{xt} < \theta^2_{xt} < \theta^3_{xt} < \ldots$ are the instants where we place orders, and where $\xi^1_{xt} , \xi^2_{xt} , \ldots \in \mathbb{R}^n$ are the amount of goods we order at time $\theta^1_{xt} , \theta^2_{xt}, \ldots$; this double sequence is deterministic or stochastic.

We denote by T the horizon, it can be finite or infinite ; we suppose here, to fix ideas, that $T < \infty$.

The cost function is defined as follows :

$$J_{xt}(v_{xt}) = E \left\{ \int_t^T f(y_{xt}(s), s) ds + N_{xt} \right\} \tag{2.2}$$

(where one should drop the Expectation if the problem is deterministic), where :

i) $y_{xt}(s)$ = state at time s = amount of goods at our disposal at time s (this is uniquely determined if $D(t,s)$ is given and if v_{xt} is chosen) ;

ii) $f(x,s)$ = holding cost by unit of time if $x > 0$,
= outage cost by unit of time if $x < 0$;

iii) N_{xt} = number of orders we place during the period $[t,T]$; we can assume here that we have to pay 1 for placing an order.

([1]) Therefore the dimension in this problem is equal to the number of items and consequently n can be very large.

This is an impulse control problem. We want to minimize (2.2) with respect to "all" possible policies (2.1). Let us define

(2.3) $$u(x,t) = \inf J_{xt}(v_{xt}) \; .$$

What we want to show (for precise statements and proofs, we refer to the works of A. Bensoussan and the Author referred to in the Bibliography), is that $u(x,t)$ can be characterized by the solution of a new free boundary problem and that this free boundary problem can be studied by using the technique of quasi variational inequalities (Q.V.I.).

2.2. Free boundary problem.

One can show that $u(x,t)$ satisfies the following set of inequalities and equalities (1) :

(2.4) $$-\frac{\partial u}{\partial t} + Au - f \leqslant 0 \; , \quad x \in \mathbb{R}^n \; , \; t \in [t_o, T]^{(2)} \; ,$$

(2.5) $$u - M(u) \leqslant 0 \; ,$$

(2.6) $$\left(-\frac{\partial u}{\partial t} + Au - f\right)(u - M(u)) = 0$$

(2.7) $$u(x,T) = 0 \; ,$$

where in (2.5)

(2.8) $$M(u)(x) = 1 + \inf . u(x+\xi), \quad \xi = \{\xi_i\}, \; \xi_i \geqslant 0.$$

Of course (2.2) obviously follows from the definition. The proof of (2.5) is immediate : if we consider the particular policy $\hat{v}_{xt}$ which consists in ordering ξ ($\xi_i \geqslant 0$) at time t, then :

$$u(x,t) = \inf J_{xt}(v_{xt}) \leqslant \inf J_{xt}(\hat{v}_{xt}) = 1 + \inf J_{x+\xi,t}(v_{x+\xi,t})$$

hence (2.5) follows. The proofs of (2.4)(2.6) are much more complicated.

(1) Of course one has to make here precise hypothesis on the nature of the Demand $D(t,s)$; cf. A. Bensoussan and the A., loc.cit.

(2) The operator A is determined by D . It is a second order (resp. first order) operator in the stochastic (resp. deterministic) case.

Let us observe now that (2.4)...(2.7) is indeed a free boundary problem. By virtue of (2.6) there are two regions in $\mathbb{R}^n_x \times [t_o, T]$; in one region, say S, one has :

$$u = M(u) \tag{2.9}$$

and in the complementary region C one has :

$$-\frac{\partial u}{\partial t} + Au - f = 0 . \tag{2.10}$$

The interface between S and C is a free boundary.

Some remarks are now in order :

Remark 2.1.

The non linear operator $u \to M(u)$ defined by (2.8) is of non local type, since to compute $M(u)$ at point x it is not sufficient to know u in a neighborhood of x.

Remark 2.2.

Many other examples or operators of the type of M as given by (2.8) arise in applications. Cf. A.Bensoussan and the Author, loc. cit.

Remark 2.3.

From the solution u one can derive the best policy (which exists) ; this is an extension of the "s – S policy", introduced in [12][14](see also the Bibliographies of these works).

Remark 2.4.

In Section 2.3. below, we briefly consider the stationary case, and more over in a bounded domain $\mathcal{O}$ of $\mathbb{R}^n$, with smooth boundary Γ (1); the problem is then to find u satisfying

$$Au - f \leqslant 0 , \text{ in } \mathcal{O} \tag{2.11}$$

$$u - M(u) \leqslant 0 , \tag{2.12}$$

$$(Au-f)(u-M(u)) = 0 , \tag{2.13}$$

and

$$\frac{\partial u}{\partial \nu} = 0 \text{ on } \Gamma \text{ (2)} . \tag{2.14}$$

We assume that A is given by (1.1) and satisfies (1.3).

(1) Cf. [3]1) 3),for the interpretation of the problem. The introduction of a bounded domain $\mathcal{O}$ eliminates some technical difficulties.

(2) Actually this boundary condition should be dropped at boundary points where $u = M(u)$.

2.3. Quasi variational inequalities (Q.V.I.).

The analogy between (2.11) ... (2.14) and (1.4) ... (1.7) is clear ; the difference is that ψ , which is given in (1.5), is now "replaced" by $M(u)$ which is not given.

One can check that (2.11) ... (2.14) is equivalent to finding u such that

(2.15)
$$\begin{cases} a(u,v-u) \geq (f,v-u) & \forall\ v \leq M(u) , \\ u \leq M(u). \end{cases}$$

The problem (2.15) is called a quasi variational inequalities (Q.V.I.).

One can show the existence and uniqueness of a maximal solution which can be obtained as the limit of solutions of V.I.

Let us introduce u^o solution of the Neumann problem :

(2.16)
$$a(u^o,v) = (f,v) \qquad \forall\ v \in H^1(\mathcal{O})$$

and let us define u^n , $n \geq 1$, as the solution of the V.I. :

(2.17)
$$\begin{cases} a(u^n,v-u^n) \geq (f,v-u^n) & \forall\ v \leq M(u^{n-1}), \\ u^n \leq M(u^{n-1}). \end{cases}$$

One can show that

(2.18)
$$u^o \geq u^1 \geq \dots \geq u^n \geq \dots$$

and obtain u , solution of (2.15), as the limit (in $L^p(\mathcal{O})$ $\forall$ p finite and $H^1(\mathcal{O})$ weakly) of u^n as $n \to \infty$.

Remark 2.5.

Algorithm (2.17) used jointly with numerical technique for solving V.I. gives numerical methods for solving Q.V.I. Cf. [2][9].

Remark 2.6.

For numerical purposes, one is of course faced with the problem of dimensionality. We shall report elsewhere on the use of decomposition methods and sub-optimal policies.

BIBLIOGRAPHY

[1] C.Baiocchi, C.R.Acad.Sc., 273 (1971), 1215-1218.

[2] A.Bensoussan, M.Goursat and J.L.Lions, C.R.Acad.Sc.Paris, séance du 2 avril 1973.

[3] A.Bensoussan and J.L.Lions,
1) C.R. Acad. Sc. Séance du 2 avril 1973,
2) C.R. Acad. Sc. Séance du 2 avril 1973,
3) Book in preparation,
4) Work to the memory of I.Petrowsky, Ousp.Mat.Nauk 1974
5) Int. Journal of Appl. Mat. and Optimization.

[4] H.Brézis and G.Duvaut, C.R.Acad. Sc. Paris, t.276 (1973), 875-878.

[5] H.Brézis and G.Stampacchia, C.R.Acad. Sc. Paris, 276, (1973), 129-132.

[6] G.Duvaut and J.L.Lions, Inéquations en Mécanique et en Physique, Dunod, 1972.

[7] G.Duvaut, C.R. Acad. Sc. Paris, June 1973.

[8] R.Glowinski, J.L.Lions and R.Trémolières, Book to appear, Dunod, 1974.

[9] M.Goursat, Report LABORIA, 1973.

[10] J.L.Lions and G.Stampacchia, Comm. Pure and Appl. Math XX (1967), 493-519.

[11] J.P.Qaudrat and M.Viot, Report LABORIA, 1973.

[12] H.Scarf Math.Meth. in the social sciences. Stanford Univ.Press, 1960.

[13] G.Stampacchia C.R.Acad. Sc. Paris. 258 (1964) 4413-4416.

[14] A.E.Veinott J. SIAM Appl. Math, 14,5, September 1966.

A CONVEX PROGRAMMING METHOD IN HILBERT SPACE AND ITS APPLICATIONS TO OPTIMAL CONTROL OF SYSTEM DESCRIBED BY PARABOLIC EQUATIONS

Kazimierz Malanowski
Polish Academy of Sciences
Institute of Applied Cybernetics
Warsaw, Poland

1. INTRODUCTION

Some optimal control problems for systems described by partial differential equations can be reduced [2, 6] to the problem of minimization of a convex functional J(y) on a closed convex and bounded set D in a Hilbert space. However, this set (the so called attainability set) usually is not given in an explicit form. Therefore the direct minimization of the functional is very difficult. On the other hand it is comparatively easy to find the points of support of D by given supporting hyperplanes. It was Gilbert who first proposed [4] to use these points of support for construction of a sequence minimizing J(y). He considered the case of a quadratic functional defined on n-dimensional Euclidean space. It turnes out that the speed of convergence of Gilbert's algorithm is comparatively law. Several attempts were made to improve the speed of convergence of Gilbert's procedure [3, 8, 9]. All these methods delt with quadratic functionals defined on n-dimensional Euclidean space.

In this paper all modifications of Gilbert's method are reduced to one general scheme, which is applied to the problem of minimization of a convex functional on closed, convex and bounded set in a Hilbert space.

The use of this scheme is proposed for solving an optimal control problem for parabolic equations. Some computations were performed and the three modifications of the method were compared on the basis of obtained numerical results.

2. CONVEX PROGRAMMING PROBLEM AND AN ITERATIVE PROCEDURE FOR SOLVING IT

In a Hilbert space H there is given a closed, convex and bounded, i.e. weakly compact [2,11] set D.

Moreover, a non-negative real convex functional J(y) is defined on the space H. It is assumed that J(y) is two times continuously differentiable and its Hessian J"(y) satisfies the following conditions:

$$n\|z\|^2 \leqslant (J''(y)z, z) \leqslant N\|z\|^2, \ \forall y \in D, \forall z \in H \tag{1}$$

where $0 < n \leqslant N < \infty$.

The condition (1) implies that $J(y)$ is strictly convex.

Our purpose is to find an element $y_{opt} \in D$ called an optimal element satisfying the relation

$$J(y_{opt}) = \inf_{y \in D} J(y). \tag{2}$$

Since $J(y)$ is lower semicontinuous functional as a convex [11] and the set D is weakly compact the point y_{opt} exists and it is unique due to the strict convexity of $J(y)$.

At the point y_{opt} the following necessary and sufficient condition of optimality has to be satisfied

$$(-J'(y_{opt}), y) \leqslant (-J'(y_{opt}), y_{opt}) \ \forall y \in D. \tag{3}$$

To determine y_{opt} an iterative procedure is used based on minimization of some quadratic approximations of the functional $J(y)$ on closed and convex subsets of D, which are constructed successively.

To this end we define functionals

$$\bar{J}_i(y) \triangleq J(y) + (J'(y_i), y - y_i) + \frac{1}{2} N(y - y_i, y - y_i) , \tag{4a}$$

$$\bar{\bar{J}}_i(y) \triangleq J(y) + (J'(y_i), y - y_i) + \frac{1}{2} n(y - y_i, y - y_i) . \tag{4b}$$

$\bar{J}_i(y)$ and $\bar{\bar{J}}_i(y)$ are quadratic functionals and

$$\bar{J}_i(y_i) = \bar{\bar{J}}(y_i) = J(y_i). \tag{5a}$$

It follows from (1) that

$$\bar{\bar{J}}_i(y) \leqslant J(y) \leqslant \bar{J}_i(y) . \tag{5b}$$

We denote by $y_i \in D_i$ the unique element satisfying the relation

$$\bar{J}_{i-1}(y_i) = \inf_{y \in D_i} \bar{J}_{i-1}(y). \tag{6}$$

Let $\bar{y}_{i+1} \in D$ by any arbitrary element such that

$$(-J'(y_i), y) \leqslant (-J'(y_i), \bar{y}_{i+1}), \ \forall y \in D, \tag{7}$$

i. e. $\bar{y}_{i+1}$ is any point at which the hyperplane M_i orthogonal to $-J'(y_i)$ supports D.

<u>Theorem</u>. If the family of closed, convex and bounded sets $D_i \subset D$ is constructed in such a way that

$$\{y_i\} \cup \{\bar{y}_{i+1}\} \subset D_{i+1} \tag{8}$$

then the sequence $\{y_i\}$ is strongly convergent to y_{opt}.

The proof of Theorem is given in Appendix 1.

As it is shown in Appendix 2 the optimal value of the functional can be estimated as follows

$$\max_{\alpha \in A} \left\{ 0,\ J(y_\alpha) + \left(\frac{n}{2} \frac{(J'(y_\alpha), \bar{y}_{\alpha+1} - y_\alpha)}{(J'(y_\alpha), J'(y_\alpha))} + 1 \right) (J'(y_\alpha), \bar{y}_{\alpha+1} - y_\alpha) \right.$$

$$\leqslant J(y_{opt}) \leqslant J(y_i), \tag{9}$$

where

$$A = \left\{ 1 \leqslant \alpha \leqslant i : (J'(y_\alpha), J'(y_\alpha) + n(\bar{y}_{\alpha+1} - y_\alpha)) \geqslant 0 \right\}. \tag{9a}$$

In the particular case where $J(y)$ is a quadratic functional of the form

$$J(y) = (z - y, z - y) \tag{10}$$

where $z \in H$ is a given element, which does not belong to D, we have $n = N = 2$, and the condition (7) takes on the form $(z-y_i, y) \leqslant (z-y_i, y_{i+1})$.

In this case the estimation (9) reduces to the following one

$$\max_{1 \leqslant \alpha \leqslant i} \left\{ 0,\ \frac{(z-y_\alpha, z-\bar{y}_{\alpha+1})^2}{(z-y_\alpha, z-y_\alpha)} \right\} \leqslant (z-y_{opt}, z-y_{opt}) \leqslant (z-y_i, z-y_i) \tag{11}$$

3. SOME METHODS OF CONSTRUCTING THE SETS D_i

Note that according to Theorem the iterative procedure is convergent to the optimal solution if the condition (8) is satisfied no matter what are the forms of the sets D_i. However the speed of convergence depends very much on the shape of D_i.

We present three methods of constructing the family D_i known from literature. In all these methods the sets D_i are characterized by finite number of parameters.

a. Gilbert's method [4]

This method is the simplest one. We start from any arbitrary point $y_0 \in D$ and at (i+1)-th iteration we choose as D_{i+1} the segment $[y_i, \bar{y}_{i+1}]$.

In this case the element y_{i+1} can be easily found from the formula $y_{i+1} = y_i + \alpha_{i+1}(\bar{y}_{i+1} - y_i)$, where

$$\alpha_{i+1} = \min \left\{ 1,\ \frac{(-J'(y_i), \bar{y}_{i+1} - y_i)}{N(\bar{y}_{i+1} - y_i, \bar{y}_{i+1} - y_i)} \right\}. \tag{12}$$

Hence the points y_{i+1} are determined in a very simple way. However the speed of convergence in terms of the number of iterations is very low [4].

This slow convergence is due to the fact that at each step we use only the information obtained in the last iteration (namely y_i) and we do not take into consideration the previous ones.

To overcome this disadvantage the following modification was proposed:

b. Barr's method [3]

In this method as in the previous one we select an arbitrary initial point y_0 and we put $D_1 = [y_0, \bar{y}_1]$, but as D_i we choose

$$D_i = \operatorname{conv}\{D_{i-1} \cup \bar{y}_i\} = \operatorname{conv}\{y_0 \cup \bar{y}_1 \cup \dots \cup \bar{y}_i\} = \\ = \operatorname{conv}\{y_i^1, y_i^2, \dots, y_i^{i+1}\}, \tag{13}$$

where y_i^j $(j = 1, \dots, i+1)$ denotes the vectors spanning the set D_i.

Every element $\hat{y} \in D_i$ can be represented in the form a convex combination of y_i^j, i.e.

$$\hat{y} = \sum_{j=1}^{i+1} \alpha^j y_i^j$$

where

$$\alpha^j \geq 0, \quad \sum_{j=1}^{i+1} \alpha^j = 1. \tag{14}$$

Hence in order to find the element y_i we have to determine the values α_i^j of the coefficients α^j at which the function $\bar{J}_{i-1}(\hat{y})$ assumes its minimum subject to the constraints (14). This functional can be rewritten in the form

$$\bar{J}_{i-1}(\hat{y}) = J(y_{i-1}) + (J'(y_{i-1}), \sum_{j=1}^{i+1} \alpha^j y_i^j) + \\ + \frac{1}{2} N\left(\sum_{j=1}^{i+1} \alpha^j y_i^j - y_{i-1}, \sum_{j=1}^{i+1} \alpha^i y_i^j - y_{i-1}\right) = \\ = c_0 - [< c_i, \alpha > + < \alpha, B_i \alpha >] \tag{15}$$

where

$c_0 = J(y_{i-1}) + \frac{1}{2} N(y_{i-1}, y_{i-1})$

$\alpha_i = \{\alpha_i^j\}^{1/i+1}$ and $c_i = \{-(J'(y_{j+1}) + Ny_{i-1}, y^j)\}^{1/i+1}$ are $(i+1)$-dimensional vectors,

$B_i = \{-\frac{1}{2} N(y_i^j, y_i^k)\}^{i+1/i+1}$ is $(i+1) \times (i+1)$-dimensional matrix,

$<.,.>$ denotes the inner product in (i+1)-dimensional Euclidean space.

Minimization of (15) subject to (14) is a typical finite dimensional quadratic programming problem and it can be solved using any of well known procedures [5].

Note that in this method the dimension of B_i usually increases after each iteration, what in turn increases the time of computations.

To avoid this difficulty some modifications of the construction of D_i were proposed [3, 7, 9], that makes it possible to limit the dimension of the vectors α_i to a given number.

c. Nahi-Wheeler's method [8]

Both methods presented up to now were general in the sense that they could be applied to any set D. The third method takes advantage of a specific form of the set D.

In linear optimal control problems as a set D we choose the so called attainability set. We consider the case of scalar control function with constrained magnitude and performance index depending on terminal state. We put

$$D = \left\{ y : y = \int_0^T A(t)\ u(t)\ dt : |u(t)| \leqslant 1,\ 0 \leqslant t \leqslant T \right\}, \tag{16}$$

where: $u(t) \in R^1$ is a scalar control function; $A(t)$ is a continuous linear operator from R^1 into H, T is a fixed time of control.

The set D defined by (16) satisfies [8] all conditions of Theorem.

The appropriate elements y will be found by means of determining control functions $u(\cdot)$ corresponding to y according to (16).

It is easy to check [7, 8] that the control $\bar{u}_{i+1}(t)$ corresponding to the element $\bar{y}_{i+1}$ satisfying (7) is given by

$$\bar{u}_{i+1}(t) = \operatorname{sgn}\left(-J'(y_i),\ A(t)\right) \tag{17}$$

As $u_0(\cdot)$ we choose any arbitrary piece-wise constant function satisfying the condition

$$|u_0(t)| \leqslant 1, \quad \forall t \in [0,T]\ . \tag{18}$$

Let

$$\left\{ t_0^j \right\} \tag{19}$$

be the set of all points of discontinuity of $u_0(\cdot)$.

The function $\bar{u}_1(\cdot)$ is derived from (17).

By $\left\{ \bar{t}_1^k \right\}$ we denote the set of all points of discontinuity of $\bar{u}_1(\cdot)$. Let us assume that this set has a finite number of elements and let us put

$$\left\{t_1^j\right\} = \left\{t_0^j\right\} \cup \left\{\bar{t}_1^k\right\} \tag{20}$$

We introduce the set U_1 of all piece-wise constant functions satisfying (18) with the points of discontinuity at t_1^j.

We put

$$D_1 = \left\{ y : y = \int_0^T A(t)\ u(t)\ dt;\ u(\cdot) \in U_1 \right\} . \tag{21}$$

It is obvious that $\{y_0\} \cup \{\bar{y}_1\} \subset D_1 \subset D$. Moreover the set D_1 is closed and convex [8], hence it satisfies all the conditions of Theorem.

By $u_1(\cdot) \in U_1$ we denote the control functions which corresponds to the element y_1 satisfying (6).

In the same way as before we denote by U_{i+1} the set of all piece-wise constant functions satisfying (18) with the points of discontinuity belonging to the set

$$\left\{t_{i+1}^j\right\} = \left\{t_i^j\right\} \cup \left\{\bar{t}_{i+1}^k\right\}, \tag{22}$$

where t_i^j and $\bar{t}_{i+1}^k$ denote the points of discontinuity of $u_i(\cdot)$ and $\bar{u}_{i+1}(\cdot)$ respectively.

It follows from this construction that usually the number of the elements t_i^j increases at each iteration. Let the number of these elements be L_i.

In the same way as D_1 we define

$$D_i = \left\{ y : y = \int_0^T A(t)\ u(t)\ dt;\ u(\cdot) \in U_i \right\} . \tag{23}$$

Now we shall show how to find the element y_i satisfying (6). Let us denote

$$l_i^j = \int_{t_i^j}^{t_i^{j+1}} A(t)\ dt, \quad j = 0, 1, \ldots, L_i \tag{24}$$

where $t_i^o = 0$, $t_i^{L_i+1} = T$, and l_i^j are given elements of the space H.

It follows from the definition of U_i and D_i that

$$D_i = \left\{ \hat{y} : \hat{y} = \sum_{j=1}^{L_i+1} \alpha^j l_i^j;\ \left|\alpha^j\right| \leqslant 1 \right\} , \tag{25}$$

where α^j are the values of the control function $u(\cdot)$ in the intervals (t_i^{j-1}, t_i^j).

Thus to find $u_i(\cdot)$ it is enough to determine the coefficients α_i^j

satisfying the condition

$$|\alpha_i^j| \leqslant 1 \tag{26}$$

that minimize the functional

$$\bar{J}_{i-1}(\hat{y}) = J(y_{i-1}) + (J'(y_{i-1}), \sum_{j=1}^{L_i+1} \alpha^j l_i^j) +$$

$$+ \frac{1}{2} N\left(\sum_{j=1}^{L_i+1} \alpha^j l_i^j - y_{i-1}, \sum_{j=1}^{L_i+1} \alpha^j l_i^j - y_{i-1}\right) =$$

$$= c_0 - \left[< c_i, \alpha > + < \alpha, B_i \alpha >\right] \tag{27}$$

Hence we obtain a quadratic form similar to (15) and to find α_i^j any procedure of quadratic programming in a finite dimensional space can be used as it was proposed in Barr's method.

4. APPLICATION TO OPTIMAL CONTROL OF SYSTEMS DESCRIBED BY PARABOLIC EQUATIONS

The method of programming presented above can be used to determine optimal control of systems described by linear parabolic equations.

Let V and H be two Hilbert spaces such that $V \subsetneq H \subsetneq V'$, and let A be a linear, continuous and coercive operator from V to V'. Let U be another Hilbert space and $G \in \mathcal{L}(U; H)$.

For $t \in [0,T]$ we consider the following equation

$$dy(t)/dt + A\, y(t) = G\, u(t) \tag{28}$$

with the initial condition

$$y(0) = y_0 \in H . \tag{28a}$$

For each $u \in L^2(0,T;U)$ equation (28) has [6] a unique solution $y(u) \in C(0,T; H)$.

A function $u \in L^2(0,T; U)$ is called an admissible control if

$$u(t) \in \mathcal{U} \quad \text{for almost all} \quad t \in [0,T], \tag{29}$$

where $\mathcal{U} \subset U$ is a closed, convex and bounded set.

The optimization problem is to find such an admissible control u_0 which minimizes the functional

$$\mathbf{J}(y(T; u)) . \tag{30}$$

Let us denote

$$D = \left\{y(T) \in H : \frac{dy(t)}{dt} + A\, y(t) = G\, u(t)\, ;\ y(0) = y_0 \quad : u(t) \in \mathcal{U}\right\}. \tag{31}$$

It can be shown [6] that D is closed, convex and bounded, therefore to find minimum of $J(y(T;u))$ we can apply the procedure described in Section 2. To this end we shall show how to find $\bar{y}_{i+1}$ satisfying (7). Let us introduce an adjoint equation

$$-(dp(t)/dt) + A^{*} p(t) = 0 , \tag{32}$$

$$p(T) = -J'(y(T;\ u_i)) . \tag{32a}$$

It is easy to show [6] that the condition (7) is equivalent to the following one

$$(G^{*} p(t),\ \bar{u}_{i+1}(t)) \geqslant (G^{*} p(t),\ u) \quad \forall u \in \mathcal{U} \text{ and for almost all } t \in [0,T]. \tag{33}$$

Thus to find $\bar{y}_{i+1} = y(T;\bar{u}_{i+1})$ we have to integrate two equations: first (32) in order to obtain $\bar{u}_{i+1}$ from (33), and then (28).

Having $\bar{u}_{i+1}(t)$ and $\bar{y}_{i+1}$ we can apply the procedure of minimalization described in Section 2.

As a numerical example we consider a system described in an oblong $(0,1) \times (0,T)$ by heat equation

$$\frac{\partial y(x,\ t)}{\partial t} - \frac{\partial^2 y(x,\ t)}{\partial x^2} = 0 \tag{34}$$

with initial condition

$$y(x,\ 0) = 0 \tag{34a}$$

and boundary conditions

$$\frac{\partial y(0,\ t)}{\partial x} = 0; \quad \frac{\partial y(1,\ t)}{\partial x} = \beta \left[u(t) - y(1,\ t)\right]. \tag{34b}$$

The cost functional is a quadratic one

$$J(u) = (z - y(T),\ z - y(T)) = \int_0^1 \left[z(x) - y(x,T)\right]^2 dx \tag{35}$$

where $z \in L^2(0,T)$ is desired final temperature distribution.Two types of constraints imposed on control function are considered

(a) $m \leqslant u(t) \leqslant M$ for almost all $t \in [0,T]$, (36)

(b) functions (u(t) are given [1, 10] as solutions of the eq.

$$\frac{du(t)}{dt} = -\frac{1}{\gamma} u(t) + \frac{1}{\gamma} v(t) , \tag{37}$$

$$u(0) = 0 , \tag{37a}$$

where

$$p \leqslant v(t) \leqslant P. \tag{37b}$$

To perform computation eq. (34) was discretized both in time and space variables. The following numerical data were taken

case (a) $\beta = 1;\ m = -1;\ M = 1;\ T = 1;\ z(x) \equiv 0.45;\ 0.5$
$\Delta t = 2.5\ 10^{-2};\ \Delta x = 10^{-1};$

case (b) $\beta = 10$; $\gamma = 0.04$; $p = 0$; $P = 1$; $T = 0.2$; 0.4;
$z(x) \equiv 0.2$; $\Delta t = 4\times 10^{-3}$; 8×10^{-3}; $\Delta x = 5\times 10^{-2}$.

The computations were performed for all three methods using a computer ODRA 1204 and obtained results are listed in Table 1.

The considered control functions are piece-wise constant and the last column of the Table 1 gives the number of discontinuities obtained in each method. This number characterizes the simplicity of obtained control.

TABLE 1

Type of constraints	Method	No. of iterations	Time sec.	J(y(T; u))	Estimation of error	No. of discont.
(a) $z(x) \equiv 0.45$	1	35	700	4.480×10^{-3}	4.320×10^{-3}	10
	2	5	374	1.576×10^{-3}	1.576×10^{-3}	5
	3	6	3911	6.163×10^{-4}	1.769×10^{-4}	5
(a) $z(x) \equiv 0.5$	1	40	800	1.599×10^{-2}	1.360×10^{-3}	6
	2	4	185	1.530×10^{-2}	8.372×10^{-4}	3
	3	5	615	1.489×10^{-2}	6.659×10^{-10}	2
(b) T = 0.2 sec.	1	14	323	4.548×10^{-2}	9.000×10^{-5}	5
	2	6	576	4.552×10^{-2}	1.296×10^{-4}	4
	3	6	4131	4.546×10^{-2}	3.204×10^{-4}	4
(b) T = 0.4 sec.	1	26	659	3.090×10^{-3}	3.090×10^{-3}	19
	2	7	540	3.414×10^{-4}	3.414×10^{-4}	7
	3	6	3647	4.989×10^{-6}	4.989×10^{-6}	6
1 - Gilbert's method; 2 - Barr's method; 3 - Nahi-Wheeler's method.						

An example of the plots of values of functional vs. the number of iterations for different method is given in Fig. 1 and the values of the functional vs. computation time in Fig. 2.

The forms of optimal control and the final temperature distributions obtained using Nahi-Wheeler method are presented in Fig.3.

It follows from the obtained results that the speed of convergence expressed in terms of the number of iterations is the lowest for Gilbert's method. On the other hand the time of each iteration for this method is comparatively low. Hence the total computation time is of the same order as in Barr's method and several times lower then in Nahi-Wheeler's . As far as the accuracy and simplicity of obtained control is concerned the Nahi-Wheeler's method is the best, Barr's is not very much inferior and Gilbert's is much worse than two others.

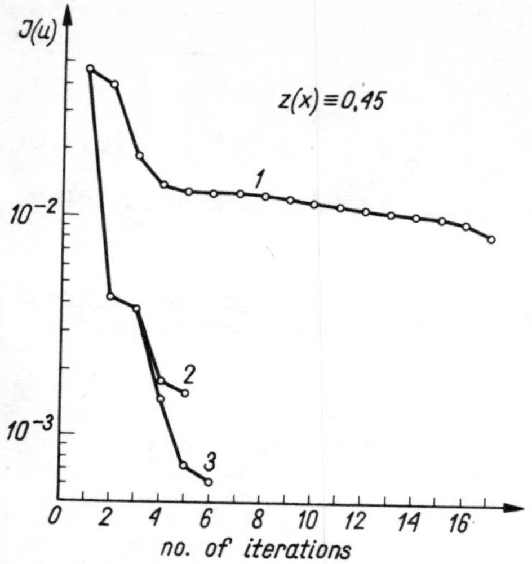

Fig.1. Value of functional J(u) vs. the number of iterations. Case (a) $z(x) \equiv 0.45$

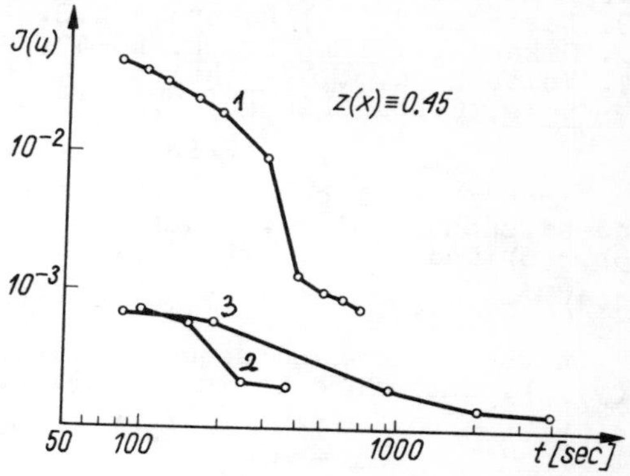

Fig.2. Value of functional J(u) vs. computation time. Case (a) $z(x) \equiv 0.45$

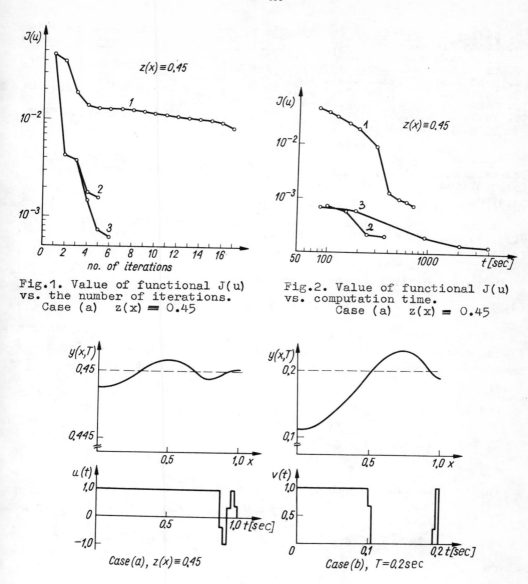

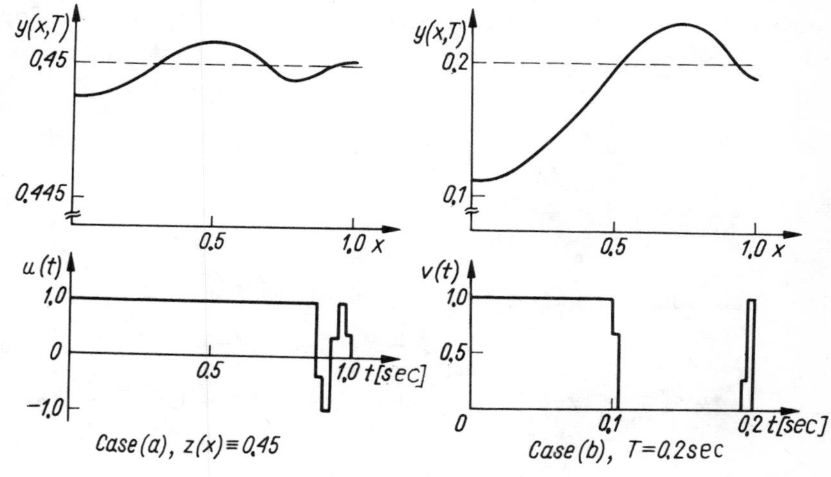

Fig. 3. Form of control and final temperature distribution

Summing up it seems that the best is Barr's method. In the case where the high accuracy is necessary Nahi-Wheeler's can be applied.

REFERENCES

1. Arienti G., Colonelli A., Kenneth P. Cahiers de IRIA 2, 81-106 (1970)
2. Balakrishnan A.V., Introduction to optimization theory in a Hilbert space, Springer Verlag, 1971.
3. Barr R.O. SIAM J. on Control 7, 415-429 (1969)
4. Gilbert E.G. SIAM J. on Control 4, 61-80 (1966)

5. Hadley G., Nonlinear and dynamic programming, Addison-Wesley, 1964
6. Lions J.L., Optimal control of systems governed by partial differential equations, Springer Verlag, 1971
7. Malanowski K., Arch. Autom. i Telemech. 18, 3-18 (1973)
8. Nahi N.E., Wheeler L.A., IEEE Trans. AC-12, 515-521 (1967)
9. Pascavaradi T., Narendra K.S., SIAM J. on Control 8, 396-402 (1970)
10. Sakawa Y., IEEE Trans. AC-11, 420-426 (1966)
11. Veinberg M.M., Variational method and method of monotone operators, Gostiechizdat, 1972 (in Russian)

APPENDIX 1. PROOF OF THEOREM

First it will be shown that the sequence $\{J(y_i)\}$ is non-increasing. Indeed, substituting in (4a) $y = y_{i+1}$, and taking into consideration (5), (6) and (8) we get

$$(1.1) \qquad J(y_{i+1}) \leqslant \bar{J}_i(y_{i+1}) \leqslant \bar{J}_i(y_i) = J(y_i).$$

On the other hand sequence $\{J(y_i)\}$ is bounded from below by $J(y_{opt})$. Therefore it is convergent. We are going to show that

$$(1.2) \qquad \lim_{i \to \infty} J(y_i) \longrightarrow J(y_{opt}).$$

To this end first we shall prove that

$$(1.3) \qquad \lim_{i \to \infty} (-J'(y_i), \bar{y}_{i+1} - y_i) = 0.$$

Let us assume that (1.3) is not satisfied. Then taking into account (7) we conclude that there exists a constant $\varepsilon > 0$ such that for every integer $m > 0$ there exists a subscript $\eta > m$ such that

$$(1.4) \qquad (-J'(y_\eta), \bar{y}_{\eta+1} - y_\eta) \geqslant \varepsilon.$$

Let us denote
$$y^\alpha = y_\eta + \alpha(\bar{y}_{\eta+1} - y_\eta), \alpha \in (0,1).$$

It follows from (6) and (8) that

$$\bar{J}_\eta(y^\alpha) \geqslant \bar{J}_\eta(y_{\eta+1}), \qquad \forall \alpha \in (0,1).$$

Taking into consideration (4) and (1.4) we get

$$(-J'(y_\eta), y_{\eta+1} - y_\eta) \geqslant \frac{1}{2} N\|y_{\eta+1} - y_\eta\|^2 + \alpha(-J'(y_\eta), \bar{y}_{\eta+1} - y_\eta) + \\ + \frac{1}{2} N\alpha^2 \|\bar{y}_{\eta+1} - y_\eta\|^2 \geqslant \frac{1}{2} N\|y_{\eta+1} - y_\eta\|^2 + \alpha\varepsilon - \frac{1}{2} N\alpha^2 \delta,$$

where

$$(1.5) \qquad \delta = \sup_{y,z \in D} \|y - z\|^2 < \infty.$$

Putting $\alpha = \min\{1, \varepsilon/N\delta\}$ we get

$$(1.6) \quad (-J'(y_\eta), y_{\eta+1} - y_\eta) \geqslant \frac{1}{2} N\| y_{\eta+1} - y_\eta\|^2 + \min\left\{\frac{1}{2}N\delta, \frac{1}{2}\frac{\varepsilon^2}{N\delta}\right\} = \\ = \frac{1}{2} N\| y_{\eta+1} - y_\eta\|^2 + \varkappa.$$

where $\varkappa > 0$ does not depend on η.

Taking into consideration (5b) and (1.6) we obtain

$$J(y_{\eta+1}) \leqslant \bar{J}_\eta(y_{\eta+1}) = J(y_\eta) + (J'(y_\eta), y_{\eta+1} - y_\eta) + \\ + \frac{1}{2} N\|y_{\eta+1} - y_\eta\|^2 \leqslant J(y_\eta) - \varkappa$$

or

$$\varkappa \leqslant J(y_{l}) - J(y_{l+1})$$

which contradicts the convergence of the sequence $\{J(y_i)\}$. This contradiction proves (1.3).

Now let us prove (1.2). Denoting

$$y^{\beta} = y_i + \beta(y_{opt} - y_i), \qquad \beta \in [0,1]$$

we get from (4) and (5)

$$(1.7) \quad \bar{\bar{J}}_i(y^{\beta}) = J(y_i) + \beta(J'(y_i), y_{opt} - y_i) + \frac{1}{2} n\beta^2 \| y_{opt} - y_i \|^2 \leqslant \leqslant J(y^{\beta}) \leqslant (1-\beta)J(y_i) + \beta J(y_{opt}).$$

From (7) and (1.7) we have

$$(1.8) \quad (-J'(y_i), \bar{y}_{i+1} - y_i) \geqslant (-J'(y_i) y_{opt} - y_i) \geqslant J(y_i) - J(y_{opt}) \geqslant 0.$$

(1.3) together with (1.8) prove (1.2).

For $J(y)$ is weakly lower semicontinuous, the set D is weakly compact and y_{opt} is the unique element satisfying (2) then (1.2) implies

$$(1.9) \quad y_i \longrightarrow y_{opt}.$$

To prove the strong convergence we note that it follows from (1) that

$$(1.10) \quad n\| y - z \|^2 \leqslant (J'(y), y - z) - (J'(z), y - z), \quad \forall y,z \in D.$$

Hence from (7) and (1.10) we get

$$(1.11) \quad n\| y_{opt} - y_i \|^2 \leqslant (J'(y_{opt}), y_{opt} - y_i) + (-J'(y_i), y_{opt} - y_i) \leqslant \leqslant (J'(y_{opt}), y_{opt} - y_i) + (-J'(y_i), \bar{y}_{i+1} - y_i).$$

For $i \longrightarrow \infty$ the first term on the right hand side of (1.10) tends to zero by (1.9) and the second by (1.3). Hence

$$(1.12) \quad \lim_{i \to \infty} \| y_{opt} - y_i \| = 0 \qquad \text{q.e.d.}$$

APPENDIX 2. ESTIMATION OF THE ERROR

It is obvious that the following estimation takes place

$$(2.1) \quad J(y_{opt}) \leqslant J(y_i).$$

To estimate $J(y_{opt})$ from below we take advantage of (5b) and find the estimation of $\bar{\bar{J}}_i(y)$ on D.

Consider the situation where the hyperplane M_i separates the set D and the point at which $\bar{\bar{J}}_i(y)$ assumes its global minimum. It takes place iff

$$\left(\bar{\bar{J}}'_i(y_i), \bar{\bar{J}}'_i(y_{i+1})\right) \geqslant 0.$$

Using (4) we rewrite this condition in the form

$$(2.2) \quad (J'(y_i), J'(y_i) + n(\bar{y}_{i+1} - y_i)) \geqslant 0.$$

In this case the minimal value of $\bar{\bar{J}}_i(y)$ on D is not greater then the minimal value of this functional on M_i.

This minimal value is assumed at the point $\widetilde{y}$, where

$$\bar{\bar{J}}_i(\widetilde{y}) = \gamma J'(y_i).$$

Taking into consideration that $\bar{y}_{i+1} \in M_i$ we get from (4b)

$$(2.3) \qquad \bar{\bar{J}}_i(\widetilde{y}) = J(y_i) + \left(\frac{n}{2} \frac{(J'(y_i), \bar{y}_{i+1} - y_i)}{(J\ (y_i), J\ (y_i))} + \right.$$

$$\left. + 1 \right) (J'(y_i), \bar{y}_{i+1} - y_i).$$

From (5b), (2.1), (2.2) and (2.3) we obtain the estimation (9) of the error of i-th iteration.

ABOUT SOME FREE BOUNDARY PROBLEMS CONNECTED WITH HYDRAULICS

CLAUDIO BAIOCCHI, Laboratorio di Analisi Numerica del C. N. R.
PAVIA, ITALY

N.1. DESCRIPTION OF THE PROBLEM

Two water reservoirs, of different levels, are separated by an earth dam, which is assumed homogeneous, isotropic and with impervious basis. We ask for the steady flow of the water between the reservoirs, and in particular for the flow region Ω (or, in other words, for the unknown part $y=\varphi(x)$ of its boundary).

Denoting by D the dam, and by y_1 and y_2 the water levels (with y_1 greater than y_2; see picture 1) the mathematical problem can be stated as follows (see for instance [8] for the general treatment of this type of problems):

Find a subset Ω of D, bounded from above by a continuous decreasing function $y=\varphi(x)$, such that there exists in Ω an harmonic function $u=u(x,y)$ which satisfies the boundary conditions: $u=y_1$ on AF; $u=y_2$ on BC; $u=y$ on CC_φ and FC_φ; $\frac{\partial u}{\partial n}=0$ on AB and FC_φ.

($\frac{\partial}{\partial n}$ denoting the normal derivative; we point out that on the "free boundary" we have both the Dirichlet and the Neumann condition).

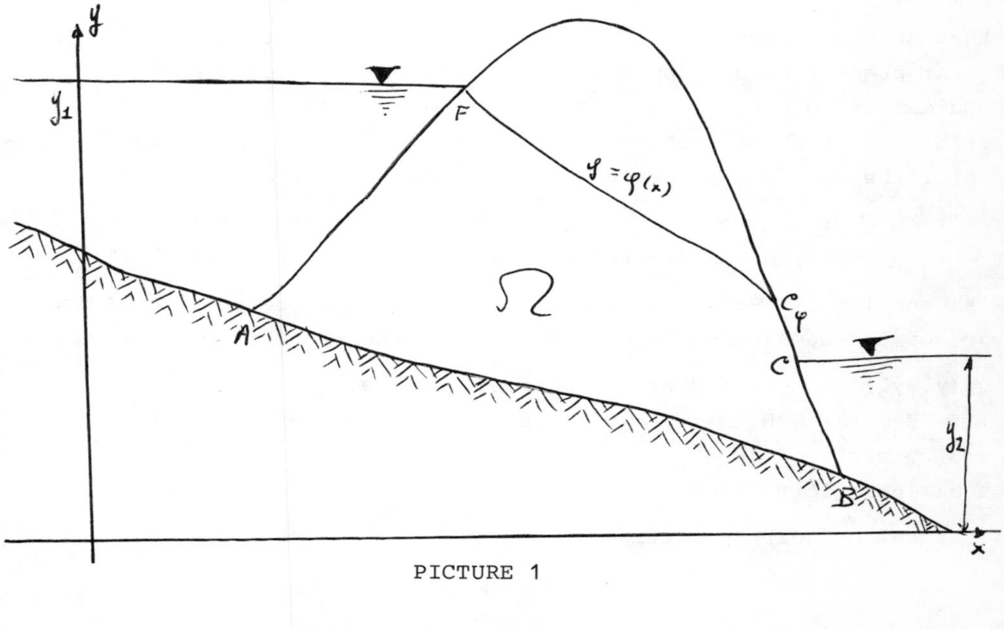

PICTURE 1

N.2. THEORETICAL RESULTS.

In a paper of 1971 (see [1],[2]) I proved that, if D is a rectangle, $D =]0,a[\times]0,y_1[$ (in picture 1 AF and BE are vertical, AB is horizontal; the origin is in A), setting:

$$w(x,y)=\int_y^{\varphi(x)}[u(x,t)-t]dt \quad \text{for } (x,y)\in\overline{\Omega}; \quad w(x,y)=0 \text{ for } (x,y)\in\overline{D}-\overline{\Omega}$$

the new unknown function w satisfies the relations:

(1) $w\geq 0;\ \Delta w\leq 1;\ w(1-\Delta w)=0$ on D; $w=g$ on ∂D

where g(x,y) is defined on ∂D by the formula:

(2) $g=\frac{(y-y_1)^2}{2}$ on AF; $g=\frac{(y-y_2)^2}{2}$ on BC; g linear on AB; g=0 elsewhere.

The system (1),(2) can be studied as a varionational inequality; precisely, denoting by $H^1(D)$ the space of square summable functions on D, whose first derivatives are also square summable, and setting:

$$K=\{v\in H^1(D);\ V\geq 0 \text{ on } D;\ v=g \text{ on } \partial D\}$$

the function w satisfies:

(3) $w\in K;\ \forall v\in K: \int_D [\text{grad}w\cdot\text{grad}(v-w)]dxdy \geq \int_D (w-v)dxdy.$

Vice-versa it is well known (see [9]) that (3) has a unique solution; and we can proof that w solution of (3) is such that, setting:

(4) $\Omega=\{(x,y)\in D;\ w(x,y)>0\}$; $u=y-D_y w$ on Ω

(and obviously graph of $\varphi=\partial\Omega-\partial D$) we get a solution of the free boundary problem. In this way we obtain an existence-uniqueness theorem, via the varionational inequalities theory; and we will see later that this approach allows us also a new numerical treatment.

At present we can apply similar methods to a wide class of free boundary problems (see [3],[4],[5]); here I will limit myself to describe one of these problems, corresponding to the case where the shape of the dam is a trapezium (see picture 2). System (1) becomes now:

(5) $\begin{cases} w\geq 0;\ \Delta w\leq 1;\ w(1-\Delta w)=0;\ w=g \text{ on FEB; } w \text{ linear on AB;} \\ D_y w=y-y_1 \text{ on AF (we reobtain } w=g \text{ on } \partial D \text{ if AF is vertical).} \end{cases}$

Denoting by $-q$ the value of $D_x w$ on AB, q represents the "discharge" of the dam (which is now unknown; in the previous case it was $q=(y_1^2-y_2^2)(2a)^{-1}$). Now we can define, for real q:

$$K_q=\{v\in H^1(D);\ v\geq 0 \text{ on } D;\ v=g \text{ on FEB };\ v_x=-q \text{ on AB}\}$$

and we must modify (3) into the formula (6) following (where a(u,v) is a bilinear form associed with $-\Delta$ (like $\int_D \text{grad}u\cdot\text{grad}v\ dxdy$) whose "natural boundary condition" on AF is $\frac{\partial}{\partial y}$):

(6) $w_q \in K_q;\ \forall v \in K_q:\quad a(w_q, v-w_q) \geq \int_D (w_q - v)\,dxdy + 2\int_{AF} (y_1 - y)(v-w_q)\,dy$

Actually we can proof (see [4] and [7]) that there exists a unique value q^* of q, such that (4) with $w = w_{q^*}$ gives the solution of the free boundary problem; and this unique value of q can be carachterized as the unique value of q such that the corresponding w_q solution of (6) is a "more regular" function, for instance $w_q \in C^1(\bar{D})$.

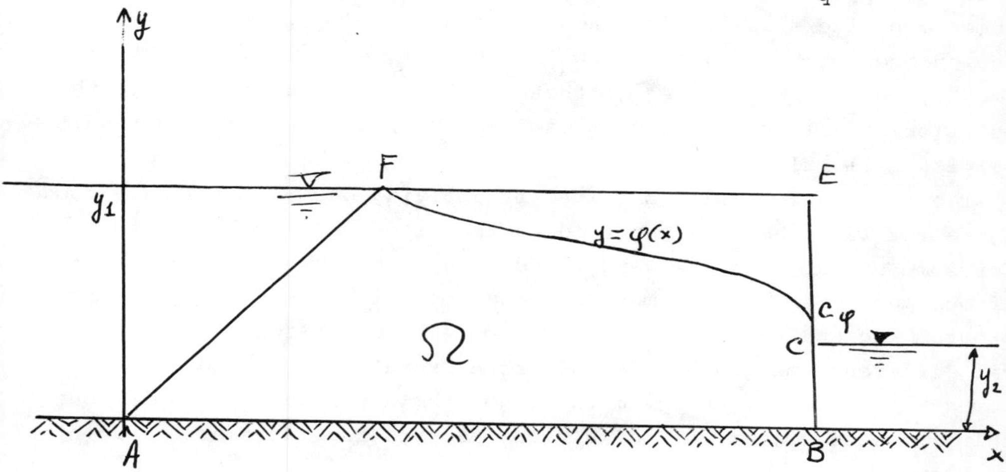

PICTURE 2

N.3. NUMERICAL APPROACH

In the cases where the shape of the dam is simple (like a rectangle, or a trapezium) we discretized the problem ((3) or (6) respectively) by a finite difference scheme. In the first case we must solve the set of inequalities:

$$(7)\quad \begin{cases} w_h \geq 0;\ \Delta_h w_h \leq 1;\ \ w_h(1-\Delta_h w_h) = 0 & \text{in the interior gridpoints;} \\ w_h = g & \text{in the gridpoints of the boundary} \end{cases}$$

where h is the mesh amplitude and Δ_h is the usual 5-points discretisation of the Laplace operator; system (7) has a unique solution and can be solved by means of an iterative scheme, with S.O.R. and projections (for the details see [3],[6]) ; Δ_h will be defined as the union of the mesh with center in the gridpoints where $w_h > 0$; setting also $u_h = Y - \delta_2 w_h$ (δ_2 discretizing D_y) we have the following convergence result (see [5]):

(8) u_h converges to u in H^1 discrete; Ω = interior of $\lim_{h\to 0} \Omega h$.

In the second case we must modify (7) into:

$$(9)\begin{cases} w_h \geq 0;\ \Delta_h w_h \leq 1;\ w_h(1-\Delta_h w_h) = 0 & \text{in the interior gridpoints} \\ w_h = g \text{ on FEB};\ \delta_1 w_h = -q \text{ on AB};\ \delta_2 w_h = y-y_1 & \text{on AF} \end{cases}$$

and we must add to the solution of (9) an algorithm for the choice of an approximation q(h) of q^*. In order to do it we look for the rooth of the equation $f_h(q) = 0$ where $f_h(q) = \delta_2 w_h(F)$; it can be proved that $q \mapsto f_h(q)$ is a convex, strictly decreasing function, which assumes opposite signe in two known points q_0 and q_1; setting q(h) as the unique rooth of $f_h(q)$ and w_h as the unique solution of (9) with q=q(h), we can proof (see [7]) that q(h) converges to q^* and (with notations similars to the ones used in (8)) the convergence result (8) is still valid.

The numerical results obteined by applying our method show a good agreement with the ones reported in the literature; we refer to [3] for some comparisons. We want to point out that the main advantages of our method from the numerical point of view are the semplicity of programming and the speed of execution (in comparison with a classical difference method we see that both execution time and Fortran statements necessary to program the algorithm are reduced to about one third); moreover our method is rigorous from the mathematical point of view.

REFERENCES

1 C. BAIOCCHI. Sur un problème à frontière libre traduisant le filtrage des liquides à travers des milieux poreux. C. R. Acad. Sc. Paris 273 (1971) 1215-1217.

2 C. BAIOCCHI. Su un problema di frontiera libera connesso a questioni di idraulica. Ann. di Mat. 97 (1972) 107-127

3 C. BAIOCCHI, V. COMINCIOLI, L. GUERRI, G. VOLPI. Free boundary problems in the theory of fluid flow through porous media: numerical approach. To appear in Calcolo

4 C. BAIOCCHI, V. COMINCIOLI, E. MAGENES, G. POZZI. Free boundary problems in the theory of fluid flow through porous media: existence and uniqueness theorems. To appear in Ann. di Mat.

5 C. BAIOCCHI, E. MAGENES. Proceedings of the Symposium "Metodi valutativi nella Fisica-Matematica", Acc. Naz. Lincei, Rome dec 1972.

6 V. COMINCIOLI, L. GUERRI, G. VOLPI. Analisi numerica di un problema di frontiera libera connesso col moto di un fluido attraverso un mezzo poroso. Publ. N. 17 of Lab. Anal. Num. Pavia (1971)

7 V. COMINCIOLI. Paper in preparation.

8 M. E. HARR. Groundwater and seepage. Mc Graw Hill, N.Y., 1962

9 G. STAMPACCHIA. Formes bilinéaires coercives sur les ensembles convexes. C. R. Acad. Sc. Paris, 258 (1964) 4413-4416.

METHODE DE DECOMPOSITION APPLIQUEE AU CONTROLE OPTIMAL DE SYSTEMES DISTRIBUES

A. BENSOUSSAN, R. GLOWINSKI, J.L. LIONS

(IRIA 78 - ROCQUENCOURT - FRANCE)

INTRODUCTION

Le but de ce travail est d'étudier l'application des techniques de décomposition-coordination à la résolution numérique des problèmes de contrôle optimal gouvernés par des équations aux dérivées partielles. L'approche retenue est celle procédant par décomposition du domaine où on ramène, par un procédé itératif, la résolution du problème de contrôle optimal initial, à celle d'une suite de sous-problèmes de même nature relatifs à des sous-domaines réalisant une partition du domaine initial. On dégagera les principes de la méthode sur un problème de contrôle où l'équation d'état est linéaire elliptique et le domaine rectangulaire, mais les techniques décrites ci-dessous s'appliquent à des géométries plus compliquées et à des équations d'état non linéaires,stationnaires ou d'évolution.

I - ETUDE DU PROBLEME GLOBAL

1.1 - NOTATIONS - HYPOTHESES.

On considère un domaine Ω de la forme

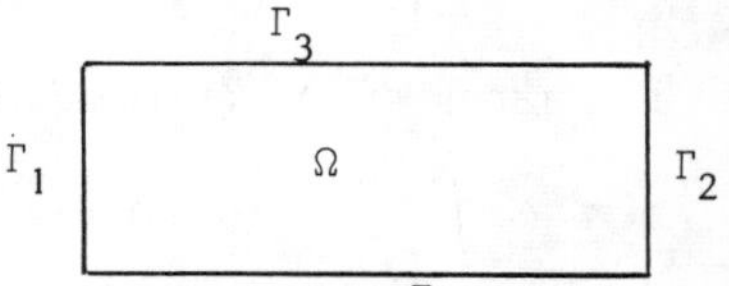

Figure 1

et on note $\partial\Omega = \Gamma_1 \cup \Gamma_2 \cup \Gamma_3 \cup \Gamma_4$ la frontière de Ω. L'état du système est donné par

$$(1.1) \qquad \begin{cases} \Delta z = 0 \\ z|_{\Gamma_3 \cup \Gamma_4} = 0 \\ \dfrac{\partial z}{\partial n}\Big|_{\Gamma_1 \cup \Gamma_2} = v \end{cases}$$

où $v = (v_1, v_2)$ est le contrôle ; $v_i \in L^2(\Gamma_i)$, $i=1,2$. Dans (1.1), $\frac{\partial z}{\partial n}$ représente la dérivée normale orientée vers l'extérieur de Ω.

Le critère est défini par (en remarquant que $z = z(v) = z(x;v)$)

$$(1.2) \qquad J(v) = \int_\Omega |z(x) - y_d(x)|^2 dx + \nu \int_{\Gamma_1 \cup \Gamma_2} v^2(x) d\Gamma \quad ; \quad \nu > 0 .$$

On considère l'ensemble (représentant les contraintes)

(1.3) $$\mathcal{U}_{ad} = \mathcal{U}^1_{ad} \times \mathcal{U}^2_{ad}\ ,\quad \mathcal{U}^i_{ad} \text{ convexe fermé de } L^2(\Gamma_i).$$

Le problème global est défini par

(1.4) $$\text{Minimiser } J(v),\ v \in \mathcal{U}_{ad}\ .$$

1.2 - CONDITIONS NECESSAIRES ET SUFFISANTES D'OPTIMALITE.

Il est standard (cf. J.L. LIONS [1]) que (1.4) possède une solution unique u, caractérisée par la condition d'Euler suivante

(1.5) $$\int_\Omega (y-y_d)(z-y)dx + \nu \int_{\Gamma_1 \cup \Gamma_2} u(v-u)d\Gamma \geqslant 0$$

$$\forall\ v \in \mathcal{U}_{ad}\ ,\ z=z(v),\ y=z(u),\ u \in \mathcal{U}_{ad}\ .$$

On introduit l'état adjoint défini par

(1.6) $$\begin{cases} -\Delta p = y-y_d & \text{dans } \Omega \\ p|_{\Gamma_3 \cup \Gamma_4} = 0 \\ \frac{\partial p}{\partial n}\Big|_{\Gamma_1 \cup \Gamma_2} = 0\ ; \end{cases}$$

alors (cf. J.L. LIONS [1]), (1.5) équivaut à

(1.7) $$\int_{\Gamma_1 \cup \Gamma_2} (p+\nu u)(v-u)d\Gamma \geqslant 0, \quad \forall\ v \in \mathcal{U}_{ad}\ ,\ u \in \mathcal{U}_{ad}\ .$$

Il sera commode de réécrire les conditions d'optimalité sous forme d'une inéquation variationnelle. On introduit l'espace de Hilbert

(1.8) $$\mathcal{V} = \{ z |\ z \in H^1(\Omega),\ z|_{\Gamma_3 \cup \Gamma_4} = 0 \}$$

où $H^1(\Omega)$ est l'espace de Sobolev d'ordre 1.
On notera que $(y,p) \in \mathcal{V} \times \mathcal{V}$. On a alors

Théorème 1.1

Le triplet (y,p,u) (où u est le contrôle optimal et y l'état optimal) est solution unique de l'inéquation variationnelle

(1.9) $$\int_\Omega \text{grad } y \text{ grad } z\ dx - \int_{\Gamma_1 \cup \Gamma_2} uz\ d\Gamma + a\left(\int_\Omega \text{grad } p \text{ grad } q\ dx - \int_\Omega (y-y_d)q\ dx\right) +$$

$$+\ b \int_{\Gamma_1 \cup \Gamma_2} (\nu u+p)(v-u)d\Gamma \geqslant 0 \quad \forall\ z,q \in \mathcal{V}\ ,v \in \mathcal{U}_{ad}\ ,\ y,p \in \mathcal{V}\ ,\ u \in \mathcal{U}_{ad}\ ,$$

où a et b sont deux constantes positives arbitraires [1]. ■

---

(1) Le rôle de a et b apparaîtra au n° 3.2

II - DECOMPOSITION

2.1 - HYPOTHESES - NOTATIONS.

Pour économiser les notations, on considère une décomposition de Ω en deux sous-domaines Ω_1, Ω_2 (la généralisation à n sous domaines s'en déduit aussitôt)

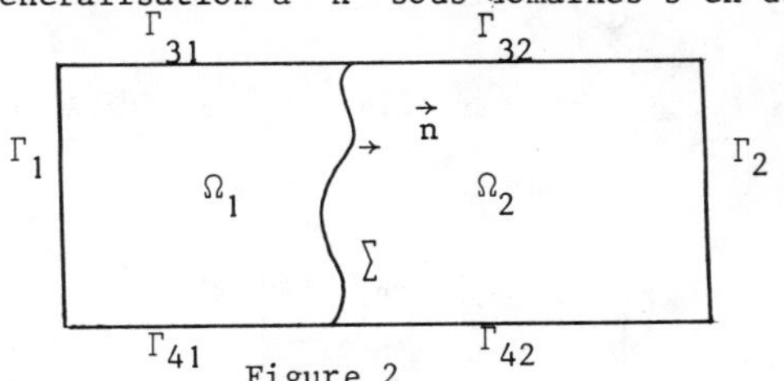

Figure 2

où, avec des notations évidentes

$$\Gamma_3 = \Gamma_{31} \cup \Gamma_{32}$$

$$\Gamma_4 = \Gamma_{41} \cup \Gamma_{42}$$

$$\Sigma = \partial\Omega_1 \cap \partial\Omega_2 \text{ , supposée régulière.}$$

Pour i=1,2, on introduit les espaces de Hilbert

$$(2.1) \qquad \mathcal{V}_i = \{ z \in H^1(\Omega_i) \mid z|_{\Gamma_{3i} \cup \Gamma_{4i}} = 0\} .$$

On pose $y_{id} = y_d|_{\Omega_i}$.

On introduit le problème suivant : trouver (y_i,p_i,u_i) dans $\mathcal{V}_i \times \mathcal{V}_i \times \mathcal{U}_{ad}^i$ vérifiant

$$(2.2) \qquad \begin{cases} y_1 = y_2 & \text{sur } \Sigma \\ p_1 = p_2 & \text{sur } \Sigma \end{cases}$$

et

$$(2.3) \begin{cases} \sum_{i=1,2} \int_{\Omega_i} \text{grad } y_i \text{ grad } z_i dx - \int_{\Gamma_i} u_i z_i \, d\Gamma_i + \\ + a\left(\sum_{i=1,2} \int_{\Omega_i} \text{grad } p_i \text{ grad } q_i \, dx - \int_{\Omega_i} (y_i - y_{id}) q_i dx\right) + \\ + b\left(\sum_{i=1,2} \int_{\Gamma_i} (\nu u_i + p_i)(v_i - u_i) d\Gamma_i\right) \geqslant 0 \quad \forall\ z_i, q_i, v_i \in \mathcal{V}_i \times \mathcal{V}_i \times \mathcal{U}_{ad}^i \end{cases}$$

et

$$(2.4) \qquad z_1 = z_2 |_\Sigma \quad , \; q_1 = q_2 |_\Sigma \; .$$

On a alors

Théorème 2.1

Il existe une solution unique de (2.2),(2.3) donnée par :

$$(2.5)\quad \begin{cases} y_i = \text{restriction de } y \text{ à } \Omega_i \\ p_i = \text{restriction de } p \text{ à } \Omega_i \\ u_i = \text{restriction de } u \text{ à } \Gamma_i \end{cases}$$

Démonstration

Définissons (y_i, p_i, u_i) par (2.5). Alors d'après (1.1) et (1.6), on a

$$(2.6)\quad \begin{cases} \Delta y_i = 0 \quad \text{dans} \quad \Omega_i \\ y_i|_{\Gamma_{3i} \cup \Gamma_{4i}} = 0 \\ \dfrac{\partial y_i}{\partial n}\Big|_{\Gamma_i} = u_i \end{cases}$$

$$(2.7)\quad \begin{cases} -\Delta p_i = y_i - y_{id} \quad \text{dans} \quad \Omega_i \\ p_i\Big|_{\Gamma_{3i} \cup \Gamma_{4i}} = 0 \\ \dfrac{\partial p_i}{\partial n}\Big|_{\Gamma_i} = 0 \end{cases}$$

$$(2.8)\quad \int_{\Gamma_i} (p_i + \nu u_i)(v_i - u_i)\, d\Gamma_i \geqslant 0.$$

Les relations (2.2) sont vérifiées (propriétés des traces dans $H^1(\Omega)$). Par ailleurs, d'après LIONS-MAGENES [1], on peut définir

$$(2.9)\quad \frac{\partial y_i}{\partial n}\Big|_{\Sigma} \in H^{-\frac{1}{2}}(\Sigma), \quad \frac{\partial p_i}{\partial n}\Big|_{\Sigma} \in H^{-\frac{1}{2}}(\Sigma)$$

et de plus, on a

$$(2.10)\quad \begin{aligned} \frac{\partial y_1}{\partial n}\Big|_{\Sigma} &= -\frac{\partial y_2}{\partial n}\Big|_{\Sigma} = \frac{\partial y}{\partial n}\Big|_{\Sigma}. \\ \frac{\partial p_1}{\partial n}\Big|_{\Sigma} &= -\frac{\partial p_2}{\partial n}\Big|_{\Sigma} = \frac{\partial p}{\partial n}\Big|_{\Sigma}. \end{aligned}$$

Soient alors z_i, $q_i \in \mathcal{V}_i$ vérifiant (2.4), $v_i \in \mathcal{U}_{ad}$. On a, d'après (2.6), (2.7), (2.8)

$$\sum_i \Big(-\int_{\Omega_i} \Delta y_i\, z_i\, dx + a \int_{\Omega_i} (-\Delta p_i - (y_i y_{id}))q_i\, dx + b \int_{\Gamma_i} (\nu u_i + p_i)(v_i - u_i)\, d\Gamma_i \Big) \geqslant 0.$$

D'après la formule de Green, il en résulte

$$\sum_i \Big(\int_{\Omega_i} \text{grad } y_i \text{ grad } z_i\, dx - \int_{\partial\Omega_i} \frac{\partial y_i}{\partial n} z_i\, d(\partial\Omega_i)\Big) + a \sum_i \Big(\int_{\Omega_i} \text{grad } p_i \text{ grad } q_i\, dx -$$

$$- \int_{\Omega_i} (y_i - y_{id}) q_i\, dx - \int_{\partial\Omega_i} \frac{\partial p_i}{\partial n} q_i\, d(\partial\Omega_i)\Big) + b \sum_i \int_{\Gamma_i} (p_i + \nu u_i)(v_i - u_i)\, d\Gamma_i \geqslant 0$$

et on vérifie aisément, grâce à (2.6), (2.7), (2.4), (2.11) que (2.11) entraîne (2.3).

Soit réciproquement une solution de (2.3). On pose

(2.12)
$$y(x) = \sum_{i=1,2} y_i(x)\ \chi_i(x)$$
$$p(x) = \sum_{i=1,2} p_i(x)\ \chi_i(x)$$

où $\chi_i(x)$ est la fonction caractéristique de Ω_i. Il est classique que (2.2) entraîne $y,p \in H^1(\Omega)$. Comme

$$y|_{\Gamma_{3i}} = y_i|_{\Gamma_{3i}} , \qquad p|_{\Gamma_{3i}} = p_i|_{\Gamma_{3i}}$$

$$y|_{\Gamma_{4i}} = y_i|_{\Gamma_{4i}} , \qquad p|_{\Gamma_{4i}} = p_i|_{\Gamma_{4i}}$$

on a bien $y,p \in \mathcal{V}$. On vérifie alors aisément que (2.3) implique (1.9). ∎

2.2 - LAGRANGIEN GENERALISE.

Bien que (2.3) soit une inéquation variationnelle non symmétrique (et donc ne provient pas d'un problème d'optimisation), on peut lui associer un lagrangien défini par

(2.13)
$$\begin{aligned}\mathcal{L} = &\sum_{i=1,2} \int_{\Omega_i} \text{grad } y_i \text{ grad } z_i dx - \int_{\Gamma_i} u_i\ z_i\ d\Gamma_i\ + \\ &+ a\ \Big(\sum_{i=1,2} \int_{\Omega_i} \text{grad } p_i \text{ grad } q_i dx - \int_{\Omega_i} (y_i-y_{id})q_i dx\Big) + \\ &+ b\ \Big(\sum_{i=1,2} \int_{\Gamma_i} (\nu\ u_i + p_i)(v_i-u_i)d\Gamma_i\Big) + \\ &+ \int_{\Sigma} \lambda(z_1-z_2)d\Sigma\ + a \int_{\Sigma} \mu(q_1-q_2)d\Sigma \qquad \text{où } \lambda\ ,\ \mu \in H^{-\frac{1}{2}}\ (\Sigma)\ .\end{aligned}$$

<u>Théorème</u> 2.2

<u>Il existe</u> $(y_i,p_i,u_i) \in \mathcal{V}_i \times \mathcal{V}_i \times \mathcal{U}_{ad}^i$ <u>et</u> $\lambda,\ \mu \in H^{-\frac{1}{2}}(\Sigma)$, <u>vérifiant les contraintes</u> (2.2) <u>et</u> $\mathcal{L} \geqslant 0$, <u>pour tout</u> $z_i,q_i \in \mathcal{V}_i,\ v_i \in \mathcal{U}_{ad}^i$. <u>Une telle solution est de plus unique</u>.

<u>Démonstration</u>

Comme (y_i,p_i,u_i) est solution de (2.2), (2.3), elle est définie de manière unique par (2.5). Soient alors

$$(y_i,p_i,u_i,\lambda,\mu) \quad \text{et} \quad (y_i,p_i,u_i,\lambda',\mu')$$

deux solutions éventuelles. On a

(2.14)
$$\int_{\Sigma} (\lambda-\lambda')(z_1-z_2)d\Sigma + \int_{\Sigma} (\mu-\mu')(q_1-q_2)d\Sigma = 0 \qquad \forall\ z_i,q_i \in \mathcal{V}_i\ .$$

Mais l'application

$$z_i \ \to\ z_i|_{\Sigma}$$

de $\mathcal{V}_i \rightarrow H^{\frac{1}{2}}(\Sigma)$ est à image dense et (2.14) implique alors

$$\lambda = \lambda' \quad , \quad \mu = \mu'$$

d'où l'unicité. Par ailleurs, on vérifie que

$$\lambda = -\frac{\partial y_1}{\partial n}\Big|_\Sigma = +\frac{\partial y_2}{\partial n}\Big|_\Sigma = -\frac{\partial y}{\partial n}\Big|_\Sigma$$

$$\mu = -\frac{\partial p_1}{\partial n}\Big|_\Sigma = \frac{\partial p_2}{\partial n}\Big|_\Sigma = -\frac{\partial p}{\partial n}\Big|_\Sigma \quad . \blacksquare$$

III - METHODE DE COORDINATION

3.1 - PRINCIPE DE LA METHODE.

La méthode proposée est itérative. Supposons que nous soyons à l'étape k, avec $\lambda = \lambda^k$, $\mu = \mu^k$. On cherche alors y_i^k, p_i^k, u_i^k réalisant $\mathcal{L} \geqslant 0$, $\forall\ z_i, q_i, v_i$ comme dans le Théorème 2.2 et pour les valeurs précédentes de λ et μ. Le problème se décompose aussitôt en

(3.1)
$$\begin{cases} \Delta y_i^k = 0 & \text{dans } \Omega_i \\ y_i^k\Big|_{\Gamma_{3i}\cup\Gamma_{4i}} = 0 \\ \dfrac{\partial y_i^k}{\partial n}\Big|_{\Gamma_i} = u_i^k \\ \dfrac{\partial y_i^k}{\partial n}\Big|_\Sigma = (-1)^i \lambda^k \end{cases}$$

(3.2)
$$\begin{cases} \Delta p_i^k = y_i^k - y_{id} & \text{dans } \Omega_i \\ p_i^k\Big|_{\Gamma_{3i}\cup\Gamma_{4i}} = 0 \\ \dfrac{\partial p_i^k}{\partial n}\Big|_{\Gamma_i} = 0 \\ \dfrac{\partial p_i^k}{\partial n}\Big|_\Sigma = (-1)^i \mu^k \end{cases}$$

(3.3)
$$\int_{\Gamma_i} (\nu\, u_i^k + p_i^k)(v_i - u_i^k)\, d\Gamma_i \geqslant 0, \qquad \forall v_i \in \mathcal{U}_{ad}^i \ .$$

Les relations (3.1), (3.2), (3.3) s'interprètent comme les C.N.S. d'optimalité du problème suivant : l'état du système est donné par (3.1) et le critère est donné par

(3.4)
$$j_i^k(u_i^k) = \int_{\Omega_i} (y_i^k - y_{id})^2 dx + \nu \int_{\Gamma_i} (u_i^k)^2 d\Gamma_i + 2(-1)^i \int_\Sigma \mu^k y_i^k \, d\Sigma \ .$$

On calcule ensuite λ^{k+1}, μ^{k+1} par les formules suivantes (où S est l'application de dualité de $H^{\frac{1}{2}}(\Sigma) \to H^{\frac{1}{2}}(\Sigma))$

$$(3.5)\quad \begin{cases} \lambda^{k+1} = \lambda^k + \rho^k\, S(y_1^k - y_2^k) \\ \mu^{k+1} = \mu^k + \rho^k\, S(p_1^k - p_2^k) \end{cases}$$

où ρ^k est une suite de nombres réels, bornée.

3.2 - LE PROBLEME DE LA CONVERGENCE.

On vérifie tout d'abord la relation suivante

$$(3.6)\quad \begin{cases} -\sum_{i=1,2} \int_{\Omega_i} |\mathrm{grad}(y_i-y_i^k)|^2 dx - \sum_{i=1,2} \int_{\Gamma_i} (u_i-u_i^k)(y_i^k-y_i)d\Gamma_i - \\ - a \sum_{i=1,2} \int_{\Omega_i} |\mathrm{grad}(p_i-p_i^k)|^2 dx - a \sum_{i=1,2} \int_{\Omega_i} (y_i-y_i^k)(p_i^k-p_i)dx - \\ - b\nu \sum_{i=1,2} \int_{\Gamma_i} (u_i-u_i^k)^2 d\Gamma_i + b \sum_{i=1,2} \int_{\Gamma_i} (p_i-p_i^k)(u_i^k-u_i)d\Gamma_i + \\ + \int_\Sigma (\lambda-\lambda^k)(y_1^k - y_2^k - y_1 + y_2)d\Sigma + a\int_\Sigma (\mu-\mu^k)(p_1^k - p_2^k - p_1 + p_2)d\Sigma \geq 0, \end{cases}$$

d'où encore (avec des notations évidentes)

$$(3.7)\quad \begin{cases} - \sum_i \left(\|y_i-y_i^k\|^2 + a\|p_i-p_i^k\|^2 + b\nu\, |u_i-u_i^k|^2\right) + \sum_i \frac{\gamma}{2}\, |u_i-u_i^k|^2 + \frac{1}{2\gamma} \int_{\Gamma_i} (y_i^k-y_i)^2 d\Gamma_i + \\ + \sum_i b\, \frac{\delta}{2}\, |u_i-u_i^k|^2 + \frac{b}{2\delta} \int_{\Gamma_i} |p_i^k - p_i|^2 d\Gamma_i + \sum_i a\, \frac{\varepsilon}{2} \int_{\Omega_i} (y_i-y_i^k)^2 dx + \\ + \frac{a}{2\varepsilon} \int_{\Omega_i} (p_i-p_i^k)^2 dx + \int_\Sigma + a\int_\Sigma \geq 0 \qquad (\text{où } \gamma, \delta, \varepsilon > 0) . \end{cases}$$

Mais d'après l'inégalité de Poincaré, et la continuité de l'application trace, il existe une constante C telle que

$$(3.8)\quad \int_{\Omega_i} |\mathrm{grad}\, z|^2 dx \geq C \left(\int_{\Omega_i} z^2 dx + \int_{\Gamma_i} z^2 d\Gamma_i \right)$$

de sorte que (3.7) et (3.8) impliquent

$$(3.9)\quad \begin{cases} - \sum_i \|y_i-y_i^k\|^2 \left(1 - \frac{1}{2C} \max\left(\frac{1}{\gamma}, a\varepsilon\right)\right) - \sum_i \|p_i-p_i^k\|^2 \left(a - \frac{1}{2C} \max\left(\frac{a}{\varepsilon},\frac{b}{\delta}\right)\right) - \\ - \sum_i |u_i-u_i^k|^2 \left(b\nu - \frac{\gamma+b\delta}{2}\right) + \int_\Sigma + a\int_\Sigma \geq 0 . \end{cases}$$

Supposons donc que

$$(3.10) \qquad 4\gamma\, C^2 > 1 .$$

On choisit

$$\frac{1}{\gamma} < 2C, \quad \frac{1}{\varepsilon} < 2C, \quad a < 2\frac{C}{\varepsilon}$$

$$\frac{b}{\delta} = \eta < a\, 2C$$

et enfin b de façon que

$$2\nu > \frac{\gamma}{b} + \frac{b}{\eta}$$

ce qui est possible si $\nu > \sqrt{\frac{\gamma}{\eta}}$, ce qui est vérifié, grâce à (3.10). Dans ces conditions, il existe une constante $C > 0$ telle que

$$(3.11) \qquad - C\Big(\sum_i \|y_i-y_i^k\|^2 + \|p_i-p_i^k\|^2 + |u_i-u_i^k|^2\Big) + \int_\Sigma + a\int_\Sigma \geq 0 .$$

Posons

$$\ell^k = \lambda^k - \lambda$$

$$m^k = \mu^k - \mu .$$

On a

$$\ell^{k+1} - \ell^k = \rho^k\, S(y_1^k-y_1-y_2^k+y_2)$$

$$m^{k+1} - m^k = \rho^k\, S(p_1^k-p_1-p_2^k+p_2)$$

d'où

$$(3.12)\begin{cases} \|\ell^{k+1}\|^2_{H^{-\frac12}(\Sigma)} = \|\ell^k\|^2_{H^{-\frac12}(\Sigma)} + (\rho^k)^2\, \|y_1^k - y_1 - y_2^k + y_2\|^2_{H^{\frac12}(\Sigma)} + \\ + 2\rho^k \int_\Sigma (\lambda^k-\lambda)(y_1^k-y_1 - y_2^k + y_2)\,d\Sigma \leq \|\ell^k\|^2_{H^{-\frac12}(\Sigma)} + (\rho^k)^2 C_1 \sum_i \|y_i^k-y_i\|^2 + \\ \qquad + 2\rho^k \int_\Sigma (\lambda^k-\lambda)(y_1^k - y_1 - y_2^k + y_2)\,d\Sigma \end{cases}$$

où C_1 est une constante. De (3.12) et de relations analogues écrites pour m^k, on déduit

$$(3.13)\begin{cases} \|\ell^{k+1}\|^2 + a^2\|m^{k+1}\|^2 \leq \|\ell^k\|^2 + a^2\|m^k\|^2 + (\rho^k)^2 C_2\Big(\sum_i \|y_i^k - y_i\|^2 + \|p_i^k - p_i\|^2\Big) - \\ - 2C\rho^k\Big(\sum_i \|y_i^k - y_i\|^2 + \|p_i^k - p_i\|^2 + |u_i-u_i^k|^2\Big). \end{cases}$$

Il est clair que l'on peut choisir la suite ρ^k de manière que $\|\ell^k\|^2 + \|m^k\|^2$ soit décroissante et convergente. Il en résulte que

$$y_i^k \to y_i , \quad p_i^k \to p_i , \quad u_i^k \to u_i .$$

De plus λ^k, μ^k demeurent dans des bornés. Un passage à la limite dans $\mathcal{L} \geq 0$ montre la convergence faible de λ^k, μ^k respectivement vers λ, μ . On a ainsi démontré le

Théorème 3.1

Sous l'hypothèse (3.10), on a

(3.14) $$y_i^k \to y_i, \quad p_i^k \to p_i \quad \underline{\text{dans}} \quad \mathcal{V}_i$$

(3.15) $$u_i^k \to u_i \quad \underline{\text{dans}} \quad L^2(\Gamma_i)$$

(3.16) $$\lambda^k, \mu^k \to \lambda, \mu \quad \underline{\text{dans}} \quad H^{-\frac{1}{2}}(\Sigma) \ \underline{\text{faible}}.$$

IV - APPLICATION NUMERIQUE

4.1 - POSITION DU PROBLEME.

On a pris $\Omega =]0,4[\times]0,1[$ et décomposé Ω en quatre sous-domaines Ω_i, i=1,2,3, 4, comme indiqué sur la figure 3, les notations étant des variantes évidentes de celles du N°2.1. Le critère étant celui du N°1.1, relation (1.2), l'état du système est donné (avec $f \neq 0_\Gamma$ car cela facilite la construction de solutions exactes) par :

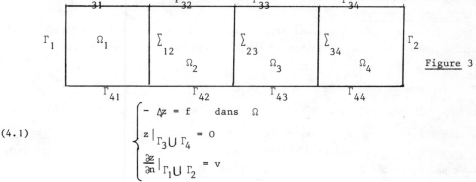

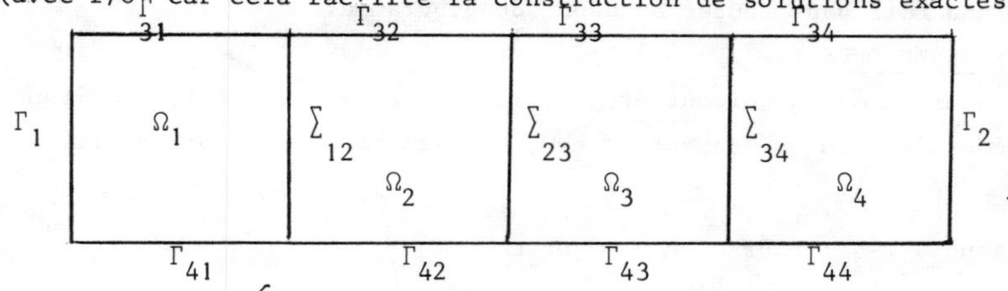

Figure 3

(4.1) $$\begin{cases} -\Delta z = f & \text{dans } \Omega \\ z|_{\Gamma_3 \cup \Gamma_4} = 0 \\ \dfrac{\partial z}{\partial n}\Big|_{\Gamma_1 \cup \Gamma_2} = v \end{cases}$$

avec

(4.2) $$v \in \mathcal{U}_{ad} = \mathcal{U} = L^2(\Gamma_1) \times L^2(\Gamma_2).$$

4.2 - CHOIX DU PROBLEME TEST.

Si dans (4.1) et dans le critère on prend f et y_d définis, respectivement, par :

(4.3) $$f(x_1,x_2) = 2\,(-x_1^2 - x_2^2 + 4x_1 + x_2)$$

(4.4) $$y_d(x_1,x_2) = -(x_2-x_2^2)(x_1^2-4x_1) - 8\,\nu$$

On vérifie facilement que le problème de contrôle optimal $\underset{v \in \mathcal{U}}{\text{Min}}\ J(v)$ admet comme solution unique :

(4.5) $$\begin{cases} v_1(x_2) = -4(x_2-x_2^2) \\ v_2(x_2) = -4(x_2-x_2^2) \end{cases}$$

l'état y et l'état adjoint p correspondant étant donnés par :

(4.6) $$y(x_1,x_2) = -(x_2-x_2^2)(x_1^2-4x_1)$$

(4.7) $$p(x_1,x_2) = 4\,\nu(x_2-x_2^2).$$

4.3 - MISE EN OEUVRE DE L'ALGORITHME DE DECOMPOSITION - COORDINATION.

L'algorithme du N°3.1, relatif à une décomposition de Ω en deux sous-domaines se généralise aisément à la décomposition du N°4.1. Au vu des relations (3.1), (3.2) il faut résoudre à chaque itération un problème de contrôle optimal, elliptique, dans chaque Ω_i, $i=1,2,3,4$; l'approximation des équations (3.1), (3.2) a été effectuée à partir d'une discrétisation par différences finies de pas $h=\frac{1}{20}$, soit environ 400 points de discrétisation par sous-domaine. Les sous-problèmes de contrôle ci-dessus ont été résolus par la méthode du gradient dont la mise en oeuvre est facilitée en remarquant que les matrices de discrétisation, associées aux équations (3.1), (3.2) sont identiques, sur chaque sous-domaine, et indépendantes de i et k; ces matrices étant symétriques et définies positives on peut alors utiliser une factorisation de CHOLESKY faite une fois pour toutes au début du programme.

4.4 - RESULTATS NUMERIQUES.

Dans les deux exemples qui ont été traités, on a initialisé l'algorithme de décomposition - coordination en prenant $\lambda^1 = 0$, $\mu^1=0$ et on a pris $S=$Identité dans (3.5).

Exemple 1

Il correspond à $\nu=5$; pour $\rho^k = \rho = 2$ on a

$$\max|y_1^1-y_2^1| = 0.651,\quad \max|y_1^4-y_2^4| = 1.5\times 10^{-2},\quad \max|y_1^{10}-y_2^{10}| = 1.8\times 10^{-4}.$$

On a amélioré la vitesse de convergence en utilisant une stratégie de ρ^k variables de type steepest descent permettant d'obtenir $\max|y_1^4 - y_2^4| = 1.8\times 10^{-4}$ d'où un gain important par rapport à la méthode avec ρ^k fixe. ■

Exemple 2

Il correspond à $\nu=0.125$; le problème a été résolu en utilisant la stratégie de ρ^k variables ci-dessus ; ν étant plus petit que dans le problème précédent la convergence est plus lente soit $\max|y_1^1 - y_2^1| = 0.506$, $\max|y_1^4 - y_2^4| = 2.10^{-3}$, $\max|y_1^9 - y_2^9| = 10^{-4}$. ■

Remarque 3.1

Dans les deux exemples précédents, la résolution à chaque itération, des sous-problèmes de contrôle optimal par la méthode du gradient n'a jamais demandé plus de trois itérations internes (pour un test d'arrêt assez sévère). ■

Remarque 3.2

Pour ν de l'ordre de 10^{-2} on n'a pu obtenir la convergence de l'algorithme de décomposition-coordination la condition (3.10) n'étant visiblement pas satisfaite. ■

Temps de calcul

Dans les deux exemples précédents, de l'ordre de 1 minute pour 5 itérations de coordination.

(1) J.L. Lions Contrôle optimal des systèmes gouvernés par des équations aux dérivées partielles, Dunod-Gauthier-Villars (1968).

(1) J.L. Lions - E. Magenes Problèmes aux limites non homogènes (T.1), Dunod (1968).

APPROXIMATION OF OPTIMAL CONTROL PROBLEMS OF SYSTEMS DESCRIBED BY BOUNDARY-VALUE MIXED PROBLEMS OF DIRICHLET-NEUMANN TYPE

P. COLLI FRANZONE

Laboratorio di Analisi Numerica del C.N.R.- Pavia, Italia

INTRODUCTION

We report here some results on optimal control of systems described by boundary-value mixed problems of Dirichlet-Neumann type for second order linear elliptic partial differential operators. These results are achieved in the framework of the deterministic theory as developed in Lions' book [1] on optimal control. We consider here the case of a boundary control on the Dirichlet condition. The state of the system is the solution of a mixed non-variational problem.

The initial control problem I) is approximated by a family I_ε) of control problems of systems described by variational problems of Neumann type and obtained by applying the penalization method. Convergence results are reported. Moreover a family I_ν) of finite dimension optimization problems, which converges to the initial problem I), is considered. Results of some numerical experiences are reported. (*)

Stating of the problem; notations and definitions.

Let Ω be a bounded open set of R^n whose boundary Γ is a(n-1) dimensional manifold and let Σ be a(n-2) dimensional manifold which separates Γ into two open non-empty disjoint sets Γ_o and Γ_1 such that:

$$\Gamma=\Gamma_o \cup \Gamma_1 \cup \Sigma \quad , \quad \Gamma_o \cap \Gamma_1 = \emptyset \quad , \quad \bar{\Gamma}_o \cap \bar{\Gamma}_1 = \Sigma$$

For $u,v \in H^1(\Omega)$ (1), set:

$$a(u,v)=\sum_{i,j=1}^{n}\int_\Omega a_{ij}(x)\frac{\partial u}{\partial x_j}\frac{\partial v}{\partial x_i}dx+\int_\Omega a_o(x)uvdx \tag{1}$$

Hereafter, we assume the following regularity hypotheses on Ω and on the

---

(*) Completely detailed proves will appear on "Rend.Istituto Lombardo (Sc.Mat. and Nat.).

(1) $H^1(\Omega)=\{v: v\in L^2(\Omega), \frac{\partial v}{\partial x_i}\in L^2(\Omega),\ i=1,\dots,n\}$; for the definitions of real Sobolev spaces $H^s(\Omega)$ and $H^s(\Gamma)$ with $s \in R$, see Lions-Magenes [1] chap. I.

the coefficients in the form $a(u,v)$:

i) $\partial\Omega=\Gamma$ and Σ are C^2-manifolds, locally Ω is totally on one side of Γ,

ii) $a_{ij}(x) \in C^3(\bar{\Omega})$, $a_o(x) \in C^1(\bar{\Omega})$ (2)

iii) $\sum_{i,j=1}^{n} a_{ij}(x)\xi_i\xi_j \geq \alpha|\xi|^2$, $\forall \xi=(\xi_1,\xi_2,\ldots,\xi_n) \in R^n$, $\forall x \in \Omega$, $\alpha>0$

$a_o(x) \geq 0$, $\forall x \in \Omega$

The form (1) is continuous and bilinear on $H^1(\Omega)$ and defines the second-order elliptic differential operator:

$$(2) \qquad A= - \sum_{i,j=1}^{n} \frac{\partial}{\partial x_i}(a_{ij}(x)\frac{\partial}{\partial x_j})+a_o(x)$$

The adjoint form $a^*(u,v)$ is defined by: $a^*(u,v)=a(v,u)$, $\forall u,v \in H^1(\Omega)$; the operator A^* associated is:

$$(3) \qquad A^* = -\sum_{i,j=1}^{n} \frac{\partial}{\partial x_i}(a_{ji}(x)\frac{\partial}{\partial x_j})+a_o(x)$$

<u>State of the system</u>

Set $T=\{t:t\in H^1(\Omega), r_o\gamma_o t=0\}$ (3) and $\phi\in H^{-\tau}(\Omega)$, $\tau \in [0,\frac{1}{2}[$, let $w \in T$ be the unique solution of the variational equation:

$$a^*(w,t) = {}_{-\tau,\Omega}\langle\phi,t\rangle_{\tau,\Omega} \quad , \forall t \in T \quad (4)$$

which is equivalent to the homogeneous mixed problem of Dirichlet-Neumann type:

$$A^*w=\phi \text{ in } \Omega,\ r_o\gamma_o w=0,\ r_1\gamma_{A^*}w=0 \quad (5)$$

Under the assumptions i),ii) it follows from the regularity results by Shamir [1] :

$$w \in H^{3/2-\eta}(\Omega), \quad \forall \eta>0$$

The space defined by

---

(2) $C^k(\bar{\Omega})$, k positive integer, is the space of functions which are k-times continuously differentiable in $\bar{\Omega}$.

(3) We denote by r_o and r_1 the restriction operators from Γ to Γ_o and Γ_1.

(4) The bracket $\langle\ ,\ \rangle$ indicates the duality between H^s and its dual space.

(5) $\gamma_o u$=trace of u on Γ, $\gamma_A u=\sum_{i,j=1}^{n} a_{ij}(x)\frac{\partial u}{\partial x_j}\cos(n,x_i)$, conormal derivative to A on Γ, $\cos(n,x_i)$=i-th direction cosine of n, n being th normal at Γ exterior Ω. For trace theorems, see Lions-Magenes [1].

$$X^{3/2-\eta}_{A^*,\tau}(\Omega)=\{v: v\in H^{3/2-\eta}(\Omega),\ A^*v\in H^{-\tau}(\Omega),\ r_o\gamma_o v=0,\ r_1\gamma_{A^*}v=0\}$$

endowed with the norm

$$||v||^2_{X^{3/2-\eta}_{A^*,\tau}(\Omega)}=||v||^2_{3/2-\eta,\Omega}+||A^*v||_{-\tau,\Omega}$$

is a Hilbert space for $\eta\in\left]0,\frac{1}{2}\right]$ and $\tau\in\left[0,\frac{1}{2}\right[$.

The operator A^* is thus an algebraic and topological isomorphism of $X^{3/2-\eta}_{A^*,\tau}(\Omega)$ on $H^{-\tau}(\Omega)$ for each $\eta\in\left]0,\frac{1}{2}\right]$ and $\tau\in\left[0,\frac{1}{2}\right[$.

Chosen

$$f\in H^{-\rho}(\Omega),\ g_o\in H^{\sigma}(\Gamma_o),\ g_1\in H^{-1+\sigma}(\Gamma_1),\ \rho\in\left[0,\tfrac{1}{2}\right[,\ \sigma\in\left]0,\tfrac{1}{2}\right[$$

by transposition, there exists a unique $y\in H^{\tau}(\Omega)$, $\forall\,\tau\in\left[0,\frac{1}{2}\right[$ solution of:

(4) $\quad \langle y,A^*w\rangle_{\tau,\Omega\ -\tau,\Omega} = \langle f,w\rangle_{-\rho,\Omega\ \rho,\Omega} - \langle g_o,r_o\gamma_{A^*}w\rangle_{\sigma,\Gamma_o\ -\sigma,\Gamma_o} + \langle g_1,r_1\gamma_o w\rangle$ (6),

$\forall\, w\in X^{3/2-\sigma}_{A^*,\tau}(\Omega)$

By definition y is the "weak" solution of the mixed problem:

(5) $\quad Ay=f$ in Ω , $r_o\gamma_o y=g_o$, $r_1\gamma_A y=g_1$

The regularity result

$$y\in H^{1/2+\sigma}(\Omega)$$

follows from Vishik-Eskin's work [1] and therefore the boundary conditions may be interpreted according to trace theorems.

We point out that the state y of the system is solution of a boundary-value problem of non-variational type; this means that the state y does not belong to $H^1(\Omega)$.

<u>Control space</u>

In many applications, it is very frequently encountered the case in which the control is exercised through the boundary (i.e. on the boundary conditions). Usually controls are described by measurable functions; however piece-wise constant controls are often used.

We shall consider the case of a boundary control on the Dirichlet condition on Γ_o. If we want to develop our theory in a Hilbert space framework and still allow piece-wise constant controls (or controls with even stronger singularities) then we are led to choose as control space

---

(6) The mapping $w\to r_1\gamma_o w$ is a linear continuous map of $T^{\sigma}=\{v: v\in H^{3/2-\sigma}(\Omega),\ r_o\gamma_o v=0\}$, $\sigma<\frac{1}{2}$, in $H^{1-\sigma}_o(\Gamma_1)$, the bracket $\langle,\rangle$ then indicates the duality of $H^{-1+\sigma}(\Gamma_1)$ and $H^{1-\sigma}_o(\Gamma_1)$.

$\mathcal{U}$ the following Sobolev space:

$$\mathcal{U}=H^{\sigma}(\Gamma_o) \text{ where } \sigma \in]0,\tfrac{1}{2}[$$

For each $v \in \mathcal{U}$, let $y(v) \in H^{1/2+\sigma}(\Omega)$ be the unique solution of:

(6) $\quad P) \{ Ay(v)=f \text{ in } \Omega,\ r_o\gamma_o y(v)=g_o+v,\ r_1\gamma_A y(v)=g_1$

Observations space

We confine ourselves to the case in which the trace of y is observed on Γ_1. The mapping $v \to r_1\gamma_o y(v)$ is an affine continuous map of $\mathcal{U}$ in $H^{\sigma}(\Gamma_1)$; in order to avoid as much possible the use of spaces H^s, s not an integer, we may consider in particular $r_1\gamma_o y(v) \in L^2(\Gamma_1)$ and then choose as observations space:

$$Y=L^2(\Gamma_1)$$

Control problem

Given

$\mathcal{U}_{ad}$ a closed convex and bounded subset of $\mathcal{U}$

$y_d \in Y$

With every control $v \in \mathcal{U}$, we associate the "cost function":

$$J(v)=\int_{\Gamma_1} |r_1\gamma_o y(v)-y_d|^2 d\sigma+\lambda\int_{\Gamma_o} v^2 d\sigma \ , \quad \lambda \geq 0$$

We shall consider the optimization problem:

I) Find $u \in \mathcal{U}_{ad}$ minimizing $J(v)$ over $\mathcal{U}_{ad}$

We obtain the following result:

Proposition -1-

Problem I) admits a unique (7) solution u, termed optimal control which is characterized by the inequality:

(7) $$\int_{\Gamma_1}[r_1\gamma_o y(u)-y_d]\cdot r_1\gamma_o y(v-u)\,d\sigma+\lambda\int_{\Gamma_o} u(v-u)\,d\sigma \geq 0, \quad \forall v \in \mathcal{U}_{ad}$$

Let us now trasform (7) by introducing the adjoint $p(u)$, unique solution in $H^1(\Omega)$ of the problem:

(8) $\quad A^*p(u)=0 \text{ in } \Omega,\ r_o\gamma_o p(u)=0,\ r_1\gamma_{A^*}p(u)=-(r_1\gamma_o y(u)-y_d)$

By a regularity result of Shamir the solution $p(u)$ belongs to $H^{3/2-\eta}(\Omega)$ $\forall\ \eta>0$. This allows us to apply Green's formula and thus transform inequality (7) into the equivalent form:

(7) For $\lambda>0$ uniqueness follows from $J(v)$ being strictly convex; for $\lambda=0$ it follows from the uniqueness of the Cauchy problem for the elliptic operator A.

(9) $$_{\mathcal{U}'}\langle r_o\gamma_{A^*}p(u),v-u\rangle_{\mathcal{U}}+\lambda\int_{\Gamma_o}u\cdot(v-u)\,d\sigma\geq 0,\quad \forall v\in\mathcal{U}_{ad}$$ ([8])

Therefore the system (6),(8),(9) admits a unique solution $\{y,p,u\}$ where u is the optimal control.

Approximation of I)

The main task in the numerical solution of problem I) is to produce a good approximation scheme for the state equation (6) which is of non-variational type.

Set

$$a_\varepsilon(w,\psi)=a(w,\psi)+\varepsilon^{-1}\int_{\Gamma_o}r_o\gamma_o w\cdot r_o\gamma_o\psi\,d\sigma \;\;(^9),\quad w,\psi\in H^1(\Omega)$$

we approximate problem P) (6), by means of a family $P_\varepsilon)$ of variational problems:

(10) $$\begin{cases}\text{for each } \varepsilon>0, \text{ let } y_\varepsilon(v) \text{ be the unique solution in } H^1(\Omega) \text{ of the}\\ \text{variational equation:}\\ a_\varepsilon(y_\varepsilon(v),\psi)={}_{-\rho,\Omega}\langle f,\psi\rangle_{\rho,\Omega}+\varepsilon^{-1}\int_{\Gamma_o}(g_o+v)\cdot r_o\gamma_o\psi\,d\sigma+{}_{-1/2,\Gamma_1}\langle g_1,r_1\gamma_o\psi\rangle_{1/2,\Gamma_1},\end{cases}$$

$$,\forall\,\psi\in H^1(\Omega)$$

where we assume $g_1\in H^{-1/2}(\Gamma_1)$ while g_o and v belong to $\mathcal{U}$.
The variational equation (10) is **equivalent** to the following boundary value problem of Neumann type:

(11) $$P_\varepsilon)\begin{cases}Ay_\varepsilon(v)=f & \text{in } \Omega\\ \varepsilon\gamma_A y_\varepsilon(v)+\chi_{\Gamma_o}\gamma_o y_\varepsilon(v)=g_\varepsilon(v) & \text{on } \Gamma\;(^{10})\end{cases}\qquad g_\varepsilon(v)=\begin{cases}g_o+v, & \text{on } \Gamma_o\\ \varepsilon g_1, & \text{on } \Gamma_1\end{cases}$$

([8]) We note that: $\frac{1}{2}J'(u)=r_o\gamma_{A^*}p(u)+\lambda u$.

([9]) We note that: for $v\in H^1(\Omega)$, $||v||^2_{1,\Omega}\leq M(\Omega)\left[\sum_{i=1}^{n}||\frac{\partial v}{\partial x_i}||^2_{o,\Omega}+\right.$ $\left.+||r_o\gamma_o v||^2_{o,\Gamma_o}\right]$, since Γ_o is a non-empty open set of Γ, it follows that $a_\varepsilon(v,v)\geq\beta||v||^2_{1,\Omega}$, $\forall v\in H^1(\Omega)$, where $\beta>0$ is independent of v and ε (small enough); $a_\varepsilon(u,v)$ is therefore coercive on $H^1(\Omega)$.

([10]) We note that boundary condition on Γ is of "natural type".

where χ_{Γ_o} is the characteristic function of Γ_o on Γ.

Let us consider the family I_ε) of control problems hereafter described:

$$I_\varepsilon) \quad \text{Find } u_\varepsilon \in \mathcal{U}_{ad} \text{ minimizing } J_\varepsilon(v) \text{ over } \mathcal{U}_{ad}$$

where $J_\varepsilon(v) = \int_{\Gamma_1} |r_1\gamma_o y_\varepsilon(v) - y_d|^2 d\sigma + \lambda \int_{\Gamma_o} v^2 d\sigma$.

We then obtain:

Proposition-2-

For each $\varepsilon>0$, problem I_ε) admits a unique solution u_ε; moreover a necessary and sufficient condition for u_ε to be an optimal control for I_ε) is that the following equations and inequalities be satisfied:

$$P_\varepsilon) \quad a_\varepsilon(y_\varepsilon,\psi) = {}_{-\rho,\Omega}\langle f,\psi\rangle_{\rho,\Omega} + \varepsilon^{-1}\int_{\Gamma_o} (g_o+u_\varepsilon)\cdot r_o\gamma_o\psi d\sigma + {}_{-1/2,\Gamma_1}\langle g_1, r_1\gamma_o\psi\rangle_{1/2,\Gamma_1}, \quad \forall \psi \in H^1(\Omega)$$

$$P^*_\varepsilon) \quad a^*_\varepsilon(p_\varepsilon,\psi) = -\int_{\Gamma_1} [r_1\gamma_o y_\varepsilon - y_d] r_1\gamma_o\psi d\sigma, \quad \forall \psi \in H^1(\Omega)$$

$$\int_{\Gamma_o} (-\frac{1}{\varepsilon} r_o\gamma_o p_\varepsilon + \lambda u_\varepsilon)(v-u_\varepsilon) d\sigma \geq 0, \quad \forall v \in \mathcal{U}_{ad}.$$

We have the following convergence result:

Theorem -1-

As $\varepsilon \to 0$ we have:

u_ε converges to u weakly in $\mathcal{U}$ (strongly in $L^2(\Gamma_o)$)

Furthemore:

$y_\varepsilon \to y$ in $H^\tau(\Omega)$, $\forall \tau < \frac{1}{2}$, $p_\varepsilon \to p$ in $H^{1+\delta}(\Omega)$, $\forall \delta < \frac{1}{2}$

and

$$J_\varepsilon(u_\varepsilon) \to J(u).$$

Numerical approximation

Let $\mathcal{H}$ be the set R^n_+ ordered as follows:

$$(h_1,\ldots,h_n) \leq (\bar{h}_1,\ldots,\bar{h}_n) \Longleftrightarrow h_i \leq \bar{h}_i, \; i=1,\ldots,n$$

For each $h \in \mathcal{H}$ let V_h be a finite dimension subspace of $H^1(\Omega)$ endowed with the induced norm.

We now suppose that the approximation V_h satisfies the following:

Condition -1-

For each $h \in \mathcal{H}$ there exists a linear continuous operator π_h of $H^1(\Omega)$

in V_h such that:

$$\lim_{h\to o}||y-\pi_h y||_{1,\Omega}=0, \quad \forall\, y\in H^1(\Omega)$$

$$||y-\pi_h y||_{k,\Omega}\leq C|h|^{s-k}||y||_{s,\Omega}, \quad \forall\, y\in H^s(\Omega), \begin{cases} k\in[0,1[\\ s\in[k,2] \end{cases}$$

C being a constant independent of h and y.

We approximate problem P_ε) (11), with the family of problem $P_{\varepsilon,h}$):

$$P_{\varepsilon,h})\begin{cases} \text{For each } \varepsilon>0 \text{ and } h\in\mathcal{H} \text{ let } y_{\varepsilon,h}(v)\in V_h \text{ the unique solution of:} \\ a_\varepsilon(y_{\varepsilon,h},w_h)={}_{-\rho,\Omega}{<}f,w_h{>}_{\rho,\Omega}+\varepsilon^{-1}\int_{\Gamma_o}(g_o+v)\,r_1\gamma_o w_h\,d\sigma+{}_{-\frac{1}{2},\Gamma_1}{<}g_1,r_1\gamma_o w_h{>}_{\frac{1}{2},\Gamma_1}, \\ \qquad\qquad\qquad\qquad\qquad\qquad\qquad\qquad\qquad\qquad\qquad\qquad , \forall\, w_h\in V_h. \end{cases}$$

Set $\mathcal{K}=R_+^n$, for each $k\in\mathcal{K}$ let $\mathcal{U}_k$ be a finite dimension subspace of $\mathcal{U}$ endowed with the induced norm, and let q_k be a linear continuous operator of $\mathcal{U}$ in $\mathcal{U}_k$ such that:

$$\forall\, v\in\mathcal{U}, \quad \lim_{k\to o}||v-q_k v||_{\mathcal{U}}=0$$

There exist approximations $\{V_h,\pi_h\}$ and $\{\mathcal{U}_k,q_k\}$ which satisfy the staded hypotheses; see, for example Aubin [1], Bramble-Schatz [1].

Finally, for each $k\in\mathcal{K}$ let:

$$\mathcal{U}_{ad}^k \text{ a closed convex set in } \mathcal{U}_k$$

such that

$$\mathcal{U}_{ad}^k\subset\mathcal{U}_{ad}, \quad q_k\mathcal{U}_{ad}\subset\mathcal{U}_{ad}^k$$

Set $\varepsilon=M\cdot|h|^s$, M positive constant and $s>0$, $y_{\varepsilon,h}(v)=y_h(v)$ and $\nu=(h,k)$, let us consider the family I_ν) of optimization problems:

I_ν) **Find** $u_\nu\in\mathcal{U}_{ad}^k$ minimizing $J_\nu(v_k)$ over $\mathcal{U}_{ad}^k$

where $J_\nu(v_k)=\int_{\Gamma_1}|r_1\gamma_o y_h(v_k)-y_d|^2 d\sigma+\lambda\int_{\Gamma_o}v_k^2\,d\sigma$.

Proposition -3-

For any ν, problem I_ν) admits at least one solution; the set of the solutions u_ν of I_ν) is characterized by the inequality:

$$\int_{\Gamma_o}[r_o\gamma_o y_h(u_\nu)-y_d]\cdot r_o\gamma_o y_h(v_k-u_\nu)\,d\sigma+\lambda\int_{\Gamma_o}u_\nu(v_k-u_\nu)\,d\sigma\geq 0, \quad \forall\, v_k\in\mathcal{U}_{ad}^k$$

Furthemore u_ν is solution of I_ν) if and only if it satisfies the system:

$$P_h)\quad a_\varepsilon(y_h,w_h)={}_{-\rho,\Omega}\langle f,w_h\rangle_{\rho,\Omega}+\varepsilon^{-1}\int_{\Gamma_0}(g_0+u_\nu)\cdot r_0\gamma_0 w_h d\sigma+{}_{-\frac{1}{2},\Gamma_1}\langle g_1,r_1\gamma_0 w_h\rangle_{\frac{1}{2},\Gamma_1},\quad \forall w_h\in V_h$$

$$P_h^*)\quad a_\varepsilon^*(p_h,w_h)=-\int_{\Gamma_1}[r_1\gamma_0 y_h-y_d]\cdot r_1\gamma_0 w_h d\sigma,\quad \forall w_h\in V_h$$

$$\int_{\Gamma_0}[-\tfrac{1}{\varepsilon}r_0\gamma_0 p_h+\lambda u_\nu]\cdot(v_k-u_\nu)\,d\sigma\geq 0,\ \forall v_k\in\mathcal{U}_{ad}^k \qquad (^{11})$$

The above described approximation is supported by the following convergence result:

Theorem -2-

For $\varepsilon=M|h|^s$, M positive constant and $s\in]0,1[$ as $\nu\to 0$ (i.e. $h\to 0$, $k\to 0$ independently) the sequence $\{u_\nu\}$, u_ν being a solution of I_ν), converges weakly in $\mathcal{U}$ (strongly in $L^2(\Gamma_0)$) to the unique solution u of problem I).

Finally some results on numerical experiences performed by G.Gazzaniga (Laboratorio di Analisi Numerica del C.N.R.-Pavia) and myself are reported; computations were carried on an IBM 360/44 installed at the Centro di Calcoli Numerici - University of Pavia.

Let us consider the following case

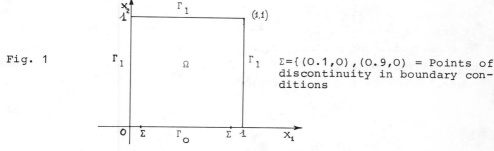

Fig. 1

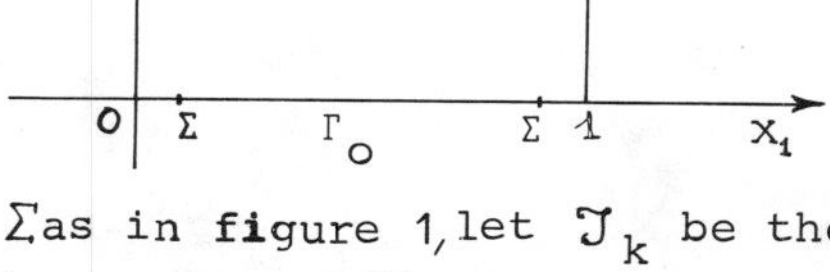

$\Sigma=\{(0.1,0),(0.9,0)\}$ = Points of discontinuity in boundary conditions

chosen Ω, Γ_0, Γ_1, Σ as in **figure** 1, let $\mathcal{T}_k$ be the partition of Γ_0 into k intervals of equal length and $\mathcal{U}_{ad}$ be the set of functions on Γ_0 piece-

(11) We observe that $\frac{1}{2}J'_\nu(u_\nu)=-\frac{1}{\varepsilon}r_0\gamma_0 p_h+\lambda u_\nu$; the described scheme presents the advantage of calculating the gradient values through the trace of the adjoint state p_h on Γ_0.

wise constant on the partition $\mathcal{T}_k$ and valued between 0 an π.

Let $h=(h_1,h_2)$ be an element of R_+^2. We denote by $\mathcal{R}_h$ the regular mesh of points $M=(m_1h_1,m_2h_2), m_i \in Z$.

We consider the approximation V_h of $H^1(\Omega)$ hereafter described:

$$\sigma_h(M)=\prod_{i=1}^{2}\left[(m_i-1)h_i,(m_i+1)h_i\right], \quad \mathcal{R}_h^1(\Omega)=\{M: M\in\mathcal{R}_h, \sigma_h(M)\cap\Omega\neq\emptyset\}$$

$$V_h=\{z_h, z_h=\sum_{M\in\mathcal{R}_h^1(\Omega)} z_h^M\theta_h^M, z_h^M\in R\}, \quad \theta_h^M(x)=\begin{cases}\prod_{i=1}^{2}\left(1-\left|\frac{x_i}{h_i}-m_i\right|\right), & x\in\sigma_h(M)\\ 0, & x\notin\sigma_h(M)\end{cases}$$

Numerical experiences are performed on the model-problem:

$$-\Delta y(v)=0 \text{ in } \Omega, \quad y|_{\Gamma_0}=v, \quad \frac{\partial y}{\partial n}\Big|_{\Gamma_1}=g_1 \quad \text{and} \quad J(v)=\int_{\Gamma_1}|y(v)|_{\Gamma_1}-y_d|^2 d\sigma$$

and we choose $y_d=z|_{\Gamma_1}$ and $g_1=\frac{\partial z}{\partial n}\Big|_{\Gamma_1}$ where:

a) $z=\text{arctg}\frac{x_2}{x_1-1/2}$ and then $y_{ott}=z$, $u_{ott}=\begin{cases}\pi, & 0.1\leq x<0.5\\ \frac{\pi}{2}, & x=0.5\\ 0, & 0.5<x\leq 0.9\end{cases}$

b) $z=\pi-\left(\text{arctg}\frac{x_2}{x_1-0.65}-\text{arctg}\frac{x_2}{x_1-0.35}\right)$ and then $y_{ott}=z$, $u_{ott}=\begin{cases}\pi, & 0.1\leq x<0.35\\ \frac{\pi}{2}, & x=0.35, x=0.65\\ 0, & 0.65<x\leq 0.9\end{cases}$

The solution of problem I_ν) has been achieved by using the direct gradient method with projection; namely, the method using as direction of descent the projection of the gradient on the linear manifold by active constraint, and a coniugate direction at the next step if no other constraint becomes active. Figure 2 represent the approximated optimal control found in the specific numeric cases considered.

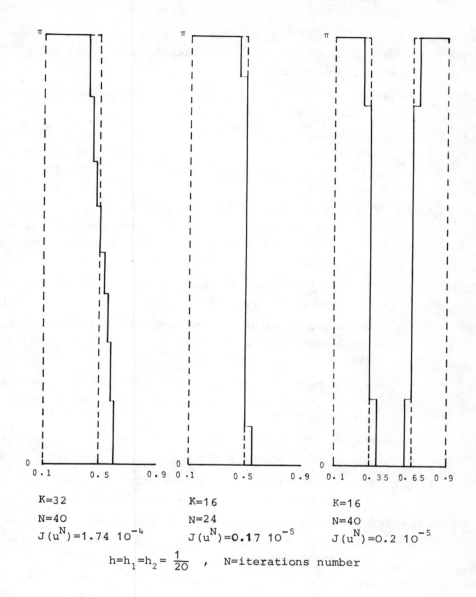

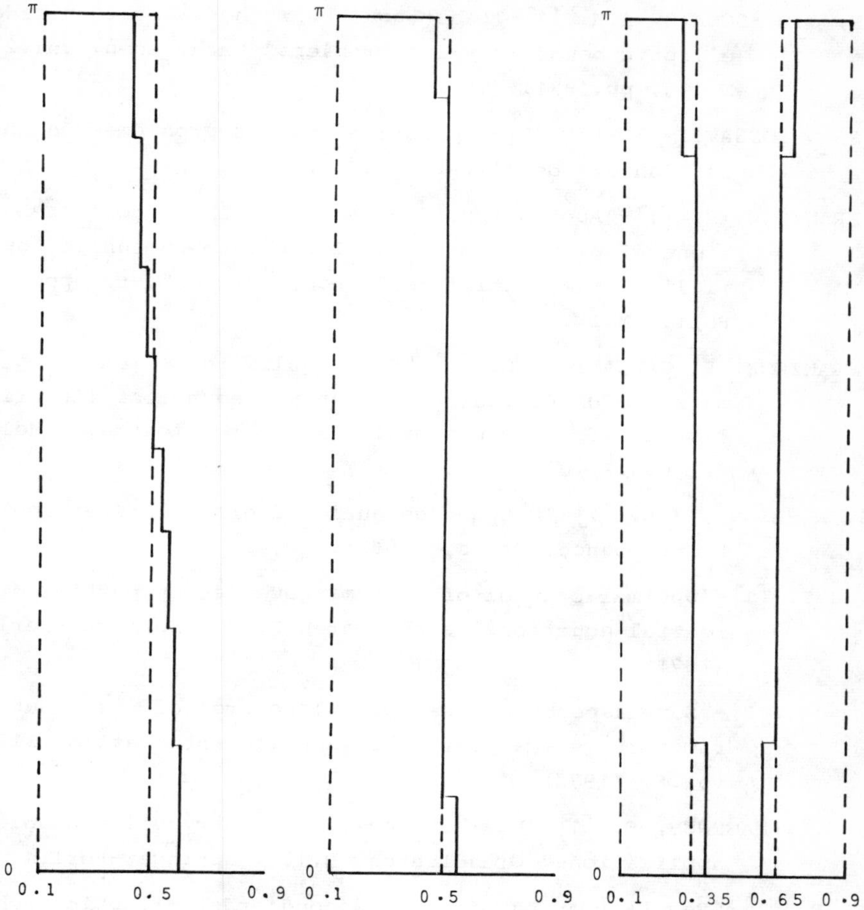

K=32
N=40
$J(u^N)=1.74\ 10^{-4}$

K=16
N=24
$J(u^N)=0.17\ 10^{-5}$

K=16
N=40
$J(u^N)=0.2\ 10^{-5}$

$h=h_1=h_2=\frac{1}{20}$, N=iterations number

Fig. 2

REFERENCES

AUBIN, J.P. [1]-"Approximations of elliptic boundary-value problems". Wiley - Interscience, vol.XXVI, (1972).

BRAMBLE, J.H.-SCHATZ, A.H. [1]-"Least squares methods for 2mth order elliptic boundary-value problems" Math. Comp. vol.25, N.113, pp.1-31, (1971).

BENSOUSSAN-BOSSAVIT-NEDELEC [1]-"Approximation des problèmes de contrôle" Cahiers de l'I.R.I.A., n.2.

COLLI FRANZONE, P. [1]-"Approssimazione mediante il metodo di penalizzazione di problemi misti di Dirichlet-Neumann per operatori lineari ellittici del secondo ordine", to appear on Boll.U.M.I..

COLLI FRANZONE, P.-GAZZANIGA, G. [1]-"Sull'analisi numerica del problema misto di Dirichlet-Neumann per equazioni lineari ellittiche del secondo ordine" Pubblicazione N.34 del L.A.N. del C.N.R. di Pavia.

LATTES, R.-LIONS, J.L. [1]-"Méthode de quasi réversibilité et applications" Dunod, Paris, (1967).

LIONS, J.L. [1]-"Optimal control of systems governed by partial differential equations" Grundlchren B.170, Springer-Berlino, (1971).

[2]-"Some aspects of the optimal control of distributed parameter systems" Series in Applied Mathematics, SIAM, vol.6, (1972).

LIONS, J.L.-MAGENES, E. [1]-"Non-homogeneous boundary value problems and applications" Grundlchren, B.181, Springer-Berlin, (1971)

SHAMIR, E. [1]-"Regularization of mixed second order elliptic problems" Israel J. of Math., 6, pp.150-168, (1968).

VISHIK, I.M.-ESKIN, I.G. [1]-"Elliptic convolution equations in bounded domains and their applications" Uspehi Math.Nauk 22, 1 (133), pp.15-76, (1967).

YVON, J.P. [1]-"Application de la pénalization à la résolution d'un problème de contrôle optimal" Cahier de l'IRIA n.2, (1970).

CONTROL OF PARABOLIC SYSTEMS WITH BOUNDARY CONDITIONS INVOLVING TIME-DELAYS

P.K.C.Wang

Department of System Science

University of California

Los Angeles,California

U.S.A.

In this paper, we consider various problems in the optimal control and stability of parabolic systems with boundary conditions involving time-delays. These problems arise physically in the control of diffusion processes in which time-delayed feedback signals are introduced at the boundary of a system's spatial domain. To illustrate the basic ideas, only the results for parabolic systems with the simplest forms of boundary conditions involving time-delays will be presented. Results of a more general nature are discussed in detail in reference [6].

I. PRELIMINARIES

Let Ω be a bounded open set in R^n with an infinitely differentiable boundary Γ and I denote a given finite time interval $]0,T[$. We consider a parabolic system of the form:

$$\frac{\partial y}{\partial t} - \Delta y = f \quad \text{in } Q=\Omega\times I, \tag{1}$$

where Δ is the Laplacian operator. The function f corresponds to either a distributed control or a specified function defined in Q.

Let Γ^1 and Γ^2 be given disjoint subsets of Γ such that $\Gamma=\Gamma^1\cup\Gamma^2$. Let $\Sigma=\Gamma\times I$; $\Sigma^i=\Gamma^i\times I, i=1,2$. We consider the following Neumann boundary condition involving a time-delay:

$$\frac{\partial y}{\partial \upsilon}(x,t) \triangleq \sum_{i,j=1}^{n} \cos(\eta,x_i)\,\frac{\partial y}{\partial x_j}(x,t) = q(x,t) \quad \text{on } \Sigma, \tag{2}$$

where

$$q(x,t) = \Phi(x)\{y(w(x),t-\tau) + u(x,t)\}, \tag{3}$$

where Φ is a given C^∞ function defined on Γ with compact support in Γ^1; $\cos(\eta,x_i)$ is the i-th directional cosine of the outward normal η at a point $x\in\Gamma$; u represents either a boundary control or a given function defined in Σ; the time-delay τ is a specified positive number; w is a continuously differentiable bijection of Γ onto Γ such that $w(x)=x$ if $x\in\Gamma^2$ and $w(x)\in\Gamma^2$ if $x\in\Gamma^1$; moreover, its Jacobian does not vanish on Γ.

The initial data for (1) are given by

$$\left.\begin{array}{ll} y(x,0) = y_o(x), & x\in\Omega, \\ y(x,t') = \phi_0(x,t'), & (x,t')\in\Gamma\times[-\tau,0[, \end{array}\right\} \quad (4)$$

where y_o and ϕ_0 are specified functions. Note that only $\tilde{\phi}_0$, the restriction of ϕ_0 to $\Gamma^2\times[-\tau,0[$, is of importance here. In what follows, we shall first give sufficient conditions for the existence of a unique solution of the mixed initial-boundary value problem (1)-(4). Then, various optimal control problems will be discussed.

For simplicity, let the final time $T=K\tau$, where K is a given positive integer. We introduce the following notations: $I_j=](j-1)\tau,j\tau[$, $Q_j=\Omega\times I_j$, $\Sigma_j=\Gamma\times I_j$ and $\Sigma_j^i=\Gamma^i\times I_j$ for $i=1,2$ and $j=0,1,\ldots,K$. Let $H^r(\Omega)$, $r\geqslant 0$, denote the Sobolev space of order r on Ω. For any pair of real numbers $r,s\geqslant 0$, the Sobolev space $H^{r,s}(Q)$ is defined by

$$H^{r,s}(Q) = H^0(I;H^r(\Omega))\cap H^s(I;H^0(\Omega)), \quad Q=\Omega\times I,$$

where $H^s(I;X)$ denotes the Sobolev space of order s of functions defined on I and with values in X.

For optimal control problems, it is of importance to consider the cases where the control f or u belongs to $L^2(Q)$ or $L^2(\Sigma)$ respectively. For these cases, we have the following results:

<u>Theorem 1</u>: Let y_o,ϕ_0,u and f be given with $y_o\in H^1(\Omega)$ {resp. $H^{\frac{1}{2}}(\Omega)$}, $\tilde{\phi}_0\in H^{\frac{1}{2},\frac{1}{4}}(\Sigma_0^2)$ {resp. $L^2(\Sigma_0^2)$}, $u\in H^{\frac{1}{2},\frac{1}{4}}(\Sigma)$ {resp. $L^2(\Sigma)$} and $f\in L^2(Q)$ {resp. $(H^{\frac{1}{2},\frac{1}{4}}(Q))'$, dual of $H^{\frac{1}{2},\frac{1}{4}}(Q)$}. Then, there exists a unique solution $y\in H^{2,1}(Q)$ {resp. $H^{\frac{3}{2},\frac{3}{4}}(Q)$} for problem (1)-(4). Moreover, $y(\cdot,j\tau)\in H^1(\Omega)$ {resp. $H^{\frac{1}{2}}(\Omega)$} for $j=1,\ldots,K$.

The above theorem can be established by first solving the problem on Q_1 using the basic results of Lions and Magenes [3;p.33,p.81] specialized to the case of (1)-(4). In a similar manner, the existence of a unique solution on Q_2 can be established by using the solution on Q_1 to generate the initial data at $t=\tau$ and boundary data on Σ_2. This advancing process is repeated for $Q_3,Q_4,\ldots$, until the final cylinder set Q_K is reached. Note that the solution y on Q_j (denoted hereafter by y_j), $j=1,\ldots,K$, is required to satisfy the initial data at $t=(j-1)\tau$:

$$y_j(x,(j-1)\tau) = y_{j-1}(x,(j-1)\tau), \quad x\in\Omega, \quad (5)$$

and boundary condition (2) whose right-hand-side is given by

$$q_j(x,t) = \Phi(x)\{y_{j-1}(w(x),t-\tau) + u(x,t)\}. \quad (6)$$

In order to apply the same results of Lions and Magenes to any Q_j, we must verify that y_{j-1} and $y_{j-1}|_{\Sigma_{j-1}}$, $j=2,\ldots,K$, satisfy the same conditions as required for y_o and q_1. This can be shown by making use of the trace theorem [3;p.9] and a result pertaining to the continuity of $y_{j-1}(\cdot,t)$ on $[(j-2)\tau,(j-1)\tau]$ [2;p.19].

II. OPTIMAL CONTROL

We shall consider optimal control problems for system (1)-(4) in which the control corresponds to either f or u belonging to a specified closed convex set $\mathcal{U}_Q \in L^2(Q)$ or $\mathcal{U}_\Sigma \in L^2(\Sigma)$ respectively.

Problem 1: Let y_o, ϕ_0 and u be given functions satisfying the hypotheses of Theorem 1 so that for a given control $f \in \mathcal{U}_Q$, a unique solution $y(f) \in H^{2,1}(Q)$ exists and $y(\cdot,T;f) \in H^1(\Omega)$. The problem is to find a $f^o \in \mathcal{U}_Q$ such that $J(f^o) \leqslant J(f)$ for all $f \in \mathcal{U}_Q$, where J is a cost functional given by

$$J(f)=\lambda_1 \int_Q |y(x,t;f)-y_d|^2 dxdt + \lambda_2 \int_\Omega |y(x,T;f)-y_{dT}|^2 dx + \lambda_3 \int_Q (Nf)f\ dxdt, \tag{7}$$

where $\lambda_i \geqslant 0$ and $\lambda_1+\lambda_2+\lambda_3>0$; y_d and y_{dT} are given in $L^2(Q)$ and $L^2(\Omega)$ respectively;and N is a positive linear operator on $L^2(Q)$ into $L^2(Q)$.

For the above problem, it is known [1] that for $\lambda_3>0$, a unique optimal control f^o exists, moreover, f^o is characterized by

$$\begin{aligned} J'(f^o)\cdot(f-f^o) &= \lambda_1 \int_Q (y(f^o)-y_d)(y(f)-y(f^o))\ dxdt \\ &+ \lambda_2 \int_\Omega (y(x,T;f^o)-y_{dT})(y(x,T;f)-y(x,T;f^o))dx \\ &+ \lambda_3 \int_Q (Nf^o)(f-f^o)\ dxdt \geqslant 0 \quad \text{for all } f \in \mathcal{U}_Q. \end{aligned} \tag{8}$$

The above condition can be simplified by introducing the following adjoint equation. For each $f \in \mathcal{U}_Q$, we define $p=p(f)=p(x,t;f)$ as the solution of

$$-\frac{\partial p(f)}{\partial t} - \Delta p(f) = \lambda_1(y(f)-y_d) \quad \text{in } Q, \tag{9}$$

with terminal condition

$$p(x,T;f) = \lambda_2(y(x,T;f)-y_{dT}), \quad x \in \Omega, \tag{10}$$

and boundary condition:

$$\frac{\partial p(f)}{\partial \upsilon}(x,t) = 0 \quad \text{for } (x,t) \in ([\Gamma - w(\text{supp}(\Phi))] \times I) \cup (w(\text{supp}(\Phi)) \times I_K), \tag{11}$$

$$\frac{\partial p(f)}{\partial \upsilon}(x,t) = \Phi(w^{-1}(x))|J_w(x)|\, p(w^{-1}(x),t+\tau;f) \quad \text{for } (x,t) \in w(\text{supp}(\Phi)) \times]0,T-\tau[, \tag{12}$$

where J_w denotes the Jacobian of w; $w(\text{supp}(\Phi))$ is the image of the support of under the mapping w and

$$\frac{\partial p(f)}{\partial \upsilon}(x,t) = \sum_{i,j=1}^{n} \cos(\eta,x_i) \frac{\partial p(f)}{\partial x_j}(x,t). \tag{13}$$

We observe that for given y_d, y_{dT} and f, problem (9)-(13) can be solved backward in time starting from t=T by first obtaining the solution $p=p_K$ on Q_K and terminal condition (19) and boundary condition

$$\frac{\partial p_K(f)}{\partial \upsilon}(x,t) = 0 \qquad \text{for } (x,t)\in \Sigma_K. \tag{14}$$

Having found p_K, we may proceed to solve the problem on Q_{K-1} backward in time with terminal data at t=(K-1) :

$$p_{K-1}(x,(K-1)\tau) = p_K(x,(K-1)\tau), \qquad x\in\Omega, \tag{15}$$

and with boundary conditions:

$$\frac{\partial p_{K-1}(f)}{\partial \upsilon}(x,t) = 0, \quad (x,t)\in[\Gamma-w(\mathrm{supp}(\Phi))]\times I_{K-1}, \tag{16}$$

$$\frac{\partial p_{K-1}(f)}{\partial \upsilon}(x,t) = \Phi(w^{-1}(x))\,|J_w(x)|\,p_K(w^{-1}(x),t+\tau;f), \quad (x,t)\in w(\mathrm{supp}(\Phi))\times I_{K-1}. \tag{17}$$

Note that the right-hand-side of (17) is completely determined once p_K is known. This backward process is repeated until the solution on the initial cylinder set Q_1 is determined. For $f\in L^2(Q)$, the existence of a unique solution $p_K(f)\in H^{2,1}(Q_K)$ with $p_K(\cdot,(K-1)\tau)\in H^1(\Omega)$ can be established by applying Theorem 1 to (9)-(13) with obvious change of variables and with reversed sense of time t'=T-t. The result can be extended to Q_j, $1\leqslant j\leqslant K-1$, in the same way, since the right-hand-side of (17) is in $H^{\frac{1}{2},\frac{1}{4}}(\Sigma_{K-1})$ (by trace theorem). Thus, we have the result:

Lemma 1: Let the hypotheses of Theorem 1 be satisfied. Then for given $y_d\in L^2(Q)$, $y_{dT}\in H^1(\Omega)$ and any $f\in L^2(Q)$, there exists a unique solution $p(f)\in H^{2,1}(Q)$ to problem (9)-(13).

Now, in view of Lemma 1, we can proceed to simplify (8) using the adjoint equation. It can be shown [6] by setting $f=f^o$ in (9)-(13), multiplying both sides of (9) by $(y(f)-y(f^o))$ and integrating over Q that (8) reduces to

$$\int_Q (p(f^o)+\lambda_3 N f^o)(f-f^o)\,dxdt \geqslant 0 \text{ for all } f\in \mathcal{U}_Q. \tag{18}$$

This result can be summarized as:

Theorem 2: For Problem 1 with cost functional (7) with $y_d\in L^2(Q)$, $y_{dT}\in H^1(\Omega)$ and $\lambda_3>0$, there exists a unique optimal control f^o which is determined by the solution of (1),

(9) with boundary conditions (2),(11)-(12), initial condition (4) and terminal condition (10) (all with $f=f^o$). Moreover, f^o satisfies (18).

For the special case where $\mathcal{U}_Q=L^2(Q)$, (18) is satisfied when

$$f^o = -\lambda_3^{-1}N^{-1}p(f^o). \tag{19}$$

In particular, if N is the identity operator on $L^2(Q)$, then, in view of Lemma 1, we have $f^o \in H^{2,1}(Q)$. To obtain the optimal control (19) in feedback form, we follow Lions' approach [1] by first considering the following set of equations with $s \in I$:

$$\left.\begin{aligned} &\frac{\partial y}{\partial t} - \Delta y + \lambda_3^{-1}N^{-1}p = 0, \\ -&\frac{\partial p}{\partial t} - \Delta p - \lambda_1 y = -\lambda_1 y_d, \end{aligned}\right\} (x,t) \in \Omega\times]s,T[, \tag{20}$$

with boundary conditions:

$$\frac{\partial y}{\partial \upsilon}(x,t) = \left\{\begin{aligned} &\Phi(x)\{y(w(x),t-\tau)+u(x,t)\} \quad \text{if } t-\tau \geqslant s \\ &\Phi(x)\{\phi_s(w(x),t-\tau)+u(x,t)\} \quad \text{if } t-\tau < s \end{aligned}\right\} (x,t) \in \Omega\times]s,T[, \tag{21}$$

$$\frac{\partial p}{\partial \upsilon}(x,t) = \left\{\begin{aligned} &0, \quad (x,t) \in \hat{\Sigma}_s \triangleq (\Gamma - w(\mathrm{supp}(\Phi))\times]s,T[\cup (w(\mathrm{supp}(\Phi))\times I_K), \\ &\Phi(w^{-1}(x))|J_w(x)|p(w^{-1}(x),t+\tau), \quad (x,t) \in w(\mathrm{supp}(\Phi))\times]s,T-\tau[, \end{aligned}\right\} \tag{22}$$

and with initial and terminal conditions:

$$\left.\begin{aligned} y(x,s) &= y_s(x), \\ p(x,T) &= \lambda_2(y(x,T)-y_{dT}), \end{aligned}\right\} x \in \Omega, \tag{23}$$

where y_s is given in $H^1(\Omega)$ and ϕ_s is a given function defined in $\Gamma\times[s-\tau,s[$, whose restriction $\tilde{\phi}_s$ to $\Gamma^2\times[s-\tau,s[$ is in $H^{\frac{1}{2},\frac{1}{4}}(\Gamma^2\times[s-\tau,s[)$. Note that (20)-(23) provide the solution to the optimal control problem associated with (1) for $t \in]s,T[$, $\mathcal{U}_Q = L^2(Q)$ and with a cost functional given by

$$J_s(f)=\lambda_1 \int_s^T \int_\Omega |y(x,t;f)-y_d|^2 \,dxdt + \lambda_2 \int_\Omega |y(x,T;f)-y_{dT}|^2 \,dx + \lambda_3 \int_s^T \int_\Omega (Nf)f \,dxdt. \tag{24}$$

The problem with $\lambda_3>0$ has a unique optimal control in the form of (19). Consequently, (20)-(23) has a unique solution $\{y,p\}$ also. In fact, for any given pair $(y_s,\tilde{\phi}_s) \in H^1(\Omega)\times H^{\frac{1}{2},\frac{1}{4}}(\Gamma^2\times[s-\tau,s[)$, the solution $y,p \in H^{2,1}(\Omega\times]s,T[)$. Moreover, the following property can be readily established.

Proposition: Let $\{y,p\}$ be the solution of (20)-(23) with $s=0$. Define σ_s, the system "state" at time s, by the pair $(y(\cdot,s),\tilde{\phi}_s)$, where

$$\tilde{\phi}_s(\cdot,t') = \begin{cases} \tilde{\phi}_0(\cdot,t') & \text{for } t'\in \hat{I}_s=[-\tau,0[\cap[s-\tau,s[, \\ y(\cdot,t')|_{\Gamma^2} & \text{for } t'\in[s-\tau,s[-\hat{I}_s. \end{cases} \tag{25}$$

Then, for all pairs $s\leqslant t$ in I,

$$p(\cdot,t) = P(t,s)\sigma_s + r_s(\cdot,t), \tag{26}$$

where $P(t,s)$ and $r_s(\cdot,t)$ are defined as follows:

(i) we solve the equations:

$$\left.\begin{aligned} &\frac{\partial\beta}{\partial t} - \Delta\beta + \lambda_3^{-1}N^{-1}\gamma = 0, \\ -&\frac{\partial\gamma}{\partial t} - \Delta\gamma - \lambda_1\beta = 0, \end{aligned}\right\} (x,t)\in\Omega\times]s,T[, \tag{27}$$

with boundary conditions

$$\frac{\partial\beta}{\partial\upsilon}(x,t) = \begin{cases} \Phi(x)\beta(w(x),t-\tau) & \text{if } t-\tau\geqslant s, \\ \Phi(x)\tilde{\phi}_s(w(x),t-\tau) & \text{if } t-\tau<s, \end{cases} \quad (x,t)\in\Omega\times]s,T[, \tag{28}$$

$$\frac{\partial\gamma}{\partial\upsilon}(x,t) = \begin{cases} 0, & (x,t)\in\hat{\Sigma}_s, \\ \Phi(w^{-1}(x))|J_w(x)|\gamma(w^{-1}(x),t+\tau), & (x,t)\in w(\mathrm{supp}(\Phi))\times]s,T-\tau[, \end{cases} \tag{29}$$

and initial and terminal conditions:

$$\left.\begin{aligned} \beta(x,s) &= y(x,s), \\ \gamma(x,T) &= \lambda_2\beta(x,T), \end{aligned}\right\} x\in\Omega, \tag{30}$$

then

$$P(t,s)\sigma_s = \gamma(\cdot,t); \tag{31}$$

(ii) we solve the equations:

$$\left.\begin{aligned} &\frac{\partial\eta}{\partial t} - \Delta\eta + \lambda_3^{-1}N^{-1}\xi = 0, \\ &\frac{\partial\xi}{\partial t} - \Delta\xi - \lambda_1\eta = -\lambda_1 y_d, \end{aligned}\right\} (x,t)\in\Omega\times]s,T[, \tag{32}$$

with boundary conditions:

$$\frac{\partial\eta}{\partial\upsilon}(x,t) = \begin{cases} \Phi(x)\{\eta(w(x),t-\tau)+u(x,t)\} & \text{if } t-\tau\geqslant s, \\ \Phi(x)u(x,t) & \text{if } t-\tau<s, \end{cases} \quad (x,t)\in\Omega\times]s,T[, \tag{33}$$

$$\frac{\partial\xi}{\partial\upsilon}(x,t) = \begin{cases} 0, & (x,t)\in\hat{\Sigma}_s, \\ \Phi(w^{-1}(x))\,|J_w(x)|\,\xi(w^{-1}(x),t+\tau), & (x,t)\in w(\mathrm{supp}(\Phi))\times]s,T-\tau[, \end{cases} \tag{34}$$

and with initial and terminal conditions:

$$\left.\begin{aligned} \eta(x,s) &= 0, \\ \xi(x,T) &= \lambda_2(\eta(x,T)-y_{dT}), \end{aligned}\right\} \quad x\in\Omega, \tag{35}$$

then

$$r_s(x,t) = \xi(x,t). \tag{36}$$

Now, the optimal feedback control can be obtained by setting s=t in (26) and substituting the result into (19):

$$f^o(\cdot,t) = -\lambda_3^{-1}N^{-1}(P(t,t)\sigma_t+r_t(\cdot,t)), \quad t\in I. \tag{37}$$

By making use of Schwartz's kernel theorem [4], it can be verified that the optimal feedback control (37) (with N being the identity operator on $L^2(Q)$) can be represented in the form:

$$f^o(x,t)=-\lambda_3^{-1}\left\{\int_\Omega K_0(x,x',t)y(x',t)dx' + \int_{t-\tau}^{t}\int_{\Gamma^2} K_1(x,x',t,t')\phi_t(x',t')d\Gamma^2dt'+r_t(x,t)\right\}, \tag{38}$$

where $\{K_0,K_1\}$ is the kernel of $P(t,t)$.

Problem 2: Let y_o,ϕ_0 and f be given functions satisfying the hypothesis of Theorem 1 with $u\in L^2(\Sigma)$. Let $y(x,t;u)$ denote the solution of (1)-(4) at (x,t) corresponding to a given control $u\in\mathcal{U}_\Sigma$. We wish to find a $u\in\mathcal{U}_\Sigma$ such that the following cost functional is minimized over $\mathcal{U}_\Sigma$:

$$J(u)=\lambda_1\int_Q |y(x,t;u)-y_d|^2dxdt + \lambda_2\int_\Omega |y(x,T;u)-y_{dT}|^2dx + \lambda_3\int_0^T\int_{\mathrm{supp}(\Phi)} (\tilde{N}u)u\, d\Gamma dt, \tag{39}$$

where the λ_i's, y_d and y_{dT} are as in Problem 1, and $\tilde{N}$ is a positive linear operator on $L^2(\Sigma)$ into $L^2(\Sigma)$. If y_o,ϕ_0 and f satisfy the conditions of Theorem 1, then for each $u\in\mathcal{U}_\Sigma$, $y(u)\in H^{\frac{3}{2},\frac{3}{4}}(Q)$ and $y(\cdot,T;u)\in H^{\frac{1}{2}}(\Omega)$. Hence $J(u)$ is defined. Similar to Problem 1, this problem has a unique optimal control $u^o\in\mathcal{U}_\Sigma$ if $\lambda_3>0$. Also, u^o can be characterized by

$$\lambda_1\int_Q(y(u^o)-y_d)(y(u)-y(u^o))\, dxdt + \lambda_2\int_\Omega(y(x,T;u^o)-y_{dT})(y(x,T;u)-y(x,T;u^o))dx$$
$$+\lambda_3\int_0^T\int_{\mathrm{supp}(\Phi)}(\tilde{N}u^o)(u-u^o)\, d\Gamma dt \geqslant 0 \quad \text{for all } u\in\mathcal{U}_\Sigma. \tag{40}$$

The foregoing inequality can be simplified by introducing an adjoint equation whose form is identical to (9)-(13). From Theorem 1, for any $u \in L^2(\Sigma)$, there exists a unique solution $y(u) \in H^{\frac{3}{2},\frac{3}{4}}(Q)$ with $y(\cdot,T;u) \in H^{\frac{1}{2}}(\Omega)$. If $y_{1d} \in L^2(Q)$ and $y_{dT} \in H^{\frac{1}{2}}(\Omega)$, then the right-hand-side of (9) and (10) are in $L^2(Q)$ and $H^{\frac{1}{2}}(\Omega)$ respectively. Similar to Problem 1, we can establish the existence of a unique solution $p(u) \in H^{\frac{3}{2},\frac{3}{4}}(Q)$ for (9)-(13). Moreover, (40) can be simplified as

$$\int_0^T \int_{\mathrm{supp}(\Phi)} (p(u^o)\Phi(x)+\lambda_3 Nu^o)(u-u^o)\, d\Gamma dt \geqslant 0 \quad \text{for all } u \in \mathcal{U}_\Sigma. \tag{41}$$

In Problems 1 and 2, the cost functionals involve only deviations of the solutions from their desired values averaged over the interior of the spatial domain Ω. Similar results can be obtained for optimal control problems in which the spatial averaging in the cost functional is taken over both Ω and its boundary Γ[6].

III. STABILITY

In applications, it is of interest to establish stability conditions for systems with specified forms of feedback controls. Here, we consider system (1) with f=0 and with a Dirichlet boundary condition of the form:

$$y(x,t)=\Phi(x)\{y(w(x),t-\tau)+u(x,t)\}, \qquad (x,t) \in \Gamma\times[0,\infty[, \tag{42}$$

where Φ and w are as in (3). In view of (38), we consider a feedback control which is a linear function of the state σ_t (defined by (25)) and has a representation of the form:

$$u(x,t)=\int_\Omega F_0(x,x',t)y(x',t)\,dx' + \int_{t-\tau}^t \int_{\Gamma^2} F_1(x,x',t,t')\phi_t(x',t')\,d\Gamma^2 dt',$$
$$(x,t) \in \Gamma\times[0,\infty[. \tag{43}$$

We shall establish a sufficient condition for the boundedness of solutions of system (1) with boundary condition (43) and feedback control (43).

Let Q_{t_1} denote the cylinder set $\Omega\times]0,t_1[$ for $0<t_1<\infty$. Let $y_o \in C^0(\bar{\Omega})$ and y be a continuous function in $\bar{Q}_{t_1}$ satisfying (1) in $Q_{t_1} \cup S_{t_1}$ where $S_t=\Omega\times\{t\}$. Then the weak maximum principle [5] asserts that the maximum of $|y|$ in $\bar{Q}_{t_1}$ is attained on the part $\bar{S}_0 \cup (\Gamma\times]0,t_1])$ of the boundary of Q_{t_1}, i.e.

$$\max_{Q_{t_1}} |y(x,t)| \leqslant \max\{\max_{\bar{\Omega}} |y_o(x)|, \sup_{\Gamma\times]0,t_1]} |y(x,t)|\}. \tag{44}$$

We shall show by contradiction that if

$$\eta(x,t)=|\Phi(x)| \left\{1 + \int_\Omega |F_0(x,x',t)|\,dx' + \int_{t-\tau}^t \int_{\Gamma^2} |F_1(x,x',t,t')|\,d\Gamma^2 dt'\right\} < 1 \tag{45}$$

for all $(x,t)\in\Gamma\times[0,\infty[$, then the maximum of $|y|$ is attained on $\bar{S}_0$ or

$$\max_{x\in\bar{\Omega}}|y(x,t)| \leqslant \max_{\bar{\Omega}}|y_o(x)| \qquad \text{for all } t\geqslant 0. \tag{46}$$

First, assume that there exists a point $(x^*,t^*)\in\Gamma\times]\tau,t_1]$ such that

$$|y(x^*,t^*)| = \max_{\bar{Q}_{t_1}} |y(x,t)|. \tag{47}$$

Then, in view of (42) and (43), we have

$$|y(x^*,t^*)|\leqslant|\Phi(x^*)|\Bigg\{|y(w(x^*),t^*-\tau)|+\int_\Omega |F_0(x^*,x,t^*)||y(x,t^*)|dx$$
$$+\int_{t^*-\tau}^{t^*}\int_{\Gamma^2}|F_1(x^*,x,t^*,t)||y(x,t)|dxdt\Bigg\} \leqslant \eta(x^*,t^*)|y(x^*,t^*)|, \tag{48}$$

where η is defined in (45). It is evident that if (45) is satisfied, then (48) leads to a contradiction. Consequently, such a point (x^*,t^*) does not exist under condition (45).

Now, assume that a point $(x^*,t^*)\in\Gamma\times[0,\tau]$ exists such that (47) holds. Then, from (42) and (43), we have

$$|y(x^*,t^*)|\leqslant|\Phi(x^*)|\Bigg\{|\phi_0(w(x^*),t^*-\tau)|$$
$$+\left(\int_\Omega |F_0(x^*,x,t^*)|dx + \int_{t^*-\tau}^{t^*}\int_{\Gamma^2}|F_1(x^*,x,t^*,t)|dxdt\right)|y(x^*,t^*)|\Bigg\}. \tag{49}$$

If we impose the condition that ϕ_0 is continuous on Σ_0 and satisfies

$$\sup_{\Sigma_0} |\phi_0(x,t)| \leqslant \max_{\bar{\Omega}} |y_o(x)|, \tag{50}$$

then

$$|\phi_0(w(x^*),t^*-\tau)| \leqslant \max_{\bar{\Omega}} |y_o(x)| \leqslant |y(x^*,t^*)| \tag{51}$$

and

$$|y(x^*,t^*)| \leqslant \eta(x^*,t^*)|y(x^*,t^*)|. \tag{52}$$

Again, under condition (45),(52) leads to a contradiction. Thus, such (x^*,t^*) does not exist. Finally, we note that the foregoing results remain valid for arbitrarily large t_1. Consequently, (46) holds. Thus, we have established:

Theorem 3: Let y be a classical solution of (1) with f=0 satisfying boundary condition (42),(43) and initial data (4). Let $y_o\in C^0(\bar{\Omega})$ and $\phi_0\in C^0(\Sigma_0)$ such that (50) is satisfied. Then, under condition (45), $\max_{x\in\bar{\Omega}}|y(x,t)| \leqslant \max_{\bar{\Omega}}|y_o(x)|$ for all $t\geqslant 0$.

The conditions in the above theorem represent restrictions on the initial data, the parameter in the boundary conditions and the kernels of the feedback control operators. Physically speaking, the result simply states that if the feedback gain and the maximum magnitude of the past boundary data are sufficiently small, then the solution will not grow with time.

IV. CONCLUDING REMARKS

In this paper, only parabolic systems with the simplest forms of boundary conditions involving time-delays have been considered. One may consider optimal control and stability problems for parabolic system (1) with a variety of more complex boundary conditions involving time-delays. A few examples are given below:

1. Boundary Condition Involving Multiple Time-Delays: This a generalization of the Neumann boundary condition (2):

$$\frac{\partial y}{\partial \upsilon}(x,t) + \gamma(x,t)y(x,t) = \Phi(x)\left\{\sum_{m=0}^{M} b_m(x,t)y(w(x),t-\tau_m) + \sum_{m=1}^{M} c_m(x,t)\frac{\partial y}{\partial \upsilon}(w(x),t-\tau_m)+u(x,t)\right\} \quad \text{on } \Sigma, \tag{53}$$

where $0 \leq \tau_0 < \tau_1 < \ldots < \tau_M$; ,$b_m$ and c_m are specified coefficients. If $\gamma, c_m, m=1,\ldots,M$ are identically zero on Γ, the results of this paper can be extended to this case without difficulty.

2. Boundary Condition with Indirect Control: Here, the boundary condition is identical to (2) and (3) except that u is generated indirectly by

$$u(x,t) = g(x)^T z(t), \qquad (x,t) \in \Sigma, \tag{54}$$

where $(\cdot)^T$ denotes transposition; g is a given mapping from Γ into R^r; $z(t) \in R^r$ is the solution of the following system of linear ordinary differential-difference equations:

$$\frac{dz(t)}{dt} = \sum_{m=0}^{M} F_m(t)z(t-\tau_m) + G(t)v(t) \tag{55}$$

with given initial data

$$z(t') = \omega(t') \in R^r, \qquad t' \in [-\tau_m, 0], \tag{56}$$

where $0 \leq \tau_0 < \tau_1 < \ldots < \tau_M$; $F_m(t)$ and $G(t)$ are given matrices. The control $v(\cdot) \in \mathcal{V}$, a specified closed convex subset of $L^2(I;R^s)$. One may consider the optimal control problem involving the minimization of a convex cost functional of the solution y and the control v. Also, instead of (55), one may replace it by a functional differential equation. Finally, one may consider similar problems for general parabolic and hyperbolic systems with boundary conditions involving time-delays. These problems will be discussed elsewhere.

ACKNOWLEDGEMENT: This work was supported by U.S.Air Force-Office of Scientific Research, Grant No. AFOSR-72-2303.

REFERENCES:

[1] J.L.Lions, Contrôle Optimal de Systèmes Gouvernés par des Équations aux Dérivées Partielles, Dunod,1968.

[2] J.L.Lions and E.Magenes, Non-homogeneous Boundary Value Problems and Applications, Vol.1, Springer-Verlag, N.Y. 1972 (Translated by P.Kenneth).

[3] J.L.Lions and E.Magenes, Non-homogeneous Boundary Value Problems and Applications, Vol.2, Springer-Verlag, N.Y. 1972 (Translated by P.Kenneth).

[4] L.Schwartz, "Theorie des Noyaux", Proc. Int. Congress of Mathematicians, Vol.1, 1950, pp.220-230.

[5] O.A.Ladyženskaja,V.A.Solonnikov and N.N.Ural'ceva,Linear and Quasilinear Equations of Parabolic Type, Translations of Math. Monographs No.23, Am.Math.Soc.R.I.1968.

[6] P.K.C.Wang, "Optimal Control of Parabolic Systems with Boundary Conditions Involving Time-Delays", Univ. of Calif. School of Engineering and Applied Science Rpt. No. UCLA-ENG-7346, June,1973.

CHARACTERIZATION OF CONES OF FUNCTIONS ISOMORPHIC TO CONES OF CONVEX FUNCTIONS

Jean-Pierre AUBIN
University of Paris-9 Dauphine - Paris

INTRODUCTION

The aim of this paper is the characterization of cones of functions $\Delta(U)$ defined on a topological space U isomorphic with the cone $\Gamma(S)$ of lower semi-continuous convex functions defined on a convex subset S of a locally convex vector space F in the following sense :

There exists a continuous map π *from* **U** *onto* S *such that any function* $\varphi \in \Delta(U)$ can be written in a unique way in the following form

$$\varphi(u) = g(\pi u) \qquad \text{where } g \in \Gamma(S)$$

The existence of such an isomorphism implies that all the properties of lower semi-continuous convex functions hold for the functions φ of $\Delta(U)$. Before stating the theorem characterizing such cones of functions, we shall describe two examples.

This paper summarizes the report [1].

1. EXAMPLE : *Functions defined on a family of convex compact subsets.*

1\. 1 Définitions

Let

(1\. 1) X and X' be two paired vector spaces supplied with their weak topologies

and let us denote by

(1\. 2) $\mathcal{A}$ the set of convex compact subsets of X

If $L = \{p_1, \ldots, p_n\}$ ranges over the family of finite subsets of X' and $\mathbb{P}_L(x) = \max_{i=1,\dots n} |\langle p_i, x\rangle|$ denotes the semi-norms of the weak topology of X, we shall supply A with the topology defined by the semi distances

(1. 3) $$\delta_L(A, B) = \max\left\{ \sup_{x\in A} \inf_{y\in B} \mathbb{P}_L(x - y), \sup_{y\in B} \inf_{x\in A} \mathbb{P}_L(x - y)\right\}$$

when L ranges over the finite subsets of X'.

<u>Definition</u> 1. 1

We shall say that $A_o \subset A$ is "convex" if

(1. 4) $$\forall\ A, B \in A_o,\ \forall\ \lambda \in [0, 1],\quad \lambda A + (1 - \lambda)B \in A_o$$

and that a function φ from A_o into $\mathbb{R}$ is "convex" if

(1. 5) $$\forall\ A, B \in A_o,\ \forall \lambda \in [0, 1],\ \varphi(\lambda A + (1-)B) \leqslant \lambda\varphi(A) + (1-\lambda)\varphi(B).$$

We shall denote by $\Delta(A_o)$ the cone of lower semi-continuous "convex" functions defined on A_o

<u>Remark</u> 1. 1

. If $\varphi : X \longmapsto \mathbb{R}$ is a convex function, then

(1. 6) $$A \in A \longmapsto \tilde{\varphi}(A) = \inf_{x\in A} \varphi(x) \text{ is "convex" on } A.$$

In particular, the support functions

(1. 7) $$A \in A \longmapsto \sigma(A, p) = \sup_{x\in A} \langle p, x\rangle \quad \text{where } p \in X'$$

are "affine".

We can prove the following result.

<u>Proposition</u> 1. 1

Any function $\varphi \in \Delta(A_o)$ can be written in a unique way

(1. 8) $$\varphi(A) = g\ (\sigma(A))$$

where

(1. 9)
$$\begin{cases} \text{i)} & \sigma(A) : \{ p \longmapsto \sigma(A, p) \} \in \mathbb{R}^{X'} \\ \text{ii)} & g \text{ is a lower semi-continuous function on } \mathbb{R}^{X'} \end{cases}$$

1. 2 Some applications

Let us introduce

(1. 10)
i) a function $\varphi \in \Delta(A_o)$

ii) a lower semi-continuous convex function $g \in \Gamma(S)$ where S is a convex subset of a locally convex vector space Y

iii) a continuous linear map $L \in L(X, Y)$

and the "conjugate functions" defined by

(1. 11)
$$\begin{cases} \text{i)} & \varphi^*(p) = \sup_{A \in \mathcal{A}_o} [\sigma(A, p) - \varphi(A)] \quad \text{where } p \in X' \\ \text{ii)} & g\ (q) = \sup_{y \in S} [\langle q, y \rangle - g(y)] \quad \text{where } q \in Y' \end{cases}$$

Proposition 1. 2

If $\{ a_o \} \in A_o$ *is such that* g *is continuous at* $L a_o$ *, then there exists* $y \in Y'$ *such that*

(1. 12)
$$\inf_{A \in A_o} [\varphi(A) + \inf_{x \in A} g(-L x)] =$$
$$= - [\varphi^*(L' \bar{q}) + g\ (\bar{q})]$$
$$= - \min_{q \in Y'} [\varphi^*(L' q) + g^*(q)]$$

We can deduce from this proposition the following "duality result" for minimization problems

Corollary 1. 1

Let $P \subset Y$ *be a closed convex cone with non-empty interior* $\overset{\circ}{P}$ (for the Mackey topology $\tau(Y, Y')$)

If $u \in Y$, let us introduce the subset

(1. 13) $\mathcal{A}_u = \{A \in \mathcal{A}_\circ$ such that there exists $x \in A$ with $L\,x - u \in P\}$

If there exists $\{a_\circ\} \in \mathcal{A}_\circ$ *such that*

(1. 14) $$L\,a_\circ - u \in \overset{\circ}{P}$$

then there exists $\bar{q} \in P^+$ *such that*

(1. 15) $$\inf_{A \in \mathcal{A}u} \varphi(A) = \inf_{A \in \mathcal{A}_\circ} [\varphi(A) - \sigma(A, L'\,\bar{q}) + \langle \bar{q}, u\rangle]$$

$$= - \varphi^*(L'\,\bar{q}) - \langle \bar{q}, u\rangle = - \min_{q \in P^+} [\varphi^*(L'\,q) - \langle q, u\rangle]$$

Brief economic interpretation

Let $X = \mathbb{R}^n$ denote a commodity space, $X' = \mathbb{R}^n$ the space of price systems, $P^+ = \mathbb{R}^n_+$ the cone of positive prices, $u \in \mathbb{R}^n$ a "demand" vector.

We denote by $\mathcal{A}_\circ$ a "convex" set of convex compact subsets containing o.

We interpret $A \in \mathcal{A}_\circ$ as a *firm* whose production set is $A - \mathbb{R}^n_+$ and whose "maximum profit function" is

$$\sigma(A, p) = \sup_{x \in A} \langle p, x\rangle \quad \text{defined on } P^+ = \mathbb{R}^n_+ .$$

We take $Y = X$ and L to be the identity mapping.

Then the set $\mathcal{A}_u$ is the set of firms which can satisfy the "demand" vector u.

If $\varphi : \mathcal{A}_\circ \longmapsto \mathbb{R}$ is a cost function defined on the set of firms, then the minimization of φ on $\mathcal{A}_u$ amounts to the minimization on $\mathcal{A}_\circ$ of the perturbed cost function

$$A \longmapsto \varphi(A) - \sigma(A, \bar{q}) + \langle \bar{q}, u\rangle$$

2. EXAMPLE : *Functions defined on σ algebras*

2. 1. Définitions

Let

(1. 1) $\begin{cases} \text{i) } \mathcal{A} \text{ be a } \sigma\text{-algebra defined on a set } \Omega \\ \text{ii) } m : \mathcal{A} \longmapsto \mathbb{R}^m \text{ be an atomless bounded vector-valued measure.} \end{cases}$

We shall set $A \approx B$ if and only if $m(A) = m(B)$ and $\mathcal{A}_\circ$ the factor set $\mathcal{A}/_{\approx}$, supplied with the distance

(2. 2) $$\delta(A, B) = \| m(A) - m(B) \|$$

By the Lyapounov theorem, $m(\mathcal{A})$ is convex.

Therefore,

(2. 3) $\begin{cases} \forall A, B \in \mathcal{A}_\circ , \forall \lambda \in [0, 1] \\ \gamma(\lambda, A, B) = m^{-1}(\lambda m(A) + (1-\lambda) m(B)) \text{ belongs to } \mathcal{A}_\circ . \end{cases}$

<u>Definition</u> 2. 1

We shall denote by $\Delta(\mathcal{A}_\circ)$ the cone of lower semi-continuous functions defined on $\mathcal{A}_\circ$ satisfying

$\forall$ $A, B \in \mathcal{A}_\circ$, $\forall \lambda \in [0, 1]$

(2. 4) $$\varphi(\gamma(\lambda, A, B)) \leq \lambda \varphi(A) + (1-\lambda)\varphi(B)$$

<u>Remarks</u> 2. 1

. If $g : \mathbb{R}^m \longmapsto \mathbb{R}$ is a lower semi-continuous convex function defined on $m(\mathcal{A})$, then the function φ defined by

(2. 5) $$\varphi(A) = g(m(A))$$

belongs to $\Delta(\mathcal{A}_\circ)$. Conversely, any function $\varphi \in \Delta(\mathcal{A}_\circ)$ can be written in a unique way in the form (2. 5) where g is convex.

. In particular, the functions $A \longmapsto \langle p, m(A) \rangle$ and $A \longmapsto -\langle p, m(A) \rangle$ (where $p \in \mathbb{R}^m$) belong to $\Delta(\mathcal{A}_\circ)$.

. Let $\Theta = \{ \theta \in L^{\infty}(\Omega, A, |m|)$ such that $\theta(\omega) \underset{p.p}{\in} [0, 1] \}$
and $\varphi \in L^1(\Omega, A, |m|)$.

Then the function $\tilde{\varphi} : A_o \longmapsto \mathbb{R}$ defined by

(2. 6) $$\tilde{\varphi}(A) = \inf \{ \int \varphi(\omega) \theta(\omega) d|m|(\omega) \mid \theta \in \Theta \text{ and } \int \theta(\omega) d|m|(\omega) = m(A) \}$$

belongs to $\Delta(A_o)$.

2. 2. Some Applications

(2. 7) Let us introduce

i) a function $\varphi \in \Delta(A_o)$

ii) a lower semi-continuous convex function $g \in \Gamma(S)$ where S is a convex subset of a locally convex vector space Y

iii) a continuous linear map $L \in L(\mathbb{R}^n, Y)$

and the "conjugate functions" defined by

$$\begin{cases} \text{i)} & \varphi^*(p) = \sup_{A \in A_o} [\langle p, m(A) \rangle - \varphi(A)] \quad \text{where } p \in \mathbb{R}^m \\ \text{ii)} & g^*(q) = \sup_{y \in S} [\langle q, y \rangle - g(y)] \quad \text{where } q \in Y' \end{cases}$$

Proposition 2. 1

If there exists $A_o \in A_o$ *such that* g *is continuous at* $L\, m(A_o)$, *then there exists* $\bar{q} \in Y'$ *such that*

(2. 9) $$\begin{cases} \inf_{A \in A_o} [\varphi(A) + g(-L\, m(A))] \\ = - [\varphi^*(L' \bar{q}) + g^*(\bar{q})] \\ = - \min_{q \in Y'} [\varphi^*(L' q) + g^*(q)] \end{cases}$$

We can deduce from this proposition the following "duality result" for minimization problems.

Corollary 2. 1

Let $P \subset Y$ *be a closed convex cone with a non-empty interior* $\mathring{P}$ *(for the Mackey topology* $\tau(Y, Y')$*)*

If $u \in Y$, *let us introduce the subset*

(2. 10) $\mathcal{A}_u = \{A \in \mathcal{A}_0$ *such that* $L\, m(A) - u \in P\}$

If there exists $A_0 \in \mathcal{A}_0$ *such that*

(2. 11) $L\, m(A_0) - u \in \mathring{P}$

then there exists $\bar{q} \in P^+$ *such that*

$$(2.\ 12)\quad \begin{cases} \inf_{A \in \mathcal{A}_u} \varphi(A) = \inf_{A \in \mathcal{A}_0} [\varphi(A) - \langle \bar{q}, L\, m(A)\rangle + \langle \bar{q}, u\rangle] \\ = - \varphi^*(L'\, \bar{q}) + \langle \bar{q}, u\rangle = - \min_{q \in P^+} [\varphi^*(L'\, q) - \langle q, u\rangle] \end{cases}$$

Brief economic interpretation

For instance, we can regard $\mathcal{A}$ as a set of "decisions" defined on the set Ω of "elementary decisions" and the measure m as the map associating with any decision $A \in \mathcal{A}_0$ an "act" $m(A) \in \mathbb{R}^m$ resulting from the decision A. We take $Y = \mathbb{R}^m$ and we interpret $u \in \mathbb{R}^m$ as an objective. Then the set $\mathcal{A}_u$ is the set of decisions, whose acts are greater than the objective u.

If $\varphi : \mathcal{A}_0 \longmapsto \mathbb{R}$ is a cost function defined on the set $\mathcal{A}_0$ of decisions, then the minimization of φ on $\mathcal{A}_u$ amounts to the minimization on $\mathcal{A}_0$ of the perturbed cost function

$A \longmapsto \varphi(A) - \langle \bar{q}, m(A)\rangle + \langle \bar{q}, u\rangle$

3. STATEMENT OF THE THEOREM OF ISOMORPHISM

The two above cones of functions are examples of cones of "γ-vex" functions.

3. 1. The cone of γ-vex functions

Let

$$(3.1)\quad \begin{cases} \text{i) } U \text{ be a set} \\ \text{ii) } \mathcal{S}'_1(U) \text{ be the set of probability discrete measures} \\ \quad \mu = \sum_{\text{finite}} \alpha_i\, \delta(u_i) \text{ on } U \end{cases}$$

(where $\delta(u_i)$ is the Dirac measure at u_i, $\alpha_i \geq 0$, $\sum_{\text{finite}} \alpha_i = 1$)

Let us introduce

$$(3.2)\quad \begin{cases} \gamma \text{ be a correspondance with non-empty values from} \\ \quad \mathcal{S}'_1(U) \text{ into } U \end{cases}$$

<u>Definition</u> 3. 1

We shall say that a function $\varphi : U \mapsto \mathbb{R}$ *is "*γ*-vex" if*

(3. 3) $\quad \varphi(\gamma\mu) \leq \sum \alpha_i\, \varphi(u_i)$ for any $\mu = \sum \alpha_i\, \delta(u_i) \in \mathcal{S}'_1(U)$

and "γ-affine"

(3. 4) $\quad \varphi(\gamma\mu) = \sum \alpha_i\, \varphi(u_i)$ for any $\mu = \sum \alpha_i\, \delta(u_i) \in \mathcal{S}'_1(U)$

<u>Remark</u> 3. 1

. If U is a convex subset of vector space X, then usual convex functions are γ-vex functions where γ is the map defined by

(3. 5) $\quad \gamma\mu = \sum \alpha_i\, u_i$ whenever $\mu = \sum \alpha_i\, \delta(u_i) \in \mathcal{S}'_1(U)$.

. If A is the family of convex compact subsets of X, then the "convex" functions $\varphi \in \Delta(A)$ are γ-vex functions where γ is the map defined by

(3. 6) $\quad \gamma\mu = \sum \alpha_i\, A_i$ whenever $\mu = \sum \alpha_i\, \delta(A_i) \in \mathcal{S}'_1(A)$

. If A is a σ-algebra and $m : A \longmapsto \mathbb{R}^m$ is a bounded atomless vector-valued measure, then the functions $\varphi \in \Delta(A_\circ)$ are γ-vex functions where γ is the map defined by

(3. 7) $\gamma\mu = m^{-1}(\Sigma\ \alpha_i\ m(A_i))$ whenever $\mu = \Sigma\ \alpha_i\ \delta(A_i) \in \mathcal{S}'_1(A_\circ)$

3. 2. The cone of lower semi-continuous γ-vex functions

Let us denote by

(3. 8) G the vector space of γ-affine functions defined on U

We shall supply U with the topology defined by the semi-distances

(3. 9) $\delta_L(u, v) = \max\limits_{i=1,\dots,n} | f_i(u) - f_i(v) |$

when $L = \{f_1, \dots, f_n\}$ ranges over the finite subsets of G.

We denote by

(3. 10) $\Delta_\gamma(U)$ the cone of lower semi-continuous γ-vex functions

Theorem 3. 1

There exist

(3. 11) i) a locally convex vector space F

ii) a continuous map π from U into F

such that

(3. 12) $S = \pi(U)$ is a convex subset of F

and such that the map

(3. 13) $g \in \Gamma(S) \longmapsto \varphi = g \circ \pi \in \Delta_\gamma(U)$

is an isomorphism from the cone $\Gamma(S)$ *of lower semi-continuous convex functions on* S *onto the cone* $\Delta\ (U)$ *of lower semi-continuous* γ*-vex functions on* U.

The proof of this theorem, based on the the minimax properties of the γ-vex functions, together with other results and the references, can be found in [1]

[1] . Jean-Pierre AUBIN

Existence of saddle points for classes of convex and non-convex functions.

Mathematics Research Center Technical Summary Report # 1289 University of Wisconsin (1972).

NECESSARY CONDITIONS AND SUFFICIENT CONDITIONS FOR PARETO OPTIMALITY IN A MULTICRITERION PERTURBED SYSTEM*

JEAN-LOUIS GOFFIN ALAIN HAURIE

Ecole des Hautes Etudes Commerciales - Montréal

1. INTRODUCTION

Vector valued optimization problems have recently attracted a renewed attention. Decision making in engineering and economic systems is confronted by a multiplicity of goals, and it seems necessary to take it into account directly, rather than finding an a priori mix of goals.

The concept of Pareto optimality has been extensively studied in the realm of economic theory and more recently in the field of optimal control theory ([1] - [6]).

In addition to the complexity due to multiple goals, the decision-maker is often faced with perturbations or uncertainties which influence the consequences of his actions. These disturbances, if their probability distribution cannot be assessed can be modelled as set-constrained perturbations following the lines given in Refs [7] - [9]. This description is quite fitting if the perturbations are caused by the existence of other decision makers - or players - which can affect the consequence of the decision of a coalition of players[10].

Formally the system is described by :

(i) A decision set X

(ii) A perturbation set Y

(iii) A vector valued cost criterion

$$\psi : X \times Y \to R^p \ , \ (x, y) \mapsto \psi(x, y) \triangleq (\psi_j(x, y))_{j=1,\dots,p}$$

In section 2, the optimality criterion is precisely defined as a generalization of Pareto-optimality.

In section 3, the scalarization process is studied for this class of system.

In section 4, a Lagrange multiplier theorem is proved.

---

*This work was supported by the Canada Council under Grants S72-0513 and S73-0268.

Notation

$$\psi(.) \stackrel{\Delta}{=} (\psi_j(.))_{j=1,\dots,p}$$

$$\forall x, y \in R^p \quad x \leqq y \quad \text{if} \quad x_j \le y_j \quad j = 1, \dots, p$$

$$R^p_+ \stackrel{\Delta}{=} \{z \in R^p : 0 \leqq z\}$$

$\psi'_x(x, y)$ will denote the gradient w.r.t. x of the fonction $\psi(., y)$ at point x.

$\varphi'_j(x; h)$ will denote the directional derivative in the direction h of the function $\varphi_j(x)$.

2. DEFINITION OF OPTIMALITY

To every decision x in X is associated a set of possible outcomes :

$$\psi(x, Y) \stackrel{\Delta}{=} \{\psi(x, y) : y \in Y\}$$

Any preorder given on $\{\psi(x, Y) : x \in X\}$ will induce a preorder on X. Given a preorder $(X, \prec)$ a minimal element will be called a Pareto-minimum in X.

In this paper the following preorder will be used :

$$x \prec x' \quad \text{if} \quad \text{Sup}\ \psi(x, Y) \leqq \text{Sup}\ \psi(x', Y)$$

where Sup $\psi(x, Y)$ is the L.U.B. of $\psi(x, Y)$ in $(R^p, \le)$. Note that :

$$\text{Sup}\ \psi(x, Y) = (\text{Sup}\ \psi_j(x, Y))_{j=1,\dots,p}$$

Thus the following definition of Pareto optimality is adopted.

Definition 2.1 : x* in X is Pareto-optimal if, for all x in X :

$$\text{Sup}\ \{\psi_j(x, y) : y \in Y\} \le \text{Sup}\{\psi_j(x^*, y) : y \in Y\} \quad \forall j \in \{1,\dots,p\}$$

implies that :

$$\text{Sup}\ \{\psi_j(x, y) : y \in Y\} = \text{Sup}\{\psi_j(x^*, y) : y \in Y\} \quad \forall j \in \{1,\dots,p\}$$

We can define the auxiliary cost function $\varphi : X \to R^p$ by :

$$\varphi(x) \stackrel{\Delta}{=} \text{Sup}\ \psi(x, Y).$$

Definition 2.1 is then equivalent to :

x* in X is Pareto optimal if for all x in X :

$$\varphi(x) \leqq \varphi(x^*) \Rightarrow \varphi(x) = \varphi(x^*).$$

In order to characterize Pareto optimal elements, two approaches will be used : The scalarization process [1] - [6] and a direct method based on an extension of the theorem of Kuhn and Tucker [11].

3. THE SCALARIZATION PROCESS

The main results in optimization with a vector-valued criterion revolve around the process of scalarization [1] - [6]. In the presence of perturbations the scalarization cannot, in general, be performed directly on the cost functions as for $\alpha_j > 0$, $j = 1,\dots, p$ the condition : $\hat{x} \in X$ and :

$$\mathrm{Sup}\ \{\sum_{j=1}^{p} \alpha_j\ \psi_j(\hat{x}, y) : y \in Y\} \leq \mathrm{Sup}\{\sum_{j=1}^{p} \alpha_j\ \psi_j(x, y) : y \in Y\}$$

$\forall\ x \in X$

does not imply that $\hat{x}$ is Pareto-optimal.(†)

The scalarization must be applied to the auxiliary cost function . First we state the following well known result.

Theorem 3.1 : Let $\alpha_j > 0$, $j = 1,\dots, p$ and x^* in X be such that :

$$\forall\ x \in X \qquad \sum_{j=1}^{p} \alpha_j\ \varphi_j(x^*) \leq \sum_{j=1}^{p} \alpha_j\ \varphi_j(x)$$

then x^* is Pareto optimal.

Corollary 3.1 : Let $\alpha_j > 0$, $j = 1,\dots, p$ and x^* in X be such that :

$$\forall\ x \in X \quad \sup_{\substack{y_1 \in Y \\ y_2 \in Y \\ \vdots \\ y_p \in Y}} \sum_{j=1}^{p} \alpha_j\ \psi_j(x^*, y_j) \leq$$

$$\sup_{\substack{y_1 \in Y \\ y_2 \in Y \\ \vdots \\ y_p\ Y}} \sum_{j=1}^{p} \alpha_j\ \psi_j(x, y_j)$$

(†) Consider as an example : $X \triangleq R$, $Y \triangleq \{-1, 1\}, \psi(x, y) \triangleq (x^2+y(x-1), x^2 - y(x-1))$. Thus $\varphi(x) = (x^2 + |x - 1|, x^2 + |x - 1|)$ and the only Pareto optimal element is $x^* = \frac{1}{2}$. But $\psi_1(x, y)+\psi_2(x, y) = 2x^2$ which attains its unique minimum at $\hat{x} = 0$.

then x* is Pareto-optimal.

Proof :

$$\sum_{j=1}^{p} \alpha_j \sup_{y \in Y} \psi_j(x, y) = \sum_{j=1}^{p} \sup_{y_j \in Y} \alpha_j \psi_j(x, y_j)$$

$$= \sup_{y_1 \in Y, \dots, y_j \in Y} \sum_{j=1}^{p} \alpha_j \psi_j(x, y_j) \qquad \blacksquare$$

For a necessary scalarization condition to hold some convexity assumptions are needed. The following lemma indicates what kind of convexity must be met.

Lemma 3.1 : x^* in X is Pareto-optimal iff the following holds :

$$(\varphi(x^*) - R_+^p) \cap (\varphi(X) + R_+^p) = \{\varphi(x^*)\} . \tag{3.1}$$

Proof : Clearly x^* is Pareto-optimal iff

$$(\varphi(x^*) - R_+^p) \cap \varphi(X) = \{\varphi(x^*)\} \tag{3.2}$$

Obviously (3.1) implies (3.2). Conversely let ω be an element of the L.H.S. of (3.1) then, for some x in X, ω' and ω'' in R^p one has :

$$\omega = \varphi(x^*) - \omega' = \varphi(x) + \omega''$$

thus $\varphi(x) = \varphi(x^*) - (\omega' + \omega'') \in \varphi(x^*) - R_+^p$ and the L.H.S. of (3.2) contains $\varphi(x)$.

Thus (3.2) implies $\varphi(x^*) = \varphi(x)$, $\omega' = \omega'' = 0$ and finally $\omega = \varphi(x^*)$, that is (3.1). $\blacksquare$

Remark 3.1 : If X is a convex subset of a linear space and each φ_j is convex on X, then $\varphi(X) + R_+^p$ is convex, (but φ_j quasi-convex is not sufficient)

Remark 3.2 : $\varphi(X) + R_+^p$ is convex iff :

$$\forall\, x, x' \in X, \ \forall \lambda \ [0, 1] \ \exists\, z \in X \quad \text{s.t.}$$

$$\varphi(z) \leqq \lambda \varphi(x) + (1 - \lambda) \varphi(x')$$

i.e. there exists a mapping $T : X \times X \times [0, 1] \to X$ such that :

$$\forall x, x' \in X, \ \forall \lambda \in [0, 1] \ \varphi(T(x, x', \lambda)) \leqq \lambda \varphi(x) + (1 - \lambda) \varphi(x').$$

Theorem 3.2 : If $\varphi(x) + R_+^p$ is convex and x^* is Pareto-optimal then there exist $\alpha_j \geq 0$ for all j and $\alpha_\ell > 0$ for some ℓ such that :

$$\forall\, x \in X \quad \sum_{j=1}^{p} \alpha_j \varphi_j(x^*) \leq \sum_{j=1}^{p} \alpha_j \varphi_j(x) .$$

Proof : Direct application of lemma 3.1 and the separation theorem. ■

Assumption 3.1 :

X is a convex subset of a Banach space U

Y is a compact space

$\psi : U \times Y \to R^p$ is continuous and convex in $x \in U$.

$x \to \psi'_x(x, y)$ is an equicontinuous family.

Then, we can prove the following theorem using results of Danskin [12] Dem'Yanov [13], and Lemaire [14].

Theorem 3.3 : Under assumption 3.1 :

(i) If $x^* \in X$ is Pareto-optimal then there exist $\alpha_j \geq 0$ for all j, $\alpha_\ell > 0$ for some ℓ such that :

$$\underset{x \in X}{\text{Min}} \sum_{j=1}^{p} \alpha_j \underset{y \in Y_j(x^*)}{\text{Max}} < \psi'_{jx} (x^*, y), x - x^* > = 0 \tag{3.3}$$

where $Y_j(x) \triangleq \{\bar{y} \in Y : \psi_j(x, \bar{y}) = \underset{y \in Y}{\text{Max}}\ \psi_j(x, y)\}$

(ii) Conversely, if there exist $\alpha_j > 0$ for each j such that (3.3) holds, then x^* is Pareto-optimal.

Proof : For each j, φ_j is directionally differentiable with (see Refs [12] - [14]) :

$$\varphi'_j(x ; h) = \underset{y \in Y_j(x)}{\text{Max}} < \psi'_{jx} (x,y), h >$$

Since :

$$\left(\sum_{j=1}^{p} \alpha_j \varphi_j\right)' (x ; h) = \sum_{j=1}^{p} \alpha_j \varphi'_j(x ; h)$$

We can use theorem 2.4 of Dem'Yanov [13] and theorems 3.1 and 3.2 to get (i) and (ii). ■

It is clear that the scalarization approach is the simpler one, but to be applicable it requires convexity assumptions, even to get necessary conditions.

In the next section, we will develop necessary conditions for Pareto-optimality, which do not require convexity.

4. A LAGRANGE MULTIPLIER THEOREM

In this section, the results of Danskin [12] and Bram [15] will be used extensively.

Assumption 4.1 :

Y is a compact space

$\psi : R^n \times Y \to R^p$ is continuous and its derivative $\psi'_x(x, y)$ is continuous.

$X \subset R^n$ is defined by a set of constraints :

$$X = \{x \in R^n : g_i(x) \geq 0 \quad i = 1, \dots , m\}$$

$$g_i : R^n \to R \text{ is } C^1, i = 1, \dots , m.$$

Theorem 4.1 : Under Assumption 4.1, if x* in X is Pareto-optimal and if the Kuhn - Tucker constraint qualification is satisfied at x*, then there exist $\lambda_i \geq 0$ for $i = 1,\dots, m$ and $\alpha_j \geq 0$ for $j = 1,\dots, p$ $\alpha_\ell > 0$ for some ℓ, such that :

$$\sum_{j=1}^{p} \alpha_j \max_{y \in Y_j(x^*)} < \psi'_{jx}(x^*, y), h > \;\geq\; \sum_{i=1}^{m} \lambda_i < g'_i(x^*), h > \qquad (4.1)$$

and $\lambda_k > 0 \Rightarrow g_k(x^*) = 0.$ (4.2)

Furthermore, if the g_i's are concave and the ψ_j's are convex in x then (4.1) with $\lambda_i \geq 0$, $i = 1,\dots, m$ and $\alpha_j > 0$, $j = 1,\dots, p$ implies that x* is Pareto-optimal.

Proof : Let $x^* \in X$ be a Pareto-optimal point. This means that for every $x \in X$ | either $\exists\, j$ s.t. $\varphi_j(x) > \varphi_j(x^*)$
| or $\forall\, j \quad \varphi_j(x) = \varphi_j(x^*)$

this implies that :

$$\forall x \in X \quad \max_j \left[\varphi_j(x) - \varphi_j(x^*)\right] \geq 0 \qquad (4.3)$$

(Note that this is a necessary condition for Pareto-optimality, but it is a necessary and sufficient condition for weak Pareto-optimality).

Let Γ be the set of vectors admissible at x, that is the set of $h \in R^n$ such that there exist an arc issuing from x, and lying in X, with a tangent vector at x equal to h.

Thus (4.3) becomes :

$$\forall h \in \Gamma \quad \max_j \varphi_j'(x^*;\ h) \geq 0 \tag{4.4}$$

Under the constraint qualification condition, one has :

$$\Gamma = \{h \in R^n : \langle g_i'(x^*),\ h \rangle \geq 0, \quad i \in I\}$$

where $\quad I = \{i : g_i(x^*) = 0\}$.

Let $\quad W_j \triangleq \psi_{jx}'(x^*,\ Y_j(x^*))$, then (4.3) becomes :

$$\max_{j=1,\dots,p} \ \max_{W \in W_j} \langle w,\ h \rangle \geq 0 \qquad \forall h \in \Gamma$$

or

$$\max_{\substack{w \in \cup W_j \\ j=1,\dots,p}} \langle w,\ h \rangle \geq 0 \qquad \forall h \in \Gamma$$

and, if W is the convex hull of $\bigcup_{j=1,\dots,p} W_j$,

$$\max_{w \in W} \langle w,\ h \rangle \geq 0 \qquad \forall h \in \Gamma$$

Consider the cone $\Gamma^* \triangleq \{z \in R^n : \langle z,\ h \rangle \geq 0 \quad \forall h \in \Gamma\}$ then following Bram's proof [15] it can be shown that $\Gamma^* \cap W$ is not empty.

Let $\hat{w} \in \Gamma^* \cap W$, then there exist $\lambda_i \geq 0$, $i \in I$ such that :

$$\hat{w} = \sum_{i \in I} \lambda_i \ g_i'(x^*)$$

and there exist $w_j \in W_j$, $\alpha_j \geq 0$, $j = 1,\dots,p$, $\sum_{j=1}^{p} \alpha_j = 1$

such that :

$$\hat{w} = \sum_{j=1}^{p} \alpha_j \ w_j$$

therefore, setting $\lambda_i = 0$ if $i \notin I$ one has :

$$\forall h \in R^n \quad \sum_{j=1}^{p} \alpha_j \max_{w_j \in W_j} \langle w_j,\ h \rangle$$

$$\geq \sum_{j=1}^{m} \lambda_i \langle g_i(x^*),\ h \rangle$$

that is (4.1).

Let us remark that if each set W_j reduces to a single element, then each φ_j is differentiable at x* and (4.1) yields the generalized Kuhn - Tucker result :

$$\sum_{j=1}^{p} \alpha_j \psi'_{jx}(x^*, Y_j(x^*)) = \sum_{i=1}^{m} \lambda_i g'_i(x^*).$$

Finally, the proof of the sufficient condition is the same as in the classical case. ■

5. CONCLUSION

The notion of Pareto-optimality has been extended to perturbed systems. The scalarization process and the extended Lagrange multiplier rule which have been obtained can be used to characterize and compute optimal decisions.

A promising field of application of these results is "n-player" game theory without side payments [16] ; the reader could verify that the boundary of the Auman's characteristic function for a given coalition corresponds to the set of all Pareto-optimal outcomes in an adequatly defined perturbed system.

REFERENCES

[1] A.W. STARR et Y.C. HO :
Nonzero-Sum Differential Games, JOTA, 3 (1969), 184-206.

[2] ______________________
Further Properties of Nonzero-Sum Differential Games, JOTA (1969), 207-219.

[3] T.L. VINCENT & G. LEITMANN :
Control Space Properties of cooperative games, JOTA, 4 (1970), 91-113.

[4] A. BLAQUIERE :
Sur la géométrie des surfaces de Pareto d'un jeu différentiel à N joueurs, C.R. Acad. Sc. Paris Sér. A, 271 (1970), 744-747.

[5] A. BLAQUIERE, L. JURICEK & K.E. WIESE :
Sur la géométrie des surfaces de Pareto d'un jeu différentiel à N joueurs; théorème du maximum, C.R. Acad. Sc. Paris Sér. A, 271 (1970), 1030-1032.

[6] A. HAURIE :
Jeux quantitatifs à M joueurs, doctoral dissertation, Paris 1970.

[7] M.C. DELFOUR & S.K. MITTER :
Reachability of Perturbed Linear Systems and Min Sup Problems, SIAM J. On control, 7 (1969), 521-533

[8] D.P. BERTSEKAS & I.B. RHODES :
On the Minimax Reachability of Targets and Target Tubes, Automatica, 7 (1971), 233-247.

[9] J.D. GLOVER & F.C. SCHWEPPE :
Control of Linear Dynamic Systems with Set Constrained Disturbances, IEEE Trans. on Control, AC-16 (1971), 411-423.

[10] A. HAURIE :
On Pareto Optimal Decisions for a Coalition of a Subset of Players, IEEE Trans. on Automatic Control, avril 1973.

[11] H.W. KUHN & A.W. TUCKER :
Non-Linear Programming, 2nd Berkeley Symposium of Mathematical Statistics and Probability, Univ. Calif. Press, Berkeley 1951.

[12] J. DANSKIN :
On the Theory of Min-Max, J.SIAM Appl. Math., Vol. 14 (1966), pp. 641-664.

[13] V.F. DEM'YANOV & A.M. RUBINOV :
Minimization of functionals in normed spaces, SIAM J. Control, Vol. 6 (1968), pp. 73-88.

[14] B. LEMAIRE :
Problèmes min-max et applications au contrôle optimal de systèmes gouvernés par des équations aux dérivées partielles linéaires, Thèse de doctorat, faculté des sciences, Université de Paris, 1970.

[15] J. BRAM :
The Lagrange Multiplier Theorem for Max-Min with several Constraints, J. SIAM App. Math. Vol 14 (1966), pp 665-667.

[16] R.J. AUMANN :
A Survey of Cooperative Games without side Payments, in Essays in Mathematical Economics, ed. M. Shubik, Princeton 1969.

A UNIFIED THEORY OF DETERMINISTIC TWO-PLAYERS ZERO-SUM DIFFERENTIAL GAMES

Christian Marchal

Office National d'Etudes et de Recherches Aérospatiales (ONERA)

92320 - Châtillon (France)

Abstract

This paper is a shorter presentation of "Generalization of the optimality theory of Pontryagin to deterministic two-players zero-sum differential games" [MARCHAL, 1973] presented at the fifth IFIP conference on optimization techniques.

The very notions of zero-sum game and deterministic game are discussed in the first sections. The only interesting case is the case when there is "complete and infinitely rapid information". When the minimax assumption is not satisfied it is necessary to define 3 types of games according to ratios between time-constant of a chattering between two or several controls and delays necessary to measure adverse control and to react to that control ; it thus emphasizes the meaning of the "complete and infinitely rapid information" concept.

In the last sections the optimality theory of Pontryagin is generalized to deterministic two-players zero-sum differential games ; it leads to the notion of extremal pencil (or bundle) of trajectories.

When some canonicity conditions generalizing that of Pontryagin are satisfied the equations describing the extremal pencils are very simple but lead to many kinds of singularities already found empirically in some simple examples and called barrier, universal surfaces, dispersal surfaces, focal lines, equivocal lines etc...

Introduction

Many authors have tried to extend to differential game problems the beautiful Pontryagin's theory used in optimization problems, but there is so many singularities in differential game problems, even in the deterministic two-players zero-sum case, that a general expression is difficult to find and to express.

A new notion is used here : the notion of "extremal pencil (or bundle) of trajectories". This notion allows to present the generalization of Pontryagin's theory in a simple way.

1. Two-players zero-sum differential games

Usually two-players zero-sum differential games are presented as follow :

A) There is a parameter of description t that we shall call the time.

B) The system of interest used n other parameters $x_1, x_2, \dots x_n$ and we shall put :

$$\overrightarrow{X} = (x_1, x_2, \dots, x_n) = \text{state vector.} \tag{1}$$

We shall assume that I, the performance index of interest (called also cost function or pay off) is a only function of the final values $\overrightarrow{X_f}$, t_f ; if necessary, if for instance I is related to an integral taken along the described trajectory $\overrightarrow{X(t)}$, we must add into $\overrightarrow{X}$ a component related to that integral.

C) There is two players that we shall call the maximisor M and the minimisor m, each of them chooses a measurable control $\overrightarrow{M(t)}$ and $\overrightarrow{m(t)}$ (respectively in the control domains $\mathcal{D}_M(t)$ and $\mathcal{D}_m(t)$ Borelian functions of t) and the velocity vector $\overrightarrow{V} = d\overrightarrow{X}/dt$ is a given Borelian function of $\overrightarrow{X}$,t and the two controls :

$$\overrightarrow{V} = d\overrightarrow{X}/dt = \overrightarrow{V}(\overrightarrow{X}, \overrightarrow{M}, \overrightarrow{m}, t) \tag{2}$$

D) There is a "playing space" $\mathcal{E}$ subset of the R^{n+1} $\vec{X},t$ space and a "terminal surface" $\mathcal{C}$, or more generally a "terminal subset" $\mathcal{C}$ along the boundaries of $\mathcal{E}$. We shall assume that the set $\mathcal{E}$ is open.

E) The control equation (2) is defined everywhere in $\mathcal{E}$ and the performance index $I(X_f, t_f)$ is defined everywhere in $\mathcal{C}$; the game starts in at a given initial point $\vec{X}_o$, t_o; the maximisor tries to maximises I at the first arrival at $\mathcal{C}$ and the minimisor tries to minimizes it.

2. Zero-sum games

Since there is only one performance index $I(X_f, t_f)$ it seems that the above defined game is zero-sum, however it is possible that both players have interest to avoid the termination of the game (e.g. the cat, the mouse and the hole around a circular lake, when the cat blocks the hole) and in order to avoid difficulties of non zero-sum games the value of the performance index must also be defined in non-finite cases.

3. Deterministic cases of two-players zero-sum differential games

A first condition of determinism is that both players have complete informations on the control function $\vec{V}(\vec{X}, \vec{M}, \vec{m}, t)$, the control domains $\mathcal{D}_M(t)$ and $\mathcal{D}_m(t)$, the performance index $I(X_f, t_f)$, the playing space $\mathcal{E}$, the terminal subset $\mathcal{C}$ and the initial conditions $\vec{X}_o$, t_o.

It is possible to imagine some particular cases of deterministic games such as :

A) One of the two players, for instance M, has more or less complete informations on the present and past state vector $\vec{X}(t)$ and choose a pure strategy based on these various informations :

$$\vec{M} = \vec{M}\left(\vec{X}(\theta), \vec{M}(\theta), t\right) \quad ; \quad t_o \leq \theta < t \tag{3}$$

B) He must indicate his choice to the other player.

C) The second player choose then its own control function $\vec{m}(t)$.

Hence : When (3) is given the problem of m is an ordinary problem of minimization and then M chooses its strategy (3) in order that this minimum be as large as possible.

However, if the informations of the first player are incomplete, the conditions A and B are generally unrealistic : the choice of a good mixed strategy improves very often the possibilities of the first player and the real problems is thus not deterministic.

The only realistic and deterministic cases are the following :

A) Both players can measure the present value of $\vec{X}$ at an infinite rate ; we shall call T_M and T_m the infinitely small delays necessary to obtain this measure and to react to it.

B) In some cases the optimal control requires a chattering between two or several controls, we shall assume that these chatterings can be made at an infinite rate and we shall call τ_M and τ_m the corresponding infinitely small intervals of time.

C) There is then 3 deterministic cases :

C1) Case when $\tau_m + T_m \ll \tau_M$

It is the maximin case or Mm-case, everything happens as if the minimisor m could choose its control after the maximisor M at any instant.

C2) $\tau_M + T_M \ll \tau_m$. It is the minimax case or mM-case symmetrical to the previous one.

C3) $\tau_M \ll \tau_m + T_m$ and $\tau_m \ll \tau_M + T_M$

It is the neutral case or N-case, both players choose their own control independently of the opponent choice.

The determinism of that last case requires some more conditions (see in chapter 5.1 the condition of equality of H_1 and H_2).

Of course the assumption of determinism implies that both players know if the game is a mM, Mm or N-game.

A simple example of these 3 tapes of game is given in the following example :

Initial conditions : $x_o = t_o = 0$; terminal subset $t_f = 1$; performance index $I = x_f$; control function $dx/dt = 2M^2 + Mm - 2m^2$; control domains : $|M| \leq 1$; $|m| \leq 1$.

Hence :

Maximin case : $M = \pm 1$; $m = -\text{sign } M$; $x_f = -1$

Minimax case : $m = \pm 1$; $M = \text{sign } m$; $x_f = +1$

Neutral case : M and m chatter equally at very high rate as in a Poisson process between +1 and −1, it gives $x_f = 0$.

We shall see that these 3 types of games are equivalent (and the comparisons of T_M, T_m, τ_M, τ_m are not necessary) if the control function has the form :

(4) $$\vec{V}(\vec{X}, \vec{M}, \vec{m}, t) = \vec{V}_M(\vec{X}, \vec{M}, t) + \vec{V}_m(\vec{X}, \vec{m}, t)$$

and if $\vec{V}_M$ and/or $\vec{V}_m$ are bounded.

4. The upper game and the lower game

Another reason of undeterminism appears when there is discontinuities of the performance index (e.g. : $I = 0$ if $x_f \neq 0$ and $I = 1$ if $x_f = 0$) or when the terminal subset has particular forms (such as the two sheets of a cone) and when an infinitely small change of the control, especially near the final instant, gives a large change of the performance index : it is indeed impossible to follow the opponent reactions with an infinite accuracy.

In order to avoid that kind of undeterminism it is sufficient to give to one player an infinitesimal right to cheat, i.e. to add to the velocity vector $\vec{V} = d\vec{X}/dt$ a component $\vec{\varepsilon}$ whose integral $\int \|\vec{\varepsilon}\| . dt$ is as small as desired by its opponent in any sufficiently large bounded set of the $\vec{X}, t$ space.

We obtain thus the upper game and the lower game according to the player who has the right to cheat (classification independant of the minimax, maximin and neutral types).

The upper and lower values of a game will be major elements of appreciation of that game in a given situation since an infinite accuracy is never possible.

5. Extension of the Pontryagin's theory to deterministic two-players zero-sum differential games

5.1. The adjoint vector $\vec{P}$ and the Generalized Hamiltonian $H^*(\vec{P}, \vec{X}, t)$

We shall use the ordinary notations and first the adjoint vector of Pontryagin $\vec{P}$ which will be closely related to the strategy of each player.

As usual the Hamiltonian will be the scalar product :

(5)
$$H = \vec{P} \cdot \overrightarrow{dX}/dt = \vec{P} \cdot \vec{V}$$

and by a direct generalization of the notion of "Generalized Hamiltonian" (MARCHAL 1971, ROCKAFELLAR 1970) the new "Generalized Hamiltonian " is :

A) For a game of "maximin type" :

(6)
$$H^*(\vec{P},\vec{X},t) = \sup_{\vec{M} \in \mathcal{D}_M(t)} \left[\inf_{\vec{m} \in \mathcal{D}_m(t)} \vec{P} \cdot \vec{V}(\vec{X},\vec{M},\vec{m},t) \right]$$

B) For a game of "minimax type" :

(7)
$$H^*(\vec{P},\vec{X},t) = \inf_{\vec{m} \in \mathcal{D}_m(t)} \left[\sup_{\vec{M} \in \mathcal{D}_M(t)} \vec{P} \cdot \vec{V}(\vec{X},\vec{M},\vec{m},t) \right]$$

C) For a game of "neutral type" let us define H_1 and H_2, the λ_i 's being positive and their sum being one :

(8)
$$\begin{cases} H_1(\vec{P},\vec{X},t) = \sup_{\lambda_i} \left\{ \sup_{\vec{M}_i \in \mathcal{D}_M(t)} \left[\inf_{m \in \mathcal{D}_m(t)} \vec{P} \cdot \sum_i \lambda_i \vec{V}(\vec{X},\vec{M}_i,\vec{m},t) \right] \right\} \\ H_2(\vec{P},\vec{X},t) = \inf_{\lambda_i} \left\{ \inf_{\vec{m}_i \in \mathcal{D}_m(t)} \left[\sup_{\vec{M} \in \mathcal{D}_M(t)} \vec{P} \sum_i \lambda_i \vec{V}(\vec{X},\vec{M},\vec{m}_i,t) \right] \right\} \end{cases}$$

It is easy to verify that always :

(9)
$$H_1(\vec{P},\vec{X},t) \leq H_2(\vec{P},\vec{X},t)$$

but these two quantities are note necessarily equal (if for instance, for given $\vec{P}$, $\vec{X}$, t, the scalar product $\vec{P}.\vec{V}$ is equal to M – m, M and m being arbitrary real positive numbers).

The determinism of a game of "neutral type" requires that the two functions $H_1(\vec{P}, \vec{X}, t)$ and $H_2(\vec{P}, \vec{X}, t)$ be identical, they of course are then equal to the Generalized Hamiltonian $H^*(\vec{P}, \vec{X}, t)$ of the game.

A sufficient condition of equality of H_1 and H_2 is that at least one of the two control domains $\mathcal{D}_M(t)$ and $\mathcal{D}_m(t)$ be a compact set of a R^q space and that for any $(\vec{X}, t)$ the control function $\vec{V} = \vec{V}(\vec{X}, \vec{M}, \vec{m}, t)$ be uniformly continuous with respect to the corresponding control parameter.

Of course the maximin type being the most favourable to the minimisor the corresponding Generalized Hamiltonian is always the smallest and conversely the Generalized Hamiltonian of the minimax type is always the largest.

It is easy to verify that the 3 Generalized Hamiltonian are identical if the control has the form (4) and if the velocities $\vec{V}_M$ and/ or $\vec{V}_m$ are bounded.

It is possible to see now how the adjoint vector $\vec{P}$ is related to the strategies of both players, let us assume for instance that locally, between the instants t and $t + \delta t$, the maximisor want to maximises the scalar product $\vec{K}.\vec{X}$ and the minimisor want to minimizes it, they will both choose the control corresponding to $\vec{P} = \vec{K}$ in (6), (7) or (8) according to the type of the game (with a chattering if necessary) and, if H* is continuous in terms of $\vec{P}$, $\vec{X}$ and t, they will obtain :

(10)
$$\vec{K} \cdot \overrightarrow{\delta X} = \vec{P} \cdot \overrightarrow{\delta X} = H^*(\vec{P},\vec{X},t) \cdot \delta t + o(\delta t)$$

5.2. The conditions of canonicity

In order to avoid the difficulties coming from the discontinuities of trajectories $\vec{X}(t)$ we shall assume that the control function $\vec{V} = \vec{V}(\vec{X}, \vec{M}, \vec{m}, t)$ (with $\vec{M} \in \mathcal{D}_M(t)$ and $\vec{m} \in \mathcal{D}_m(t)$ is bounded in any bounded set of the $\vec{X}$, t space, it implies that the trajectories $\vec{X}(t)$ are Lipchitz functions of t and that the Generalized Hamiltonian $H^*(\vec{P}, \vec{X}, t)$ is bounded in any bounded set of the $\vec{P}, \vec{X}$, t space.

On the other hand let us note that the part of the terminal subset where the performance index is very small is considered as a forbidden zone by the maximisor and conversely the part of the terminal subset where the performance index is very large is considered as a forbidden zone by the minimisor.

Thus, in order to obtain a generalization of the Pontryagin theory to differential games, it is necessary that the conditions of application of that theory to problems with forbidden zones be satisfied, that is (MARCHAL 1971, page 151).

A) The problem must be canonical in the generalized meaning of Pontryagin for the admissibility of the discontinuous type, i.e. here, since velocities $\vec{V}$ are locally bounded :

The Generalized Hamiltonian $H^*(\vec{P}, \vec{X}, t)$ must be a locally Lipschitzian function of $\vec{P}, \vec{X}$ and t.

This severe condition has the advantage to involve the equivalence between chattering and relaxation, which is necessary especially for neutral type games.

B) The terminal subset $\mathcal{C}$ must be union or intersection of a finite number of closed or open "smooth" sets (i.e. manifolds with everywhere a tangent subspace Lipschitzian function of the position) these manifolds must be limited by a finite number of "smooth" hypersurface themselves limited by a finite number of "smooth" hyperlines et c....

For instance the terminal subset can be the surface of a polyhedron or of a cylinder and this condition is satisfied in almost all ordinary problems, however it is not satisfied at the origin for the function $y = x^n$ when either $0.5 < n < 1$ or $1 < n < 2$.

C) For any value I_o the parts of the terminal subset defined either by $I > I_o$ or by $I \geqslant I_o$ or by $I < I_o$ or by $I \leqslant I_o$ must satisfy the same condition of "smoothness".

5.3. The Generalized Pontryagin's formulas

In ordinary problems of optimization, when the Generalized Hamiltonian $H^*(\vec{P}, \vec{X}, t)$ is defined and when the conditions of canonicity are satisfied it is possible to adjoin to each extremal trajectory $\vec{X}(t)$ an absolutely continuous adjoint function $\vec{P}(t)$ different from zero and such that, with $H^*[\vec{P}(t), \vec{X}(t), t] = H^*(t)$, either :

$$d\vec{X}(t)/dt = \partial H^*/\partial \vec{P} \; ; \; d\vec{P}(t)/dt = -\partial H^*/\partial \vec{X} \; ; \; dH^*(t)/dt = \partial H^*/\partial t \tag{11}$$

or, more generally (when $H^*(\vec{P}, \vec{X}, t)$ is locally Lipchitzian but not continuously differentiable in terms of $\vec{P}, \vec{X}$ and t).

(12) The vector $\dfrac{d}{dt}\left(\vec{X}(t), -\vec{P}(t), H^*(t)\right)$ belongs for almost all t to the domain $D_{Ht}\left(\vec{P}(t), \vec{X}(t), t\right)$

This domain $D_{Ht}(\vec{P}, \vec{X}, t)$ being the smallest closed and convex set (of the R^{2n+1} space) containing the gradient vectors $\partial H^*/\partial(\vec{P}, \vec{X}, t)$ obtained at points $(\vec{P} + \vec{\delta P}, \vec{X} + \vec{\delta X}, t)$ where :

A) H^* is differentiable in terms of $\vec{P}, \vec{X}$ and t ;

B) $\vec{\delta P}$ and $\vec{\delta X}$ are infinitely small (i.e. of course D_{Ht} is the limit for $\varepsilon \to 0$ of domains $D_{Ht\varepsilon}$ obtained when $\|\vec{\delta P}\|$ and $\|\vec{\delta X}\|$ can vary from 0 to ε ; D_{Ht} is a particular "convex hull").

When there is forbidden zones with "smooth" boundaries the adjoint function $\vec{P}(t)$ becomes the sum of an absolutely continuous function and a "jump function" (i.e. a function with a finite or denumerable number of discontinuities and which is constant between these discontinuities) and the equations (11) and (12) become more complex.

Let us now try to generalize these equations to differential game problems.

We shall only consider upper game problems with a bounded playing space and we shall decompose them into the different "games of kind" corresponding to either $I \geq I_o$ or $I < I_o$.

Let us note that some people consider that the part of the terminal subset $\mathcal{C}$ corresponding to $I \geq I_o$. in the upper game problem, part that we shall call $\mathcal{C}_+$, is only the closure of the corresponding part of the initial game, some other people add to that closed set the points where the local upper limit of the performance index is I_o and thus obtain all points where that local upper limit is larger than or equal to I_o. Anyhow in both cases the set $\mathcal{C}_+$ is closed and thus the two cases are similar.

We shall call $\mathcal{C}_-$ the subset $\mathcal{C} - \mathcal{C}_+$.

It is easy to demonstrate that the part of the playing space corresponding to $I < I_o$ (in the upper game problem) is open, we shall call that subset O_{Io} and we shall call $\mathcal{F}_{Io}$ the remaining part of the playing space. We are especially interested by the closed set $\mathcal{B}$ intersection of the boundaries of O_{Io} and $\mathcal{F}_{Io}$.

The generalization of Pontryagin's theory (see demonstration in MARCHAL, to appear) leads to :

If the playing space is bounded, to each point $(\vec{X}_o, t_o)$ of the boundary $\mathcal{B}$ corresponds a pencil (or bundle) of absolutely continuous trajectories $\vec{X}_i(t)$ belonging entirely to $\mathcal{B}$, each of them being associated to an adjoint function $\vec{P}_i(t)$ defined on the same interval of time and sum of an absolutely continuous function of t and a jump function of t (with a bounded total variation).

We shall call extremal pencil or extremal bundle the union of the trajectories $\vec{X}_i(t)$; this extremal pencil satisfy the following generalized conditions :

A) The pencil begins at the point $(\vec{X}_o, t_o)$ of interest with at least one trajectory.

B) Each trajectory $\vec{X}_i(t)$ is defined on a closed interval of time (t_{io}, t_{if}) and ends at the terminal subset $\mathcal{C}$ (and not only at $\mathcal{C}_+$ as written erroneously in MARCHAL 1973).

C) A point $(\vec{X}, t)$ of the extremal pencil can belong to one or to several trajectories $\vec{X}_i(t)$ and there is always at least one corresponding adjoint vector $\vec{P}_i(t)$ different from zero.

D) With $H_i^*(t) = H^*(\vec{P}_i(t), \vec{X}_i(t), t)$, the equations (11) become :

$$
(13) \quad \begin{cases} d\vec{X}_i(t)/dt = \partial H^*/\partial \vec{P} \\ d\vec{P}_i(t)/dt = -\partial H^*/\partial \vec{X} + \sum_j \lambda_{ij}\left(\vec{P}_i(t) - \vec{P}_j(t)\right) \\ dH_i^*(t)/dt = \partial H^*/\partial t + \sum_j \lambda_{ij}\left(H_i^*(t) - H_j^*(t)\right) \end{cases}
$$

with :

(14) $$\vec{X}_j(t) \neq \vec{X}_i(t) \Rightarrow \lambda_{ij} = 0 \quad ; \quad \vec{X}_j(t) = \vec{X}_i(t) \Rightarrow \lambda_{ij} \geq 0$$

E) In the same way the generalized equations (12) becomes :

For any given i we have for almost all t :

(15) $$\frac{d}{dt}\left(\vec{X}_i(t), -\vec{P}_i(t), H_i^*(t)\right) \in D_{Ht}\left(\vec{P}_i(t), \vec{X}_i(t), t\right) + \sum_j \lambda_{ij}\left[0, \vec{P}_j(t) - \vec{P}_i(t), H_i^*(t) - H_j^*(t)\right]$$

F) The functions $\vec{P}_i(t)$ and $H_i^*(t)$ can also have jumps in the directions given by the infinite values of the positive factors λ_{ij}.

G) As usual when there is forbidden zones, if a point $(\vec{X}, t)$ of the pencil belong to the terminal zone $\mathcal{C}$ one can add to the derivatives of the vector $\overrightarrow{(P, -H^*)}$ given in (13) and (15) a component normal to $\mathcal{C}$ and directed outward (if $(\vec{X}, t) \in \mathcal{C}_-$) or toward the playing space (if $(\vec{X}, t) \in \mathcal{C}_+$), one can even add a jump provided that exist connecting absolutely continuous functions $\vec{P}_i(\varphi)$, $H_i^*(\varphi)$ leading from $\vec{P}(t-)$, $H^*(t-)$ to $\vec{P}(t_+)$, $H^*(t_+)$, verifying for any φ the relation (6)(or (7) or (8)) and having their derivatives with respect to φ in the directions given by these outer or inner normal components.

H) Finally for each trajectory $\vec{X}_i(t)$ of the extremal pencil the final values $[\vec{P}_i(t_{if}), H_i^*(t_{if})]$ must satisfy the ordinary final conditions of Pontryagin. (also called "transversality conditions").

- - - - - - - - - - - - - - - - - -

A simple way to obtain the directions normal to $\mathcal{C}$ is to use a Lipschitzian penalty function $f(\vec{X}, t)$ equal to zero on $\mathcal{C}$ and negative in the playing space $\mathcal{E}$; the local gradient of $f(\vec{X}, t)$ with respect to $(\vec{X}, t)$ gives the outer normal direction (or directions, for instance at a corner of $\mathcal{E}$).

On the same way, for the condition H. the final vector $[\vec{P}_i(t_{if}) ; -H_i^*(t_{if})]$, with (n + 1) components, must be parallel to and in the direction of the local gradient of a Lipschitzian function $f_+(\vec{X}, t)$ equal to zero on $\mathcal{C}_+$ and negative anywhere else (or antisymmetrically with respect to $\mathcal{C}_-$ if $(X_{if}, t_{if}) \in \mathcal{C}_-$).

Let us note that around points $(\vec{X}, t)$ belonging to $\mathcal{C}_+$ but not to $\mathcal{C}_-$ it is useless to consider the function $f_+(\vec{X}, t)$ out of $\mathcal{E} + \mathcal{C}$ (and conversely for $f_-(\vec{X}, t)$ if $(\vec{X}, t) \in \mathcal{C}_-$).

On the other hand if the direction of grad $f_+(\vec{X}, t)$ (or grad $f_-(\vec{X}, t)$), it is not continuous at the final point $(\vec{X}_{if}, t_{if})$ of interest, the vector $[\vec{P}_{if}, -H^*_{if}]$ may be into any direction of the corresponding conic convex hull.

Conclusion

The generalization of the optimization theory of Pontryagin to deterministic two-players zero-sum differential games leads to the notion of extremal pencil and to the above equations and rules which are sometimes sufficient to determine these pencils (see for instance the two examples of MARCHAL 1973). The main remaining question is to improve the conditions of backward construction of extremal pencils outlined in that reference.

References

ATHANS, M. - The status of optimal control theory and applications for deterministic systems, A, 8 differential games. IEEE trans. on automatic control, (April 1966).

BEHN, R.D. and HO, Y.C. - On a class of linear stochastic differential games. IEEE trans. on auto-control, (June 1968).

CHYUNG, D.H. - On a class of pursuit evasion games. IEEE trans. on auto-control, (August 1970).

HO, Y.C. and BARON, S. - Minimal time intercept problem. IEEE trans. on auto-control, (April 1965).

HO, Y.C. , BRYSON, A.E. and BARON, S. - Differential games and optimal pursuit-evasion strategics. IEEE trans. on auto-control, (October 1965).

ISAACS, R. - Differential games. John Wiley and Sons, (1965).

JACOB, J.P. and POLAK, E. - On a class of pursuit-evasion problems. IEEE trans. on auto-control, (December 1967).

MARCHAL, C. - Theoretical research in deterministic optimization. ONERA publication n° 139, (1971).

MARCHAL, C. - The bi-canonical systems. Techniques of optimization. A.V. Balakrishnan Editor, Academic Press, New York and London, (1972).

MARCHAL, C. - Generalization of the optimality theory of Pontryagin to deterministic two-players zero-sum differential games. ONERA, tiré à part.n° 1233, (1973).

MARCHAL, C. - Theoretical research in deterministic two-players zero-sum differential games. ONERA publication, to appear.

MESCHLER, P.A. - On a goal-keeping differential game. IEEE trans. on auto-control, (February 1967).

MESCHLER, P.A. - Comments on a linear pursuit-evasion game. IEEE trans. on auto-control, (June 1967).

PONTRYAGIN, L.S. , BOLTYANSKII, V.G. , GAMKRELIDZE, R.V. and MISCHENKO, E.F. - The mathematical theory of optimal processes. Interscience Publishers, John Wiley and Sons, Inc. New York, (1962).

ROCKAFELLAR, R.T. - Generalized Hamiltonian equations for convex problems of Lagrange. Pacific J. of Math. 33, 411-427 (1970).

ROCKAFELLAR, R.T. - Dual problems of optimal control. Techniques of optimization. A.V. Balakrishnan editor. Ac. Presse, New York, London, (1972).

WARGA, J. - Minimax problems and unilateral curves in the calculus of variations. SIAM Journal on Control A, 3, 1, (1965).

ABOUT OPTIMALITY OF TIME OF PURSUIT

M.S.NIKOL'SKII

STEKLOV MATHEMATICAL INSTITUTE,

MOSCOW, USSR

I'll tell about results of me, Dr. P.B.Gusjatnikov and V.I.Uhobotov in the problem optimality of pursuit time. I have studied with Dr. P.B.Gusjatnikov this problem in 1968. There is the article about this results in Soviet Math. Dokl. (vol. 184, N 3, 1969). After this article was the article of N.N.Krasovskii and A.I.Subbotin about this question. Their result is more general. I'll not tell about result of N.N.Krasovskii and A.I.Subbotin (see Soviet Journal of Applied Math. and Mechan., N 4, 1969). Recently Dr. P.B.Gusjatnikov and I have got some results in this field in cooperation with Dr.V.I.Uhobotov.

Let the motion of a vector z in Euclidean space R^n be described by the linear vector differential equation

$$\dot{z} = Cz - u + v, \tag{1}$$

where $z \in R^n$, C is a constant square matrix, $u \in P \subset R^n$, $v \in Q \subset R^n$. u and v are control vectors. The vector u corresponds to the pursuer, the vector v to the evader. P and Q are convex compact sets. Let a terminal subspace M be assigned in R^n. The pursuit is assumed to be completed when the point z reaches M for the first time. It is in the interest of the pursuer to complete the pursuit. The information of pursuer is the equality (1) and $z(t)$, $v(t)$ for all present $t \geq 0$. The functions $u(t)$, $v(t)$ are meserable functions. The pursuer don't know $v(t)$ in advance, $v(t)$ can be arbitrary meserable function.

Pontrjagin have constructed the method of pursuit from initial point z_0 and gave estimate $\tau(z_0)$ of time of pursuit.

I'll say some words about this method.

Let L is complemental subspace of M and π is operator of projection of R^n onto L parallely M.

Pontrjagin considers the compact sets $\pi e^{\tau C} P$, $\pi e^{\tau C} Q$, their geometrical difference $\hat{w}(\tau) = \pi e^{\tau C} P \overset{*}{-} \pi e^{\tau C} Q$ and integral $W(t) = \int_0^t \hat{w}(\tau)\, d\tau$, which is a compact convex set in R^n. The time of pursuit from z_0 in the theory of Pontrjagin is the least root of inclusion:

$$\pi e^{tC} z_0 \in W(t). \tag{2}$$

Let $\tau(z_0)$ is such root.

<u>Definition 1</u>. The optimal time of pursuit from z_0 is the least time within pursuer can complete the pursuit from z_0.

The Pontrjagin time $\tau(z_0)$ is optimal if the following Condition A is fulfilled.

<u>Condition A</u>. For all extreme point $u \in P$ exists $v = \underbrace{v(u)}_{\in Q}$ such that for all $\tau \in [0,T]$

$$\pi e^{\tau C} u - \pi e^{\tau C} v(u) \in \hat{w}(\tau).$$

<u>Theorem 1</u>. If the Condition A is fulfilled and $\tau(z_0) \leq T$, then $\tau(z_0)$ is optimal time of pursuit.

We shall give some sufficient conditions for fulfilment the Condition A.

Let us $0 \in P$, 0 is interior point of Q in its support subspace and U, V their support subspaces in R^n, $f(\tau)$ is restriction of mapping $\pi e^{\tau C}$ on U, $g(\tau)$ is restriction of mapping $\pi e^{\tau C}$ on V.

Let us $f(\tau)$ can be factored in such way

(3) $$f(\tau) = h(\tau)\mathcal{D}, \quad 0 \le \tau \le T,$$

where $\mathcal{D}$ is linear mapping $\mathcal{U}$ into $\mathcal{L}$, $\dim \mathcal{D}\mathcal{U} \geq 1$, $h(\tau)$ is linear mapping $\mathcal{D}\mathcal{U}$ into $\mathcal{L}$, $h(\tau)$ is analytical for $\tau \in [0,T]$ and homeomorphic for $\tau \in [0,T]$ with the exception finite set K; the set $g(\tau)Q$ sweeps completely the set $f(\tau)P$ for all $\tau \in [0,T]$:

(4) $$f(\tau)P \equiv \hat{w}(\tau) + g(\tau)Q,$$

$\dim V \geq 1$.

<u>Definition 2</u>. A convex closed set S from space $\mathcal{N}$ $(\dim \mathcal{N} \geq 1)$ is strictly convex, if each its support plane has only one common point with S.

<u>Definition 3.</u> A convex closed set S from space $\mathcal{N}\,(\dim \mathcal{N} \geq 1)$ is regular, if it is strictly convex and each its boundary point has only one support plane.

<u>Theorem 2</u>. Let $g(\tau)$ can be factored in the form

(5) $$g(\tau) = h(\tau)\alpha(\tau)F, \quad 0 \le \tau \le T,$$

here F is linear mapping V into $\mathcal{D}\mathcal{U}$, $\alpha(\tau)$ - nonnegative analytical function. Let FQ is strictly convex in $\mathcal{D}\mathcal{U}$. In these conditions Condition A is fulfilled.

<u>Theorem 3</u>. Let Q is regular in V. In these conditions the equality (5) is necessary for fulfilment the Condition A.

The another sufficient conditions for fulfilment the Condition A are given by the following Theorem 4.

<u>Theorem 4</u>. If for $t \geq 0$ π and e^{tC} are commutative on

$\mathcal{U}$, V and πQ sweeps completely the set πP , then the Condition A is fulfilled.

Example. The Pontrjagin's test example:

$$\ddot{x} + \alpha \dot{x} = \rho u ,$$

$$\ddot{y} + \beta \dot{y} = \sigma v ,$$

$M = \{x = y\}$, $\alpha, \beta, \rho, \sigma$ are positive constants,
$x, \dot{x}, y, \dot{y} \in R^k$, $k \geq 1$, $|u| \leq 1$, $|v| \leq 1$.

If $\rho > \sigma$, $\frac{\rho}{\alpha} > \frac{\sigma}{\beta}$, then the condition A is fulfilled and the time $\tau(z_0)$ is optimal.

ALGEBRAIC AUTOMATA AND OPTIMAL SOLUTIONS IN PATTERN RECOGNITION

E.ASTESIANO - G.COSTA

Istituto di Matematica - Università di Genova -

Via L.B. Alberti,4 16132 GENOVA (ITALY)

INTRODUCTION U. Grenander (1969) proposed a formalization of the linguistic approach in pattern recognition (see also Pavel (1969)). Though the main interest in this method is undoubtely for its pratical application to the construction of particular grammars of patterns,we think the theoretical questions worthwhile further insight. Therefore we devote our attention to the abstract formulation of the problem; however, in order to give a detailed model,we restrict ourselves to a definite class of decision rules. In the recognition system we consider here, the objects to be recognized are (represented by) terms on a graded alphabet and the decision rules are (implemented by) algebraic automata; moreover we assume that sample classes (in some sense the "training sets") and rules of identifications of objects in images are given. In this context we investigate the problems related to the definition,the existence and the effective construction of optimal grammars of patterns.
This paper is a generalization and improovement of a previous work of one of the authors. The results can also be considered (but the point of wiew is completely different) a generalization of well known classical results in algebraic theory of automata; these can in fact be obtained as a particular case of our results (when the sample classes are a partition). The first part is devoted to set up substantially well known definitions and results, in a language more apt to treat our problem. In the second the recognition model and a definition of the optimal solution are proposed and explained. In the last section,conditions of existence for the optimal solution are given and the problem of the effective construction is solved. Finally a few examples are presented to show that some seemingly incomplete results cannot in fact be substantially improoved.

1. PRELIMINARY NOTIONS.

We refer to Cohn (1965) and Grätzer (1968) for the algebraic concepts and to Thatcher and Wright (1968) and Arbib and Give'on (1968) for what concerns algebraic automata.

If X is a non empty set, DF(X) is the set of all families of non empty,mutually disjoint,subsets of X; we denote its elements by $\mathcal{A},\mathcal{B},\ldots,\mathcal{F},\ldots$ If $\mathcal{A}=\{A_i,\ i\in I\}$ then $A_o := X - \bigcup_{i\in I} A_i$ and $\mathring{\mathcal{A}} := \mathcal{A}\cup\{A_o\}$. Consider on DF(X) the relation $\vdash$ defined by: $\mathcal{A}\vdash\mathcal{B}$ iff a (unique)map $\rho:\mathcal{A}\to\mathcal{B}$ exists s.t. $\forall A\in\mathcal{A}$, $A\subseteq\rho(A)$; we also write $\mathcal{A}\vdash_{\rho}\mathcal{B}$.

Proposition 1.1 $\vdash$ is an order relation on DF(X) and DF 1 (X) $:=(DF(X),\vdash)$ is a complete lattice.//

Remark. We indicate by // the end of the prof; the proof is omitted whenever it is

---

This work was partially supported by a G.N.A.F.A. (C.N.R.) grant.

straightforward or substantially known.
Denote now by E(X) the complete lattice of equivalences on the set X, ordered by $\subseteq$ as a subset of $X \times X$. If $\mathcal{A} \in DF(X)$ and $\Theta \in E(X)$, then Θ is an <u>$\mathcal{A}$-equivalence</u> iff, $\forall A \in \mathcal{A}$, $A = \bigcup_{x \in A} [x]_\Theta$. For any $P \subseteq E(X)$, let DL(X,P) be the set $\{(\Theta, \mathcal{B}) / \mathcal{B} \in DF(X), \Theta \in P, \Theta$ is a $\mathcal{B}$-equivalence and $<$ be the relation on DL(X,P) defined by : $(\Theta, \mathcal{B}) < (\Psi, \mathcal{D})$ iff $\Theta \subseteq \Psi$ and $\mathcal{B} \vdash \mathcal{D}$.

<u>Proposition 1.2.</u> $<$ is an order relation and $DL1(X,P) := (DL(X,P), <)$ is a complete lattice iff P is a (complete) sublattice of E(X). //

A <u>graded (ranked) set</u> is a pair (Σ, σ), where Σ is a non empty set and σ a map from Σ into $\mathbb{N}$ (the set of non negative integers); if $n \in \mathbb{N}$, then $\Sigma_n := \sigma^{-1}(n)$. From now on we shall simply write Σ instead of (Σ, σ).
A <u>Σ-algebra</u> (or simply: an algebra) is a pair $\mathfrak{A} = (A, \alpha)$, where A is a non empty set, the carrier of $\mathfrak{A}$, and α is a map that assigns to each element ω in Σ an operator α_ω on A. If $\omega \in \Sigma_n$, α_ω is an n-ary operator, that is : if $n \geqslant 1$, then $\alpha_\omega : A^n \to A$; if $n = 0$, then α_ω is an element of A. We use capital German letters, $\mathfrak{A}, \mathfrak{B}, \ldots, \mathfrak{F}, \ldots, \mathfrak{Q}, \ldots$ to indicate Σ-algebras.
If X is a set disjoint from Σ, then we denote <u>free Σ-algebra of Σ-terms on X</u> (or simply :terms) by $\mathfrak{F}_\Sigma(X)$ and its carrier by $F_\Sigma(X)$; if $X = \emptyset$ we write $\mathfrak{F}_\Sigma$ and F_Σ. Given $\mathfrak{A} = (A, \alpha)$, we denote by $rp_\mathfrak{A}$ the unique homomorphism from $\mathfrak{F}_\Sigma$ into $\mathfrak{A}$; we say that <u>$\mathfrak{A}$ is connected</u> iff $rp_\mathfrak{A}$ is onto. The set of all congruences on $\mathfrak{A}$ will be denoted by $C(\mathfrak{A})$; $C(\mathfrak{A})$ is a (complete) sublattice of E(A), hence $DL1(A, C(\mathfrak{A}))$ is a complete lattice. If $\mathcal{A} \in DF(A)$, then an <u>$\mathcal{A}$-congruence</u> is simply an $\mathcal{A}$-equivalence on A which is a congruence on $\mathfrak{A}$; $C(\mathfrak{A}, \mathcal{A})$ is the set of all $\mathcal{A}$-congruences on $\mathfrak{A}$. We write : C, $C(\mathcal{A})$, DL1(P), instead of $C(\mathfrak{F}_\Sigma)$, $C(\mathfrak{F}_\Sigma, \mathcal{A})$, $DL1(F_\Sigma, P)$.
A <u>graded (ranked) alphabet</u> is a finite graded set Σ, s.t. $\Sigma_0 \neq \emptyset$. From now on Σ will be a graded alphabet that we consider as fixed and given.
A <u>Σ-automaton</u> is a pair $M = (\mathfrak{Q}, \mathcal{F})$, where : $\mathfrak{Q} = (Q, \alpha)$ is a Σ-algebra and $\mathcal{F} \in DF(Q)$. The <u>response function of M</u> is just $rp_\mathfrak{Q}$; we shall denote it also by rp_M. <u>M is connected</u> iff $\mathfrak{Q}$ is connected, and <u>M is finite</u> iff Q is finite. The <u>behaviour of M</u> is $\mathcal{B}^M := \{B / B = rp_M^{-1}(F), F \in \mathcal{F}\}$; the <u>equiresponse congruence of M</u> ER(M), is the canonical congruence on $\mathfrak{F}_\Sigma$ associated with rp_M; eventually, $\mu(M) := (ER(M), \mathcal{B}^M)$.
<u>Lemma 1.3.</u> For any Σ-automaton M, $\mu(M) \in DL1(C)$.//
If $(\Theta, \mathcal{B}) \in DL1(C)$, then $\pi((\Theta, \mathcal{B}))$ is the Σ-automaton $(\mathfrak{F}_\Sigma/\Theta, \mathcal{B}/\Theta)$, where $\mathcal{B}/\Theta := \{B/\Theta, B \in \mathcal{B}\}$.
<u>Lemma 1.4.</u> If $(\Theta, \mathcal{B}) \in DL1(C)$, then $\pi((\Theta, \mathcal{B}))$ is connected and $\mu(\pi((\Theta, \mathcal{B}))) = (\Theta, \mathcal{B})$.//

Let $M_i = (\mathfrak{A}_i, \mathfrak{F}_i)$ be a Σ-automaton, $i = 1,2$; an <u>homomorphism from M_1 into M_2</u> is a pair (φ, ρ), where $\varphi : \mathfrak{A}_1 \to \mathfrak{A}_2$ is an algebras homomorphism, $\rho : \mathfrak{F}_1 \to \mathfrak{F}_2$ is a map and, $\forall F \in \mathfrak{F}_1$, $\varphi(F) \subseteq \rho(F)$; (φ, ρ) is an isomorphism, and we write $M_1 \simeq M_2$, iff : φ is an algebras isomorphism, ρ is one to one and onto and $\varphi(F) = \rho(F)$. If M_1 and M_2 are connected and there is an homomorphism $(\bar\varphi, \bar\rho)$ from M_1 into M_2, then $(\bar\varphi, \bar\rho)$ is uniquely determined by the properties : $rp_{M_2} = \bar\varphi \cdot rp_{M_1}$ and $\bar\varphi(F) \subseteq \bar\rho(F)$ $\forall F \in \mathfrak{F}$, moreover $\bar\varphi$ is onto, we indicate this by writing $M_1 \twoheadrightarrow M_2$. These definitions and properties allow us to state the following lemma.

<u>Lemma 1.5.</u> If M and M' are connected automata, then : i) $\pi(\mu(M)) \simeq M$; ii) $M \twoheadrightarrow M'$ iff $\mu(M) < \mu(M')$.//

We denote by $\mathfrak{M}^c_\Sigma$ the class of connected Σ- automata mod. the isomorphism equivalence, by $[M]_\simeq$ the class of $\mathfrak{M}^c_\Sigma$ corresponding to M and by $\measuredangle$ the relation on $\mathfrak{M}^c_\Sigma$ defined by : $[M_1]_\simeq \measuredangle [M_2]_\simeq$ iff $M_2 \twoheadrightarrow M_1$. By the above lemma, $\measuredangle$ is correctly defined and it is order relation.

<u>Theorem 1.6.</u> $(\mathfrak{M}^c_\Sigma, \measuredangle)$ is a poset (i.e. partially ordered set) anti-isomorphic to DL1(C), therefore $(\mathfrak{M}^c_\Sigma, \measuredangle)$ is a complete lattice anti-isomorphic to DL1(C).//

We recall that similar results, for monadic algebra automata without final states, can be found in Büchi (1966).

Given $\mathfrak{A} = (A, \alpha)$, a symbol x not in $A \cup \Sigma$ and $S \subseteq A$, then for each τ in $F_\Sigma(S \cup \{x\})$ we can define the unary operator on $\mathfrak{A}$ $\|\tau\|_\mathfrak{A}$, as follows : $\forall a \in A$,
i) if $\tau = s \in S$, then $\|\tau\|_\mathfrak{A}(a) = s$; ii) if $\tau = x$, then $\|\tau\|_\mathfrak{A}(a) = a$;
iii) if $\tau = \tau_1 \dots \tau_n \omega$, then $\|\tau\|_\mathfrak{A}(a) = \alpha_\omega(\|\tau_1\|_\mathfrak{A}(a), \dots, \|\tau_n\|_\mathfrak{A}(a))$; clearly if $\tau = \omega \in \Sigma_0$, then $\|\tau\|_\mathfrak{A}(a) = \alpha_\omega$.

<u>Remarks.</u> i) the above operators correspond to the "polynomials" in Grätzer (1968); ii) One can verify that if $\mathfrak{A}$ is connected, then, $\forall S \subseteq A$, $\{\|\tau\|_\mathfrak{A}, \tau \in F_\Sigma(S \cup \{x\})\} = \{\|\tau\|_\mathfrak{A}, \tau \in F_\Sigma(\{x\})\}$; iii) if $\mathfrak{A} = \mathfrak{F}_\Sigma(Y)$, for any set Y , then $\|\tau\|_\mathfrak{A}(a)$ is the Σ-term on Y obtained from τ by replacing each occurrence of x in τ with the term a.

Consider $\mathcal{A} \in DF(A)$ and $\mathfrak{A}$ connected.

<u>Definition 1.1.</u> $N_\mathcal{A}$ is the relation on A s.t., if $a,b \in A$, $(a,b) \in N_\mathcal{A}$ <u>iff</u> $\forall \tau \in F_\Sigma(\{x\})$, $\forall A_i \in \mathcal{A}$, $\|\tau\|_\mathfrak{A}(a) \in A_i$ iff $\|\tau\|_\mathfrak{A}(b) \in A_i$.

<u>Theorem 1.7.</u> i) $N_\mathcal{A}$ is the maximum of $C(\mathfrak{A}, \mathcal{A})$; ii) $C(\mathfrak{A}, \mathcal{A}) = \{\Theta \in C(\mathfrak{A}) / \Theta \subseteq N_\mathcal{A}\}$; iii) $C(\mathfrak{A}, \mathcal{A})$ is a (complete) sublattice of the complete lattice $C(\mathfrak{A})$.

<u>Proof.</u> Hint : use a modified, but equivalent (see above remark), definition of $N_\mathcal{A}$, considering $F_\Sigma(A \cup \{x\})$ instead of $F_\Sigma(\{x\})$.//

For any $\mathcal{A} \in DF(F_\Sigma)$, set $C(\mathcal{A}) := C(\mathfrak{F}_\Sigma, \mathcal{A})$; by theorems 1.6 and 1.7 a class of minimal Σ-automata for $\mathcal{A}$ (i.e. having $\mathcal{A}$ as their behaviour) exists: the class $[\pi((N_\mathcal{A}, \mathcal{A}))]_\simeq$.

2. <u>THE RECOGNITION MODEL.</u> We need a few other definitions before we can give our recognition model. Consider on DF(X), X non empty set, the relation $\Vdash$ defined by : $\mathcal{B} \Vdash \mathcal{D}$ iff $\mathcal{B} \xrightarrow{\varphi} \mathcal{D}$ and the map φ is a bijection. It is soon verified that $\Vdash$ is an order relation. If $\mathfrak{A} = (A, \alpha)$ and $\Theta \in C(\mathfrak{A})$ then Θ is an <u>$\mathcal{A}$-separating congruence</u> iff each congruence class intersects at most one element of $\mathcal{A}$. We denote by SC $(\mathfrak{A}, \mathcal{A})$ the set of all $\mathcal{A}$-separating congruences on $\mathfrak{A}$.

Consider now the free algebra $\mathfrak{F}_\Sigma$, and a set $H = \{(t_i, t'_i) \mid (t_i, t'_i) \in F_\Sigma \times F_\Sigma$, $i = 1, \ldots, N\}$; we denote by $\Theta(H)$ the congruence on $\mathfrak{F}_\Sigma$ generated by H, see Grätzer (1968). One can verify that $\Theta(H)$ coincides with the reflexive, symmetric and transitive closure of the relation R(H), defined by : $(t,t') \in R(H)$ <u>iff</u> there is a pair (t_i, t'_i) in H s.t. t' is obtained from t replacing in t t_i by t'_i.

We can now give in detail our recognition model; see Grenander (1969) and Pavel (1969) for the motivations, for part of the terminology and a wider discussion on the subject.

- The "objects" to be recognized are coded on a graded alphabet .
- What we actually recognize are (<u>structural</u>)<u>descriptions</u> of the "objects" (analogous to the "configurations" in Grenander (1969)) i.e. -terms. One "object" may correspond to many different descriptions.
- We have some information about the corrispondence descriptions - "objects", i.e. we are given a finite set H F x F : $(t,t') \in H$ means that t and t' correspond to the same "object" and can thus be identified. If appears quite natural that we identify also all the pairs in R(H) and that we extend the process by reflexivity, symmetry and transitivity. This eventually amounts to consider on F the identifications given by the congruence $\Theta(H)$. We can restate all this, by saying that we are given a finitely generated congruence ${}^I\Theta$ on $\mathfrak{F}_\Sigma$; we call <u>images</u> the classes (mod ${}^I\Theta$) and the images now become, for us, the objects to be recognized.
- We are given $\mathcal{A} \in DF(F_\Sigma)$, <u>the family of examples</u>, i.e. a family of sets of already classified descriptions. $\mathcal{A}$ and ${}^I\Theta$ must be such that ${}^I\Theta$ is $\mathcal{A}$-separating.
- An <u>admissible</u>, for ${}^I\Theta$ and $\mathcal{A}$, <u>family of patterns</u> is a family $\mathcal{B} \in DF(F_\Sigma)$ s.t. $\mathcal{A} \Vdash \mathcal{B}$ and ${}^I\Theta$ is a $\mathcal{B}$-congruence.
- An <u>admissible</u>, for ${}^I\Theta$ and $\mathcal{A}$, <u>decision rule</u> is a map $r : F_\Sigma \longrightarrow {}^o\mathcal{A}$, such that : $\mathcal{B}_r := \{r^{-1}(A), A \in \mathcal{A}\}$ is an admissible family of patterns and a (connected) Σ-automaton M exists such that $\mathcal{B}_r = \mathcal{B}^M$ (we say that r is implemented by M).

Usually a decision rule can be implemented by several (even non isomorphic) Σ- aut.

Definition 2.1. A solution of the recognition problem, with ${}^I\Theta$ and $\mathcal{A}$-given, is a Σ-automaton implementing an admissible, for ${}^I\Theta$ and $\mathcal{A}$, decision rule.

Consider $\mathcal{U}=(A,\alpha)$, $\mathcal{A}\in DF(A)$ and $P\subseteq SC(\mathcal{U},A)$, then : $EDL(\mathcal{A},P):=\{(\Psi,\mathcal{B})\ (\Psi,\mathcal{B})\in DL(A,P)$,

Remark. The order $\ll$ induced on $EDL(\mathcal{A},P)$ by the order $<$ we have on $DL(A,P)$ is s.t. $(\Psi,\mathcal{B})\ll(\Omega,\mathcal{D})$ iff $\Psi\leq\Omega$ and $\mathcal{B}\Vdash\mathcal{D}$.

We call quasi-complete lower semilattice (q.c.l.s.-lattice) any ordered set in which all non empty subsets have a g.l.b. Thus, if we consider the set $EDF(\mathcal{A}):=$ $=\{\mathcal{B}\in DF(F_\Sigma)\ |\mathcal{A}\Vdash\mathcal{B}\}$, the set of extensions of $\mathcal{A}$, ordered by $\Vdash$; it is easy to see that $(EDF(\mathcal{A}),\Vdash)$ is a q.c.l.s.-lattice but not a lattice. From this we have also that, if P is a q.c.l.s.-lattice, then $EDL(\mathcal{A},P)$ is a q.c.l.s.-lattice, but, even if P is a complete lattice, $EDL(\mathcal{A},P)$ is not in general a lattice. This fact is of great importance as we shall see now.

Set $SC(\mathcal{A}):=SC(\mathcal{F}_\Sigma,\mathcal{A})$ and, if $P\subseteq SC(\mathcal{A})$, ${}^IP:=\{\Psi|\Psi\in P\ /\Psi\geq{}^I\Theta\}$, then from theorem 1.6 and the remark about $\ll$ we have the following theorem.

Theorem 2.1. M is a solution of the recognition problem, for given ${}^I\Theta$ and $\mathcal{A}$, iff $\mu(M)\in EDL(\mathcal{A},{}^ISC(\mathcal{A}))$.//

Remark. We can use this theorem for a new definition of "solution of the recognition problem". Indeed from now on we shall refer to $EDL(\mathcal{A},{}^ISC(\mathcal{A}))$ as the set of solutions of the recognition problem, (for given ${}^I\Theta$ and $\mathcal{A}$).

We can now characterize different kinds of solutions. Let : $\mathcal{A}\in DF(F_\Sigma)$, ${}^I\Theta\in SC(\mathcal{A})$ and $P\subseteq SC(\mathcal{A})$ be given. First of all we observe that considering $P\subseteq SC(\mathcal{A})$ instead of $SC(\mathcal{A})$ means that we are using only a subclass of all the admissible decision rules: exactly the class of rules implemented by automata which, for their algebraic structure, correspond to P (see theorem 1.6). For instance if $P=FSC(\mathcal{A}):=$ $=\{\Psi\in SC(\mathcal{A}),\ ind(\Psi)<+\infty\}$, then we consider only the decision rules implemented by finite Σ-automata whose equiresponse congruence is $\mathcal{A}$-separating (now we are not taking account of ${}^I\Theta$).

Definition 2.2. $EDLE(\mathcal{A},{}^IP):=\{(\Psi,\mathcal{B}):(\Psi,\mathcal{B})\in EDL(\mathcal{A},{}^IP),\ \Psi=\max.\text{of}\ ({}^IP\cap C(\mathcal{B}))\}$.

This is the set of "economical" solutions: for each admissible family of patterns $\mathcal{B}$, we consider the minimal, i.e. "more economical", Σ-automata for $\mathcal{B}$ in the class of automata corresponding to P (if they exist).

<u>Definition 2.3.</u> If $\mathfrak{A}=(A,\alpha)$, $\mathcal{B}\in DF(A)$ and $\Psi\in SC(\mathfrak{A},\mathcal{B})$, then $\mathcal{B}_\Psi:=\{B_\Psi, B\in\mathcal{B}\}$, where $B_\Psi:=\bigcup_{a\in B}[a]_\Psi$

<u>Definition 2.4.</u> $EDLJ(\mathcal{A},{}^IP):=\{(\Psi,\mathcal{A}_\Psi)/\Psi\in{}^IP\}$.

This is the set of <u>justified solutions</u>: for each Ψ in ${}^IP$ we extend each A in $\mathcal{A}$ by adding to A only elements of F_Σ which are Ψ-congruent to at least one element of A. It is now quite reasonable the following definition.

<u>Definition 2.5.</u> The set of <u>good solutions</u> of the recognition problem, for given $\mathcal{A}$, ${}^I\Theta$ and P is $GS(\mathcal{A},{}^IP):=EDLE(\mathcal{A},{}^IP)\cap EDLJ(\mathcal{A},{}^IP)=\{(\Psi,\mathcal{A}_\Psi)/\Psi=\max({}^IP\cap C(\mathcal{A}_\Psi))$.

The three above sets are ordered by $\ll$, as subsets of $EDL(\mathcal{A},{}^IP)$.

<u>Definition 2.6.</u> The <u>optimal solution</u>, for given $\mathcal{A}$, ${}^I\Theta$ and P, is the maximum of $GS(\mathcal{A},{}^IP)$, if it exists. We denote it by o.s. $(\mathcal{A},{}^IP)$.

Consider now the two conditions :

$[\alpha]$ $\forall(\Psi,\mathcal{A}_\Psi)$, $\Psi\in{}^IP$, $\text{l.u.b.}_{/C(\mathcal{A}_\Psi)}({}^IP\cap C(\mathcal{A}_\Psi))\in{}^IP$.

$[\beta]$ $\forall(\Psi,\mathcal{B})\in EDL(\mathcal{A},{}^IP)$, $\text{l.u.b.}_{/C(\mathcal{B})}({}^IP\cap C(\mathcal{B}))\in{}^IP$.

<u>Proposition 2.2.</u> i) $EDLJ(\mathcal{A},{}^IP)\simeq{}^IP$ (as posets); ii) if max $EDLJ(\mathcal{A},{}^IP)$ exists, then max GS $(\mathcal{A},{}^IP)$ exists and the two coincide; if condition $[\alpha]$ holds, the converse is also true. //

<u>Proposition 2.3. (Stability property)</u> If condition $[\beta]$ holds and $(\bar\Psi,\bar{\mathcal{B}})=$ o.s.$(\mathcal{A},{}^IP)$ then, $\forall\mathcal{A}'$ s.t. $\mathcal{A}\Vdash\mathcal{A}'\Vdash\bar{\mathcal{B}}$, o.s. $(\mathcal{A}',{}^IP\cap SC(\mathcal{A}'))=(\bar\Psi,\bar{\mathcal{B}})$.

<u>Proof.</u> $\mathcal{A}'\Vdash\bar{\mathcal{B}}=\mathcal{A}_{\bar\Psi}\Rightarrow\bar\Psi=SC(\mathcal{A}')$ and, as $\bar\Psi\geq{}^I\Theta$, ${}^I\Theta\in SC(\mathcal{A}')$. Since by prop.2.2 $\bar\Psi=\max{}^IP$ and clearly $\mathcal{A}\Vdash\mathcal{A}'\Vdash\bar{\mathcal{B}}:=\mathcal{A}_{\bar\Psi}$ implies $\mathcal{A}'_{\bar\Psi}=\mathcal{A}_{\bar\Psi}:=\bar{\mathcal{B}}$, then $(\bar\Psi,\bar{\mathcal{B}})=$ $=\max EDLJ(\mathcal{A}',{}^IP\cap SC(\mathcal{A}'))$. Thus by prop.2.2 $(\bar\Psi,\bar{\mathcal{B}})=\max G.S(\mathcal{A}',{}^IP\cap SC(\mathcal{A}'))$.//

<u>Remark.</u> One can easily verify that condition $[\beta]$, and therefore $[\alpha]$, is met, for example, when $P=SC(\mathcal{A})$ or $P=FSC(\mathcal{A})$.

3. <u>EXISTENCE OF OPTIMAL SOLUTIONS.</u> Let $M=(\mathfrak{Q},\mathfrak{F})$ be a Σ-automaton; if $\Phi\in SC(\mathfrak{Q},\mathfrak{F})$, then M/Φ is the Σ-automaton $(\mathfrak{Q}/\Phi,\mathfrak{F}/\Phi)$, where $\mathfrak{F}/\Phi:=\{F'/F'=\{[q]_\Phi, q\in F\}$, $F\in\mathfrak{F}\}$. Recalling def.2.3, we shall write ${}^I\mathcal{A}$ instead of $\mathcal{A}_{{}^I\Theta}$.

<u>Proposition 3.1.</u> If $ER(M)={}^I\Theta$ and $\mathcal{B}^M={}^I\mathcal{A}$, then : o.s. $(\mathcal{A},{}^ISC(\mathcal{A}))$ exists iff max $SC(\mathfrak{Q},\mathfrak{F})$ exists; if max $SC(\mathfrak{Q},\mathfrak{F})=\bar\Psi$ then o.s.$(\mathcal{A},{}^ISC(\mathcal{A}))=\mu(M/\bar\Psi)$.

<u>Proof.</u> Set $\Theta:={}^I\Theta$; from well known isomorphism theorems (see, for instance, Grätzer (1968)), we have the lattice isomorphisms : $C(\mathfrak{Q})\simeq C(\mathfrak{F}_\Sigma/\Theta)\simeq\{\Psi/\Psi\in C(\mathfrak{F}_\Sigma)$, $\Psi\geq\Theta\}$; this implies the poset isomorphisms: $SC(\mathfrak{Q},\mathfrak{F})\simeq SC(\mathfrak{F}_\Sigma/\Theta$, $\mathcal{B}^M/\Theta)\simeq\{\Psi/\Psi\in{}^ISC(\mathcal{B}^M)={}^ISC({}^I\mathcal{A})={}^ISC(\mathcal{A})$. Thus by Prop.2.2 the first part of the proposition is proved. As for the second, if $\bar{\bar\Psi}:=\max{}^ISC(\mathcal{A})$, remarking that $\bar{\bar\Psi}/\Theta$ corresponds

to $\overline{\Psi}$ in the isomorphism $SC(\mathfrak{A}, \mathfrak{F}) \simeq {}^{I}SC(\mathcal{A})$, we have : $\mathfrak{A}/\Psi \simeq \mathfrak{F}_{\Sigma}/\Theta/\overline{\overline{\Psi}}/\Theta \simeq \mathfrak{F}_{\Sigma}/\overline{\overline{\Psi}}$.
Hence and from the definitions of $M/\overline{\Psi}$ and μ, we have : $\mu(M/\overline{\Psi}) = (\overline{\overline{\Psi}}, \mathcal{B}^{M}_{\overline{\overline{\Psi}}}) =$
$= (\overline{\overline{\Psi}}, {}^{I}\mathcal{A}_{\overline{\overline{\Psi}}}) = (\overline{\overline{\Psi}}, \mathcal{A}_{\overline{\overline{\Psi}}}) = o.d. (\mathcal{A}, {}^{I}SC(\mathcal{A}))$.//

Proposition 3.2. If o.s. $(\mathcal{A}, {}^{I}SC(\mathcal{A}))$ exists and $\mu(M) \in EDLJ(\mathcal{A}, {}^{I}SC(\mathcal{A}))$, then $\overline{\Psi}$ exists s.t. $\overline{\Psi} = \max SC(\mathfrak{A}, \mathfrak{F})$ and $\mu(M/\overline{\Psi}) =$ o.s. $(\mathcal{A}, {}^{I}SC(\mathcal{A}))$.

Proof. Setting $\Psi := ER(M)$, by propo.2.2, o.s. $(\mathcal{A}, {}^{I}SC(\mathcal{A})) = \max EDLJ(\mathcal{A}, {}^{I}SC(\mathcal{A}))$ and thus : $({}^{I}\Theta, {}^{I}\mathcal{A}) \ll \mu(M) = (\Psi, \mathcal{B}^{M}) \ll$ o.s. $(\mathcal{A}, {}^{I}SC(\mathcal{A}))$. Since $\mathcal{A} \vdash \mathcal{B}^{M}$ implies $SC(\mathcal{B}^{M}) \subseteq SC(\mathcal{A})$, recalling the stability property (Prop.2.3), we have :
o.s.$(\mathcal{A}, {}^{I}SC(\mathcal{A})) =$ o.s. $(\mathcal{B}^{M}, {}^{I}SC(\mathcal{A}) \cap SC(\mathcal{B}^{M})) =$ o.s.$(\mathcal{B}^{M}, {}^{I}SC(\mathcal{B}^{M}))$. Prop.2.2 yields : $\exists$ o.s. $(\mathcal{B}^{M}, {}^{I}SC(\mathcal{B}^{M})) \Rightarrow \exists \max {}^{I}SC(\mathcal{B}^{M}) \Rightarrow \exists$ o.s.$(\mathcal{B}^{M}, {}^{\Psi}SC(\mathcal{B}^{M})) =$
$=$ o.s. $(\mathcal{B}^{M}, {}^{I}SC(\mathcal{B}^{M}))$. Thus by Prop.3.1 $\overline{\Psi}$ exists and $\mu(M/\overline{\Psi}) =$ o.s.$(\mathcal{B}^{M}, {}^{\Psi}SC(\mathcal{B}^{M}))$
$=$ o.s. $(\mathcal{A}, {}^{I}SC(\mathcal{A}))$.//

Consider $\mathfrak{A} = (A, \alpha)$ and $\mathcal{B} \in DF(A)$; we now investigate on the nature of $SC(\mathfrak{A}, \mathcal{B})$.
If $x \notin \Sigma \cup A$, recalling the definition of the operators $\|\tau\|_{\mathfrak{A}}$, $\tau \in F_{\Sigma}(\{x\})$, we have the following definition.

Definition 3.1. V' is the relation on A defined by : if $a, b \in A$, $(a,b) \in V'_{\mathcal{B}}$ iff $\forall \tau \in F_{\Sigma}(\{x\})$, $\forall B \in \mathcal{B}$, $\|\tau\|_{\mathfrak{A}}(a) \in B \Rightarrow \|\tau\|_{\mathfrak{A}}(b) \in B \cup B_{o}$ and $\|\tau\|_{\mathfrak{A}}(b) \in B \Rightarrow \|\tau\|_{\mathfrak{A}}(a) \in B \cup B_{o}$.

This relation is a modification and generalization of a relation defined in Verbeek (1967). It is clear that $V'_{\mathcal{B}}$ is reflexive and symmetric but examples show that it is not in general transitive (e.g. when $\mathfrak{A} = \mathfrak{F}_{\Sigma}$ and $\bigcup_{B \in \mathcal{B}} B$ is a finite set).
Let $V_{\mathcal{B}}$ be the transitive closure of $V'_{\mathcal{B}}$; we have then the following proposition .

Proposition 3.3. i) $(a,b) \in V'_{\mathcal{B}} \Rightarrow (\|\tau\|_{\mathfrak{A}}(a), \|\tau\|_{\mathfrak{A}}(b)) \in V'_{\mathcal{B}}$ $\forall \tau \in F_{\Sigma}(\{x\})$; ii) $V_{\mathcal{B}}$ is a congruence on $\mathfrak{A}$; iii) $SC(\mathfrak{A}, \mathcal{B}) = \{\Psi \in C(\mathfrak{A}) / \Psi \subseteq V'_{\mathcal{B}}\}$; iv) $SC(\mathfrak{A}, \mathcal{B})$ is a (quasi-complete) lower sub-semilattice of the complete lattice $C(\mathfrak{A})$; if $V'_{\mathcal{B}} = V_{\mathcal{B}}$ then $V_{\mathcal{B}} = \max SC(\mathfrak{A}, \mathcal{B})$ and $SC(\mathfrak{A}, \mathcal{B})$ is a complete lattice.

Proof. i) follows from the property : $\forall \tau, \tau' \in F(\{x\})$, $\forall a \in A$, $\|\tau'\|_{\mathfrak{A}}(\|\tau\|_{\mathfrak{A}}(a)) =$ $= \|\tau''\|_{\mathfrak{A}}(a)$, where $\tau'' := \|\tau'\|_{\mathfrak{F}}(\tau)$ and $\mathfrak{F} := \mathfrak{F}_{\Sigma}(\{x\})$. ii) consider $\Theta(V'_{\mathcal{B}})$; by i) and the definition of $\Theta(V'_{\mathcal{B}})$, it is easy to prove that $\Theta(V'_{\mathcal{B}}) = V_{\mathcal{B}}$. The proof of iii) again is based on property i) and the fact that the same property holds for any congruence. Eventually it is quite easy to verify the first part of iv), while the second is rather trivial given ii) and iii). //

Remark. If $V'_{\mathcal{B}} \neq V_{\mathcal{B}}$ it is not true, in general, that $V_{\mathcal{B}}$ is the l.u.b. of $SC(\mathfrak{A}, \mathcal{B})$, as we shall see with an example later on.

The following proposition is strictly related, as we shall see, to the stability property.

Proposition 3.4. If $V'_{\mathcal{B}} = V_{\mathcal{B}}$, set $\bar{\mathcal{B}} = \mathcal{B}_{V_{\mathcal{B}}}$; then : $V_{\bar{\mathcal{B}}} = N_{\bar{\mathcal{B}}} = V'_{\mathcal{B}} = V_{\mathcal{B}}$ and $C(\mathcal{U},\bar{\mathcal{B}}) = SC(\mathcal{U},\bar{\mathcal{B}}) = SC(\mathcal{U},\mathcal{B})$, hence $SC(\mathcal{U},\mathcal{B})$ is a (complete) sublattice of $C(\mathcal{U})$.

Proof. It is easy to verify that $\mathcal{B} \Vdash \bar{\mathcal{B}}$ implies $V'_{\mathcal{B}} \supseteq V'_{\bar{\mathcal{B}}}$; hence : $V_{\mathcal{B}} \supseteq V_{\bar{\mathcal{B}}} \supseteq N_{\bar{\mathcal{B}}}$. If $B \in \mathcal{B}$, denote $B_{V_{\mathcal{B}}}$ by $\bar{B}$, and $\|\tau\|_{\mathcal{U}}$ by $\|\tau\|$, $\forall \tau \in F(\{x\})$. As $V_{\mathcal{B}}$ is a congruence $(a,b) \in V_{\mathcal{B}} \Rightarrow (\|\tau\|(a), \|\tau\|(b)) \in V_{\mathcal{B}}$; now, $\forall \bar{B} \in \bar{\mathcal{B}}$, we have : $\|\tau\|(a) \in \bar{B} \Leftrightarrow \exists\, z \in B$ s.t. $(\|\tau\|(a),z) \in V_{\mathcal{B}} \Leftrightarrow (\|\tau\|(b),z) \in V_{\mathcal{B}} \Leftrightarrow \|\tau\|(a) \in \bar{B}$, $\forall \tau \in F_{\Sigma}(\{x\})$, i.e. $(a,b) \in N_{\bar{\mathcal{B}}}$; so $N_{\bar{\mathcal{B}}} \supseteq V_{\mathcal{B}}$, hence : (+) $V_{\mathcal{B}} = V_{\bar{\mathcal{B}}} = N_{\bar{\mathcal{B}}}$ which implies also $V'_{\mathcal{B}} = V_{\mathcal{B}}$. Moreover : $\mathcal{B} \Vdash \bar{\mathcal{B}}$ yields $SC(\mathcal{U},\mathcal{B}) \supseteq SC(\mathcal{U},\bar{\mathcal{B}}) \supseteq C(\mathcal{U},\bar{\mathcal{B}})$; by (+), Theor.1.7 ii) and Prop. 3.3, iii) , we have the coincidence of the three sets.//

Consider now $\Phi^{(1)}, \Phi^{(2)} \in SC(\mathcal{U},\mathcal{B})$, where $\Phi^{(1)} \subseteq \Phi^{(2)}$; if $\mathfrak{A}^{(i)} := \mathcal{U}/\Phi^{(i)}$ and $\mathfrak{F}^{(i)} := \mathcal{B}^{(i)}/\Phi^{(i)}$, i = 1,2, then the canonical epimorphism $\varphi : \mathfrak{A}^{(1)} \to \mathfrak{A}^{(2)}$ is such that $\forall B \in \mathcal{B}$, $B/\Phi^{(i)} \xrightarrow{\varphi} B/\Phi^{(2)}$ and $\varphi^{-1}(B/\Phi^{(2)}) = B/\Phi^{(1)}$.

Proposition 3.5. If $u,v \in A/\Phi^{(1)}$, the carrier of $\mathfrak{A}^{(1)}$, then : $(u,v) \in V'_{\mathfrak{F}^{(1)}}$ iff $(\varphi(u), \varphi(v)) \in V'_{\mathfrak{F}^{(2)}}$ (where obviously $V'_{\mathfrak{F}^{(i)}}$ is on $A/\Phi^{(i)}$).

Proof. It follows from the definition of the V'-relation and of the epimorphism and from the following property, (see Grätzer (1968)): $\forall \tau \in F_{\Sigma}(\{x\}), \forall u \in Q^{(1)}$ $\varphi(\|\tau\|_{\mathfrak{A}^{(1)}}(u)) = \|\tau\|_{\mathfrak{A}^{(2)}}(\varphi(u))$.//

Setting $\mathcal{U} = \mathfrak{F}_{\Sigma}$, if $M = (\mathfrak{A},\mathfrak{F})$, we have the following results.

Corollary 3.6. If $\mathfrak{F} = \{F_i,\ i \in I\}$ and $\mathcal{B} = \mathcal{B}^M = \{B_i \mid B_i = rp_M^{-1}(F_i),\ i \in I\}$, then $V'_{\mathcal{B}} = V_{\mathcal{B}}$ iff $V'_{\mathfrak{F}} = V_{\mathfrak{F}}$ (where $V'_{\mathcal{B}}$ is on $\mathfrak{F}_{\Sigma}$ and $V'_{\mathfrak{F}}$ on $\mathfrak{A}$).//

Cor.3.6. and Prop.3.3, together with Prop.3.2, give the following theorem.

Theorem 3.7. If $\mu(M) \in EDLJ(\mathcal{A}, {}^{I}SC(\mathcal{A}))$ and $V'_{\mathfrak{F}} = V_{\mathfrak{F}}$ then o.s.$(\mathcal{A}, {}^{I}SC(\mathcal{A}))$ exists and o.s. $(\mathcal{A}, {}^{I}SC(\mathcal{A})) = (V_{\mathcal{A}}, \mathcal{A}_{V_{\mathcal{A}}}) = \mu(M/V_{\mathfrak{F}})$.//

This theorem gives a sufficient condition for the existence of the optimal solution, and can be used to obtain an effective construction for it. This problem is investigated in what follows.

Consider $\mathfrak{A} = (Q,\alpha)$ and $\mathfrak{F} \in DF(Q)$; on Q we define the following relation $Z'_{\mathfrak{F}} := \bigcap_{n \geq 0} Z'_n$, where, if $a,b \in Q$, i) $(a,b) \in Z'_0$ iff $a \in F \Rightarrow b \in F \cup F_0$ and $b \in F \Rightarrow a \in F \cup F_0$ $\forall F \in \mathfrak{F}$; ii) $(a,b) \in Z'_{m+1}$ iff $(a,b) \in Z'_m$ and $\forall \omega \in \Sigma$, if $\omega \in \Sigma_{p+1}$ then, $\forall a_1,\dots,a_p \in Q$, the elements $\alpha_{\omega}(a_1,\dots,a_j,a,a_{j+1},\dots,a_p)$ and $\alpha_{\omega}(a_1,\dots,a_j,b,a_{j+1},\dots,a_p)$ are Z'_m - related, j = 0,...,p.

Lemma 3.8. i) $Z'_{\mathfrak{F}} \supseteq V'_{\mathfrak{F}}$; ii) $\Psi \leq SC(\mathfrak{A},\mathfrak{F}) \Rightarrow \Psi \leq Z'_{\mathfrak{F}}$; iii) if $Z'_{\mathfrak{F}}$ is transitive, then $Z'_{\mathfrak{F}} = V'_{\mathfrak{F}} = V_{\mathfrak{F}}$.

Proof. We use a modified, but equivalent, definition of $V'_{\mathfrak{F}}$, considering $F_{\Sigma}(\{x\} \cup Q)$

instead of $F_{\Sigma}(\{x\})$ and set $\|\tau\| := \|\tau\|_{\mathfrak{A}}$.

i). Obviously $(a,b) \in V'_{\mathfrak{F}} \Rightarrow (a,b) \in Z'_0$; if $(a,b) \in V'_{\mathfrak{F}} \Rightarrow (a,b) \in Z'_k$, $\forall k \leq m$, then $\forall \omega \in \Sigma_{p+1}$, $\forall a_1, \ldots, a_p \in Q$ consider, for $j = 0, \ldots, p$ and $u = a, b$: $\hat{u}_j := \alpha_\omega(a_1, \ldots, a_j, u, a_{j+1}, \ldots, a_p)$ and $\tau_j := a_1 \ldots a_j \; x \; a_{j+1} \ldots a_p \omega \in F_{\Sigma}(\{x\} \cup Q)$. Then, also recalling Prop. 3.3 , we have : $(a,b) \in V'_{\mathfrak{F}} \Rightarrow (\|\tau_j\|(a), \|\tau_j\|(b) = = (\hat{a}_j, \hat{b}_j) \in V'_{\mathfrak{F}}$ and thus by induction hypothesis $(\hat{a}_j, \hat{b}_j) \in Z'_m$. This shows that $(a,b) \in V'_{\mathfrak{F}} \Rightarrow (a,b) \in Z'_{m+1}$.

ii) is quite obvious as $Z'_{\mathfrak{F}}$ is an $\mathfrak{F}$-separating relation.

iii). We show that if $Z'_{\mathfrak{F}}$ is transitive then it is an $\mathfrak{F}$-sepatating congruence, so $Z'_{\mathfrak{F}} \subseteq V'_{\mathfrak{F}}$;hence, by i), $Z'_{\mathfrak{F}} = V'_{\mathfrak{F}}$ .If $Z'_{\mathfrak{F}}$ is transitive, clearly it is an $\mathfrak{F}$-separating equivalence; we have only to prove it is a congruence. First of all we observe that : $(a,b) \in Z'_{\mathfrak{F}} \Rightarrow \forall \omega \in \Sigma_{p+1}, \forall p \geq 0$, $\forall a_1, \ldots, a_p \in Q$, $(\hat{a}_j, \hat{b}_j) \in Z'_{\mathfrak{F}}$, $j = 0, \ldots, p$. Then, if $\omega \in \Sigma_p$ and $(u_j, v_j) \in Z'_{\mathfrak{F}}$, $j=1, \ldots, p$ and $\sim := \sim (\text{mod } Z'_{\mathfrak{F}})$, $\alpha_\omega(u_1, \ldots, u_p) \sim \alpha_\omega(v_1, u_2, \ldots, u_p) \ldots \sim \alpha_\omega(v_1, \ldots, v_p)$.//

<u>Remark.</u> It is also quite clear that : a) if $Z'_m = Z'_{m+1}$ then $Z'_m = Z'_{m+p}$, $\forall m,p \geq 0$; b) if Q is finite then $\bar{m}$ exists, $\bar{m} \leq |Q|^2$, s.t. $Z'_{\mathfrak{F}} = Z'_{\bar{m}}$. (The transitivity is not involved here.)

<u>Theorem 3.9.</u> If a finite automaton $M = (\mathfrak{A}, \mathfrak{F})$ is given s.t. $\mu(M) = ({}^I\Theta, {}^I\mathcal{A})$, there is an algorithm for deciding whetner o.s. $(\mathcal{A}, {}^I SC(\mathcal{A}))$ exists and, if so, for finding it.

<u>Proof.</u> By the above remark $Z'_{\mathfrak{F}}$ can be actually computed. If $Z'_{\mathfrak{F}}$ is transitive, then by lemma 3.8 $Z'_{\mathfrak{F}} = V'_{\mathfrak{F}} = V_{\mathfrak{F}}$, o.s. $(\mathcal{A}, {}^I SC(\mathcal{A})) = (V_{\mathcal{A}}, \mathcal{A}_{V_{\mathcal{A}}}) = \mu(M/V_{\mathfrak{F}})$ and $M/V_{\mathfrak{F}}$ is obtained by an effective construction. Otherwise, we can by inspection on all the congruences contained in $Z'_{\mathfrak{F}}$ (and they are a finite number) verify whether $SC(\mathfrak{A}, \mathfrak{F})$ has a maximum or not and so decide whether o.s $(\mathcal{A}, {}^I SC(\mathcal{A}))$ exists or not ; if it exists, by Prop.3.1., o.s. $(\mathcal{A}, {}^I SC(\mathcal{A})) = \mu(M/\bar{\Psi})$, where $\bar{\Psi} := = \max SC(\mathfrak{A}, \mathfrak{F})$.//

<u>Theorem 3.10.</u> If a finite automaton $M = (\mathfrak{A}, \mathfrak{F})$ is given , s.t. $\mu(M) \in EDLJ(\mathcal{A}, {}^I SC(\mathcal{A}))$ and $Z'_{\mathfrak{F}}$ is transitive, then o.s. $(\mathcal{A}, {}^I SC(\mathcal{A}))$ exists and there is an algorithm for finding it.

<u>Proof.</u> It follows directly from Prop.3.7 and lemma 3.8//

<u>Theorem 3.11.</u> Let $\Theta := \Theta(H) \in C(\mathfrak{F}_{\Sigma})$, H finite set, and a regular family $\mathcal{A}$ (i.e. $\mathcal{A} = \mathcal{B}^{\bar{M}}$, for a finite Σ-automaton $\bar{M}$) be given. There is an algorithm for deciding whether o.s. $(\mathcal{A}, {}^{\Theta} SC(\mathcal{A}))$ exists and if it exists, for finding it.

Proof. Recalling from section 2 the definition of $\Theta(H)$ we can see that if D is the maximal depth of the terms in H, then ind $(\Theta) < +\infty$ iff each term od depth D+1 is Θ-congruent to some term of depth $\leq D$. If ind $(\Theta) < +\infty$ it is not difficult to construct a finite Σ-automaton $\tilde{M}$ s.t. $ER(\tilde{M}) = \Theta$. If $\tilde{M} = (\tilde{\mathfrak{A}}, \tilde{\mathfrak{F}})$ and $\bar{M} = (\bar{\mathfrak{A}}, \bar{\mathfrak{F}})$, $\bar{M}$ being the automaton for $\mathcal{A}$ and $\bar{\mathfrak{F}} := \{\bar{F}_i, i \in I\}$ let $M := (\bar{\mathfrak{A}} \times \tilde{\mathfrak{A}}, \bar{\mathfrak{F}} \times \tilde{\mathfrak{F}})$ and M^c be the connected subautomaton of M (whose construction is effective). $M^c = (\mathfrak{A}^c, \mathfrak{F}^c)$, where : $\mathfrak{F}^c = \{F_i^c / F_i^c := \bar{\bar{F}}_i \times \tilde{\tilde{F}}_i , i \in I$ and $\tilde{\tilde{F}}_i := \{q \in \tilde{Q} / rp_{\tilde{M}}^{-1}(q) \cap A_i \neq \emptyset\}$, $\mathcal{A} \ni A_i := rp_{\bar{M}}^{-1}(F_i)$. Θ is -separating iff $DF(\tilde{Q}) \ni \mathcal{P} := \{\tilde{\tilde{F}}_i , i \in I\}$; $\mathcal{P}$ is finite, then we can decide whether its elements are non empty and mutually disjoint. If it is so, define $\bar{\bar{M}} := (\tilde{\mathfrak{A}}, \mathcal{P})$; then $\mathcal{B}^{\bar{\bar{M}}} = \mathcal{A}_\Theta$ and $ER(\bar{\bar{M}}) = \Theta$. Hence by Theorem 3.9 we have proved our assertions. //

CONCLUSIONS AND EXAMPLES. The above results show not only that, as it was to be expected, the existence and the nature of the optimal solution for a recognition problem depend on the devices we use to recognize, bout also that when the o.s. does not exist every maximal element in the set of good solutions is an o.s. in respect to a restricted class of recognizers. Thus supplementary "goodness" criteria are needed to define a unique o.s. . Moreover we have shown the essential role of the images ; for instance, if we are able to recognize the images, then for a regular family of examples $\mathcal{A}$ we know all about the optimal solution. Finally if $\mathcal{A}$ is a partition, the $V'_{\mathcal{A}} = N_{\mathcal{A}}$, $Z'_{\mathfrak{F}} = V'_{\mathfrak{F}} = N_{\mathfrak{F}}$ and the algorithmfor computing $Z'_{\mathfrak{F}}$ is exactly the one used by Brainerd (1968) for minimizing tree automata.

Example 1. Let $\Sigma = \Sigma_0 \cup \Sigma_1 = \{\lambda\} \cup \{a,b\}$; $Q = \{q_0, q_1, q_2, q_3\}$ and $\mathfrak{A} = (Q, \alpha)$, where $\alpha_\lambda = q_0$; $\alpha_a : q_0 \mapsto q_1, q_1 \mapsto q_2, q_2 \mapsto q_3, q_3 \mapsto q_3$; $\alpha_b : q_0 \mapsto q_3, q_1 \mapsto q_3, q_2 \mapsto q_1, q_3 \mapsto q_3$.

$\mathfrak{A}$ is therefore a monadic-algebra ; it can be seen that in a monadic algebra the Z' and V' relations coincide. If $\mathfrak{F} = \{\{q_1\}, \{q_2\}\}$, then we have $V'_{\mathfrak{F}} = Z'_{\mathfrak{F}} = Z'_1$ and the "relation classes" are $\{q_1, q_3\}$, $\{q_2, q_3, q_0\}$; hence $V'_{\mathfrak{F}} \neq V_{\mathfrak{F}}$. One can verify by inspection that the only $\mathfrak{F}$-separating congruence on $\mathfrak{A}$ is the diagonal congruence Δ .Therefore : max $SC(\mathfrak{A}, \mathfrak{F}) = \Delta$ and abviously $\Delta = N_{\mathfrak{F}} \neq V_{\mathfrak{F}}$.

Example 2. Let Σ be as above; then the terms in F_Σ can be viewed simply as elements of Σ_1^* . So consider the congruence Θ on $\mathfrak{F}_\Sigma$ whose classes are : $s_0 = \{\Lambda\}$, $s_1 = \{a\}$, $s_2 = a^2b(ab)^*$, $s_3 = a(ab)^*a$, $s_4 = ab(ab)^*a$, $s_5 = ab(ab)^*$, $s_6 = b(ab)^*$; $s_7 = \Sigma_1^* - \bigcup_{i=1}^{6} s_i$.

Set $\mathcal{A} = \{ s_1 \cup s_2 , s_3 \}$; then $\mathcal{B} := \mathcal{A}/\Theta = \{ \{[a]_\Theta, [a^2 b]_\Theta\}, \{[a^2]_\Theta\}\}$. It is easy to verify that on $\mathfrak{F}_\Sigma/\Theta$: the "relation classes" of $V'_\mathcal{B}$ are : $\{s_1, s_2, s_4, s_5, s_6, s_7\}$, $\{s_3, s_o, s_4, s_5, s_6, s_7\}$; we also have : l.u.b. $SC(\mathfrak{F}_\Sigma/\Theta, \mathcal{B})$ is the (non $\mathcal{B}$-sep.) congruence whose classes are $\{s_o\}$ and $\{s_1, \ldots, s_7\}$, and which is obviously different from $V_\mathcal{B}$. Moreover, setting $M = (\mathfrak{A}, \mathfrak{F})$ - see ex.1- , $\mu(M) \in EDL(\mathcal{A}, {}^\Theta SC(\mathcal{A}))$; then , as we have seen, max, $SC(\mathfrak{A}, \mathfrak{F})$ exists, while max $SC(\mathfrak{F}_\Sigma/\Theta, \mathcal{B})$ does not exist, which implies that also o.s. $(\mathcal{A}, {}^\Theta SC(\mathcal{A}))$ does not exist.

Example 3. Let $\Sigma = \Sigma_o \cup \Sigma_1 = \{\lambda\} \cup \{a,b,c\}$; then we can consider Σ_1^* instead of F_Σ. Set : $D := \{a\} \cup \{b\} \cup cc^*a \cup cc^*b$, $A_1 := c^*bD^*$, $\bar{A}_1 := c^*bD^*cc^*$, $A_2 := c^*aD^*$, $\bar{A}_2 := c^*aD^*cc^*$, $A_3 := c^*$, $\mathcal{A} = A_1, A_2$, ${}^I\Theta = \Delta$. Then $V'_\mathcal{A} = V_\mathcal{A}$, with classes $A_3, A_1 \cup \bar{A}_1, A_2 \cup \bar{A}_2$, so that o.s.$(\mathcal{A}, SC(\mathcal{A})) = (V_\mathcal{A}, \{A_1 \cup \bar{A}_1 , A_2 \cup \bar{A}_2\})$.

REFERENCES

ARBIB,M.A. and GIVE'ON, Y.(1968) "Algebra automata I : parallel programming as a prolegomena to the categorical approach" Inform. and Control 12,331-345.

ASTESIANO,E. (1973) "A recognition model", to appear in Pubblicazioni dell'Istituto di Matematica dell'Università di Genova.

BRAINERD,W.S. (1968) "The minimalization of tree automata", Inform.and Control 13, 484-491.

BÜCHI,J.R. (1966) "Algebraic Theory of feedback in discrete systems- Part I " in Automata theory ,edited by E.R. Caianello ,Academic Press.

COHN,P.M. (1965) "Universal algebra", Harper,New York.

GRÄTZER,G. (1968) "Universal algebra" ,Von Nostrand,New York.

GRENANDER,U.(1969) "Foundations of patterns analysis" ,Quart.Appl.Math.27, 1-55

PAVEL,M. (1969) "Fondements mathématiques de la reconnaissance des structures", Hermann, Paris

THATCHER,J.W. and WRIGHT,J.B.(1968) "Generalized finite automata theory with an application to a decision problem of second order logic",Math.Systems Theory,Vol.2. N.1,57-81.

VERBEEK, L.A.M. (1967) " Congruence separation of subsets of a monoid with application to automata " , Math. Systems Theory, Vol.1, N.4, 315-324

A NEW FEATURE SELECTION PROCEDURE FOR PATTERN RECOGNITION BASED ON SUPERVISED LEARNING

by

Josef Kittler

Control and Systems Group, Department of Engineering

University of Cambridge, England

1. Introduction

Linear methods of feature selection are characterised by a linear transformation or 'mapping' of a pattern vector $\underline{x}$ from an N dimensional space X into a $n < N$ dimensional space Y. The feature vector $\underline{y} \in Y$ which is obtained from $\underline{x}$ by transformation T, i.e. $\underline{y}^T = \underline{x}^T T$, has a reduced number of components and should, if successful, contain all of the information necessary for discriminating between classes present in the original vector $\underline{x}$.

Many methods have been suggested for determining the transformation matrix T required for linear feature selection. But most of these can be classified in one of the following two categories:

a) methods based on the Karhunen-Loeve expansion

b) methods using discriminant analysis techniques.

In the first part of the present paper a new method of feature selection based on the Karhunen-Loeve (K-L) expansion is proposed. Subsequently a relationship between the superficially different K-L expansion and discriminant analysis approaches is established and in so doing a more unified approach to the problem of feature selection is introduced.

2. A Method of Feature Selection for Pattern Recognition Based on Supervised Learning

The method of feature selection discussed in this paper is based on the properties of the K-L expansion. Since the detailed treatment of the Karhunen-Loeve expansion of discrete and continuous processes can be found elsewhere (Mendel and Fu, Fukunaga, Kittler and Young) only a brief description of the method will be given here. Also for simplicity, we shall confine our discussion to the case of discrete data.

Consider a sample of random N dimensional pattern vectors $\underline{x}$. Each vector is associated with one of m possible classes ω_i. Let the mean of $\underline{x} \varepsilon \omega_i$ be $\underline{\mu}_i$ and denote the noise on $\underline{x}_i$ by $\underline{z}_i$, i.e.

$$\underline{x}_i = \underline{z}_i + \underline{\mu}_i \tag{1}$$

Without loss of generality we can assume that the overall mean $\underline{\mu} = E\{\underline{x}\} = 0$ since it is clearly possible to centralize the data prior to analysis by removal of the overall mean. Suppose that the probability of occurence of i-th class is $P(\omega_i)$ and let membership of patterns in their corresponding classes be known.

Suppose that we would like to expand the vector $\underline{x}$ linearly in terms of some deterministic functions $\underline{t}_k$ and associated coefficients y_k, i.e.

$$\underline{x} = \sum_{k=1}^{N} y_k \underline{t}_k \tag{2}$$

subject to the conditions:

α) $\underline{t}_k$ are orthonormal

β) y_k are uncorrelated

γ) the representation error $\bar{e}$,

$$\bar{e} = \sum_{i=1}^{m} P(\omega_i) \, E\{|\underline{x}_i - \hat{\underline{x}}_i|^2\} \tag{3}$$

incurred by approximating $\underline{x}$ with $\hat{\underline{x}}$ composed of $n < N$ terms in the expansion (2), i.e.

$$\hat{\underline{x}} = \sum_{k=1}^{n} y_k \underline{t}_k$$

is minimised.

Fu and Chien have shown that the deterministic functions satisfying the property α through β are the eigenvectors T

$$T = [\underline{t}_1 \; \underline{t}_1 \; \ldots \; \underline{t}_N]$$

of the sample covariance matrix C defined as

$$C = E\{\underline{x}\underline{x}^T\} = \sum_{i=1}^{m} P(\omega_i) \, E\{\underline{x}_i \underline{x}_i^T\}$$

In order to satisfy condition γ , it is necessary to arrange the eigenvectors $\underline{t}_1, \ldots, t_N$ in the descending order of their associated eigenvalues,

$$\lambda_1 \geq \lambda_2 \geq \ldots \lambda_n \geq \ldots \geq \lambda_N$$

Chien and Fu also showed that the eigenvalues λ_k are the variances of the transformed features y_k and that the expansion has some additional favourable properties, in particular, the total entropy and residual entropy associated with the transformed features are minimised.

The transformation T of the pattern vectors into the K-L coordinate system results in compression of the information contained in the original N-dimensional pattern vectors $\underline{x}$ into $n < N$ terms of the K-L expansion. This latter property has been utilised for feature selection in pattern recognition by various authors and a few of the possibilities are listed below.

i) If the features y_k are to be uncorrelated then T should be chosen as the matrix of eigenvectors associated with the mixture covariance matrix C_1 defined as (Watanabe)

$$C_1 = E\{\underline{x} \, \underline{x}^T\}$$

ii) If the transformation is required to decorrelate the components of the noise vector $\underline{z}_i$ then the matrix of eigenvectors T should correspond to the averaged within class covariance matrix C_2, i.e.

$$C_2 = \sum_{i=1}^{m} P(\omega_i) \; E\{\underline{z}_i \underline{z}_i^T\}$$

It should be noted that method ii) selects features irrespective of the discriminatory information contained in the class means and the utilization of information about the mean vectors in method i) is not optimal in any sense. This can be seen from a detailed analysis of variances λ_k of the new features y_k, i.e.

$$\lambda_k = \sum_{i=1}^{m} P(\omega_i) \; E(y_{ki}^2) = \sum_{i=1}^{m} P(\omega_i) \; \sigma_{ki}^2 + \sum_{i=1}^{m} P(\omega_i) \; \gamma_{ki}^2 = \tilde{\rho}^2 + R_{kk} \tag{4}$$

where

$$\sigma_{ki}^2 = E\{(y_{ki} - \gamma_{ki})^2\} \tag{5}$$

and

$$\gamma_{ki} = E\{y_{ki}\} \tag{6}$$

In an earlier paper, Kittler, J. and Young, P.C. (1973) have shown that the discriminant power of a feature against class means is related to the ratio $R_{kk}/\tilde{\rho}^2$ provided that the averaged within class covariance matrix is in the diagonal form.

The magnitude of the first term in (4), however, contains no discriminatory information at all. It is therefore desirable to normalise the averaged within class variances to unity and thus to allow selection of the features on the basis of the magnitude of the eigenvalues λ_k. It is this normalising transformation which is the essence of the proposed new feature selection technique.

iii) In order to normalise the noise, we first have to diagonalise the averaged within class covariance matrix C_2,

$$C_2 = \sum_{i=1}^{m} P(\omega_i) \; E\{\underline{z}_i \; \underline{z}_i^T\} \tag{7}$$

by transforming $\underline{x}$ into a new feature vector $\underline{y}$ using the system of eigenvectors U associated with C_2. Thus the feature vector $\underline{y}$ where

$$\underline{y}^T = \underline{x}^T U \tag{8}$$

will have a diagonal covariance matrix

$$C_y = \sum_{i=1}^{m} P(\omega_i) \; E\{U^T \underline{z}_i \underline{z}_i^T U\} = \begin{bmatrix} \lambda_i & & & \\ & \lambda_2 & & 0 \\ & 0 & \ddots & \\ & & & \lambda_N \end{bmatrix} \tag{9}$$

where λ_k's are eigenvalues associated with C_2.

The matrix C_y can now be transformed into identity matrix by multiplying each component of the feature vector $\underline{y}$ by the inverse of its standard deviation. In matrix form this operation can be written as

$$\underline{g}^T = \underline{y}^T S \tag{10}$$

where

$$S = \begin{vmatrix} \lambda_k^{-\frac{1}{2}} & & 0 \\ & \ddots & \\ 0 & & \lambda_{NN}^{-\frac{1}{2}} \end{vmatrix} \tag{11}$$

Once the averaged within class covariance matrix is in the identity form, by solving the eigenvalue problem, $C_g B = B\Lambda_g$ (12)
we can obtain a new K-L coordinate system, B, which is optimal with respect to the mixture covariance matrix C_g, where

$$C_g = E\{\underline{g}\,\underline{g}^T\} \tag{13}$$

Λ_g is the matrix of eigenvalues of the matrix C_g.
Note that the class means are included in this case. Thus B is a coordinate system in which the square of the projections of the class means onto the first coordinates averaged over all classes is maximised.

The eigenvalues λ_{gk} which are the diagonal elements of Λ_g can now be expressed as

$$\lambda_{gk} = 1 + \sum_{i=1}^{m} P(\omega_i)\,(E\{\underline{b}_k^T\,\underline{g}_i\})^2 = 1 + \lambda_{kk} \tag{14}$$

It follows that the features selected according to the magnitude of λ_{gk} will now be ordered with respect to their discriminatory power.

The feature selection procedure can be summarised in the following steps:

1) Using K-L expansion diagonalise the mixture covariance matrix $\tilde{C} = E\{\underline{\tilde{x}}\,\underline{\tilde{x}}^T\}$ of the original $\tilde{N}$-dimensional data $\underline{\tilde{x}}$. Disregard those features which are associated with negligable eigenvalues and generate a feature vector $\underline{x}^T = \tilde{x}^T W$ of dimension $N < \tilde{N}$ where W is the system of eigenvectors of $\tilde{C}$.
2) Find the K-L coordinate system U in which the averaged within class covariance matrix C_y defined in (9) is diagonal.
3) Normalise the features y_k to transform the matrix C_y into identity matrix form.
4) Determine the final K-L coordinate system B which is associated with the mixture covariance matrix C_g.

If we ignore the first K-L expansion, which does not affect feature selection but only reduces the computational burden of the following two K-L analyses, we can view the overall linear transformation $T = USB$ as one that decorrelates the mixture covariance matrix C_1 subject to the condition that

$$T^T C_2 T = I \tag{15}$$

The resulting feature vector $\underline{f}$ is then given as

$$\underline{f}^T = \underline{x}^T T \tag{16}$$

Note that the proposed feature selection technique is applicable to supervised pattern recognition problems only because the membership of the training patterns in the individual classes must be known a priori so that the averaged within class covariance matrix C_2 can be computed.

In order to show the relationship between the K-L expansion techniques i), ii), iii), and the discriminant analysis techniques which we describe in the next section it is necessary to derive the results outlined above in an alternative manner. This requires that the problem is viewed as one of simultaneous diagonalization of the matrices C_2 and C_1. From the previous discussion, we know that

$$SU^T C_2 US = I \tag{17}$$

Now by utilising (17), the eigenvalue problem (12) can be written as

$$SU^T (C_1 - \lambda_g C_2) US\underline{t} = 0 \tag{18}$$

But since SU^T is nonsingular, the condtion (18) will be satisfied only if

$$|C_1 - \lambda_g C_2| = 0 \tag{19}$$

It follows that λ_g are the eigenvalues of the matrix $C_1 C_2^{-1}$, i.e.

$$(C_1 C_2^{-1} - \lambda_g I)\underline{t} = 0 \tag{20}$$

2.1 Experimental Results

The results of an experimental comparison of the three K-L procedures outlined above is given in Fig. 1. Data for the experiment were generated digitally according to the rule,

$\underline{x}$ (class A): $x_1 \sim N(2,2), x_2 \ldots x_9 \sim N(0,1), x_{10} \sim N(0,0.25)$

$\underline{x}$ (class B): $x_1 \sim N(1.95,2), x_2 \ldots x_9 \sim N(0,1), x_{10} \sim N(0.5,0.25)$

where $N(\mu,\sigma)$ defines a normal distribution with standard deviation σ and mean μ. From these results, it will appear that the method iii) performs substantially better than the other two K-L procedures.

3. Discriminant Analysis

Let us now formulate the problem of feature selection in a different way. Suppose that we wish to find a linear transformation matrix T which maximises some distant criterion d defined over the sample of random vectors in the transformed space. Two of the most important distance measures are as follows:

a) The intraset distance d_{1n} between the n-th feature of all the pattern vectors in one class averaged over all classes, which is defined by

$$d_{1n} = \tfrac{1}{2}\sum_{i=1}^{m} P(\omega_i) \frac{1}{N_i^2} \sum_{j}^{N_i} \sum_{l}^{N_i} \underline{t}_n^T (\underline{x}_{ij} - \underline{x}_{il})(\underline{x}_{ij} - \underline{x}_{il})^T \underline{t}_n \tag{21}$$

where N_i is the number of vectors $\underline{x} \in \omega_i$ and $\underline{t}_n$ is the n-th column of the transformation matrix T.

b) The interset distance d_{2n} between the n-th feature of all patterns, which is defined by

$$d_{2n} = \sum_{i=2}^{m} P(\omega_i) \sum_{h=1}^{i=1} P(\omega_h) \frac{1}{N_i N_h} \sum_{j}^{N_i} \sum_{l}^{N_h} \underline{t}_n^T (\underline{x}_{ij} - \underline{x}_{hl})(\underline{x}_{ij} - \underline{x}_{hl})^T \underline{t}_n \tag{22}$$

It has been shown, Kittler, J. (1973), that these distance critera can be expressed in terms of sample covariance matrices. In particular the distances d_{1n} and d_{2n} become

$$d_{1n} = \underline{t}_n^T C_2 \underline{t}_n \tag{23}$$

with C_2 given by (4) and

$$d_{2n} = \underline{t}_n^T \psi \, \underline{t}_n$$

where

$$\psi = C_1 - \sum_{i=1}^{m} P^2(\omega_i) \, E\{\underline{z}_i \underline{z}_i^T\} \tag{24}$$

By analogy, the sum of the interset and the intraset distances d_{3n}, is

$$d_{3n} = \tfrac{1}{2} \sum_{i=1}^{m} P(\omega_i) \sum_{h=1}^{m} P(\omega_h) \frac{1}{N_i N_h} \sum_{j=1}^{N_i} \sum_{l=1}^{N_h} \underline{t}_n^T (\underline{x}_{ij} - \underline{x}_{hl})(\underline{x}_{ij} - \underline{x}_{hl})^T \underline{t}_n \tag{25}$$

can be written as

$$d_{3n} = \underline{t}_n^T (C_2 + M)\underline{t}_n = \underline{t}_n^T C_1 \underline{t}_n \tag{26}$$

where

$$M = \sum_{i=1}^{m} P(\omega_i) \, \underline{\mu}_i \underline{\mu}_i^T$$

Clearly a number of different distance criteria could be constructed; but, in all cases it would be possible to express the distance in terms of a covariance-type matrix C, i.e.

$$d_n = \underline{t}_n^T C \underline{t}_n \tag{27}$$

Using these results the maximisation of a chosen distance criterion d can be carried out by maximising with respect to the transformation vector $\underline{t}$, subject to some additional constraints, e.g. holding constant some distance s where

$$s = \underline{t}_n^T C_s \underline{t}_n \tag{28}$$

which is irrelevant for classification purposes. s might, for example, be defined as the intraset distance. The solution for this kind of problem can be obtained by method of Lagrange multipliers. In case of a simple constraint, maximisation of d subject to $s = \text{const}$ can be written as the maximisation of f, where

$$f = d - \lambda(s - \text{const}) = \underline{t}_n^T C \underline{t}_n - \lambda(\underline{t}_n^T C_s \underline{t}_n - \text{const}) \tag{29}$$

Setting the first derivatives of f with respect to the components of $\underline{t}_n$ equal zero yields

$$(C - \lambda C_s)\underline{t}_n = 0 \tag{30}$$

But if (30) is then postmultiplied by the inverse of C_s, we get an eigenvalue problem

$$(CC_s^{-1} - \lambda I)\underline{t}_n = 0 \tag{31}$$

When there is more than a single constraint, the solution is more complicated since the function under consideration now becomes

$$f = d - \sum_l \nu_l (S_l - \text{const}) \tag{32}$$

the solution must be obtained using general optimisation techniques. However, in the special case when there are only two constraints and one of these is simply the condition of orthonormality of the matrix T, i.e.

$$T^T T = I \tag{33}$$

the problem of optimisation can be posed as the eigenvalue problem

$$(C - \nu C_s - \lambda I)T = 0 \tag{34}$$

And since in the design of pattern recognition systems our chief interest is in classification performance, we can determine ν experimentally to achieve the minimum error rate.

3.1 Experimental Results

This latter approach for two constraints has been tested on four class, artificially generated data. The classes were generated according to the rule,

$\underline{x}$ (class A): $x_1,\ldots\ldots,x_{10} \sim (0,1)$

$\underline{x}$ (class B): $x_1,\ldots\ldots,x_8 \sim (0,1), x_9 \sim (4.2,1), x_{10} \sim (-4.2,1)$

$\underline{x}$ (class C): $x_1, \ldots, x_8 \sim (0,1), x_9 \sim (2.2,1.3), x_{10} \sim (2.2,1.3)$

$\underline{x}$ (class D): $x_1, \ldots, x_8 \sim (0,1), x_9 \sim (2.2,1.3), x_{10} \sim (-2.2,1.3)$

The summary of the results obtained using classifier with linear discriminant function is in Fig. 2. The best error rate corresponds to $\nu = 0.7$.

4. Discussion

There are many possible ways of defining the distance between the elements of a sample and there are even more combinations of constraints and distance criteria that could be maximised in any particular feature selection problem. Consequently we shall restrict our discussion here to a few specific methods that have been suggested in the past.

First, let us consider the method proposed by Sebestyen. In this procedure, the distance criterion to be maximised is d_{3n} , subject to the condition that the sum of the intraset distances remains constant. In this case the transformation matrix T' is the solution of the equation (31), i.e.

$$(C_1 C_2^{-1} - \lambda I)\underline{t}_n' = 0 \tag{35}$$

Comparing the relationship (35) with (20) we see that the column vectors $\underline{t}_n'$ are colinear with the coordinate system obtained by the method iii) described in section 2, i.e.

$$\underline{t}_n' = \zeta_n \underline{t}_n \tag{36}$$

But in contrast to the method iii), where T was chosen to satisfy

$$T^T C_2 T = I \tag{37}$$

the columns of the transformation matrix T' in the present case must be such that

$$\sum_{n=1}^{N} \underline{t}_n'^T C_2 \underline{t}_n' = \text{const} = K \tag{38}$$

And from (36) it follows that

$$\sum_{n=1}^{N} \zeta_n^2 \underline{t}_n^T C_2 \underline{t}_n = K \tag{39}$$

Thus, we can choose the N-1 coefficients ζ_n and evaluate the last one. But is there any particular choice of the coefficients ζ_n which would give better features? From (35) it follows that

$$\underline{t}_n'^T C_1 \underline{t}_n' - \lambda_n \underline{t}'^T C_2 \underline{t}_n' = \zeta_n^2 + \zeta_n^2 \lambda_{M_n} - \lambda_n \zeta_n^2 = 0 \tag{40}$$

where

$$\lambda_{M_n} = \underline{t}'^T_n M \underline{t}_n' \tag{41}$$

and $T'^T M T'$ is diagonal matrix. Using (40) and (36) the maximised distance d_{3n}

can be written

$$d_{3n} = \lambda_n \zeta_n^2 \tag{42}$$

However, from (40), the eigenvalue λ_n can be expressed in terms of λ_{M_n} as

$$\lambda_n = 1 + \lambda_{M_n} \tag{43}$$

Thus we can conclude that although the distance d_{3n} is proportional to ζ_n^2, the discriminatory power inherent in d_{3n} is a function of λ_n and therefore independent of ζ_n. Thus ζ_n can be chosen

$$\zeta_n^2 = \zeta_i^2 = \text{const} \qquad n,i. \tag{44}$$

and T' becomes

$$T' = \text{const } T \tag{45}$$

Apart from some constant of proportionality the features obtained in this manner will be exactly the same as those obtained by the method iii) and will satisfy the ordering criterion (13). It is interesting to note that the distance d_{3n} is only proportional to the discriminatory power of the n-th feature. If d_{3n} is used as an ordering criterion, therefore, any ill-chosen coefficients could result in a suboptimal ordering of the features.

If the interset distance d_{2n} is maximised instead of d_{3n}, with the same constraints, then the situation is rather more complicated since the matrix C in (31) is now replaced by ψ. It can be shown that the matrix T obtained as the solution of (31) diagonalizes both matrices ψ and C_2 and this means that

$$T^T \psi T - \Lambda\Lambda_O = O \tag{46}$$

where $\Lambda_O = T^T C_2 T$ and Λ is the matrix of eigenvalues of ψC_2^{-1}. Substituting for ψ from (24) the first term on the left hand side of (46) can be rewritten to yield

$$\Lambda_O + T^T(M - \sum_{i=1}^{m} P^2(\omega_i)E\{\underline{z}_i\underline{z}_i^T\})T - \Lambda\Lambda_O = O \tag{48}$$

The term in the middle of the left hand side of the equation must also be diagonal and it is, in fact, this term which determines the optimal coordinate system T. Depending on the relative dominance of M or $\sum_{i=1}^{m} P^2(\omega_i)E\{\underline{z}_i\underline{z}_i^T\}$ the axes T may coincide with the coordinate system in the previous case or lie in the direction determined by the term $\sum_{i=1}^{m} P^2(\omega_i)E\{\underline{z}_i\underline{z}_i^T\}$. However, in general, T will be a compromise between these two.

Let us denote the second term of (48) by $\tilde{\Lambda}_M$, i.e.

$$\tilde{\Lambda}_M = T^T(M - \sum_{i=1}^{m} P^2(\omega_i)\ E\{\underline{z}_i\underline{z}_i^T\})T \tag{49}$$

Then by analogy to (43) the elements λ_n of the matrix of eigenvalues Λ can be expressed as

$$\lambda_n = 1 + \tilde{\lambda}_{M_n} \tag{50}$$

Now even if we select the features according to the magnitude of the eigenvalues λ_n, instead of the distance d_{2n} the ordering may not necessarily be satisfactory since the magnitude of λ_{M_n} is proportional not only to the discriminatory power of the n-th feature but also to noise, represented by the second term of $\tilde{\lambda}_{M_n}$, i.e.

$$\tilde{\lambda}_{M_n} = \underline{t}_n^T M\underline{t}_n - \underline{t}_n^T (\sum_{i=1}^{m} P^2(\omega_i)E\{\underline{z}_i\underline{z}_i^T\})\underline{t}_n \tag{51}$$

Thus the method might in certain circumstances yield inferior features to those obtained by maximising the criterion d_{3n}.

Finally, a few remarks are necessary in connection with the special case of feature selection with two constraints discussed in the previous section. From the experimental results obtained where the distance d_{2n} was maximised subject to the condition

$$\sum_{n=1}^{N} t_n^T C_2 t_n = \text{const}, \tag{52}$$

we can conclude that the optimal coordinate axes T almost coincide with the eigenvectors of the matrix M (since ν is such that the constraint matrix cancels out the within class scatter matrix $C_2 - \sum_{i=1}^{m} P^2(\omega_i)E\{\underline{z}_i\underline{z}_i^T\})$. This fact is even more obvious from the experimental results of Nieman, who originally suggested this approach. He assumed that the a priori class probabilities associated with the classification of ten numerals were equal. The minimum error rate was then obtained for $\nu = 0.9$. Now if $P(\omega_i) = 0.1, \forall_i$, then the eigenvalue problem defined by (34) becomes

$$(C_2 - C_2/10 + M - \nu C_2 - \lambda I)T = (M - \lambda I)T = 0 \tag{53}$$

when $\nu = 0.9$. This result is only to be expected since we cannot possibly decorrelate both within and between class scatter matrices by a single orthonormal transformation. And it is quite reasonable therefore, that the most important features will be selected with some degree of confidence only in the coordinate system where their means are decorrelated. Thus, in practice, we can only hope that our choice will not be affected by the projected noise.

These same remarks apply when the distance d_{3n} is maximised. But in this case when $\nu = 0$ the problem reduces to the K-L method i). However, from the above discussion we see that better results can be obtained for $\nu \to +1$. Consequently Niemann's method will yield superior features.

5. Conclusion

The most important result of the comparative study described in this paper is the correspondence and, in particular cases, the direct equivalence that has been established between some statistical feature selection techniques developed from the Karhunen-Loeve expansion and some alternative techniques obtained by maximisation of a distance criterion. This allows us to extend the properties of features obtained by linear transformation derived from the K-L expansion to the distance optimisation methods, and vice versa. Thus we know, for instance, that the features obtained by Sebestyen's method will be not only maximally separated but also uncorrelated.

In a previous paper we have shown analytically and confirmed experimentally that the K-L procedure iii) is particularly favourable for feature selection applications. Some additional experimental results supporting this conclusion are presented in section 2 of the present paper. Moreover the comparative study described here reveals that this procedure retains its advantages for an even larger class of linear transformation techniques which include methods based on separability measures.

The coordinate system used by the K-L method iii) can be obtained by a successive application of the K-L expansion, as described in section 2. Alternatively it can be obtained from the system of eigenvectors associated with the product of two matrices, one of them being the within class covariance matrix as described in section 3. However, the designer may have some difficulties with the latter approach, particularly if the matrix being inverted is not well defined. Both for this reason, and also in order to have a greater control of the analysed data, it seems better to use the first method. Although this implies two eigenvalue analyses, the matrices involved are symmetric and the problem is computationally quite simple.

References

1. Chien, Y.T., Fu, K.S., IEEE Trans. Inf. Theory, IT-15, 518 (1967)
2. Fukunaga, K.: Introduction to statistical pattern recognition, The Macmillan Company, New York, (1972)
3. Kittler, J.: On Linear Feature Selection. Technical Report of Cambridge University Eng. Dept., CUED/B-Control/TR54 (1973).
4. Kittler, J., Young, P.C.: A new approach to feature selection based on the Karhunen-Loeve expansion, Jnl. Pattern Recognition (to be published, 1973).
5. Mendel, J.M., Fu, K.S.: Adaptive, Learning and Pattern Recognition Systems, Academic Press, New York (1970).
6. Niemann, H.: An Improved Series Expansion for Pattern Recognition, Nachrichkentechn, Z, pp. 473-477, (1971).

7. Sebestyen, G.S.: Decision Making Processes in Pattern Recognition, the Macmillan Company, New York (1962).

8. Watanabe, S.: Computers and Information Science II, Academic Press, New York (1967).

Fig. 1

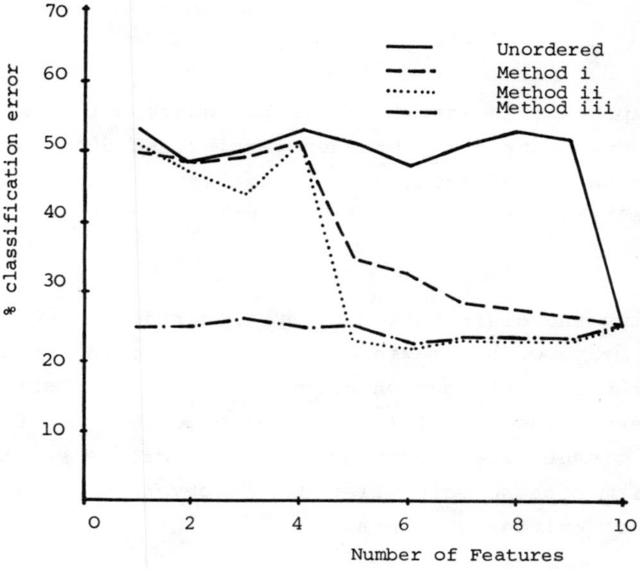

Fig. 2

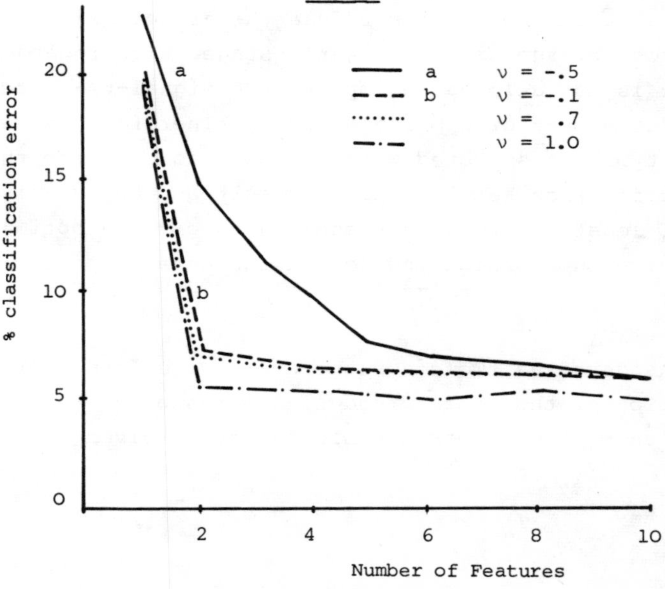

A CLASSIFICATION PROBLEM IN MEDICAL RADIOSCINTIGRAPHY

Georg Walch
IBM Heidelberg Scientific Center

1. INTRODUCTION

In this paper we present a classification procedure which we have developed in the framework of hypotheses testing and decision theory to do the classification of scintigraphic images into the class of normal pattern and that of pattern with anomalies in an interactive way.

In nuclear medicine the distribution of applied radioactive compounds in human body is measured by imaging devices like moving scanner and gamma-camera to receive information about tumours and metastases on the base of different storage effects in those lesions and the surrounding healthy tissue. The accumulated data is called scintigraphic image or scintigram because scintillations in crystals are involved in the detection of emitted γ-quanta.

These scintigrams are characterized by high statistical noise, both radioactive decay and γ-detection being of statistical nature, and low spatial resolution. Both facts reduce the detectability of small anomalies produced by tumours in an early stage. With regard to radiation damage it is not allowed to improve the signal-to-noise ratio by increasing the amount of applied activity. Therefore, digital filters of Wiener type, constructed with the goal to minimize the expectation of the difference between filter result and object distribution in the sense of least squares, were applied to give an optimal compromise between noise suppression and resolution enhancement (Pistor et al).

They may be considered as two stage operators, the first part for the optimal estimation of the undisturbed signal response, the second part for the inversion of the linear transformation involved in the imaging process.

In the filtered images the resolution, the contrast, the signal-to-noise ratio, and thereby the detectability of small anomalies are improved. But the classification between normal and abnormal pattern, the decision whether an observed fluctuation is of statistical or biological nature, is still left to the human inspector. To do this classification automatically or at least to base the human decision on quantitative measures, a likelihood ratio test is adapted to this special case.

2. LIKELIHOOD RATIO TEST

2.1 General Remarks

In our case the null hypothesis H_0 is:

The image distribution is a known normal intensity pattern.

The alternative H_1 is:

The normal pattern is changed at known position by an anomaly of known shape.

The paramters size and strength of the anomaly are unknown and may vary. Therefore, the alternative is a composite hypothesis. It is reduced to a single one by estimating those parameters according to the maximum likelihood method.

The likelihood ratio test rejects the null hypothesis H_0 if the ratio

$$\Lambda = \frac{L_0(X)}{L_1(X)} < K \qquad (1)$$

or

$$T = -2 \log \Lambda > K' \qquad (2)$$

where X is the given sample of observations, L_0 and L_1 denote the joint probability functions of the sample under H_0 and H_1, respectively. According to the Neyman-Pearson lemma a likelihood ratio test is most powerful, i.e. there is no test with equal or smaller type I error which has a smaller type II error (Lindgren).

2.2 Specialized Formulae

An image area containing n observed quanta or counts is divided in k image cells. If p_i is the probability for a single count to be in the i-th cell, the expected frequency in this cell is np_i. The observed frequencies x_i are Poisson distributed with means np_i which distributions may be approximated by normal distributions with the same means and variances. The x_i being statistically independent, the joint probability L of the sample $X = (x_1, \ldots, x_k)$ is

$$L(X) = \prod_{i=1}^{k} \frac{1}{(2\pi np_i)^{1/2}} e^{-\frac{(x_i-np_i)^2}{2np_i}} \tag{3}$$

and

$$-2 \log L(X) = k \log(2\pi n) + \sum_{i=1}^{k} \log p_i + \sum_{i=1}^{k} \frac{(x_i-np_i)^2}{2np_i} \tag{4}$$

If H_0 is true we write p_{0_i} for p_i and $(\chi^2)_0$ for the last sum in eq. (4), if H_1 is true we write p_{1_i} and $(\chi^2)_1$, respectively. Then T is expressed by

$$T = \sum_{i=1}^{k} \log \left(\frac{p_{0_i}}{p_{1_i}}\right) + (\chi^2)_0 - (\chi^2)_1 \tag{5}$$

To make use of this formula we have to formulate the hypotheses H_0 and H_1, i.e. to give explicitely the p_{0_i} and p_{1_i}, and we have to fix the critical value K'. Furthermore we want to know the probabilities of the errors of type I and type II.

2.3 Distribution of the Test Statistic and Decision Strategie

To fix K' we have to know the distribution of T under H_0. Asymptotically it is χ^2 distribution if the test area is chosen randomly. But applying the test only at those areas where the difference between filtered image and expected normal image is maximum, the probability to get greater values for T increases, therefore, the distribution of T is changed. This is gained by simulation of images with given intensity distribution, calculating T at the points of those maximum differences (the first term in (5), which is small compared to $(\chi^2)_0 - (\chi^2)_1$ for small anomalies, is neglected).

Figure 1 shows in the left curve the probability P_0 that the statistic T is greater than the given abscissa X if H_0 is true. It allows to determine the type I error α for a given critical value K' or vice versa to fix K' for given significance level α.

To determine type II error β further simulation of images containing anomalies of known size and strength are needed. The right curve of figure 1 shows the probability P_1 that T is less than X in case of H_1. We can read from it the probability that T is less than a fixed K' in spite of H_1 being true which is the type II error. (This curve is valid only for fixed parameters mean count rate, size, and strength of the anomaly).

For definitively fixing the critical value K' different strategies are possible, e.g. to take K' for a fixed α, or to take K' which minimizes the total error $\alpha + \beta$, or to choose K' to minimize the total risk R which is the sum of the two errors weighted by risk functions R_0 and R_1 and by the a-priori probabilities p (H_0) and p (H_1) for H_0 and H_1 being true, respectively:

$$R = R_0 \, p(H_0) \, \alpha + R_1 \, p(H_1) \, \beta \qquad (6)$$

The power of the test, which is the probability of rejecting H_0 if H_0 is false, is shown in the receiver operating characteristic in figure 2 as function of α for a given parameter set.

Figure 3 gives the power as function of the strength of the anomaly for a confidence level α = 5 %.

2.4 Interactive Application

To apply the test procedure for clinical images we have to overcome two difficulties:

1) We do not know the shape of the anomaly. But we assume that a sphere is a good approximation for the shape of tumours in an early stage.

2) We do not know the normal pattern with sufficient precision

because there are large individual differences. But we have to make use of this knowledge only at the area where the test is executed. Therefore we approximate the trend in the neighbourhood of this area as quadratic polynom by least squares fitting.

There remains the problem to define the test area and its surrounding. The best way to do this is an interactive one because size and shape of the surrounding depend on the position of the anomaly within the organ under examination. This interactive procedure is demonstrated with the help of some figures.

The first example in figure 4 is a simulated image where we know the shape of the trend and the positions of anomalies, but we take the action as in the case without this knowledge. We first display intensity profiles (figure 5) crossing the suspicious region, marked with arrows 1 in figure 4. The crossing point of the profiles is marked with a star in these curves. We define with the light pen at each of these curves two inner points which border the test area and two outer points which limit the neighbourhood used for approximation of the trend. The test is executed at the position where the difference between filtered image and trend is maximal. The next display (figure 6) then shows, crossing the position of this maximum two profiles of observed values as points, two profiles of the filtered image as wavy lines, and two profiles of the trend as parabolae. In addition position coordinates and the value of the test statistic are displayed at the right.

While the anomaly at the first test region is obvious, the two following cases at the position of arrows 2 and 3 in figure 4 are doubtful. The appearance of the profiles is very similar, but the test statistic is greater than the critical value $K' = 16$ for $\alpha = 5$ % at the position of the true anomaly (figure 7) and the statistic is very small at the position without true anomaly (figure 8).

The next example shows a clinical case of a liver scintigram (figure 9) where the search is made for metastases with negative storage effects. Figure 10 shows the profiles through the position marked in figure 9 and the result in figure 11 is acceptance of an anomaly.

3. CONCLUSION

While digital image processing in nuclear medicine until recently consisted in a quality improvement by various filter methods, but left the recognition and classification fully to the human inspection and experience, the proposed test is a step towards automatic pattern recognition and classification. Its practical application is still in the beginning stage, but the success with simulated images lets us assume that it will be helpful in the clinical work.

REFERENCES

(1) B.W. Lindgren: Statistical Theory, Macmillan, London (1968)

(2) P.Pistor, G.Walch, H.G.Meder, W.A.Hunt, W.J.Lorenz, A.Amann, P.Georgi, H.Luig, P.Schmidlin, H.Wiebelt: Digital Image Processing in Nuclear Medicine, Kerntechnik 14, 299 - 306, and 353 - 359 (1972)

(3) P.Pistor, P.Georgi, G.Walch: The Heidelberg Scintigraphic Image Processing System, Proc.2. Symposium on Sharing of Computer Programs and Technology in Nuclear Medicine, Oak Ridge National Laboratory (1972)

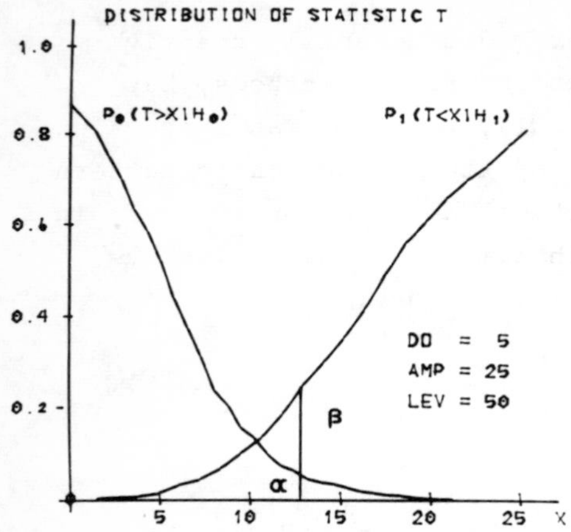

Fig. 1: Distribution of test statistic

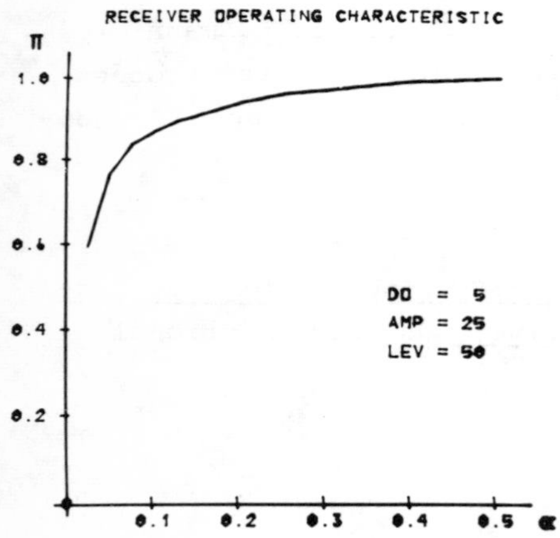

Fig. 2: Receiver operating characteristic

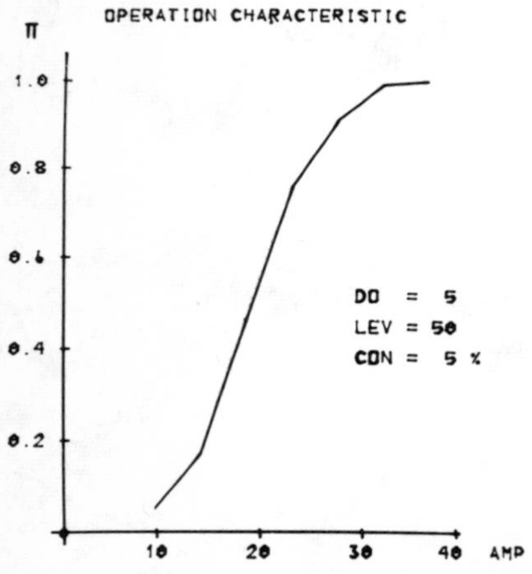

Fig. 3: Operation characteristic

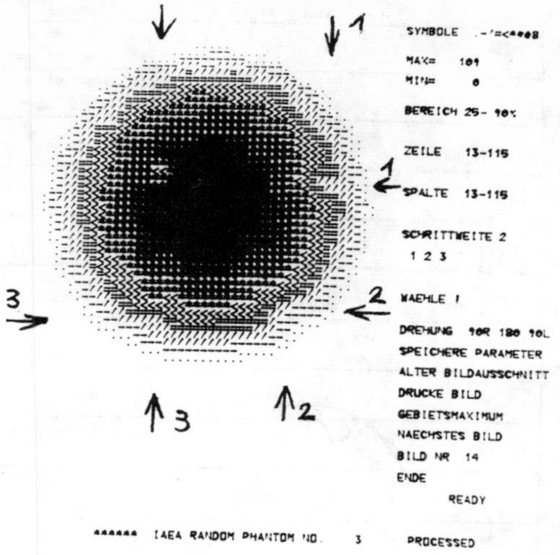

Fig. 4: Simulated image after filtering

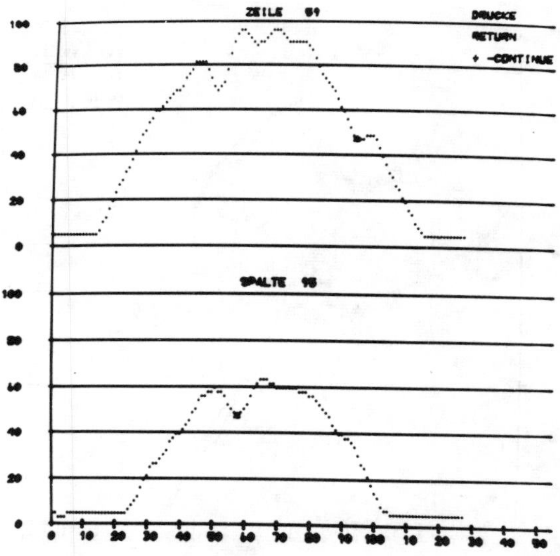

Fig. 5: Profiles at 1

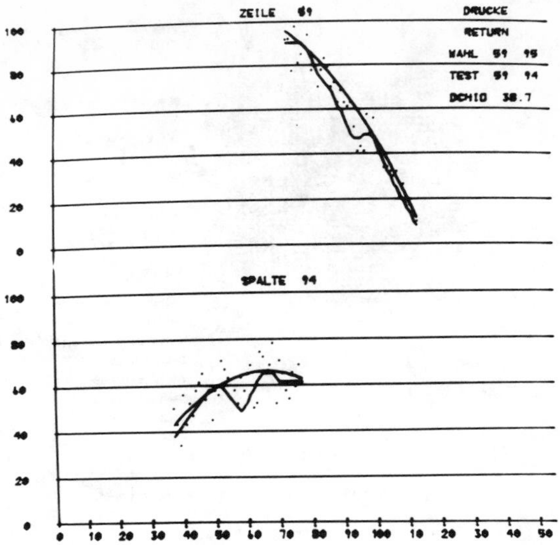

Fig. 6: Result at 1

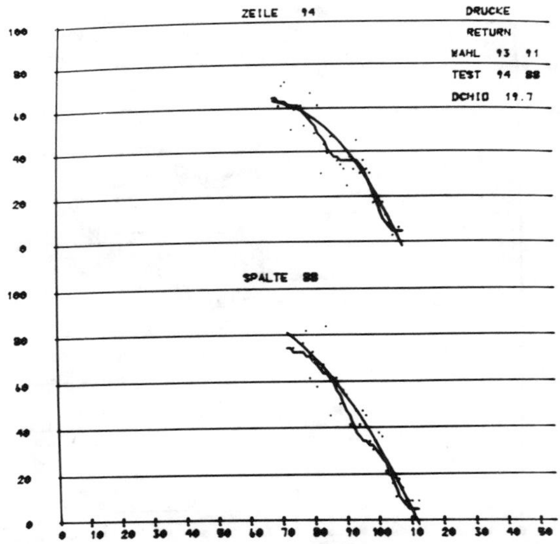

Fig. 7: Result at 2

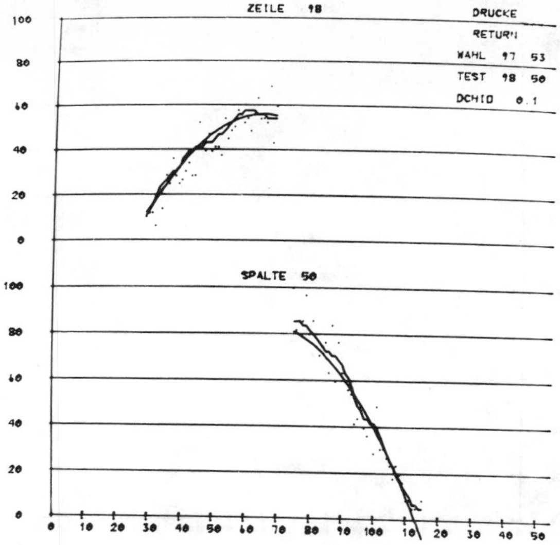

Fig. 8: Result at 3

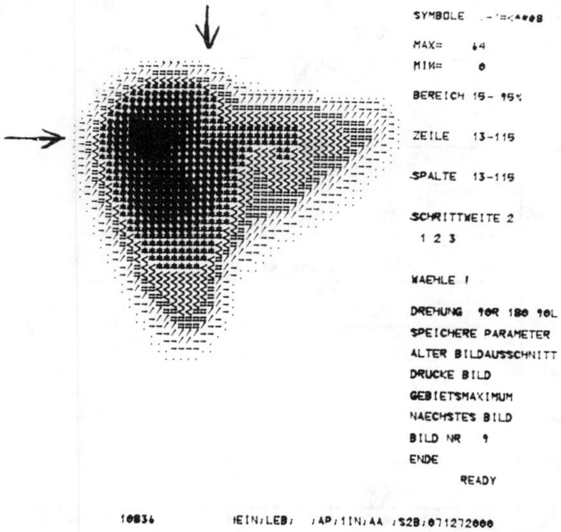

Fig. 9: Liver scintigram after filtering

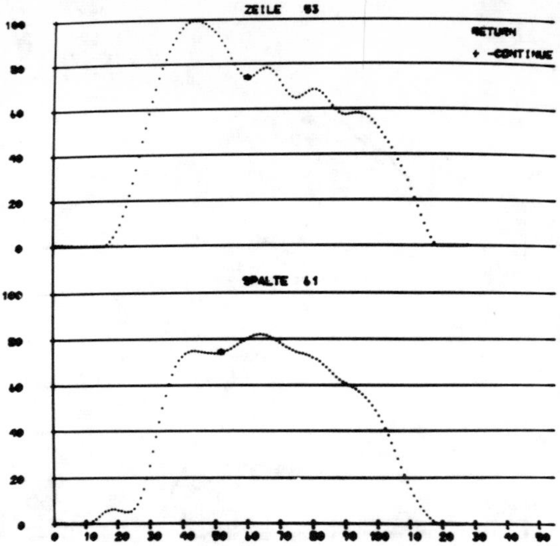

Fig. 10: Profiles of scintigram at marked position

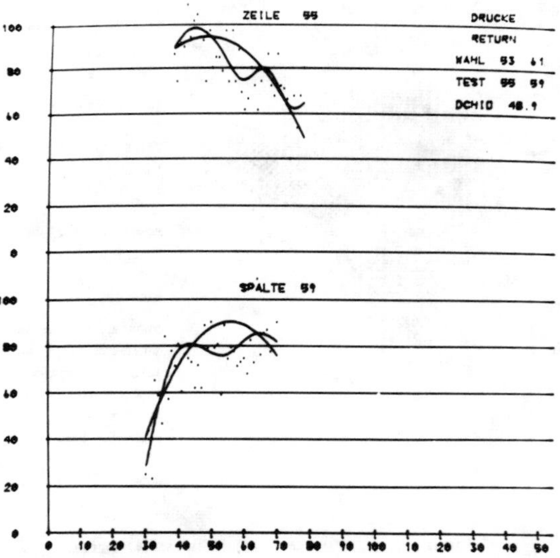

Fig. 11: Results of trend fitting and testing

THE DYNAMIC CLUSTERS METHOD AND OPTIMIZATION IN NON HIERARCHICAL-CLUSTERING

E. DIDAY

I.R.I.A.* Rocquencourt (78) FRANCE

Abstract

Algorithms which are operationnally efficient and which give a good partition of a finite set, produce solutions that are not necessarily optimum. The main aim of this paper is a synthetical study of properties of optimality in spaces formed by partitions of a finite set. We formalize and take for a model of that study a family of particularily efficient techniques of "clusters centers" type. The proposed algorithm operates on groups of points or "kernels"; these kernels adapt and evolve into interesting clusters.

After having developed the notion of "strong" and "weak" patterns, and the computer aspects, we illustrate the different results by an artificial example.

1/ Introduction

1.1 - The problem

In various scientific areas (medecine, biology, archeology, economy, etc) vastsets of objects represented by a finite number of parameters frequently appear ; for the specialist, obtaining the groupings "natural and homogeneous" together with the most representative elements of such a set constitutes an important stage in the understanding of his data. A good approach for the solution of the problem is provided by clustering techniques which consist of finding a partition for a finite set such that each object resembles more to the objects within its group than the objects outside. In mathematical terms the problem can be elaborated under one of the following forms : considering a certain W criterion :

A - Find the partition of E which optimizes W.

B - Find the partition of E which optimizes W among all the partitions in K classes.

The family of methods to which we will refer concerns mainly problem B,

---

(*) Institut de Recherche d'Informatique et d'Automatique.

but it will also be helpful for the user in resolving the following C problem.

C - Find among all the partitions in K classes, the partition for which each class will have the most representative kernel (kernel is a group of points from the population to be classified) .

In paragraph 1.2., we shall briefly give the main properties of the dynamic clusters method . This family of methods will be used as a model for the true purpose of this study which will be developed in & 1.3.

1.2 - The Dynamic Clusters method

One takes a function g which permits the transformation of a partition of E in a finite set of kernels and a function f permitting the passage of several kernels to a partition. The principle of this method is simple, it consists of applying in an alternative manner the f and g functions from an initial choice of kernels ; providing some hypothesis which will be given, the decreasing of criterion W is insured . The formalism which we are giving allows us to obtain numerous variations for this technique and notably, as particular cases, the methods of HALL and BALL (1965), FREEMAN (1969), and DIDAY (1970). We took this family of methods as a model of our study for numerous reasons :

a) They allow us to avoid storing the table of $N.\frac{(N-1)}{2}$ (where N=card(E)) similarities of objects two by two. This permits the processing of a population much more important than by other more classical techniques. (SOKAL and SNEATH (1963), JOHNSON (1969), ROUX (1968), LERMAN (1970).

b) These techniques are very fast. For instance, the variant studied in DIDAY (1970) allows the processing on an IBM 360/91 of a population of 900 items each characterized by 35 parameters in three and a half minutes.

c) These techniques do not suffer from the chain effect (cf. JOHNSON (1967)). In other words, they do not tend to bring two points that are apart closer to each other if these two points are joined by a line of points closer to each other.

d) It is not necessary to define arbitrary thresholds to determine the classes nor to stop the process (cf. SEBESTIEN (1966), Bonner (1964), HILL (1967) etc...).

e) The use of kernels favors the realization of partitions around the agglomerations with a high density and attenuates the effect of the marginal points (cf. figs 14 & 15). It favors also the apparition of empty classes. And finally, let us underline that the use of kernels permits us to provide problem c) with "local optima".

1.3 - Synthetic study of the obtained solutions

All of the realizable techniques, having as their goal to minimize the criteria W, provide solutions for which nothing proves that they are optimal. Yet, the various studies recently carried out on the present status of the research in "clustering" (see BOLSHEV (1969), FISHER and VAN NESS (1971), WATTANABE (1971), BALL (1970), CORMACK (1971)) emphasize the nonexistence of a synthetic study of the solutions obtained for a given algorithm.

The present paper is devoted to this study. We have limited ourselves to a particular type of algorithm ; but naturally, this analysis could be extended to other techniques. The set of solutions will be called V_k. Each solution attained by an algorithm is optimal with respect to a certain part of V_k which is a particularly rooted tree. This leads to attribute a structure to the V_k space. It is particularly shown that, under some hypothesis, this space can be partitioned into a finite number of rooted trees which have for roots so-called "non biased" stable solutions and for leaves some "impasse members". The various attained results are applied as follows :

a) One builds a random variable which permits the attaining an idea of the structure of V_k. An invariant is thus obtained which is interesting for multiple reasons : notably for the data evolving timewise and a comparison of the efficiency of the techniques.

b) We define several types of "fuzzy sets" which will truly provide the user with the various facets of reality that he wants to grasp.

c) We are herewith publishing a new kind of technique which will allow, by switching from one rooted tree to another, an approach to the global optimum.

The example will particularly underline the interests of the "strong forms" which are a very useful tool for the practitioner allowing him to extract from his population the most significant groups of points.

Finally, let us mention that we have skipped the theoretical developments, restraining ourselves to results that are both interesting for understanding and for the computer techniques employed.

2/ A Few Notations and Definitions

E : the set of objects to be classified, it will be assumed to be finite.

$\mathbb{P}(E)$: the set of the subset of E.

$\mathbb{P}_k$: the set of partitions of E in a number $n \leqslant k$ parts.

$\mathbb{L}_k \subset \{L = (A_1,\dots,A_k)/A_i \subset A\}$ where, A will represent, for instance, E or R^n.

$V_k = \mathbb{L}_k \times \mathbb{P}_k$.

W an injective application : $V_k \to \mathbb{R}^+$.

A local optimum on $C \subset V_k$ will be an element v^* such that $W(v^*) = \min_{v \in C} W(v)$

If $C = V_k$ one has a global optimum.

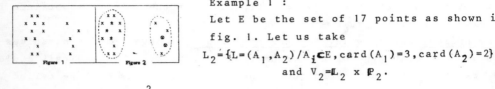

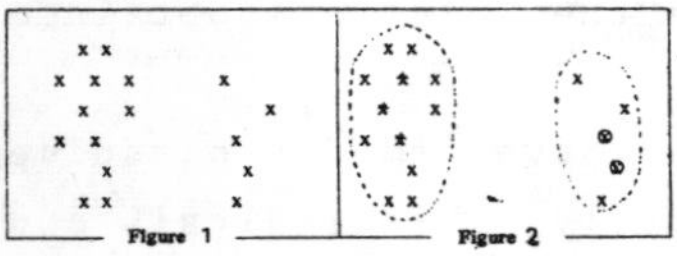

Figure 1 Figure 2

Example 1 :

Let E be the set of 17 points as shown in fig. 1. Let us take
$L_2 = \{L = (A_1, A_2)/A_i \subset E, \mathrm{card}(A_1) = 3, \mathrm{card}(A_2) = 2\}$
and $V_2 = \mathbb{L}_2 \times \mathbb{P}_2$.

Let us choose $W(v) = \sum_{i=1}^{2} \frac{1}{\mathrm{card}\, P_i} \sum_{x \in A_i} \sum_{y \in P_i} d(x,y)$ where d is the Euclidian distance. The global optimum $v^* = (L^*, P^*)$ where $L^* = (A_1^*, A_2^*)$ is shown in fig. 2. The dotted line indicates the points of E that form P_1^* and P_2^* : the 3 points identified by the ✱ form A_1 ; the ⊠ is used to represent the two points which constitute A_2.

3/ Constructing the Triplets (f,g,w)

3.1- General formulation

We shall write $v = (L,P) \in V_k$ where $L \in \mathbb{L}_k$: $L = (A_1,\dots,A_k)$ with $A_i \subset A$ and $P \in \mathbb{P}_k$: $P = (P_1,\dots,P_k)$ where the P_i's are the classes of the partition P of E. We shall also take the following mappings :

$$D : E \times \mathbb{P}(E) \to R^+$$

which in practice will express the similarity of one element of E with one component of E.

$R : A \times T \times \mathbb{P}_k \to \mathbb{R}^+$ (where T is the set of integers between 1 and k). This mapping will be used to aggregate and separate the classes. For instance, $R(x,i,P) = D(x,P_i)$ can be chosen.

The triplet (f,g,w) is constructed as follows :

$$W : V_k \to \mathbb{R}^+ : v = (L,P) \Rightarrow W(v) = \sum_{i=1}^{k} \sum_{x \in A_i} R(x,i,P)$$

$$f : \mathbb{L}_k \to \mathbb{P}_k : f(L) = P \quad \text{with}$$

$$P_i = \{x \in E / D(x,A_i) \leqslant D(x,A_j) \text{ for } j \neq i\},$$

in case of equality, x is attributed to the part of the smallest index.

$$g : \mathbb{P}_k \to \mathbb{L}_k : g(P) = L \quad \text{with}$$

A_i = {the n_i elements $a \in A$ which minimize $R(a,i,P)$} . The value of n_i will depend upon the variant chosen (cf 2).
In [9] we took as a convention to call A_i the "kernel" of the ith class.

Remark : If $R : A \times T \times V_k \to \mathbb{L}_k$, then $g : V_k \to \mathbb{L}_k$ must be chosen.

The dynamic clusters method consists in applying alternatively function f followed by the function g on the attained result and this from $L^{(0)} \in \mathbb{L}_k$ either estimated or drawn at random.

3.2-The different variants and a comparison of interests

It is not our intention to explore all the possible variants. Rather we shall explore those which appear to be interesting by simply varying the choice of g and R (allowing the reader to dream up others).

a) For this variant, one has : $A \equiv R^n$, $n_i = 1$ $\forall i$;
If furthermore $R(x,i,P) = D(x,P_i)$, $g(P) = L$ is such that A_i is the center of gravity (*) of P_i in the sense of D.
WATANABE gives an history of this kind of method in [25] .
b) $A \equiv E$ and $n_i = \text{card}(F_i)$ where :

$F_i = \{ x \in E / R(x,i,P) \leqslant R(x,j,P)\ j \neq i \}$, if $i<j$ and $R(x,i,P) = R(x,j,P)$,

x is affected to F_i. It is obvious that the A_i's are identical to the F_i's and constitute a partition of E.

A thorough study of this case can be found in [9] (which is a generalization of the method proposed by FREEMAN in [12] where $\mathbb{L}_k = \mathbb{P}_k$ and g is replaced by f). Let us note that an interesting variant of this method consists in chosing : $i \in \{1,2,\ldots,k\}$, $n_i = \alpha.\text{card}(F_i)$ with $\alpha = \frac{1}{3}$ as an example.
c) $A \equiv E$ and n_i is fixed once and for all $i \in \{1,\ldots,k\}$; n_i will be chosen by the user if he has an idea of the contents of his data, otherwise, he can let :

$$n_i = \frac{\alpha.\text{card } E}{k} \quad \text{for all } i \text{ (see [9] and [10]).}$$

d) $A \equiv E$, n_i is fixed or equal to $\alpha.\text{card } P_i$ with $0<\alpha<1$; A_i is defined as being the n_i elements of P_i which minimize $R(x,i,P)$. When n_i is fixed and in the case where the number of elements per kernel become superior to the number of elements of the corresponding class, one will take for instance, $n_i = \text{card } P_i$ if class P_i is concerned.

(*) x is called the center of gravity of P_i in the sense of D if

$$D(x,P_i) = \inf_{x \in R^n} D(x,P_i)$$

3.3- Construction of triplets that make the sequence u_n decreasing

Definition of the sequences u_n and v_n : let h be the $V_k \to V_k$ mapping such that $v = (L,P) \in V_k \Rightarrow h(v) = (g(P), f(g(P)))$.

A sequence $\{v_n\}$ is defined by v_o and $v_{n+1} = h(v_n)$
A sequence $\{u_n\}$ is defined from the sequences $\{v_n\}$ by $u_n = W(v_n)$

Definition of S :

Let $S : \mathbb{L}_k \times \mathbb{L}_k \to \mathbb{R}^+$:

$$S(L,M) = \sum_{i=1}^{k} \sum_{x \in A_i} R(x,i,Q) \text{ where } Q = f(M).$$

Definition of a square function (*)

R will be called square if :

$$S(L,M) \leqslant S(M,M) \Rightarrow S(L,L) \leqslant S(L,M)$$

Theorem 1** : If R is square, the (f,g,W) triplet makes the u_n sequence decrease for those cases where the number of elements per kernel is fixed.

4/ The Structure of $\mathbb{L}_k, \mathbb{P}_k, V_k$ and Optimality Properties.

Let us consider the graph $\Gamma = (V_k, h)$. Then appear particular elements in V_k :

4.1- The non-biased elements :

The following properties are equivalent and characterize a non-biased element (***). $v = (L,P) \in V_k$

a) v is a root of a looped tree (****) of Γ.

b) $L = h(v)$.

c) $L = g(P); f(L) = P$.

The d) and e) properties respectively allow the characterization of the non-biased elements of $\mathbb{L}_k$ (resp. $\mathbb{P}_k$).

(*) We have shown an example of a square function in [9] and [10]

(**) The demonstrations of all the results of this paper are in the "Thèse d'Etat" of the author.

(***) This name comes from the fact that the kernels corresponding to such an element are in the center (in the meaning of g) of the class they determine (in the meaning of f).

(****) cf. Appendix 1 for a definition of a "looped tree".

d) $g(f(L)) = L$.

e) $f(g(P)) = P$

4.2- The impasse elements :

The properties a) and b) are equivalent and characterize an impasse element $v = (L,P) \in V_k$.

a) v is a leaf of Γ .

b) $P \neq f(L)$ or $f^{-1}(g^{-1}(L)) = \Phi$

Let us point out that the c) and d) properties which follow respectively, permit us to characterize the impasse elements of $\mathbb{L}_k$ (respectively $\mathbb{P}_k$).

c) $g^{-1}(L) = \Phi$ or $f^{-1}(g^{-1}(L)) = \Phi$.

d) $f^{-1}(P) = \Phi$ or $g^{-1}(f^{-1}(P)) = \Phi$ or $f^{-1}(g^{-1}(f^{-1}(P))) = \Phi$.

The following theorem can immediately be deduced from the definitions of proposition 2, (cf. appendix 1) and from theorem 1.

Theorem 2 :

If R is square, then :

a) Each connected component of $\Gamma = (V_k, h)$ is a looped tree.

b) There exists in V_k at least one non-biased element.

c) If a non-biased element $v \in V_k$ is the vertex of a tree C, then v is a local optimum with respect to the set of vertices of C.

d) If $w \in V_k$ is not a non-biased element, w belongs to a looped tree with the root w^* , and $W(w) > W(w^*)$.

e) The global optimum is a non-biased element.

5/ Searching For Invariants

5.1 - Measure of the rooted trees :

One will first assume that the triplet (f,g,W) makes u_n decreasing $\forall u_o$, (in other words $\forall x \in V_k$, $W(h(x)) < W(x)$). The probality space $(\Omega, \mathcal{A}, P)$ of the family of the looped trees is defined as follows :

$\Omega = V_k$; $\mathcal{A}$ = { the algebra generated by the partition of Ω in looped trees} (i.e. the set of the parts of Ω which are the unions of looped trees).

$P : \mathcal{A} \to (0,1)$ is such that if $C \in \mathcal{A}$ is the union of n looped trees $C_1, \ldots, C_n$ then $P(C) = \frac{1}{\text{card}(\Omega)} \cdot \sum_{i=1}^{n} \text{card } C_i$.

The random variable X (so called of the family of the trees) of $(\Omega, \mathcal{A}, P)$ in $(\mathbb{R}, B)$ where B is the Borelean tribu, is the mapping $\Omega \to \mathbb{R}$ such that $X(v) = W(w)$ where w is a non-biased element of the looped tree containing v. X is actually a random variable, for if $I \in B, X^{-1}(I)$ is the union of trees of V_k with elements such that $X(v) \in I$ for the vertices. The distribution function $F(x = pr(X < x)$ expresses the probability of obtaining an element $v \in V_k$ in a looped tree or a loop containing a non-biased element w such that $W(w) < x$. In 7.1, an example of an empirical distribution function corresponding to a n-sample of V_k is given. In the case where $x : W(f(x)) > W(x)$ (in other words, one does not assume that the series u_n is decreasing $\forall u_o$), one can also define a random variable on the connected components of V_k. The random variable of $(V_k, \mathfrak{a}, P)$ in (R,B) is such that $X(v) = \inf_{y \in C} W(y)$ where C is the connected part of V_k to which v belongs. The introduction of these random variables permits us to give an idea of the connected components and of their respective size by means of the empirical distribution functions. This also gives us a tool to make a comparison of the different techniques ; the best one being those where the roots of the largest trees correspond to the smallest values taken by W (see 6.1.).

5.2 - Strong forms, fuzzy-sets :

5.2.1 - Characterization of the various types of forms :

Let $C_1, \ldots, C_n$ be the n connected parts of the graph (V_k, h) and $C = C_1 \times C_2 \times \ldots \times C_n$; one defines the mapping $Z : C \to \mathbb{R}^+$ by :

$$Z(V) = W(v_1) + \ldots + W(v_n) \text{ where : } V = (v_1, \ldots, v_n) \in C \text{ and } v_i \in C_i$$

Let $V : Z(V^*) = \underset{V \in C}{\text{Min}}\, Z(V)$. Let $V^* = (v_1, \ldots, v_n)$ and $v_i^* = (L^{i*}, P^{i*})$

(If R is square, C_i is a looped tree or a loop and v_i^* is the non-biased element of C_i). Let us denote P_j^i as the jth class of the partition P^i. Let H be the mapping $E \to \mathbb{N}^n$ which, to each element $x \in E$ associates the vector $(\alpha_1, \ldots, \alpha_n)$ where α_i is the number of the class to which the element α belongs in P^{i*}. Let $H(y) = (\beta_1, \ldots, \beta_n)$ and $\delta(x,y)$ be the number of indices $i = 1,2,\ldots,n$ such that $x_i - y_i = 0$. Let F_n and F_1 be two multi-valued functions defined on E such that :

$$F_n(x) = \{y \in E / \delta(x,y) = n\} \text{ and } F_1(x) = \{y \in E / \delta(x,y) \geq 1\}$$

Definition of strong forms (*)

The following properties are equivalent and characterize the partition P^* of E for which each class is a strong form.

1) $P^* = P^{1*} \cap P^{2*} \cap \ldots \cap P^{n*}$

2) P^* is the less thin (**) of the partitions which are thinner than $P^{1*}, \ldots P^{n*}$.

(*) The intersection of two partitions is the set of the parts obtained in taking the intersection of each class of one by all the classes of the other.

(**) A participation P is said to be thinner than a partition P' of E if every class of P' is the union of classes of P.

3) P^* is the partition defined by the quotient space E/H.
4) P^* is the partition defined by the connected parts of the graph $\Gamma_n = (E,F_n)$.

Definition of weak forms :

The following properties are equivalent(*) and characterize the partition Q^* of E for which each class is a weak form.
1) Q^* is the thinner of the partitions which are less thin than $P^{1*},\dots,P^{n*}$.

2) Q^* is the partition defined by the set of the connected parts of the graph $\Gamma_1 = (E,F_1)$.

More generally, if we impose $F_p(x) = \{y\in E/\delta(x,y)\geqslant p\}$ and $\Gamma_p = (E,F_p)$, the set of the connected parts of Γ_p for $p = 0,1,2,\dots,n$ constitutes a hierarchy which induces the subdominant ultra-metric of a certain distance (cf Appendix 3).

Remark :

It follows from these definitions that P^* is a thinner partition than Q^*.

Definition of the overlapping points and of the isolated points :
They are characterized by the fact that they are strong forms reduced to a single point. They are distinguished by the following properties :
- a point $a\in E$ is isolated if $\delta(a,x) = 0$; $\forall x\in E$.
- a point $a\in E$ is overlapping if $\forall x\in E$: $0<\delta(a,x)<n$.

5.2.2 - Fuzzy-sets

The interest of "fuzzy sets" of Zadeh that we are introducing here is that they permit :
a) To obtain new forms from set operations on the strong forms (union, intersection, etc.).
b) To characterize these new forms without having to define their profile types (for instance, by the calculating of means) and even without knowing the elements of which they are constituted.
Each strong form A can be considered as a "fuzzy set" characterized by the mapping $h_A : E\to\{0,1\}$ such that $h_A(x) = \frac{\delta(x,a)}{n}$ where $a\in A$. One see from this definition (3rd property) that $h_A(a) = 1 \quad \forall a\in A$. One can use h_A in order to have an idea of the degree of similarity with A of an overlapping point or of another strong form. One can also use the

(*) For a demonstration of this equivalence cf. Appendix 2.

mapping $F : \mathcal{F} \to \{0,1\}$ where $\mathcal{F}$ is the set of the weak forms of E and

$$F(B) = \frac{1}{card(B)} \sum_{x \in B} \left(\frac{1}{m} \sum_{j=1}^{m} h_{Aj}(x)\right)$$

where the Aj are the m strong forms which constitute B. This mapping F expresses the degree of weakness of B but the more the strong forms Aj are dissimilar, the more F(B) will be smaller.

5.4 - Approaching the global optimum by changing trees :

It is a matter of constructing with the aid of two non-biased elements v_1 and v_2 of $\mathbb{V}_k$ a third non-biased element v_3 to improve the criterion Suppose R is square.

We will denote $v^i = (L^i, P^i) \in V_k$ with $L^i = (L^i_1, \ldots, L^i_k)$ and $P^i = (P^i_1, \ldots, P^i_k)$; $v^i_j = (L^i_j, P^i_j)$.

We are presuming that W is additive (which is practically often true) otherwise that there exists a mapping z :

$z : \mathbb{P}(E) \times \mathbb{P}(E) \to \mathbb{R}^+$ such that

$$W(v^i) = \sum_{j=1}^{k} z(v^i_j)$$

Let us assume that v^1 and v^2 are two non-biased solutions and that $\{v^i_{j_1}, \ldots, v^i_{j_k}\}$, are the k smallest values taken by z(x) with $x \in \{v^i_j / i = 1,2$ and $j = 1,2,\ldots k\}$. Let us denote $P = (P^i_{j_1}, \ldots, P^i_{j_k})$ and $L = (L^i_{j_1}, \ldots, L^i_{j_k})$. One can easily prove then thefollowing proposition.

Proposition :

If $L \neq L^1$, $L \neq L^2$ and $P \in \mathbb{P}_k$, then the looped tree containing $v = (L,P)$ has as a root a non-biased element v_3 such that $W(v_3) < \inf(W(v_1), W(v_2))$.

6/ Examples of Applications

6 .1 - The artificial example of Ruspini

We have applied the case c) on the inputs of Ruspini (cf fig.6). First of all, this permitted us to observe the speed of the method in comparison to that of Ruspini ; therefore, in taking $K=4, n_1=n_2=n_3=n_4=n_5$, $R(x,i,L) = D(x,C_i) = \sum_{y \in C_i} d(x,y)$ where d is the Euclidian distance, we ran 50 passes of the method (each time changing the drawing of $L^{(0)}$)in 2,57 minutes on the CII 10 070. These 50 passes have pointed out the existence of 6 looped trees. The frequency of the appearance of each of these 6 solutions are indicated in figure 12. This graph is, in fact, a histogram of the random variables which have been defined in 5.1. The abcissa is represented by $u = \text{Lim}\, u_n$(cf 3.3.) ; the convergence is generally obtained at about 4 iterations. The most frequent solution is the one that corresponds to the four best classes ; the value of u for this solution is clearly better than for the other solutions, which shows that it really corresponds to the best partition. The best solution corresponds to the root of the biggest looped tree, which is satisfactory for the method. The most frequent solutions are indicated in figures 7,8,9 and 10. One can easily see that the solutions given in figures 9,10 and 11 do not carry any information (cf 5.2) to the solution given in figure 7. From these solutions one obtains 4 "strong forms" corresponding exactly to the four classes of the best solution.

Let us remark that in applying proposition 3 to the solution given in figures 8 and 9, one brings out the looped tree whose root is the solution correspondong to that of figure 7.

One gives (table 1), the table of the strong forms obtained by taking this time $K=6, n_5=n_6=5$, without changing the other parameters and in taking 5 passes of the method (n=5). This table brings out the existence of 6 strong forms and 3 weak forms. Let B_1, B_2, B_3 denote the weak forms and A_1 the strong forms (cf Fig.13). One can measure the "weakness" of B_i by using the F functions (cf 5.2.2). As

$$\sum_{j=1}^{3} h_{A_j}(x) = 1+\frac{4}{5}+\frac{4}{5} \; \forall x \in B_1 \; ; \; \sum_{j=4}^{5} h_{A_j}(x) = 1+\frac{1}{5} \; \forall x \in B_2 ; \text{ and } h_{A_6}(x) = 1 \; \forall x \in B_3$$

One has :

$$F(B_1) = \frac{13}{15},\; F(B_2) = \frac{3}{5} \quad \text{and} \quad F(B_3) = 1$$

One sees that B_3 is a strong form, that B_1 is almost a strong form and that B_3 is a relatively weak form. These strong and weak forms that appear in fig. 13 express quite well these values. Finally, let us point out that in 4 of the 5 solutions, there appear empty classes which signify that the number of classes that actually do exist must be smaller than 6.

Conclusion

A wide field of research is still open : in practice one should : develop the choice of f and g (e.g. by means of the learning methods) ; develop the techniques allowing the choice of k (the number of classes required a priori) ; realize an exhaustive comparison of the various variations of the clusters centers method ; develop in depth the techniques of the passage from one tree to another ; make a statistical survey of the structure of the space V_k in relation with E, particularly as for as the relative number of impasse elements, non-biased elements, size of the trees, levels, etc... is concerned ; set up techniques allowing a clearer vision of the strong forms table(e.g.of the minimum spanning tree type) ; use the weak forms in order to detect among the strong forms the low density zones (obtaining "holes" leading to numerous practical applications). Let us also point out the fact that "strong forms table" allow us to obtain at once the three types of classification procedure :

a) Partitioning (by taking the partition corresponding to the best value of W) ;

b) Clumping (using the overlapping points) ;

c) Hierarchical classification (using the "connected descendant" method)

BIBLIOGRAPHY

(1) BALL G.H., 1970 - Classification Analysis - Technical Note, Stanford Research Institute.Menlo Park,California 94025 USA

(2) BARBU M.,1968 - Partitions d'un ensemble fini : leur treillis M.S.H. n° 22

(3) BENZECRI J.P., 1971 - Algorithmes rapides d'agrégation.Sup.Class. n°9 Laboratoire de Statistique Mathématique. Université de Paris-6

(4) BENZECRI J.P., 1970 - Représentation Euclidienne d'un ensemble muni de masses et de distances. Université de Paris-6

(5) BERGE C., 1967 - Théorie des graphes et ses applications - Dunod Ed. Paris.

(5)'BOLSHEV L.N., 1969 - Cluster Analysis - I.S.I.R.S.S.'69

(6) BONNER R.E.,1964 - On some clustering technics - IBM Journal of Research and Development

(7) CORMACK R.M.,1971 - A review of Classification - The Journal of the Royal Statistical Society,Serie A, vol 134, Part 3

(8) DIDAY E,BERGONTM, BARRE J, 1970-71-72 - Differentes notes sur la programmation de la Méthode des nuées dynamiques Note I.R.I.A., Rocquencourt 78.

(9) DIDAY E, 1970 - La méthode des nuées dynamiques et la reconnaissance des formes. Cahiers de l'I.R.I.A., Rocquencourt 78

(10)DIDAY E, 1971 - Une nouvelle méthode en classification automatique et reconnaissance des formes - Revue de Statistique appliquée, vol. XIX,n° 2.

(11)FISHER L,VAN NESS J.W. 1971 - Admissible Clustering Procedures. Biometrika,58,1,p. 91.

(12)FREEMAN N, 1969 - Experiments in discrimination and classification. Pattern Recognition J.vol.1,n° 3.

(13) HALL D.J, HALL G.H, 1965 - Isodata a Novel Method of Data Analysis and Pattern Classification. Technical Report, 5R I Project 5533, Stanford Research Institute, Menlo Park, California U.S.A.

(14) HILL D.R, 1967 - Mechanized Information Storage, retrieval and dissemination. Proceedings of the F.I.D/I.F.I.P. Joint Conference Rome.

(15) JOHNSON S.C, 1967 - Hierarchical clustering schemes. Psychometrica 32, 241-45

(16) LERMAN H, 1970 - Les bases de la classification automatique - Gauthiers-Villars, 1970

(17) PICARD J, 1972 - Utilisation des méthodes d'analyse de données dans l'étude de courbes expérimentales. Thèse de 3è cycle Laboratoire de Statistique Mathématique Université Paris-6

(18) ROMEDER J.M, 1969 - Methodes de discrimination. Thèse de 3è cycle Statistique Mathématique. Faculté des Sciences de Paris-6

(19) ROUX M, 1968 - Un algorithme pour construire une hiérarchie particulière - Thèse de 3è cycle. Laboratoire de Statistique Mathématique, Université de Paris-6.

(20) RUSPINI H.R, 1970 - Numerical Methods for fuzzy clustering. Information Science 2, p.319-350

(21) SANDOR G, LENOIR P, KERBAOL M, 1971 - Une étude en ordinateur des corrélations entre les modifications des protéines sériques en pathologie humaine. C.R. Acad. Sc. Paris, t.272, p.331-334

(22) SANDOR G, DIDAY E, LECHEVALLIER y, BARRE J, 1972 - Une étude informatique des corrélations entre les modifications des protéines sériques en pathologie humaine. C.R. Acad. Sc. Paris, t. 274, d.p. 464-467

(23) SEBESTIEN G.S, 1966 - Automatic off-line Multivariate Data Analysis Proc. Fall Joint Computer Conference pp.685-694

(24) SOKHAL R.R, SNEATH P.H.R, 1963 - Numerical Tasconomy. W.H. Freeman & Co, San Francisco and London.

(25) WATANABE M.S, 1971 - A unified view of clustering algorithms. IFIP Congress 71, Ijubiana, Booklet TA-2

(26) ZADEH L.A, 1965 - Fuzzy sets. Inf. Control 8, pp. 338-353

(27) ZAHN C.I, 1971 - Graph Theoretical methods for detecting and describing Gestalt Clusters. I.E.E.E. trans. on Computers, vol. C-20, n° 1, January.

(28) McQUEEN j, 1967 - Some Methods for Classification and Analysis of Multivariate Observations. 5th Berkeley Symposium on Mathematics, Statistics and probability, vol. 1, n° 1, pp. 281-297

APPENDICES

Appendix 1

Let B be a finite set and a function h : $B \to B$. The graph defined by B and the set of the arcs $(h(x),x)$ will be noted by $\Gamma = (B,h)$. One knows that the set of the connected components of Γ constitute a partition of B ; each of these components have a particular form :

Proposition 1 : Each connected component of Γ contains at maximum one circuit.

One gives (fig.4) an example of a connected component of Γ . We shall say that x is a fixed point if $x=h(x)$. Atree having for its root a fixed poin will be called a looped tree(see fig.5). Let W be a mapping $B \to \mathbb{R}^+$.

Proposition 2 : If W is injective on the entire sequence v_n and verifies the property: $W(h(x))<W(x)$ then :

1) Each connected component of Γ contains a loop and only one and does not contain another circuit.

2) Each connected component of Γ is a looped tree or a loop.

3) If $y \in B$ is not a fixed point, there exists a fixed point x such that $W(x)<W(y)$

Appendix 2

The problem is to show that the following two properties are equivalent to characterize a weak form.

1) Q^* is the finest of the partitions which are less fine than $P^{1*},\ldots,P^n$

2) Q^* is the parittion defined by the set of the connected parts of the graph $\Gamma_1 = (E,F_1)$.

Appendix 3

Theorem : Let Δ be the mapping (*) $E \times E \to \mathbb{N}$ such that $\Delta(x,y) = n-\delta(x,y)$ and let E' be the quotient space (**) E/H. If F_p is the multi-mapping E' $E' \to \mathbb{P}(E')$ such that $F_p(x) = \{y \in E' / \delta(x,y) \geq P\}$ and Γ_p is the graph (E',F_p), then :

1) The set of the connected parts of Γ_p for $p = 0,1,2,\ldots,n$ constitute a hierarchy on E'.

2) This hierarchy induces the subdominant ultrametric of Δ .

(*) See definition of δ in 5.2.1.

(**) H is the mapping as defined in 5.2.1.

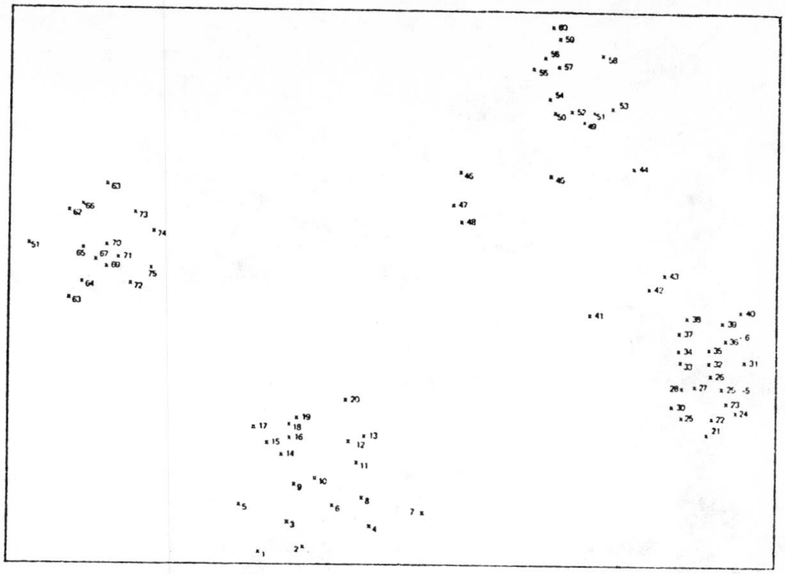

Fig. 6

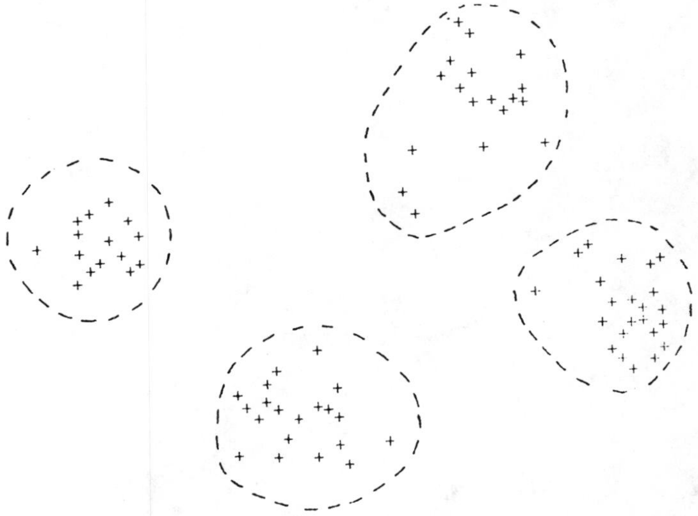

Fig. 7

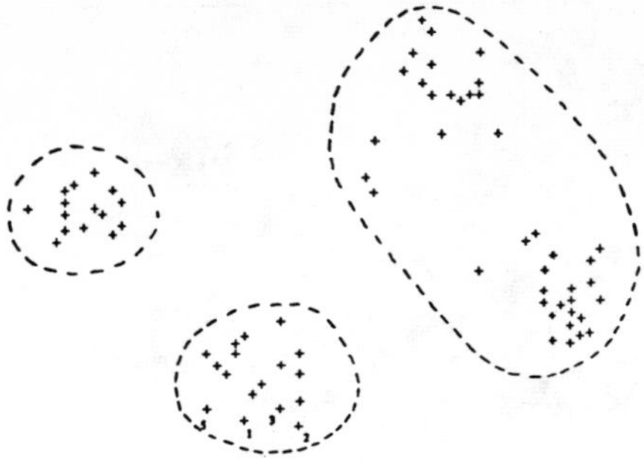

Fig. 8

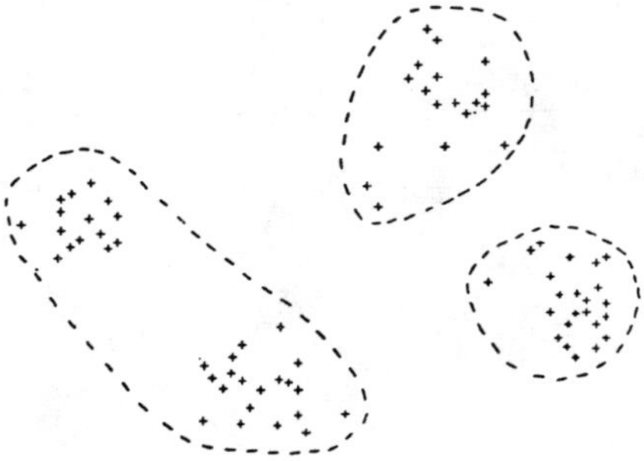

Fig. 9

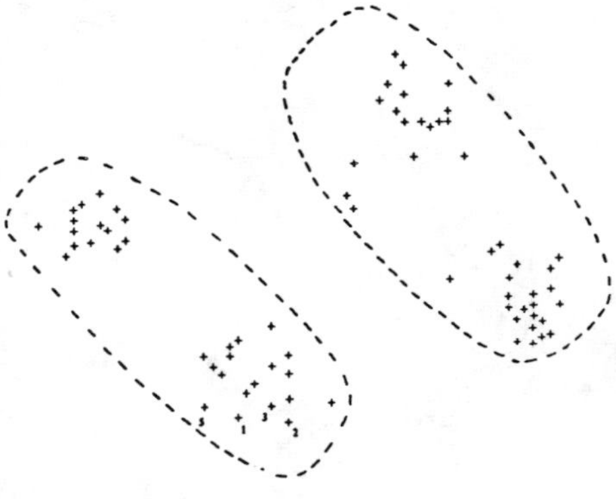

Fig. 10

Frequency of appearance of each solution after 50 drawings L (0).

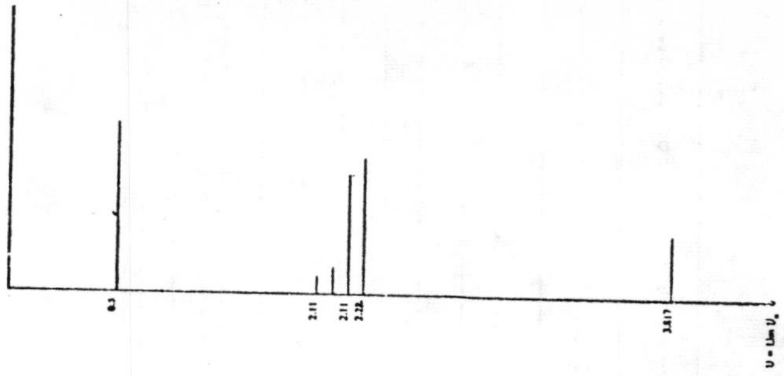

Fig. 12

The value

U= 0.5 corresponds to the solution in Fig. 7
U= 2.11 corresponds to the solution in Fig. 8
U= 2.22 corresponds to the solution in Fig. 9
U= 3.817 corresponds to the solution in Fig. 10

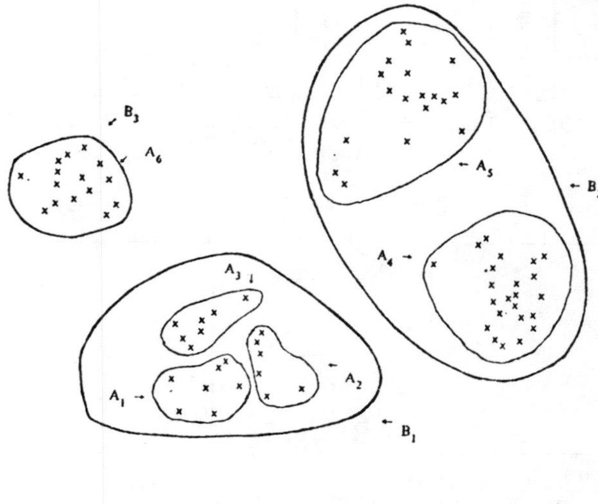

Fig. 13

Nº of points	obtained solutions 1st	2nd	3rd	4th	5th	δ_{ij}		
1	1	3	3	1	4	0		
2	1	3	3	1	4	0		
3	1	3	3	1	4	0	A_1	
5	1	3	3	1	4	0		
6	1	3	3	1	4	0		
9	1	3	3	1	4	0		
10	1	3	3	1	4	0		
8	1	3	3	1	6	1		
4	1	3	3	1	6	0		
7	1	3	3	1	6	0	A_2	B_1
11	1	3	3	1	6	0		
12	1	3	3	1	6	0		
13	1	3	3	1	6	0		
14	1	3	3	1	5	1		
15	1	3	3	1	5	0		
16	1	3	3	1	5	0		
17	1	3	3	1	5	0	A_3	
18	1	3	3	1	5	0		
19	1	3	3	1	5	0		
20	1	3	3	1	5	0		
21	2	2	1	2	2	5		
22	2	2	1	2	2	0		
23	2	2	1	2	2	0		
24	2	2	1	2	2	0		
25	2	2	1	2	2	0		
26	2	2	1	2	2	0		
27	2	2	1	2	2	0		
28	2	2	1	2	2	0		
29	2	2	1	2	2	0		
30	2	2	1	2	2	0		
31	2	2	1	2	2	0	A_4	
32	2	2	1	2	2	0		
33	2	2	1	2	2	0		
34	2	2	1	2	2	0		
35	2	2	1	2	2	0		
36	2	2	1	2	2	0		
37	2	2	1	2	2	0		
38	2	2	1	2	2	0		
39	2	2	1	2	2	0		
40	2	2	1	2	2	0		B_2
41	2	2	1	2	2	0		
42	2	2	1	2	2	0		
43	2	2	1	2	2	0		
44	3	4	1	4	1	4		
45	3	4	1	4	1	0		
46	3	4	1	4	1	0		
47	3	4	1	4	1	0		
48	3	4	1	4	1	0		
49	3	4	1	4	1	0		
50	3	4	1	4	1	0		
51	3	4	1	4	1	0		
52	3	4	1	4	1	0		
53	3	4	1	4	1	0	A_5	
54	3	4	1	4	1	0		
55	3	4	1	4	1	0		
56	3	4	1	4	1	0		
57	3	4	1	4	1	0		
58	3	4	1	4	1	0		
59	3	4	1	4	1	0		
60	3	4	1	4	1	0		
61	4	1	2	3	3	5		
62	4	1	2	3	3	0		
63	4	1	2	3	3	0		
64	4	1	2	3	3	0		
65	4	1	2	3	3	0		
66	4	1	2	3	3	0		
67	4	1	2	3	3	0		
68	4	1	2	3	3	0	A_6	B_3
69	4	1	2	3	3	0		
70	4	1	2	3	3	0		
71	4	1	2	3	3	0		
72	4	1	2	3	3	0		
73	4	1	2	3	3	0		
74	4	1	2	3	3	0		
75	4	1	2	3	3	0		
U =	1.2135	1.2135	4.8910	1.2135	1.0000			

Table 1: Strong forms for the inputs of Ruspini

Fig. 14

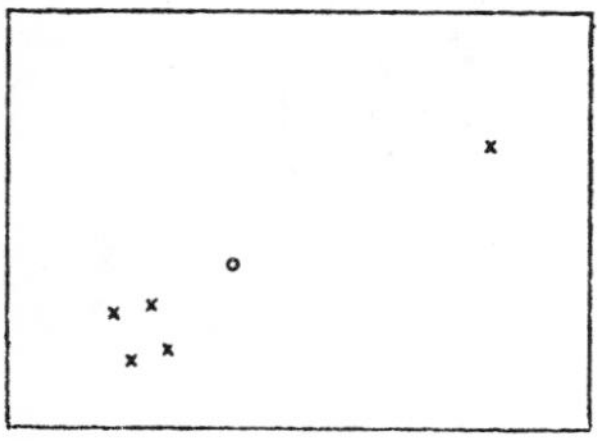

The signs "x" represent the elements to be classified whereas the sign "0" represent the center of gravity of the 5 elements.

Fig. 15

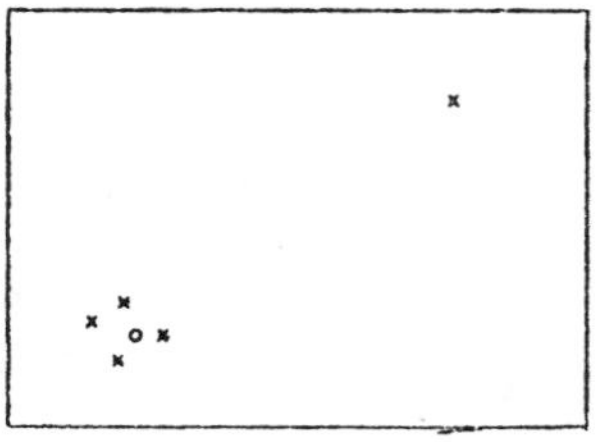

The three closest elements of the population are represented by the sign "x"; the center of gravity of these three elements attenuate the effect of the marginal element.

A MAXIMUM PRINCIPLE FOR GENERAL CONSTRAINED OPTIMAL CONTROL PROBLEMS - AN EPSILON TECHNIQUE APPROACH

Jerome W. Mersky

1215 S. Leland St.
San Pedro, CA 90731
U.S.A.

We wish to present an extension of the Epsilon Technique (Reference 1) to general constrained optimal control problems with systems governed by ordinary differential equations. The use of the Epsilon Technique provides a straightforward constructive approach to the maximum principle, and, in particular, to the lagrange multipliers for a very general constrained optimal control problem which subsumes the so-called "bounded phase coordinate" problem.

Before formally stating the problem, we establish some definitions and notations. We shall be working in the class of generalized controls in the sense of McShane (Reference 2) and L. C. Young (Reference 4):

Definition Let U be a compact set in R^k. A function u which, to almost every t, assigns a probability measure $\mu(\cdot, t)$, defined on the lebesque subsets of U, is said to be a generalized control.

Definition If $f(\xi, \cdot)$ is a function defined on U, it may be extended to be a function of generalized controls $\tilde{f}$ by the following: if μ_o is the probability measure associated with the generalized control u_o then

$$\tilde{f}(\xi, u_o) = \int_u f(\xi, u) d\, \mu_o(u, t)$$

Since in the following all controls are assumed to be generalized there will be no confusion in dropping the tilde over the symbol for generalized controls.

Generalized controls may be considered as elements of the dual space $\mathscr{C}^*_o(\Omega)$. Therefore, we shall use the weak * topology for the topology of U_g, the class of generated by the compact set U.

We are now in a position to state the problem we wish to consider: viz.

Problem P   In the class of generalized controls, minimize

$$\int_0^T g(t, x(t), u)dt$$

subject to

$$\dot{x}(t) = f(t, x(t), u) \qquad \text{a.e.}$$

$$\varphi(t, x(t), u) = 0 \qquad \text{a.e.}$$

$$\phi(t, x(t), u) \leq 0 \qquad \text{a.e.}$$

$$\psi(t, x(t)) \leq 0 \qquad \text{a.e.}$$

where $x(t) \in R^n$, $\varphi(t, x(t), u) \in R^p$, $\phi(t, x(t), u) \in R^r$, $\psi(t, x(t)) \in R^q$, f, g, φ and ϕ are continuous in (t, x, u) and continuously differentiable in x and, ψ is C'' in (t, x). In addition, we require that, for all admissable x(t),

$$\int_0^T \|\dot{x}(t)\|^2 dt \leq N < \infty$$

where N is a fixed constant, and that

$$[x, f(t, x, u)] \leq C(1 + \|x\|^2) \qquad t \in [0, T].$$

We replace this with the "epsilon-problem".

Problem P_ϵ In the class of generalized controls minimize $h(\epsilon, x(\cdot), u) =$

$$\frac{1}{2\epsilon} \int_0^T \|\dot{x}(t) - f(t, x(t), u)\|^2 dt + \frac{1}{2\epsilon} \int_0^T \|\varphi(t, x(t), u)\|^2 dt$$

$$+ \frac{1}{2\epsilon} \int_0^T [m(t, x(t), u), \phi(t, x(t), u)]dt + \frac{1}{2\epsilon} \int_0^T [n(t, x(t)), \psi(t, x(t))]dt$$

$$+ \int_0^T g(t, x(t), u)dt$$

where $\epsilon > 0$, and

$$m(t, x(t), u) = \begin{cases} 0 & \phi(t, x(t), u) < 0 \\ \phi(t, x(t), u) & \phi(t, x(t), u) \geq 0 \end{cases}$$

$$n(t, x(t)) = \begin{cases} 0 & \psi(t, x(t)) < 0 \\ \psi(t, x(t)) & \psi(t, x(t)) \geq 0 \end{cases}$$

with the same conditions as in Problem P.

The following theorem gives the existence of, and necessary conditions for, solutions to Problem P_ϵ.

Theorem 1 Under the above conditions, there exists a solution pair $x_\epsilon(\cdot)$, u_ϵ for problem P_ϵ. This pair satisfies the "ϵ-maximum principle":

1.1) Let

$$\mathscr{H}(\epsilon, t, x(t), u) = [Z(\epsilon, t), f(t, x(t), u)] - [L(\epsilon, t), \varphi(t, x(t), u)]$$

$$-[M(\epsilon, t), \phi(t, x(t), u)] - [N(\epsilon, t), \Psi(t, x(t), u)] - g(t, x(t), u)$$

where

$$Z(e, t) = \frac{1}{\epsilon}\left(\dot{x}_\epsilon(t) - f(t, x_\epsilon(t), u_\epsilon) + \frac{1}{2}\frac{\partial\psi^*}{\partial x}\int_t^T n(s, x_\epsilon(s))ds\right)$$

$$L(\epsilon, t) = \frac{1}{\epsilon}\varphi(t, x_\epsilon(t), u_\epsilon)$$

$$M(\epsilon, t) = \frac{1}{\epsilon}m(t, x_\epsilon(t), u_\epsilon)$$

$$N(\epsilon, t) = \frac{1}{2\epsilon}\int_t^T n(s, x_\epsilon(s))ds$$

$$\Psi(t, x(t), u) = \frac{\partial\psi}{\partial t}(t, x(t)) + \frac{\partial\psi}{\partial x}(t, x(t))f(t, x(t), u)$$

Then,

$$\mathscr{H}(\epsilon, t, x_\epsilon(t), u_\epsilon) = \max_{u \in U_g} \mathscr{H}(\epsilon, t, x_\epsilon(t), u)$$

1.2) If we write $Y(\epsilon, t) = \frac{1}{\epsilon}(\dot{x}_\epsilon(t) - f(t, x_\epsilon(t), u_\epsilon))$ then we have

$$\dot{Y} = -\nabla f^* Y + \nabla\varphi^* L + \nabla\phi^* M + \nabla\psi^*(\frac{1}{2\epsilon} n) + \nabla g^* \qquad \text{a.e.}$$

$$Y(\epsilon, T) = 0$$

evaluated along $(t, x_\epsilon(t), u_\epsilon)$.

1.3) The multipliers $N_i(\epsilon, t)$ are monotonic non-increasing and constant on intervals along which $\psi_i(t, x_\epsilon(t)) < 0$.

Comments The proof is omitted here and may be obtained from Reference 3. The basic idea is to observe first that

$$\int_o^T [n(t, x(t)), \psi(t, x(t))]dt =$$

$$\int_o^T \left[\int_t^T n(s, x(s))ds, \frac{d\psi}{dt}(t, x(t)) \right] dt -$$

$$\left[\int_t^T n(s, x(s))ds, \psi(t, x(t)) \right]_o^T$$

Then if $r(\epsilon, t, \dot{x}, \chi)$ is used to denote the sum of the integrands in the definition of $h(\epsilon, t, x(t), u)$ as a function of

$$\chi \in \overline{C(t, x)} = \overline{co} \left\{ f(t, x, u), g(t, x, u), \varphi(t, x, u), \phi(t, x, u) \,\middle|\, u \in U \right\}$$

then

$$\frac{dr}{d\theta}(\epsilon, t, \dot{x}_\epsilon, \chi_\epsilon + \theta(\chi - \chi_\epsilon)) \Big|_{\theta=0} \geq 0$$

This gives 1.1). To obtain 1.2), let d(t) be an element of the Schwarz space of infinitely smooth functions with compact support, and $x_d(t) = x_\varepsilon(t) + \theta\, d(t)$. Then

$$\left.\frac{dh(\varepsilon, x_d(\cdot), u_\varepsilon)}{d\theta}\right|_{\theta=0} = 0$$

gives 1.2). Condition 1.3) is immediate.

We may proceed now to the limiting form of Theorem 1 which gives us a new existence theorem and maximum principle for Problem P.

<u>Theorem 2</u> As ε converges to zero, $x_\varepsilon(\cdot)$ converges uniformly to $x_o(\cdot)$ and u_ε converges weak * to u_o with $x_o(\cdot)$, u_o being a solution pair to Problem P. If, furthermore, the matrix

$$\begin{pmatrix} \dfrac{\partial\varphi}{\partial u} \\ \dfrac{\partial\phi}{\partial u} \\ \dfrac{\partial\Psi}{\partial u} \end{pmatrix}$$

has full rank along $(t, x_o(t), u_o)$ then the following limits exist:

$$\zeta(t) = \lim_{\varepsilon\to 0} Z(\varepsilon, t)$$

$$\lambda(t) = \lim_{\varepsilon\to 0} L(\varepsilon, t)$$

$$\mu(t) = \lim_{\varepsilon\to 0} M(\varepsilon, t)$$

$$\nu(t) = \lim_{\varepsilon\to 0} N(\varepsilon, t)$$

and we have the maximum principle:

2.1) Let

$$\mathscr{H}(t, x(t), u) = [\zeta(t), f(t, x(t), u)] - [\lambda(t), \varphi(t, x(t), u)]$$

$$- [\mu(t), \phi(t, x(t), u)] - [\nu(t), \Psi(t, x(t), u)]$$

$$- g(t, x(t), u)$$

Then

$$\mathcal{H}(t, x_o(t), u_o) = \max_{u \in U_g} \mathcal{H}(t, x_o(t), u)$$

2.2) $$\dot{\zeta} = -\nabla f^* \zeta + \nabla \varphi^* \lambda + \nabla \phi^* \mu + \nabla \Psi^* \nu + \nabla g^* \qquad \text{a.e.}$$

$$\zeta(T) = 0$$

evaluated along $(t, x_o(t), u_o)$.

2.3) The multipliers $\mu_i(t)$ are non negative and $\mu_i(t) = 0$ when $\phi_i(t, x_o(t), u_o) < 0$; the multipliers $\nu_i(t)$ are monotonic non increasing functions of t which are constant along intervals along which $\psi_i(t, x_o(t)) < 0$ with $\nu_i(T) = 0$.

Comments Again a detailed proof may be found in Reference 3. The main idea here is that the matrix above, having full rank, has an inverse which allows us to write $u_\varepsilon - u_o$ as a function of the constraint terms $\varphi(t, x_\varepsilon(t), u_\varepsilon)$, etc. so that the convergence arguments proceed in a straight-forward manner.

References

1. Balakrishnan, A. V.,"The Epsilon Technique - A Constructive Approach to Optimal Control" in Control Theory and the Calculus of Variations, A. V. Balakrishnan (ed), Academic Press, New York, 1969
2. McShane, E. J.,"Optimal Controls, Relaxed and Ordinary," in Mathematical Theory of Control, A. V. Balakrishnan and L. W. Neustadt (eds), Academic Press, New York, 1967
3. Mersky, J. W., An Application of the Epsilon Technique to Control Problems with Inequality Constraints, Dissertation, University of California, Los Angeles, 1973
4. Young, L. C., Lectures on the Calculus of Variations and Optimal Control Theory, W. B. Saunders Co., Philadelphia, 1969

OPTIMAL CONTROL OF SYSTEMS GOVERNED BY VARIATIONAL INEQUALITIES

J.P. YVON

(IRIA 78 - ROCQUENCOURT - FRANCE)

SUMMARY

In many physical situations, systems are not represented by equations but by variationnal inequalities : a typical case is systems involving semi-porous mediums but there are many other examples (cf. e.g. Duvaut-Lions [4]). This paper (1) is devoted to the study of optimal control problems for such systems. As in the case of partial differential equations we are led to consider the analogous separation between elliptic and parabolic systems ; this is studied first and then we give two algorithms with application to a biochemical-example.

I - ELLIPTIC INEQUALITIES

Let us define

- V Hilbert space, K closed convex subset of V,
- $a(y,z)$ bilinear form on V, continuous and positive definite,
- $j(z)$ convex ℓ.s.c. functional on V,

and

- $\mathcal{U}$ Hilbert space, $\mathcal{U}_{ad}$ a closed convex subset of $\mathcal{U}$
- $B \in \mathcal{L}(\mathcal{U}; V')$.

Then we consider the following problem :

Problem E_o

For each $v \in \mathcal{U}$ find $y \in K$ solution of

(1.1) $$a(y, \varphi - y) + j(\varphi) - j(y) \geqslant (f+Bv, \varphi - y) \quad \forall \varphi \in K$$

where f is given in V'.

Theorem 1.1

Under the following hypothesis

(1.2) $$\begin{cases} - \ a(.,.) \ \text{ is coercive : } a(\varphi,\varphi) \geqslant \alpha \|\varphi\|^2 \quad \alpha > 0 \quad \forall u \in V \\ \quad \text{or} \\ - \ j(.) \ \text{ is strictly convex} \end{cases}$$

there is a unique solution $y(v)$ to Problem E_o.

For the proof of this theorem cf. Lions-Stampacchia [6]. We introduce now :

- $\mathcal{H}$ Hilbert space and $C \in \mathcal{L}(V;\mathcal{H})$
- z_d given in $\mathcal{H}$

and we consider

Problem E_1

Find $u \in \mathcal{U}_{ad}$ solution of

(1.3) $$J(u) \leqslant J(v)$$

(1) More details about definitions proofs etc... will be given in Yvon [8] .

where

(1.4) $$J(v) = \|C\, y(v) - z_d\|_{\mathcal{H}}^2 + \nu \| v\|_{\mathcal{U}}^2 \qquad \nu \geqslant 0$$

Theorem 1.2

If we assume

(1.5)
- i) $\mathcal{U}_{ad}$ bounded (or $\nu > 0$ in (1.4)) and
- ii) If $\varphi_n \to \varphi$ weakly in V and $v_n \to v$ weakly in $\mathcal{U}$, then $\lim\, (B\, v_n, \varphi_n) \leqslant (Bv, \varphi)$,

then there is at least one solution to (1.3).

For the proof one uses a minimizing sequence of $J(v)$ and the hypothesis (1.5)-ii) allows to take the limit in (1.1).

Remark 1

The assumption ii) of (1.5) is a property of compacity of B. For instance in the case

(1.6) $$H \hookrightarrow V \hookrightarrow V'$$

(each space dense in the next with continuous injection) we may take

$$B\, v = \bar{B}\, v \qquad \bar{B} \in \mathcal{L}(\mathcal{U};H)$$

so that

$$(B\, v,\varphi)_{VV'} = (\bar{B}\, v,\varphi)_H$$

If the injection from V into H is compact then we obtain property (1.5)-ii).

II - PARABOLIC SYSTEMS

We suppose now that we have (as in (1.6))

- V,H Hilbert spaces, $V \subset H$ dense with continuous injection so that
$$V \hookrightarrow H = H' \hookrightarrow V'$$
- $a(y,z)$ bilinear form on V, symmetric and coercive,
- f given in $L^2(0,T;H)$,
- $j(z)$ a convex ℓ.s.c. functional on V with domain
$$D(j) = \{\varphi \in V \mid j(\varphi) < +\infty\}$$
- y_o given in the closure of $D(j)$ in H.

Then we have

Theorem 2.1

With the previous data, there exists a unique function y such that

(2.1) $$y \in C\,([0,T]\,;V)\ ,\quad \frac{dy}{dt} \in L^2(0,T;H)$$

and satisfying

(2.2) $$\left(\frac{dy}{dt}\, , z-y\right) + a(y,z-y) + j(z) - j(y) \geqslant (f,z-y) \quad \text{a.e. in } (0,T),$$

$$\forall\, z \in D(j).$$

(2.3) $$y(0) = y_o$$

Demonstration in Brezis [2] .

Now let us define
- $\mathcal{U}$ Hilbert space, $\mathcal{U}_{ad}$ closed convex subset of $\mathcal{U}$
- $B \in \mathcal{L}(\mathcal{U};L^2(0,T;H))$

and

Problem P_o

For each $v \in \mathcal{U}_{ad}$ find y satisfying (2.1) and

(2.4) $$(\frac{dy}{dt}, z-y) + a(y,z-y) + j(z) - j(y) \geqslant (f+Bv, z-y)$$

There exists a unique $y(v)$ solution of Problem P_o . Now let us introduce

- $\mathcal{H}$ Hilbert space and $C \in \mathcal{L}(L^2(0,T;V); \mathcal{H})$
- z_d given in $\mathcal{H}$

and

Problem P_1

Find $u \in \mathcal{U}_{ad}$ such that

(2.5) $$J(u) \leqslant J(v) \qquad \forall\ v \in \mathcal{U}_{ad}$$

with

(2.6) $$J(v) = \| Cy(v) - z_d\|^2_{\mathcal{H}} + \nu\| v\|^2_{\mathcal{U}} \qquad \nu \geqslant 0$$

Theorem 2.2

With the following hypothesis

(2.7) $$\begin{cases} \text{i) The injection from } V \text{ into } H \text{ is compact,} \\ \text{ii) } \mathcal{U}_{ad} \text{ is bounded (or } \nu > 0) \end{cases}$$

there exists at least one solution u to Problem P_1 .

An example

Let Ω an open set of $\mathbb{R}^n$, Γ its boundary and consider the system

(2.8) $$\begin{cases} \frac{\partial y}{\partial t} - \Delta y = v & \text{in } \Omega \times\ ]0,T[\\ \frac{\partial y}{\partial n} + \psi(y) = 0 & \text{on } \Gamma \times\ ]0,T[\\ y(0) = y_o \text{ in } \Omega & y_o \text{ given in } L^2(\Omega) \end{cases}$$

and

(2.9) $$\psi(r) = \begin{cases} 0 & \text{if } r < h \qquad h,k > 0 . \\ kr & \text{if } r \geqslant h \end{cases}$$

In order to set properly the inequality associated with relations (2.8) let

$$H=L^2(\Omega) \qquad V = H^1(\Omega) \qquad \mathcal{U} = L^2(\Omega)$$

(2.10) $$a(y,z) = \int_\Omega \sum_{i=1}^n \frac{\partial y}{\partial x_i}(x) \frac{\partial z}{\partial x_i}(x)\, dx$$

(2.11) $$j(z) = \int_\Gamma F[z(\gamma)]d\gamma \quad \text{with} \quad F(r) = \begin{cases} 0 & \text{if } r \leqslant h \\ \frac{k}{2}(r^2-h^2) & \text{if } r \geqslant h \end{cases}$$

so that $y(v)$ is the unique solution of the variationnal inequality :

$$(2.12)\begin{cases} \left(\frac{dy(v)}{dt}, z-y(v)\right) + a(y(v),z-y(v)) + j(z) - j(y(v)) \geqslant (v,z-y(v))_{L^2(\Omega)} \\ y(v)\big|_{t=0} = y_o \end{cases}$$

The cost functional is

$$(2.13) \qquad J(v) = \int_0^T \int_\Gamma \left|\frac{\partial y(v)}{\partial n} - z_d\right|^2 d\Gamma\, dt + \nu \, \|v\|^2_{\mathcal{U}} \qquad \nu \geqslant 0$$

Then if $\mathcal{U}_{ad}$ is bounded or $\nu > 0$ there exists at least one optimal control U solution of (2.5) with (2.13).

Remark 2.1

As in the elliptic case the main difficulty of the problem (theoretically and numerically) is the non differentiability of the mapping

$$(2.14) \qquad v \to y(v) \quad \text{solution of Problem } P_o \ (\text{or } E_o).$$

So that the question of necessary optimal conditions from u is, as far as we know, yet opened. For some results towards this direction cf. F. Mignot [7] .

III - DUALITY APPROACH

Using the ideas submitted in a paper from Cea-Glowinski-Nedelec [3] one can obtain a dual formulation of the variationnal inequality. For reason of simplicity we suppose that we are in the elliptic case and that the variationnal inequality comes from the minimization problem :

$$(3.1) \qquad \inf_{\varphi \in V} \frac{1}{2} a(\varphi,\varphi) + j(\varphi) - (f + Bv,\varphi)$$

Then the fundamental assumptions are :

There exists a Banach space L, a closed convex bounded set Λ of L' with $0 \in \Lambda'$ and an operator G (non necessarily linear) from V into L such that

$$(3.2) \qquad j(\varphi) = \sup_{\mu \in \Lambda} \langle \mu , G(\varphi)\rangle^{(1)}$$

Now we can rewrite the problem (3.1) in the form

$$(3.3) \qquad \inf_{\varphi \in V} \ \sup_{\mu \in \Lambda} \ \mathcal{L}(\varphi,\mu;v)$$

with

$$(3.4) \qquad \mathcal{L}(\varphi,\mu;\, v) = \frac{1}{2} a(\varphi,\varphi) + \langle \mu ,G(\varphi) \rangle - (f+Bv,\varphi).$$

The dual formulation of (3.3) is

$$(3.5) \qquad \sup_{\mu \in \Lambda} \ \inf_{\varphi \in V} \ \mathcal{L}(\varphi,\mu;v)$$

Example 3

We take the analogue of example 2. The problem is in the general form of (3.1) with a(.,.) and j(.) given by (2.10) (2.11).

So that we have

$$L = L^1(\Gamma) \qquad L' = L^\infty(\Gamma)$$

$$\Lambda = \{\mu \in L^\infty(\Gamma) \mid 0 \leqslant \mu(\gamma) \leqslant 1 \quad \text{a.e. on } \Gamma\}.$$

and

$$G(\varphi) = \frac{k}{2} (\varphi^2(\gamma) - h^2)$$

which is a non linear operator from $V = H^1(\Omega)$ into $L^1(\Gamma)$.

---

(1) Here $\langle .,. \rangle$ denotes the duality product between L' and L.

Theorem 3.1

Under the following hypothesis

$$(3.6)\qquad \begin{cases} \text{For each } \lambda \in \Lambda \text{ the mapping} \\ \varphi \to \langle \lambda, G(\varphi) \rangle \\ \text{is convex } \ell.s.c., \end{cases}$$

there exists a saddle point for $\mathcal{L}(.,.;v)$

The proof is an application of the theorem of KY Fan-Sion.

Theorem 3.2

If $G(\varphi)$ is continuous from V- weak into L'-weak* and if assumption (3.6) holds, then

$$M(\lambda) = \mathcal{L}(y(\lambda),\lambda;v) = \inf_{\varphi \in V} \mathcal{L}(\varphi,\lambda;v) \tag{3.7}$$

is Gâteaux diffenriatiable with derivative given by

$$M'(\lambda).\mu = \langle \mu, G(y(\lambda)) \rangle . \tag{3.8}$$

Corollary 3.1

A necessary and sufficient condition for $\lambda^* \in \Lambda$ solving the dual problem (3.5) is

$$\langle \lambda^* - \lambda, G(y(\lambda^*)) \rangle \geqslant 0 \quad \forall \ \lambda \in \Lambda \tag{3.9}$$

Corollary 3.2

If λ^* solve the dual (3.5) then $y(\lambda^*)$ solve the primal (3.3). Now we can state the problem corresponding to the paragraph 1 :

Optimal control problem

$$(3.10)\qquad \begin{cases} \text{A.} \quad \text{For each } v \in \mathcal{U}_{ad} \text{ compute } \lambda(v) \text{ and } y(v) = y(\lambda(v)) \text{ solution of} \\ \qquad \sup_{\mu \in \Lambda} \ \inf_{\varphi \in V} \ \mathcal{L}(\varphi,\mu;v) \\ \text{B.} \quad \text{Find } u \in \mathcal{U}_{ad} \text{ such that} \\ \qquad J(u) \leqslant J(v) \quad \text{with } J \text{ given in (1.4)} \end{cases}$$

Then using the optimality condition (3.9), we can associate to the previous problem the

Suboptimal control problem

$$(3.11)\qquad \begin{cases} \text{A.} \quad \text{For } \lambda \text{ fixed in } \Lambda \text{ compute } y(\lambda;v) \text{ solution of} \\ \qquad \inf_{\varphi \in V} \ \mathcal{L}(\varphi,\lambda;v) \\ \text{B.} \quad \text{Solve the optimal control problem} \\ \qquad \inf_{v \in \mathcal{U}_{ad}} \ \| Cy(\lambda;v) - z_a \|^2_{\mathcal{H}} + \nu \| u \|^2_{\mathcal{U}} \\ \qquad \text{this gives } u(\lambda) \text{ and } y(\lambda) = y(\lambda;u(\lambda)) \\ \text{C.} \quad \text{Then finally find } \lambda^* \in \Lambda \text{ satisfying} \\ \qquad \langle \lambda^* - \lambda \ , G(y(\lambda^*)) \rangle \geqslant 0 \quad \forall \ \lambda \in \Lambda \end{cases}$$

Remark 3.1

The previous technique which consists in permuting the determination of u and λ is due to Begis-Glowinski [1] . Notice that problem (3.11) is not equivalent to problem (3.10) this can be shown on very simple counter-examples (cf. e.g. Yvon [8]).

Theorem 3.3

Under assumptions of §1 and theorem 3.2 there exists at least one solution λ^* to problem (3.11).

IV - REGULARIZATION

An other way to avoid the non differentiability of the mapping (2.14) is to approach Problem P_1 (for instance) by a sequence of problem more regular wich ensure the differentiability of cost functions. We will expose the method in the parabolic case (§2).

Fundamental hypothesis.

(4.1) There exists a family of functionnal $j_\varepsilon(.)$ of convex functionals on V twice continuously differentiable such that

(4.2) $j_\varepsilon(\varphi) + a(\varphi,\varphi) \geqslant \beta \qquad \forall\ \varphi \in V \quad \beta$ independant of ε

(4.3) $$\lim_{\varepsilon\to 0} j_\varepsilon(\varphi) = j(\varphi) \qquad \forall\ \varphi \in V \ {}^{(1)}$$

(4.4) $$\begin{cases} \text{if } y_\varepsilon \longrightarrow y \quad \text{weakly in } L^2(0,T;V) \text{ as } \varepsilon \to 0, \text{ then} \\ \displaystyle\int_0^T j(y(t))dt \leqslant \underline{\lim} \int_0^T j_\varepsilon(y_\varepsilon(t))dt \end{cases}$$

(4.5) There exists a sequence z_ε bounded in V such that $j'_\varepsilon(z_\varepsilon) = 0$

Then we define

Problem $P_{o\varepsilon}$

Find y_ε solution of

(4.6) $$\begin{cases} \left(\dfrac{dy_\varepsilon}{dt}, z-y_\varepsilon\right) + a(y_\varepsilon, z-y_\varepsilon) + j_\varepsilon(z) - j_\varepsilon(y_\varepsilon) \geqslant (f+Bv, z-y_\varepsilon) \\ y_\varepsilon(0) = y_o \qquad \forall\ z \in V \end{cases}$$

Theorem 4.1

For each $v \in \mathcal{U}_{ad}$ there is a unique solution $y_\varepsilon(v)$ to (4.6) such that

$$y_\varepsilon(v) \in C([0,T];V).$$

Furthermore

$$y_\varepsilon(v) \to y(v) \quad \text{in } L^2(0,T;V) \quad \text{as} \quad \varepsilon \to 0$$

where $y(v)$ is the solution of Problem P_o.

With notation of §2 we introduce

(4.7) $$J_\varepsilon(v) = \|Cy_\varepsilon(v) - z_d\|^2_{\mathcal{H}} + \nu\ \|v\|^2_{\mathcal{U}}$$

and the

Problem $P_{1\varepsilon}$

Find $u_\varepsilon \in \mathcal{U}_{ad}$ such that

(4.8) $$J_\varepsilon(u_\varepsilon) \leqslant J_\varepsilon(v) \qquad \forall\ v \in \mathcal{U}_{ad}.$$

Theorem 4.2

There exists at least one u_ε solution of (4.8) and as $\varepsilon \to 0$ there exists a subsequence $\{u_{\varepsilon'}\}$ of $\{u_\varepsilon\}$ such that

$$u_{\varepsilon'} \to u \quad \text{in } \mathcal{U}$$

where u is a solution of Problem P_1.

---

(1) For simplicity we assume that $D(j) = D(j_\varepsilon) = V$. (Cf. notations of Th. 2.1)

Remark 4

As $j_\varepsilon(.)$ is in class C^2 the Problem $P_{o\varepsilon}$ may be rewritten as

$$\left(\frac{dy_\varepsilon}{dt},z\right) + a(y_\varepsilon,z) + (j'_\varepsilon(y_\varepsilon),z) = (f+Bv,z) \qquad \forall \quad z \in V$$

and Problem $P_{1\varepsilon}$ is then an ordinary optimal control problem for parabolic systems.

V - APPLICATION TO A BIOCHEMICAL-EXAMPLE

The system represents an enzymatic reaction in a membrane with semi-porous boundary. The problem is unidimensional in space and the functions $a(x,t)$, $s(x,t)$, $p(x,t)$ represent respectively the concentration of activator, substrate and product in the membrane[(1)]. In dimension-less variables the problem may be stated as

$$(5.1) \quad \begin{cases} \dfrac{\partial a}{\partial t} - \dfrac{\partial^2 a}{\partial x^2} = 0 & x \in (0,1) \quad t \in (0,T) \\ a(o,t) = v_o(t) \quad a(1,t) = v_1(t) & \\ a(x,o) = 0 & \end{cases}$$

$$(5.2) \quad \begin{cases} \dfrac{\partial s}{\partial t} - \dfrac{\partial^2 s}{\partial x^2} + \sigma \dfrac{a}{1+a} \cdot \dfrac{s}{1+s} = 0 & \sigma > 0 \\ s(o,t) = \alpha_o \quad s(1,t) = \alpha_1 & \alpha_o, \alpha_1 \in \mathbb{R} \\ s(x,o) = 0 & \end{cases}$$

$$(5.3) \quad \begin{cases} \dfrac{\partial p}{\partial t} - \dfrac{\partial^2 p}{\partial x^2} = \sigma \dfrac{a}{1+a} \cdot \dfrac{s}{1+s} \\ -\dfrac{\partial p}{\partial x}(o,t) + \psi(p(o,t)) = 0 \\ \dfrac{\partial p}{\partial x}(1,t) + \psi(p(1,t)) = 0 \\ p(x,o) = 0 \end{cases}$$

where $\psi(r)$ is real function given in (2.9).

Control variables are v_o and v_1 :

$$\mathcal{U} = L^\infty(0,T) \times L^\infty(0,T)$$

and

$$\mathcal{U}_{ad} = \{v \in \mathcal{U} \mid \quad 0 \leqslant v_i(t) \leqslant M \quad \text{a.e. on } (0,T), \qquad i=1,2\}$$

The cost function is

$$(5.4) \qquad J(v) = \int_0^T \left| -\frac{\partial p}{\partial x}(o,t) - z_o(t)\right|^2 dt + \int_0^T \left|\frac{\partial p}{\partial x}(1,t) - z_1(t)\right|^2 dt$$

Theorem 5.1

The system (5.1)...(5.3) admits a unique solution $a(v), s(v), p(v)$.

(1) For more details about enzymatic systems cf. Kernevez [5] .

Theorem 5.2

There exist at least one $u \in \mathcal{U}_{ad}$ satisfying $J(u) \leqslant J(v) \quad \forall v \in \mathcal{U}_{ad}$ with $J(v)$ given by (5.4).

VI - NUMERICAL RESULTS

Example 1

To give comparative numerical results on the two types of algorithm we have considered first the example of § II.Computation have been performed with

$$\Omega =]0,1[\qquad \text{so that} \qquad \Gamma = \{0\} \cup \{1\}$$

and z_d only function of time so that the cost function is

$$(61) \qquad J(v) = 2 \int_0^T \left|\frac{\partial y}{\partial x}(1,t) - z_d(t)\right|^2 dt + \nu \|v\|_{\mathcal{U}}^2$$

(solution is symmetric by changing x in $1-x$).

Figure 1

Represents the state corresponding to an "optimal control" computed by duality method (§III). The threshold is given by h=0.5.

Figure 2

Gives comparison between regularization and duality on the same example. The suboptimality of duality is clear on this picture. Actually the "optimal" values of cost are :

duality : $4. \quad 10^{-2}$
regularization : $0.36 \quad 10^{-2}$

Remark 6.1

In the previous examples ν in (6.1) has been taken near zero so that the optimal state may fit z_d as well as possible.

Example 2

(Bio-chemical example of §V).

As an example 1 the problem has been considered completely symmetric with a unique control $v(t)$ so that boundary conditions in (5.1) are

$$a(o,t) = a(1,t) = v(t).$$

Figure 3 and Figure 4

give optimal control and corresponding optimal value of the state computed by regularization. Figure 3 represents also values of optimal control computed for two values of the regularization parameter ε . The only active constraint is $v \geqslant 0$ in this example.

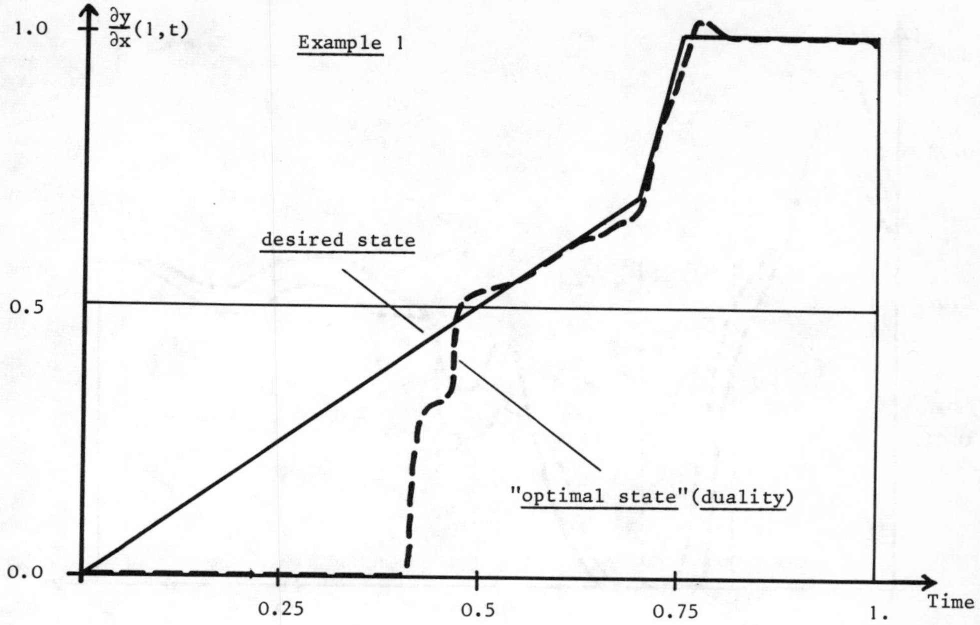

Figure 1 Optimal state

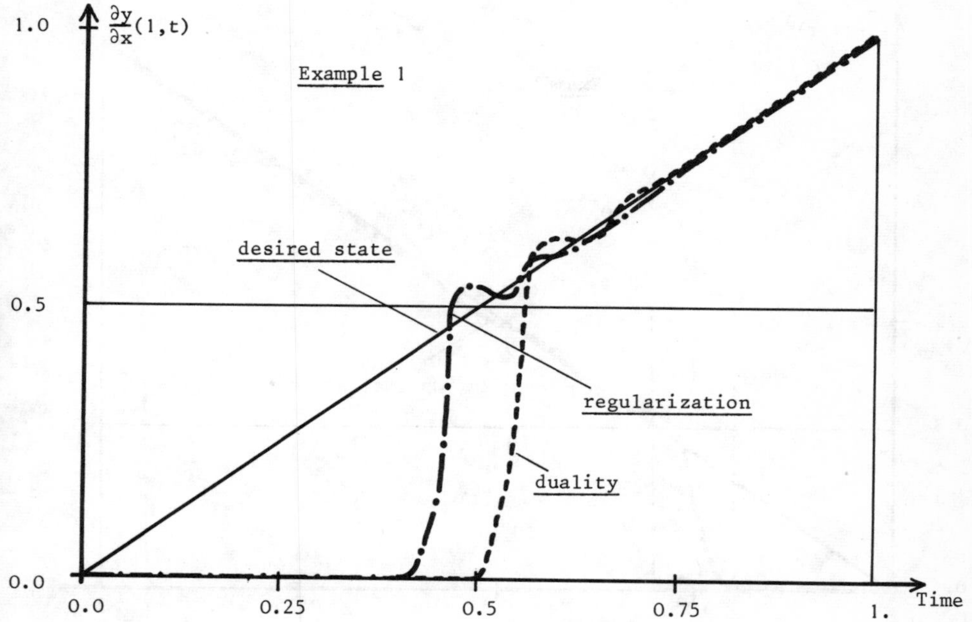

Figure 2 Optimal state

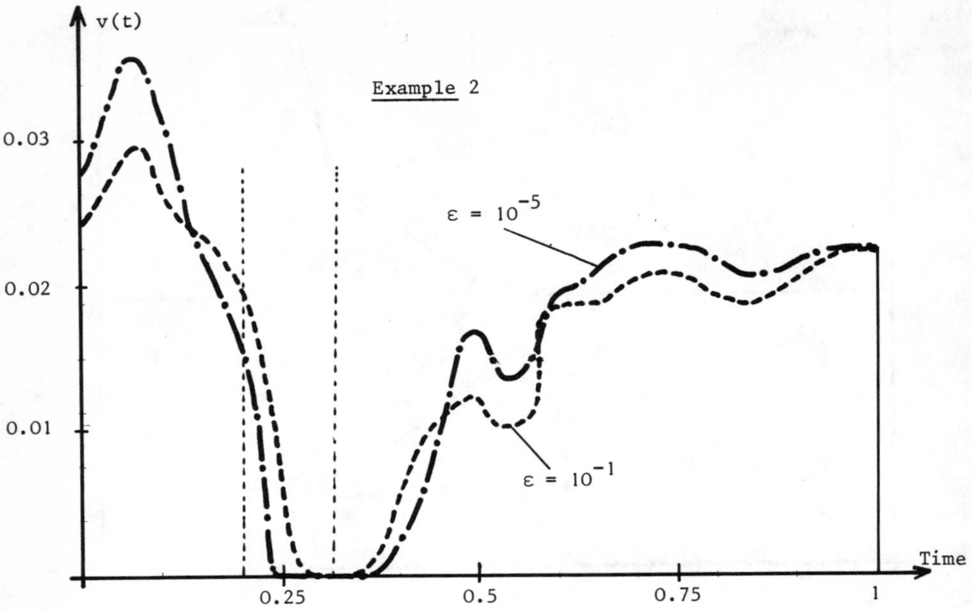

Figure 3 Optimal control

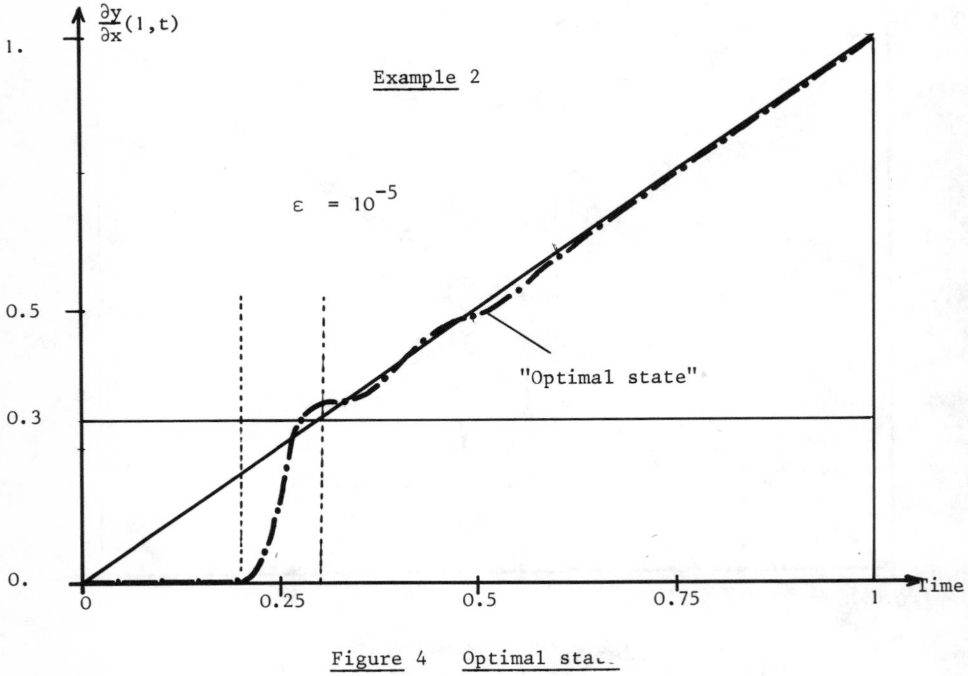

Figure 4 Optimal sta...

VII - REFERENCES

(1) D. Begis, H. Glowinski "Dual num. meth. for some variational problem..." in Techniques of optimization. Academic Press (1972).

(2) H. Brezis "Problèmes unilatéraux", Journal of Math. pures et appliquées 51, (1972).

(3) J. Cea, R. Glowinski, J.C. Nedelec "Minimisation de fonctionnelles non différentiables", Proceedings of the Dundee Num. Anal. Symp. (1972).

(4) G. Duvaut, J.L. Lions "Les inéquations en mécanique et en physique", Dunod Paris (1972)

(5) J.P. Kernevez "Evolution et contrôle des systèmes bio-mathématiques" Thèse, Paris (1972).

(6) J.L. Lions, G. Stampacchia "Variational Inequalities", Comm. on pure and app. math. vol XX, pp. 439-519 (1967).

(7) F. Mignot Séminaire Lions-Brezis Paris 1972-1973.

(8) J.P. Yvon Thèse Paris 1973.

ON DETERMINING THE SUBMANIFOLDS OF STATE SPACE WHERE THE OPTIMAL VALUE SURFACE HAS AN INFINITE DERIVATIVE

Harold L. Stalford

Radar Analysis Staff
Radar Division
Naval Research Laboratory
Washington, D. C. 20375

ABSTRACT

The problem of obtaining the optimal value surface of an optimal control process is investigated. In practice, the optimal cost surface often possesses an infinite derivative at points of certain submanifolds of the state space. A necessary condition is derived with which the equations of such submanifolds can be established without solving first the entire optimal control problem. The necessity of the condition is proved in a theorem, but only for submanifolds having one dimension less than the dimension of the state space. Three examples are provided to illustrate the utility of the condition.

I. INTRODUCTION

An objective in solving optimal control problems is to calculate the optimal cost of transfer from any initial state to the terminal set. Calculating and plotting the optimal cost values above the state space produces a surface which we shall call the optimal value surface. An optimal value surface is not necessarily smooth at every point. In general, the surface consists of points belonging to three distinct classes. First, we have the set of points where the surface is smooth. Second, there are those points where all approaching tangents to the surface remain bounded but where the surface is not smooth. The third class consists of those points where at least one approaching tangent cannot be bounded. We investigate this latter class.

Our desire is to determine where the optimal value surface has an infinite derivative without solving first the entire optimal control problem. In practice, an infinite derivative will occur along smooth submanifolds of the state space. This paper presents a necessary condition for establishing the equations of such submanifolds.

We shall now define the family of optimal control processes for which the ensuing theory holds.

II. A FAMILY OF OPTIMAL CONTROL PROCESSES

The family of optimal control processes under investigation have their dynamical behavior governed by systems of ordinary differential equations and have their evolution of state described by the motion of a point in n-dimensional Euclidean space E^n. The seven basic elements that are needed to define such optimal control processes are four functions (f, U, f_o, and g_o), two sets (X and Θ) and a function space Ω. These elements are described subsequently. The dynamical behavior of an optimal control process, hereafter, is modeled by means of the state velocity function f in the state equation

$$(1) \qquad \dot{\varphi}(t) = f(\varphi(t), u(t)), \ \varphi(t) \in E^n, \ u(t) \in E^m$$

where $f: E^n \times E^m \to E^n$ is a continuous function. The evolution of state is described by a point moving in the state space X, an open subset of E^n. The terminal set Θ is a closed set contained in the closure of X. For non-autonomous systems (that is, f an explicit function of t), one component of φ is time t itself.

The controller of the process is equipped with the two elements Ω and U. The function space Ω is the space of all Lebesgue measurable functions of time t on bounded intervals whose values have range in E^m. Constraints on the control functions in Ω are given implicitly by the set valued function

$$(2) \qquad U : X \to \text{set of all compact subsets of } E^m,$$

where U is a continuous set value function. For each state x, the set U(x) is precisely the set of control values available to the controller at the state x.

A solution of the differential equation (1) for some measurable control function and given initial conditions is called a trajectory. A trajectory $\varphi : [t_o, t_f] \to E^n$ is said to be admissible if it lies entirely in the state space X for all times t contained in the interval $[t_o, t_f)$. An admissible trajectory is said to be terminating if $\varphi(t_f)$ is contained in the terminal set Θ. The time t_f is called the terminating or final time for a terminating admissible trajectory.

The time t_f belongs to the interval $[t_o, \infty)$; t_f does not necessarily have to be the same terminating time for distinct trajectories unless it is constrained to be fixed by the terminal set Θ.

A control function $u:[t_o, t_f] \to E^m$ is said to be admissible if it has at least one corresponding admissible trajectory $\varphi:[t_o, t_f) \to X$ such that $u(t) \in U(\varphi(t))$ for all t contained in $[t_o, t_f)$. Here, the trajectory of φ corresponds to the control function u if

$$\varphi(t) = \varphi(t_o) + \int_{t_o}^{t} f(\varphi(\tau), u(\tau))\, d\tau$$

for all t contained in $[t_o, t_f]$. Note that, since f is required to be only continuous, solutions to the differential equation (1) are not necessarily unique for each control function u.

Let x_o be contained in X. Let $\mathcal{C}(x_o)$ denote the set of all admissible control functions $u:[t_o, t_f] \to E^m$ having at least one terminating admissible trajectory emanating from x_o. For the control function u contained in $\mathcal{C}(x_o)$, let $\mathcal{T}(x_o;u)$ denote the set of all terminating admissible trajectories φ emanating from x_o, corresponding to the control function u, and satisfying $u(t) \in U(\varphi(t))$ for all t contained in the domain of u.

The state is to be transferred from x_o contained in X to the terminal set Θ; the initial time t_o is fixed, while the final time t_f is not necessarily specified. The performance criterion

$$(3) \qquad J(x_o, \varphi, u) = g_o(\varphi(t_f)) + \int_{t_o}^{t_f} f_o(\varphi(\tau), u(\tau))\, d\tau$$

is to be minimized where the function g_o is a real valued continuously differentiable function defined on a neighborhood of the terminal set Θ, the function f_o is a real valued bounded continuous function with domain $E^n \times E^m$, the control function u belongs to $\mathcal{C}(x_o)$ and the trajectory φ is a member of $\mathcal{T}(x_o; u)$. The real number $J(x_o, \varphi, u)$ denotes the value of performance associated with the transfer.

In summary, a member of the family of optimal control processes is represented by the Septuple $(f, U, f_o, g_o, X, \Theta, \Omega)$ where f, U, and f_o are continuous functions, g_o is continuously differentiable, Ω is the space of Lebesgue measurable controls, X is open in E^n, and Θ is a closed set sontained in the closure of X. Let Γ denote this family of optimal control processes.

III. OPTIMAL VALUE FUNCTION

Let x_o be contained in the state space X. Let the control function u^* be contained in $\mathcal{C}(x_o)$ and let the trajectory φ^* be contained in $\mathcal{T}(x_o; u^*)$. The pair (u^*, φ^*) is said to be optimal at x_o if, for all control functions u contained in $\mathcal{C}(x_o)$ and for all trajectories φ contained in $\mathcal{T}(x_o; u)$, the following inequality is satisfied:

$$J(x_o, \varphi^*, u^*) \leq J(x_o, \varphi, u)$$

If the pair (u^*, φ^*) is optimal at x_o, then the value $J(x_o, \varphi^*, u^*)$ is arbitrarily defined to be $V(x_o)$. If an optimal pair (u^*, φ^*) exists for every x_o contained in the state space X then we have a function from X into the real numbers:

$$V : X \to E^1$$

Thus, for a state x, the value V(x) denotes the optimal transfer cost from x to the terminal set. The function V is called the optimal value function. We suppose that V is well defined on the entire state space X. A plot of the optimal value function above the state space is called the optimal value surface.

Some definitions are needed in order to describe an assumption placed on the optimal value function V.

Definition 1. A decomposition D of the state space X is a denumerable collection of disjoint subsets whose union is X. This is usually written as $D = \left\{X_o, X_j : j \in J\right\}$ where J is a denumerable index set for the members of D other than X_o.

Definition 2. A regular decomposition D of the state space X is a decomposition $D = \left\{X_o, X_j : j \in J\right\}$ such that X_o is open and dense in X and such that for each $j \in J$, X_j is a continuously differentiable submanifold of E^n.

Since X_o is open in X, it follows that X_o is a continuously differential submanifold of dimension n of E^n.

Definition 3. Let B be a subset of E^n. A mapping $F : B \to E^1$ is said to be continuously differentiable on B if there is an open set W containing B such that F may be extended to a function which is continuously differentiable on W.

We are now in a position to describe the assumption. It is an unresolved conjecture in optimal control theory for the family of optimal control processes considered herein. It is satisfied by hypothetically constructed examples as well as optimal control models of physical processes.

Assumption 1. There exists a regular decomposition D of the state space X such that the optimal value function V is continuously differentiable on the members of D.

The optimal value function of most control processes is continuous. It is, however, discontinuous in some control problems that model processes in nature. For example, Vincent [5] presents an optimal control model of an agricultural problem where the objective is to control insects that eat or destroy crops. Therein, the optimal value surface is shown to have a tear or split extending along a smooth submanifold of the state space. Since the continuity assumption can fail to be satisfied, we introduce another assumption to take its place. Incidentally, it is readily satisfied if the optimal value function is continuous.

Suppose Assumption 1 is met and let X_j be a member of the regular decomposition D. Let α be a point on the submanifold X_j.

Assumption 2. There exists an open neighborhood $\mathcal{R}$ of α in the topology of X_j and an open subset S of X_o whose closure contains $\mathcal{R}$ such that the optimal value function V when restricted to S has a continuous extension V_1 from S to a closed set $\mathcal{C}$ of X that contains $\mathcal{R}$ and such that V_1 is continuously differentiable on $\mathcal{R}$.

Here, a function $V_1 : \mathcal{C} \to E^1$ is said to be a continuous extension of V from S to $\mathcal{C}$ if V_1 is continuous on $\mathcal{C}$ and $V_1(x) = V(x)$ for all x contained in S.

Consider an optimal value surface which is continuous and satisfies Assumption 1. Take a pair of scissors and make a number of smooth cuts in the surface. Then deform the surface in a continuously differentiable manner. The resulting surface is of the general type that satisfies Assumption 2. It is introduced so that the ensuing theory encompasses problems in which the optimal value surfaces have tears running along smooth submanifolds of the state space.

IV. THE FUNDAMENTAL PARTIAL DIFFERENTIAL EQUATION OF DYNAMIC PROGRAMMING

Let $(f, U, f_o, g_o, X, \Theta, \Omega)$ be an optimal control process in Γ such that its optimal value function V satisfies Assumption 1. Let $D = \left\{X_o, X_j : j \in J\right\}$ be a regular decomposition with which V satisfies Assumption 1. It is shown in Stalford [4] that the fundamental partial differential equation of dynamic programming must be met on the open and dense member X_o of the decomposition D of X. This equation is written as

$$(4) \qquad 0 = \underset{v \in U(x)}{\text{MINIMUM}} \left\{ f_o(x,v) + \text{Grad } V(x) \cdot f(x,v) \right\}$$

and holds for all $x \in X_o$.

This equation will be utilized subsequently in determining points where the optimal value function is not smooth.

V. PROBLEM STATEMENT

Let X_j be some n-1 dimensional member of the regular decomposition D. For convenience, let M denote the continuously differentiable submanifold X_j. Let the state α denote a point on M and let $N(\alpha)$ denote a unit row vector that is normal to M at α. Finally, let $N(\alpha)^T$ denote the transpose of $N(\alpha)$.

Definition 4. The optimal value function V has an infinite derivative at α in the normal direction $N(\alpha)$ if the limit

$$\underset{h \to o^+}{\text{limit}} \ \text{Grad } V(\alpha + h \cdot N(\alpha)) \cdot N(\alpha)^T$$

cannot be bounded. The notation $h \to o^+$ denotes that h takes on only positive values.

Suppose that the optimal value function V has at each point of M an infinite derivative in at least one of the normal directions to M. Recall that an n-1 dimensional submanifold of E^n has two normal directions at each of its points. Vectorially, one is of course the negative of the other.

Problem. Determine the equation of the submanifold M without solving first the optimal control problem for optimal control feedback policies.

VI. ORTHOGONAL TRANSFORMATION OF COORDINATES

Let $T(\alpha)$ denote a matrix of tangential orthogonal vectors to M at α. We desire to transform the x coordinates linearly into new coordinates such that the new coordinate system coincides with the normal vector $N(\alpha)$ and the tangential vectors $T(\alpha)$. Such a transformation is given by the matrix equation

$$(5) \qquad y = K(\alpha) \cdot x$$

where $K(\alpha)$ is the matrix composed of the normal vector $N(\alpha)$ and the tangential vector $T(\alpha)$, that is,

$$(6) \qquad K(\alpha) = \begin{bmatrix} N(\alpha) \\ \cdots \\ T(\alpha) \end{bmatrix}$$

The normal vector $N(\alpha)$ and the tangential vectors $T(\alpha)$ can be chosen such that the equation

$$(7) \qquad K(\alpha) \cdot K(\alpha)^T = \text{Identity Matrix}$$

holds where $K(\alpha)^T$ is the transpose of $K(\alpha)$. Equation (7) implies that $K(\alpha)^T$ is equal to the inverse of $K(\alpha)$. Thus Equation (5) is an orthogonal transformation of coordinates.

Equation (4) can be rewritten as

$$(8) \qquad 0 = \underset{v \in U(x)}{\text{MINIMUM}} \left\{ f_o(x,v) + \text{Grad } V(x) \cdot K(\alpha)^T \cdot K(\alpha) \cdot f(x,v) \right\}$$

where $x \in X_o$. Substituting Equation (6) into Equation (8) and deleting zero terms we obtain

$$(9) \qquad 0 = \underset{v \in U(x)}{\text{MINIMUM}} \Big\{ f_o(x, v) + [\text{Grad } V(x) \cdot T(\alpha)^T] \cdot [T(\alpha) \cdot f(x, v)]$$
$$+ [\text{Grad } V(x) \cdot N(\alpha)^T] \cdot [N(\alpha) \cdot f(x, v)] \Big\}$$

for all $x \in X_o$.

VII. THEOREM

Let $\{x_k\}$ be any sequence in X_o that converges to the state α. The following observation can be verified by invoking Assumptions 1 and 2.

<u>Observation 1.</u> In the limit as x_k converges to α, the (1 x n-1) matrix $[\text{Grad } V(x_k) \cdot T(\alpha)^T]$ is bounded.

This observation states that the slopes of the optimal value surface in the tangent direction to M at α remain finite as α is approached. Lemma 8.2.1 of Stalford [3, p. 84] asserts this observation for the case that the optimal value function is continuous. When this is not the case, Assumption 2 can be used to amend the lemma.

Let $C(\alpha)$ be the convex closure of the set

$$\left\{N(\alpha) \cdot f(\alpha, v) \; : \; v \in U(\alpha)\right\}.$$

Note that $C(\alpha)$ is composed of scalars or one dimensional vectors since $N(\alpha)$ is a (1 x n) vector and $f(\alpha, v)$ is a (n x 1) vector. Below, when we speak of the zero vector in $C(\alpha)$ we, in essence, mean the real number zero.

<u>Theorem 1.</u> If α is a point of the submanifold M where the optimal value function has an infinite derivative in the normal direction $N(\alpha)$, then it is necessary that the zero vector belongs to the boundary of $C(\alpha)$.

<u>Proof.</u> We prove the theorem by showing that a contradiction arises if the zero vector belongs either to the interior or the exterior of $C(\alpha)$.

Suppose that the zero vector belongs to the interior of $C(\alpha)$. Let δ be some positive number such that δ and $-\delta$ are contained in the interior of $C(\alpha)$. Let $\{x_k\}$ be the sequence defined by

$$x_k = \alpha + h_k \cdot N(\alpha) \tag{10}$$

where the sequence $\{h_k\}$ is positive and converges to zero as k goes to infinity. Since the functions f and U are continuous, there exists an integer K such that if $k \leq K$ then the real numbers δ and $-\delta$ are contained in the convex closure of the set

$$\left\{ N(\alpha)\cdot f(x_k, v) \ : v \in U(x_k)\right\}.$$

In particular, for $k \geq K$ there exist controls $v_1(x_k)$ and $v_2(x_k)$ contained in the control set $U(x_k)$ such that

$$N(\alpha)\cdot f(x_k, v_1(x_k)) < -\delta \tag{11}$$

$$N(\alpha)\cdot f(x_k, v_2(x_k)) > \delta. \tag{12}$$

Let the sequence $\{R(x_k)\}$ of real numbers be defined by

$$R(x_k) = \text{Grad } V(x_k)\cdot N(\alpha)^T \tag{13}$$

for all x_k contained in $\{x_k\}$. Similarly, let the sequences $\{S1(x_k)\}$ and $\{S2(x_k)\}$ be defined by

$$S1(x_k) = N(\alpha)\cdot f(x_k, v_1(x_k)) \tag{14}$$

$$S2(x_k) = N(\alpha)\cdot f(x_k, v_2(x_k)), \tag{15}$$

respectively. In view of Definition 4, Equations (10) and (13) imply that the sequence $\{R(x_k)\}$ cannot be bounded. Thus, invoking Equations (11) and (12), we see that one of the sequences $\{S1(x_k)\cdot R(x_k)\}$ and $\{S2(x_k)\cdot R(x_k)\}$ cannot be bounded from below. This implies that the minimum of the expression in the curly brackets of Equation (9) cannot be zero for x_k sufficiently close to α. This is because the first two terms in that expression are bounded for all x_k sufficiently close to α. The first is bounded since the functions f_o and U are continuous and since U(x) is compact for each x contained in X. The second is bounded since, in addition, the function f is continuous and Observation 1 holds. We have already remarked that the third term takes on large negative values for x_k sufficiently close to α. This contradiction of Equation (9) proves the falsehood of the zero vector belonging to the interior of $C(\alpha)$. Next, we show that the zero vector cannot belong to the exterior of $C(\alpha)$.

Suppose that the zero vector belongs to the exterior of $C(\alpha)$. Since the function f is continuous and $U(\alpha)$ is compact, the set

$$\left\{ N(\alpha) \cdot f(\alpha, v) \, : \, v \in U(\alpha) \right\}$$

is a compact subset of the real numbers. Therefore, the set $C(\alpha)$ is a compact interval $[b, c]$ where b and c represent the endpoints. Since the zero vector is in the exterior of $C(\alpha)$, either b and c are both positive or they are both negative. Thus the third term in Equation (9) takes on only large positive numbers or large negative numbers for x_k sufficiently close to α. As before, the first two terms in Equation (9) are bounded for all x_k sufficiently close to α. Therefore, we have again contradicted the validity of Equation (9). In conclusion, zero must belong to the boundary of $C(\alpha)$ and our theorem is proved.

VIII. APPLICATIONS

The necessary condition of the theorem is applied in this section to three examples. In each example the necessary condition determines a family of equations for the sought submanifold M. The family of equations corresponds to a family of terminal sets. Thus, the equation selected is the one that has an intersection with the terminal set of the example being studied. Employing the knowledge that trajectories enter the terminal set rather than leave it is also a factor in determining the submanifold M. As in previous sections the submanifold M designates the points of the state space where the optimal value function has an infinite derivative.

Example 1. An antique example to illustrate the theory is given by the one-dimensional time-optimal regulator process of Leitmann [1, p. 48] and Pontryagin [2, p. 23]. In this example, a rocket travels in a straight line toward a terminal set. With the motion of the rocket controlled by a thrust program, the objective is to bring the rocket to rest at the terminal set and render the transfer time a minimum.

The equation of motion of a point mass moving horizontally above a flat earth states that the acceleration is equal to the thrust value.

In state equation form, we have

$$(16)\qquad \dot{x}_1 = x_2$$

$$\dot{x}_2 = v \qquad\qquad -1 \le v \le 1$$

where x_1 is the position of the rocket relative to the terminal set, x_2 is the speed of the rocket and the control value v represents the thrust. The maximum thrust is normalized to unity and the rocket engines are reversible. Here, the terminal set is the two-dimensional point (x_1, x_2) with $x_1 = 0$ and $x_2 = 0$. The state space X is the two-dimensional Euclidean space E^2. In the performance criterion, the function g_o is identically zero and the function f_o is identically one since we are minimizing the transfer time from an initial state to the origin. Finally, note that $U(x_1, x_2) = [-1, 1]$ for all (x_1, x_2) contained in X.

Without solving this example for optimal control feedback policies, we desire to find those points of the state space where the optimal value surface has an infinite derivative. Our approach is to apply the necessary condition of the theorem. More specially, we select an arbitrary point (α_1, α_2) in the state space and check the necessary condition of the theorem to see if it is possible for the optimal value function to have an infinite derivative at that point. This is accomplished by employing an orthogonal change of coordinates from the old coordinates (x_1, x_2) to new coordinates (y_1, y_2) such that the component y_2 serves as the normal direction in the Theorem. Therefore, we obtain a normal direction at those points where an infinite derivative may possibly exist. And, this is sufficient to determine the equation of the submanifold passing through the points sought for.

Let (α_1, α_2) be contained in X. Consider the orthogonal transformation

$$(17)\qquad \begin{bmatrix} y_1 \\ y_2 \end{bmatrix} = \begin{bmatrix} \cos(\theta) & \sin(\theta) \\ -\sin(\theta) & \cos(\theta) \end{bmatrix} \cdot \begin{bmatrix} x_1 \\ x_2 \end{bmatrix}$$

where θ is an angle yet unspecified. Suppose that (α_1, α_2) lies on a submanifold M having at (α_1, α_2) the normal vector

$$N(\alpha_1, \alpha_2) = [-\sin(\theta) \ \cos(\theta)]$$

such that the optimal value function has an infinitive derivative at all points of M. The product $N(\alpha_1, \alpha_2)\cdot f(\alpha_1, \alpha_2\ v)$ reduces to

$$N(\alpha_1, \alpha_2)\cdot f(\alpha_1, \alpha_2, v) = -\alpha_2 \sin(\theta) + v \cos(\theta)$$

where $-1\le v\le 1$. According to the theorem we seek an angle $\theta = \theta(\alpha_1, \alpha_2)$ such that zero is contained in the boundary of the set

(18) $\{-\alpha \sin(\theta) + v \cos(\theta) : -1 \le v \le 1\}$

If α_2 equals zero then this condition is met if and only if

$$\theta(\alpha_1, 0) = \pm \pi/2.$$

For a non-zero α_2, zero is contained in the boundary of the Set (18) if θ satisfies the equation

(19) $\tan(\theta) = \pm 1/\alpha_2.$

The angle θ is the angle through which the old coordinates (x_1, x_2) are rotated so that the new coordinate y_2 is normal to the submanifold M at the point (α_1, α_2). The new coordinate y_1 is, therefore, tangent to M at (α_1, α_2). The submanifold M is a curve in which locally, it can be expressed as a functional relationship between α_1 and α_2. If the equation of M is given by α_2 as a function of α_1, then the slope to M is given by

(20) $$\frac{d\alpha_2}{d\alpha_1} = \tan(\theta).$$

Substituting Equation (19) into Equation (20) and integrating, we obtain the solutions

(21) $$\tfrac{1}{2}(\alpha_2)^2 = -\alpha_1 + c_1$$

(22) $$\tfrac{1}{2}(\alpha_2)^2 = \alpha_1 + c_2$$

for the equation of M. Here, c_1 and c_2 are constants of integration. We derived these equations by utilizing the necessary condition of the theorem. Thus, if there are submanifolds of the state space on which the optimal value function has an infinite derivative then they must satisfy these equations. And, indeed, for terminal conditions of

$\alpha_1 = \alpha_2 = 0$ together with invoking that trajectories enter the terminal set rather than leave it, we obtain the submanifold M. That is, the terminal conditions of $\alpha_1 = \alpha_2 = 0$ imply that $c_1 = c_2 = 0$. Since trajectories enter the terminal set (0,0) rather than leave it, the equations for the sought submanifold are reduced to

(23) $$\tfrac{1}{2}(\alpha_2)^2 = -\alpha_1, \ \alpha_2 > 0$$

(24) $$\tfrac{1}{2}(\alpha_2)^2 = \alpha_1, \ \alpha_2 \leq 0.$$

The optimal value function for this example is given in Stalford [4]. Its derivative is easily obtained since therein the optimal value function is given in analytical form. The derivative is infinite only at the points defined by Equations (23) and (24).

Example 2. As a second example, consider the rectilinear motion of a rocket operating at constant power. In Leitmann [1, p. 29], the equations of motion are given as

(25) $$\dot{x}_1 = v$$

(26) $$\dot{x}_2 = v^2. \qquad \Bigg\} \quad -1 \leq v \leq 1$$

The terminal set is the origin (0,0). The objective is to transfer the rocket from an initial state to the terminal set and render the transfer time a minimum. Thus, the function g_o is identically zero while f_o is identically one. The state space X is the half space below the x_1 - axis. Consider the orthogonal transformation given in Equation (17) and let the y_2 - component be the normal direction to M. According to the theorem, we seek an angle θ such that zero is contained in the boundary of the set

(27) $$\{-v \sin(\theta) + v^2 \cos(\theta) : -1 \leq v \leq 1\}$$

This is only possible if $\theta = 0$ at each point of M.

Integrating Equation (20) with $\theta = 0$, we obtain

(28) $$\alpha_2 = c$$

where c is a constant of integration. Equation (28) implies that the submanifold M is a straight line parallel to the x_1 - axis. Applying the terminal conditions of $\alpha_1 = \alpha_2 = 0$, we determine the integration constant to be zero. The submanifold M coincides with the x_1 - axis.

In this example, it is not possible to reach the terminal set (0,0) from any point on M with the exception of (0,0). This is because the x_2 - coordinate is non-increasing only for the control value v = 0. Interestingly, the state space X does not contain M. This example therefore lies outside the scope of the theorem. However, as a point on M is approached normal to M, the derivative of the optimal value function does indeed become unbounded.

<u>Example 3.</u> Vincent [5] presents optimal control models of several pest management programs in agriculture where insecticides and predators are utilized to minimize the damage done to crops by insects. In particular, a model is given of a program where an insecticide spray nonharmful to the predators is used and where no biological control is introduced to change the natural population growth of the predators. The state equations of the model are

(29) $$\dot{x}_1 = x_1(1 - x_2) - v\, x_1, \quad 0 \le v \le 1$$

(30) $$\dot{x}_2 = x_2(x_1 - 1).$$

Here, x_1 represents the ratio of the actual number of insect pests and a tolerable level of such pests. The state x_2 is the ratio of the actual number of predators and a desired level of them. The control value v corresponds to the amount of insecticide used. It is desired to transfer the state of the system to the equilibrium point (1,1) and minimize the cost criterion

$$\int_0^{t_f} (5x_1 + 5v)\, d\tau.$$

This integral models the cost associated with crop loss and the insecticide used.

As before, we consider the orthogonal transformation given in Equation (17) and let the y_2 - component be normal to M. If (α_1, α_2) is a point of M, an angle $\theta = \theta(\alpha_1, \alpha_2)$ is to be determined such that zero is contained in the boundary of the set

(31) $$\left\{ -[\alpha_1(1-\alpha_2) - v\, \alpha_1] \sin(\theta) + \alpha_2(\alpha_1 - 1) \cos(\theta) \;:\; 0 \le v \le 1 \right\}$$

An inspection of Set (31) reveals that θ must be a solution of one of the equations

(32) $$\tan(\theta) = \frac{\alpha_2(\alpha_1 - 1)}{\alpha_1(1 - \alpha_2)}$$

(33) $$\tan(\theta) = (1 - \alpha_1)/_{\alpha_1}.$$

Substituting Equation (32) into Equation (20), integrating the result and applying the terminal conditions of $\alpha_1 = \alpha_2 = 1$, we obtain

(34) $$\ln(\alpha_2) + \ln(\alpha_1) = \alpha_1 - 1 + \alpha_2 - 1.$$

The inequality

(35) $$\ln x < (x-1),\ x \in (0, 1) \cup (1, \infty)$$

implies that $(\alpha_1, \alpha_2) = (1, 1)$ is the only point satisfying Equation (34).

Substituting Equation (33) into Equation (20), integrating the result and applying the terminal conditions of $\alpha_1 = \alpha_2 = 1$, we have the equation

(36) $$\alpha_2 = \ln(\alpha_1) - \alpha_1 + 2.$$

Equation (33) resulted from the Set (31) for the control value $v = 1$. Implementation of this control value into Equation (29) results in the state x_1 decreasing in time. Therefore, since trajectories enter the terminal set (1,1) rather than leave it, Equation (36) with the constraint $\alpha_1 \geq 1$ is the derived equation of the submanifold M. And, this is indeed the set of states where the optimal value function has an infinite derivative. As pointed out in Vincent [5], the optimal value function is, in addition, discontinuous across the submanifold M. But the optimal value function does satisfy Assumption 2.

REFERENCES

[1]. Leitmann, G., AN INTRODUCTION TO OPTIMAL CONTROL, McGraw Hill, (1966).

[2]. Pontryagin, L. S., et al., THE MATHEMATICAL THEORY OF OPTIMAL PROCESSING, Interscience Publishers, New York, (1962).

[3]. Stalford, H., "Sufficiency Conditions in Optimal Control and Differential Games," ORC 70-13, Operations Research Center, University of California, Berkeley, California, 1970.

[4]. Stalford, H., "An Equivalency Optimality Theorem Over a Family of Optimal Control Processes," Proc. of the 1972 Int. Conf. on Cybernetics and Society, IEEE Systems, Man and Cybernetics Society, October 9-12, 1972, Washington, D. C.

[5]. Vincent, T. L., "Pest Management Programs Via Optimal Control Theory," 13th Joint Automatic Control Conference of the American Automatic Control Council, Stanford University, Stanford, California, August 16-18, 1972.

CONTROL OF AFFINE SYSTEMS WITH MEMORY

M.C. Delfour*

(Université de Montréal)

and

S.K. Mitter**

(Massachusetts Institute of Technology)

1. INTRODUCTION

In this paper we present a number of results related to control and estimation problems for affine systems with memory. The systems we consider are typically described by linear functional differential equations or Volterra integro-differential equations.

Our results may be divided into four categories:

(i) State-space description of systems with memory.

(ii) Feedback solution of the finite-time quadratic cost problem.

(iii) Feedback solution of the infinite-time quadratic cost problem.

(iv) Optimal linear filtering.

The main difficulty in the study of the systems considered in this paper

---

* The work of the first author was supported in part by National Research Council of Canada Grant A-8730 at the Centre de Recherches Mathématiques, Université de Montréal, Montréal 101, Québec, Canada.

** The work of the second author was supported by AFOSR Grant 72-2273, NSF Grant GK-25781 and NASA Grant NGL-22-009-124 all at the Electronic Systems Laboratory, M.I.T., Cambridge, Mass. 02139.

is that the state spaces involved are infinite dimensional and that the equations describing the evolution of the state involve unbounded operators. Once an appropriate function space is chosen for the state space a fairly complete theory for the control and estimation problems for such systems can be given.

2. Affine Systems with Memory

In this paper we shall consider two typical systems: one with a fixed memory and one with a time varying memory.

Let X be the evolution space and U be the control space. We assume that X and U are finite-dimensional Euclidean spaces.

2.1. Constant Memory

Given an integer $N \geq 1$ and real numbers $-a = \Theta_N < \ldots < \Theta_1 < \Theta_o = 0$ and $T > 0$, let the system with constant memory be described by:

$$(1)\quad \begin{cases} \dfrac{dx}{dt}(t) = A_{oo}(t)x(t) + \displaystyle\sum_{i=1}^{N} A_i(t)x(t+\Theta_i) \\ \qquad + \displaystyle\int_{-a}^{o} A_{o1}(t,\Theta)x(t+\Theta)d\Theta + f(t) \\ \qquad + B(t)v(t) \quad \text{in} \quad [0,T] \\ x(\Theta) = h(\Theta) \quad , \quad -a \leq \Theta \leq 0, \end{cases}$$

where A_{oo}, A_i, A_{o1} and B are strongly measurable and bounded, $f \in L^2(0,T;\, X)$ and $v \in L^2(0,T;\, U)$.

We first need to choose an appropriate space of initial data and an appropriate state space. It was shown in DELFOUR-MITTER [1], [2], that this can indeed be done provided that (1) is rewritten in the following form:

$$(2)\quad \begin{cases} \dfrac{dx}{dt} = A_{oo}(t)x(t) + \displaystyle\sum_{i=1}^{N} A_i(t) \begin{cases} x(t+\Theta_i) & , t+\Theta_i \geq 0 \\ h^1(t+\Theta_i) & , \text{otherwise} \end{cases} \\ \qquad + \displaystyle\int_{-a}^{o} A_{o1}(t,\Theta) \begin{Bmatrix} x(t+\Theta) & , & t+\Theta \geq 0 \\ h^1(t+\Theta) & , & \text{otherwise} \end{Bmatrix} d\Theta \\ \qquad + f(t) + B(t)v(t) \quad , \text{ in } [0,T], \\ x(0) = h^o. \end{cases}$$

We can pick initial data $h=(h^o,h^1)$ in the product space $X \times L^2(-a,0;X)$, where the solution of (2) is $x : [0,T] \rightarrow X$.

We can now define the <u>state at time t</u> as an element $\tilde{x}(t)$ of $X \times L^2(-a,0;X)$ as follows:

$$(3)\quad \begin{cases} \tilde{x}(t)^o = x(t) \\ \tilde{x}(t)^1(\Theta) = \begin{cases} x(t+\Theta) & , \quad t+\Theta \geq 0 \\ h^1(t+\Theta) & , \quad \text{otherwise.} \end{cases} \end{cases}$$

For additional details see DELFOUR-MITTER [1], [2]. System (1) has a memory of fixed duration [-a,0].

2.2. Time Varying Memory

Consider the system

$$(4)\quad \begin{cases} \dfrac{dx}{dt}(t) = A_o(t)x(t) + \displaystyle\int_0^t A_1(t,r)x(r)\,dr \\ \qquad\qquad + f(t) + B(t)v(t)\ , \text{ in } [0,T] \\ x(0) = h^o \text{ in } X, \end{cases}$$

Where A_o, A_1 and B are strongly measurable and bounded. If we change the variable r to $\Theta = r-t$ and define

$$(5)\quad \begin{cases} A_{oo}(t) = A_o(t) \\ A_{o1}(t,\Theta) = \begin{cases} A_1(t,t+\Theta) & , \ -t < \Theta < 0, \\ 0 & , \ -\infty < \Theta < t, \end{cases} \end{cases}$$

equation (4) can be rewritten in the form

$$(6)\quad \begin{cases} \dfrac{dx}{dt}(t) = A_{oo}(t)x(t) + \displaystyle\int_{-\infty}^{0} A_{o1}(t,\Theta) \begin{Bmatrix} x(t+\Theta) & , \ t+\Theta\geq 0 \\ h^1(t+\Theta) & , \ \text{otherwise} \end{Bmatrix} d\Theta \\ \qquad\qquad + f(t) + B(t)v(t) \quad \text{in} \quad [0,T] \\ x(0) = h^o \ \text{ in } X,\ h^1 \text{ in } L^2(-\infty,0;X), \end{cases}$$

with $h^1 = 0$. In this form equation (6) is similar to equation (2). However here we consider the system to have a memory of infinite duration in order to accomodate the growing memory duration [-t,0]. The <u>state space</u> will be chosen to be the product $X \times L^2(-\infty,0;X)$. The <u>state</u> at time t is an element $\tilde{x}(t)$ of $X \times L^2(-\infty,0;X)$ which is defined as

$$(7)\quad \begin{cases} \tilde{x}(t)^{o} = x(t) \\ \tilde{x}(t)^{1}(\Theta) = \begin{cases} x(t+\Theta) & . -t < \Theta < 0 \\ h^{1}(t+\Theta) & , -\infty < \Theta < -t \end{cases} \end{cases}$$

3. State Equation

It will be more convenient to work with an evolution equation for the state of the system rather than equations (1) or (4). In order to obtain the state evolution equation corresponding to equation (1) let

$$(8)\quad \begin{cases} H = X \times L^{2}(-a,0;X) \\ V = \{(h(0),h) \mid h \in H^{1}(-a,0;X)\}. \end{cases}$$

The injection of V into H is continuous and V is dense in H. We identify H with its dual. Then if V' denotes the dual space of V, we have

$$V \subset H \subset V'.$$

This is the framework utilized by Lions (cf. J.L. LIONS) to study evolution equations. Define the unbounded operator $\tilde{A}(t)$: $V \rightarrow H$ by,

$$(9)\quad \begin{cases} (\tilde{A}(t)h)^{o} = A_{oo}(t)h(0) + \sum_{i=1}^{N} A_{i}(t)h(\Theta_{i}) + \int_{-a}^{0} A_{o1}(t,\Theta)h(\Theta)d\Theta \\ (\tilde{A}(t)h)^{1}(\Theta) = \frac{dh}{d\Theta}(\Theta), \end{cases}$$

and the bounded operator

$$\tilde{B}(t):\ U \rightarrow H \text{ by}$$

$$
(10)\quad \begin{cases} (\tilde{B}(t)u)^{o} = B(t)u \\ \\ (B(t)u)^{1}(\Theta) = 0 \end{cases}
$$

and $\tilde{f}(t) \in H$ by

$$
(11)\quad \tilde{f}(t)^{o} = f(t), \quad \tilde{f}(t)^{1} = 0.
$$

Then for all h in V, it can be shown that x is the unique solution in

$$
(12)\quad W(0,T) = \{z \in L^{2}(0,T;V) \mid D\, z \in L^{2}(0,T;H)\}
$$ [D denotes the distributional derivative]

of

$$
(13)\quad \begin{cases} \dfrac{d\tilde{x}}{dt}(t) = \tilde{A}(t)x(t) + \tilde{B}(t)u(t) + \tilde{f}(t) \text{ in } [0,T] \\ \\ \tilde{x}(0) = h. \end{cases}
$$

Similarly in the case of equation (4) we let

$$
(14)\quad \begin{cases} H = X \times L^{2}(-\infty,0;X) \\ \\ V = \{(h(0),h) \mid h \in H^{1}(-\infty,0;X)\}. \end{cases}
$$

We again have

$$
V \subset H \subset V'.
$$

We now define $\tilde{A}(t)\colon V \to H$ as follows:

$$
(15)\quad \begin{cases} (\tilde{A}(t)h)^{o} = A_{oo}(t)h(0) + \displaystyle\int_{-\infty}^{o} A_{o1}(t,\Theta)h(\Theta)d\Theta \\ \\ (\tilde{A}(t)h)^{1}(\Theta) = \dfrac{dh}{d\Theta}(\Theta), \end{cases}
$$

$\tilde{B}(t)$ and $\tilde{f}(t)$ be as defined in (6) and (7). For all h in V, $\tilde{x}$ is the unique solution in

(16) $$W(0,T) = \{z \in L^2(0,T;V) \mid D z \in L^2(0,T;H)\}$$

of

(17) $$\begin{cases} \frac{d\tilde{x}}{dt}(t) = \tilde{A}(t)\tilde{x}(t) + \tilde{B}(t)v(t) + \tilde{f}(t), \text{ in } [0,T], \\ \tilde{x}(0) = h. \end{cases}$$

4. Optimal Control Problem in [0,T]

We now consider a quadratic cost function,

(18) $$\begin{cases} J(v,h) = (\tilde{L}\,\tilde{x}(T),\ \tilde{x}(T))_H - 2(\tilde{\ell},\tilde{x}(T))_H \\ \quad + \int_0^T [(\tilde{Q}(t)\tilde{x}(t),\ \tilde{x}(t))_H - 2(\tilde{q}(t),\ \tilde{x}(t))_H]dt \\ \quad + \int_0^T (N(t)v(t),\ v(t))_U dt, \end{cases}$$

where $\tilde{L} \in \mathscr{L}(H)$, $\tilde{\ell} \in H$, $q \in L^2(0,T;H)$ and $\tilde{Q}$: $[0,T] \to \mathscr{L}(H)$ and N: $[0,T] \to \mathscr{L}(U)$ are strongly measurable and bounded. Moreover $\tilde{L}$, $\tilde{Q}(t)$ and $N(t)$ are self adjoint and positive and there exists $c > 0$ such that

(19) $$\forall t,\ \forall u,\ (N(t)u,u)_U > 0.$$

For this problem we know that given h in V, there exists a unique optimal control function u in $L^2(0,T;U)$ which minimizes $J(v,h)$ over all v in $L^2(0,T;U)$. Moreover this optimal control can be synthesized via the feedback law

(20) $$u(t) = -N(t)^{-1}\tilde{B}(t)^*[\pi(t)\tilde{x}(t) + r(t)],$$

where π and r are characterized by the following equations:

$$(21)\quad \begin{cases} \dfrac{d\pi}{dt}(t) = \tilde{A}(t)^*\pi(t) + \pi(t)\tilde{A}(t) - \pi(t)\tilde{R}(t)\pi(t) + \tilde{Q}(t) = 0, \text{ in } [0,T] \\ \pi(T) = \tilde{L} \end{cases}$$

$$(22)\quad \begin{cases} R(t) = B(t)^*N(t)B(t) \\ \tilde{R}(t) = \begin{bmatrix} R(t) & 0 \\ 0 & 0 \end{bmatrix} \end{cases}$$

and

$$(23)\quad \begin{cases} \dfrac{dr}{dt}(t) + [\tilde{A}(t) - \tilde{R}(t)\pi(t)]^*r(t) + [\pi(t)\tilde{f}(t) + \tilde{q}(t)] = 0, \text{ in } [0,T] \\ r(T) = \tilde{\ell}. \end{cases}$$

Here a solution of (21) is a map π: $[0,T] \to \mathscr{L}(H)$ which is weakly continuous such that for all h and k in V the map $t \to (h,\pi(t)k)_H$ is in $H^1(0,T;R)$; a solution of (23) is a map r: $[0,T] \to H$ such that $r \in L^2(0,T;H)$ and $D\, r \in L^2(0,T;V')$. For details see DELFOUR-MITTER [3] and BENSOUSSAN-DELFOUR-MITTER [1].

5. Optimal Control Problem in $[0,\infty]$

We can also give a complete theory for cost functions of the form

$$(24)\quad J_\infty(v,h) = \int_0^\infty [(\tilde{Q}\tilde{x}(t), \tilde{x}(t))_H + (Nv(t),v(t))_U]dt$$

with the following hypothesis:

1) $\tilde{Q} \in \mathscr{L}(H)$, $N \in \mathscr{L}(U)$ are positive and self adjoint and there exists $c > 0$ such that

$$\forall u\ (Nu,u)_U \geq c\ |u|_U^2 ;$$

2) $\tilde{x}$ is the solution of

$$(25)\quad \begin{cases} \dfrac{d\tilde{x}}{dt}(t) = \tilde{A}x(t) + \tilde{B}v(t) \quad \text{in} \quad [0,\infty] \\ \tilde{x}(0) = h \text{ in } V; \end{cases}$$

3) (Stabilizability hypothesis) there exists a <u>feedback operator</u> $G \in \mathscr{L}(V,U)$ of the form

$$(26)\quad Gh = G_{oo}h(0) + \sum_{j=1}^{M} G_i h(\Theta_i) + \int_{-a}^{0} G_{o1}(\Theta)h(\Theta)d\Theta$$

such that the closed loop system

$$(27)\quad \begin{cases} \frac{d\tilde{x}}{dt}(t) = [\tilde{A} + \tilde{B}\, G]\tilde{x}(t) \quad \text{in} \quad [0,\infty] \\ \tilde{x}(0) = h \in V \end{cases}$$

be L^2-stable, that is

$$(28)\quad \forall\, h \in H, \quad \int_0^\infty ||\tilde{x}(t)||_H^2\, dt < \infty.$$

For a study of the stabilizability problem see VANDEVENNE [1] [2].

When system (25) is stabilizable, there exists a unique u in $L^2_{loc}(0,\infty;U)$ which minimizes $J_\infty(v,h)$ over all v in $L^2_{loc}(0,\infty;U)$ for a given h. Moreover this optimal u can be synthesized via a constant feedback law.

$$(29)\quad u(t) = -N^{-1}\tilde{B}^*\pi\,\tilde{x}(t),$$

where π is a solution of the algebraic Riccati equation

$$(30)\quad \tilde{A}^*\pi + \pi\tilde{A} - \pi\tilde{R}\pi + \tilde{Q} = 0.$$

A solution of (30) is a positive self adjoint element of $\mathscr{L}(H)$ such that (30) is verified as an equation in $\mathscr{L}(V,V')$. The operator π in $\mathscr{L}(H)$ can be decomposed into a matrix of operators

$$\begin{pmatrix} \pi_{oo} & \pi_{o1} \\ \pi_{1o} & \pi_{11} \end{pmatrix}$$

(since H is either $X \times L^2(-a,0;X)$ or $X \times L^2(-\infty,0;X)$) where

$$\begin{cases} \pi_{oo} \in \mathscr{L}(X) \quad , \quad \pi_{o1} \in \mathscr{L}(L^2(-a,0;X),X) \\ \pi_{1o} \in \mathscr{L}(X,L^2(-a,0;X)) \;, \; \pi_{11} \in \mathscr{L}(L^2(-a,0;X)). \end{cases}$$

Moreover

$$\begin{cases} \pi_{oo}A_{oo} + A_{oo}^*\pi_{oo} + \pi_{1o}(0) + \pi_{1o}(0)^* + Q - \pi_{oo}R\pi_{oo} = 0 \\ \pi_{oo}^* = \pi_{oo} \geq 0 \end{cases}$$

$$(\pi_{1o}h^o)(\alpha) = \pi_{1o}(\alpha)h^o \quad , \quad \alpha \to \pi_{1o}(\alpha) : [-a,0] \to \mathscr{L}(X)$$

$$\begin{cases} \frac{d\pi_{1o}}{d\alpha}(\alpha) = \pi_{1o}(\alpha)\,[A_{oo} - R\pi_{oo}] + \sum_{i=1}^{N-1} A_i{}^*\pi_{oo}\,\delta(\alpha-\Theta) \\ \qquad + A_{o1}(\alpha)^*\pi_{oo} + \pi_{11}(\alpha,0)\;,\ \text{a.e. in } [-a,0] \\ \pi_{1o}(-a) = A_N{}^*\pi_{oo} \end{cases}$$

$$(\pi_{o1}h^1(\alpha) = \int_{-a}^{o} \pi_{1o}(\alpha)^* h^1(\alpha)\,d\alpha$$

$$\begin{cases} (\pi_{11}h^1)(\alpha) = \int_{-a}^{o} \pi_{11}(\alpha,\beta)h^1(\beta)\,d\beta \\ (\alpha,\beta) \to \pi_{11}(\alpha,\beta) : [-a,0] \times [-a,0] \to \mathscr{L}(X) \\ \left[\frac{\partial}{\partial\alpha} + \frac{\partial}{\partial\beta}\right]\pi_{11}(\alpha,\beta) = A_{o1}(\alpha)^*\pi_{1o}(\beta)^* + \pi_{1o}(\alpha)A_{o1}(\beta) \\ \qquad + \sum_{i=1}^{N-1} A_i{}^*\pi_{1o}(\beta)^*\delta(\alpha-\Theta_i) + \sum_{j=1}^{N-1}\pi_{1o}(\alpha)A_j\,\delta(\beta-\Theta_i) \\ \qquad - \pi_{1o}(\alpha)\,R\pi_{1o}(\beta)^* \\ \pi_{11}(-a,\beta) = A_N{}^*\pi_{1o}(\beta)^*\;,\ \pi_{11}(\alpha,-a) = \pi_{1o}(\alpha)A_N \\ \pi_{11}(\alpha,\beta) = \pi_{11}(\beta,\alpha)^*. \end{cases}$$

Under additional hypothesis on $\tilde{A}$ and $\tilde{Q}$ we can also describe the asymptotic behaviour of the closed loop system

$$(31)\quad \begin{cases} \frac{dx}{dt}(t) = [\tilde{A} - \tilde{R}\pi]\,\tilde{x}(t) & \text{in } [0,\infty] \\ \tilde{x}(0) = h & \text{in } V. \end{cases}$$

<u>Definition</u> Given a Hilbert <u>space of observations</u> Y and an <u>observer</u> $M \in \mathscr{L}(H,Y)$, System (25) is said to be <u>observable</u> by M if each initial datum h at time 0 can be determined from a knowledge of v in $L^2_{loc}(0,\infty;U)$ and the observation

$$(32)\quad z(t) = M\,x(t) \quad \text{in} \quad [0,\infty].$$

When System (15) is observable by $\tilde{Q}^{1/2}$, for each initial datum h

(33) $\tilde{x}(t) \to 0$ as $t \to \infty$,

where $\tilde{x}$ is the solution of the closed loop system (31).

In the special case where

$$\tilde{Q} = \begin{bmatrix} Q_{oo} & 0 \\ 0 & 0 \end{bmatrix}$$

and Q_{oo} is positive definite, the closed loop system (21) is L^2-stable. For further details see DELFOUR-MCCALLA-MITTER.

6. Optimal Linear Filtering and Duality

Let E and F be two Hilbert spaces. We consider the system

$$(34) \quad \begin{cases} \frac{dx}{dt}(t) = A_{oo}(t)x(t) + \sum_{i=1}^{N} A_i(t)x(t+\Theta_i) + \int_{-a}^{o} A_{o1}(t,\Theta)x(t+\Theta)d\Theta \\ \qquad + B(t)\xi(t) + f(t), \\ x(0) = h^o + \zeta^o \\ x(\Theta) = h^1(\Theta) + \zeta^1(\Theta), \quad -a \le \Theta < 0, \end{cases}$$

where $\zeta = (\zeta^o, \zeta^1)$ is the noise in the initial datum, and ξ is the input noise with values in F. We assume an observation of the form (with values in E)

$$(35) \quad z(t) = C(t)x(t) + \eta(t),$$

where η represents the error in measurement and $C(t)$ belongs to $\mathscr{L}(X,E)$. As in BENSOUSSAN [1] $\{\zeta^o, \zeta^1, \xi, \eta\}$ will be modelled as a Gaussian linear random functional on the Hilbert space.

$$(36) \quad \Phi = X \times L^2(-a,0;X) \times L^2(0,T;E) \times L^2(0,T;F)$$

with zero mean and covariance operator

$$(37) \quad \Lambda = \begin{bmatrix} P_o & & & \\ & P_1(\Theta) & & \\ & & Q(t) & \\ & & & R(t) \end{bmatrix}.$$

For each T we want to determine the best estimator of the linear random functional $x(T)$ with respect to the linear random functional $z(s)$, $0 \le s \le T$. For the solution to this problem see BENSOUSSAN [2] and BENSOUSSAN-DELFOUR-MITTER [2].

References

A. BENSOUSSAN [1] Filtrage Optimal des Systémes Lineáires, Dunod, Paris 1971.
[2], Filtrage Optimal des Systémes Lineáires avec retard, I.R.I.A. report INF 7118/71027, Oct. 1971.

A. BENSOUSSAN, M.C. DELFOUR and S.K. MITTER [1] Topics in System Theory in Infinite Dimensional Spaces, forthcoming monograph.

A. BENSOUSSAN, M.C. DELFOUR and S.K. MITTER [2] Optimal Filtering for Linear Stochastic Hereditary Differential Systems, Proc. 1972 IEEE Conference on Decision and Control, New Orleans, Louisiana, U.S.A., Dec. 13-15, 1972.

M.C. DELFOUR and S.K. MITTER [1] Hereditary Differential Systems with Constant Delays, I - General Case, J. Differential Equations, 12 (1972), 213-235.
[2], Hereditary Differential Systems with Constant Delays, II - A Class of Affine Systems and the Adjoint Problem. To appear in J. Differential Equations.
[3], Controllability, Observability and Optimal Feedback Control of Hereditary Differential Systems, SIAM J. Control, 10 (1972), 298-328.

M.C. DELFOUR, C. McCALLA and S.K. MITTER, Stability and the Infinite-Time Quadratic Cost Problem for Linear Hereditary Differential Systems, C.R.M. Report 273, Centre de Recherches Mathématiques, Université de Montréal, Montréal 101, Canada; submitted to SIAM J. on Control.

J.L. LIONS, Optimal Control of Systems Governed by Partial Differential Equations, Springer-Verlag, New York, 1971.

H.F. VANDEVENNE, [1] Qualitative Properties of a Class of Infinite Dimensional Systems, Doctoral Dissertation, Electrical Engineering Department, M.I.T. January 1972.
[2], Controllability and Stabilizability Properties of Delay Systems, Proc. of the 1972 IEEE Decision and Control Conference, New Orleans, December 1972.

Andrzej P. Wierzbicki, Andrzej Hatko*

COMPUTATIONAL METHODS IN HILBERT SPACE FOR OPTIMAL CONTROL PROBLEMS WITH DELAYS

Summary

The paper consits of two parts. The first part is devoted to basic relations in the abstract theory of optimization and their relevance for computational methods. The concepts of the abstract theory (developed by Hurwicz, Uzawa, Dubovitski, Milyutin, Neustadt and others) linked together with the notion of a projection on a cone result in an unifying approach to computational methods of optimization. Several basic computational concepts, such as penalty functional techniques, problems of normality of optimal solutions, gradient projection and gradient reduction techniques, can be investigated in terms of a projection on a cone.

The second part of the paper presents an application of the gradient reduction technique in Hilbert space for optimal control problems with delays. Such an approach results in a family of computational methods, parallel to the methods known for finite-dimmensional and other problems: conjugate gradient methods, variable operator (variable metric) methods and generalized Newton's (second variation) method can be formulated and applied for optimal control problems with delays. The generalized Newton's method is, as usually, the most efficient; however, the computational difficulties in inverting the hessian operator are limiting strongly the applications of the method. Of other methods, the variable operator technique seems to be the most promissing.

* Technical University of Warsaw, Institute of Automatic Control, Faculty of Electronics, Nowowiejska 15/19, Warsaw, Poland.

I. Basic relations in the abstract theory of optimization and computational methods

1. Basic theory.

Two basic rezults of the abstract theory of extremal solutions are needed in the sequel.

Theorem 1. (See e.g. [5]). Let E, F be linear topological spaces, D be a nonempty convex cone (positive cone) in F. Let $Q: E \to R^1$, $P: E \to F$, $p \in F$ and $Y_p = \{y \in E: p - P(y) \in D\}$.

(i) Suppose there exists $\eta \in D^*$ (called Lagrange multiplier) and $\hat{y} \in Y_p$ such that $\langle \eta, P(y) - p \rangle = 0$ and

$$Q(\hat{y}) + \langle \eta, P(\hat{y}) - p \rangle \leqslant Q(y) + \langle \eta, P(y) - p \rangle \quad \text{for all} \quad y \in E \tag{1}$$

Then

$$Q(\hat{y}) \leqslant Q(y) \quad \text{for all} \quad y \in Y_p \tag{2}$$

(ii) Let Q, P be convex (P relative to the cone D). Let the cone D have an interior point and suppose there exists $y_1 \in E$ such that $p - P(y_1) \in \mathring{D}$. Suppose there exists a point $\hat{y}$ satisfying (2). Then there exists $\eta \in D^*$ satisfying (1) and such that

$$\langle \eta, P(\hat{y}) - p \rangle = 0 \tag{3}$$

(iii) Given $p_1, p_2 \in F$ suppose there exist $\hat{y}_1, \hat{y}_2$ minimizing Q in Y_{p1}, Y_{p2} respectively. Suppose there exist η_1, η_2 satisfying (1) and (3) for $\hat{y}_1$ and $\hat{y}_2$. Then

$$\langle \eta_1, p_1 - p_2 \rangle \leqslant Q(\hat{y}_2) - Q(\hat{y}_1) \leqslant \langle \eta_2, p_1 - p_2 \rangle \tag{4}$$

Recall that the general form of a Lagrange functional is

$$L(\eta_o, \eta, y) = \eta_o Q(y) + \langle \eta, P(y) - p \rangle \tag{5}$$

with $\eta_o \geqslant 0$, whereas the normal form, with $\eta_o > 0$, is equivalent to

$$L(\eta, y) = Q(y) + \langle \eta, P(y) - p \rangle \tag{6}$$

Therefore, the theorem 1 (ii) gives a sufficient condition of normality of a convex optimization problem. The requirement of a nonempty $\mathring{D}$ is fairly severe and by no means necessary (we shall give an example of the existence of a normal Lagrange multiplier when $\mathring{D}$ is empty). However, weaker conditions of normality of convex problems have not been sufficiently investigated.

The part (iii) of the theorem is basic for sensitivity analysis of optimization problems and results in the following corollary.

Corollary 1. Suppose the space F is normed. Suppose there is an open set $\mathcal{P} \subset F$ such that the assumptions of theorem 1, part (iii), hold for each $p_1, p_2 \in \mathcal{P}$. Define the functional $\hat{Q}: \mathcal{P} \to R^1$

by $\hat{Q}(\hat{p}) = Q(\hat{y}) = \min_{y \in Y_p} Q(y)$. Suppose the normal Lagrange multipliers η are determined uniquely for each $p \in \mathcal{P}$ and the mapping $N: \mathcal{P} \to F$, $\eta = N(p)$, is continuous. Then the functional $\hat{Q}$ is differentiable and $\delta\hat{Q}(p, \delta p) = - \langle \eta, \delta p \rangle$; hence the gradient of the functional $\hat{Q}$ is $-\eta$.

The properties of the mapping N - uniqueness, continuity and Lipschitz continuity - are quite important in sensitivity analysis and ~~t~~ other computational aspects of optimization. However, we shall not investigate these properties here.

Another theorem of basic importance, which summarizes results proven in [6] , [7] , [8] , is the following.

Theorem 2. Let E be a linear topological space. Suppose the functional $Q : E \to R^1$ has a local constrained minimum at a point $\hat{y}$ in a given set $Y_p \subset E$. Suppose the set $Y_q = \{y \in E: Q(y) < Q(\hat{y})\}$ has a nonempty internal cone K_i at $\hat{y}$ (that is, a convex open cone K_i such that there is a neighbourhood $U(\hat{y})$ such that $(K_i + \hat{y}) \cap U(\hat{y}) \subset Y_q$). Suppose the set Y_p has a nonempty external cone K_e at $\hat{y}$ (that is, a convex cone K_e such that for each $k \in K_e$, for each open cone K_o containing k, and for each neighbourhood $U(\hat{y})$ the set $(K_o + \hat{y}) \cap U(\hat{y}) \cap Y_p$ contains more points than only $\hat{y}$). Then there are nonzero functionals $q_o \in K_i^*$ and $q_1 \in K_e^*$ such that $q_o + q_1 = \Theta$.

The fact that $q_o \neq \Theta$, $q_1 \neq \Theta$ does not necessarily imply that a corresponding Lagrange functional has a normal form. Actually, if $Y_p = \{y \in E: p - P(y) \in D\}$, we need additionally some assumptions resulting in a form of the Farkas lemma in order to represent the elements of K_e^* by the elements of D^*.

Corollary 2. Suppose E,F are normed spaces. Let $Q: E \to R^1$ be differentiable (with the gradient denoted by $Q_y^*(y)$); hence $K_i^* = \{\alpha Q_y^*(\hat{y}), \alpha \leq 0\}$. Let $P: E \to F$ be differentiable (with the derivative denoted by $P_y(y)$).

(i) Suppose $Y_p = \{y \in E: p - P(y) \in D\}$, where D is a nontrivial positive cone (inequality constraint). Suppose $P_y^*(\hat{y})D^*$ is weakly* closed; hence, by Farkas lemma, $K_e^* = -P_y^*(y)(D^* \cap \{\eta \in F^*: \langle \eta, P(\hat{y}) - p \rangle = 0\})$. Thus there exists $\eta \in D^*$ such that $\langle \eta, P(\hat{y}) - p \rangle = 0$ and $Q_y^*(\hat{y}) + P_y^*(\hat{y})\eta = \Theta$; therefore, the Lagrange functional has a normal form.

(ii) Suppose $Y_p = \{y \in E: p - P(y) = \Theta\}$, hence D is a trivial cone (equality constraint). Suppose $P_y(y)$ is onto; hence, by Lyusternik theorem, $K_e^* = P_y^*(\hat{y}) F^*$. Thus there exists $\eta \in F^*$ (obviously, $\langle \eta, P(\hat{y}) - p \rangle = 0$) such that $Q_y^*(\hat{y}) + P_y^*(\hat{y})\eta = \Theta$; therefore, the Lagrange functional has a normal form.

The sufficient condition of normality in (ii) - that is, the requirement that $P_y(y)$ is onto - is quite natural and useful. The sufficient condition of normality in (i) - the weak* closedness of $P_y^*(\hat{y})D^*$ - is less restrictive than the nonemptiness of $\overset{\circ}{D}$, but rather cumbersome to check.

2. Projection on cones in Hilbert space

Let D be a closed convex cone in Hilbert space $\mathcal{H}$. Given $y \in \mathcal{H}$, denote by y^D an element of D such that

$$\|y^D - y\| = \min_{d \in D} \|d - y\| \tag{7}$$

The element y^D is called the projection of y on D.

Lemma 1. (see e.g. [9]). If E is a strictly normed space (such that $\|x+y\| = \|x\| + \|y\|$ implies $x = \alpha y$, $\alpha \in R^1$; in particular, a Hilbert space) then the projection y^D of $y \in E$ on a closed convex set $D \subset E$ is determined uniquely.

Lemma 2. The element $y^D \in D$ is the projection of $y \in \mathcal{H}$ on a closed convex cone $D \subset \mathcal{H}$ if and only if

(i) $y^D - y \in D^*$ (8)

(ii) $\langle y^D, y^D - y \rangle = 0$ (9)

Proof. If (i) and (ii), then $\|d-y\|^2 = \|y^D-y\|^2 + 2\langle y^D-y,d\rangle + \|d-y^D\|^2 \geqslant \|y^D-y\|^2$ for all $d \in D$. If not (i), then there is $d \in D$ such that $\langle y^D-y,d\rangle < 0$ and there is $\varepsilon > 0$ such that $\langle y^D-y,d\rangle + \varepsilon\|d\|^2 < 0$. Hence $\|y^D + \varepsilon d - y\|^2 < \|y^D-y\|^2$; since D is convex cone, $y^D + \varepsilon d \in D$, and y^D cannot satisfy (7). If not (ii), then $\langle y^D, y^D-y\rangle > 0$ (we have always $\langle y^D, y^D-y\rangle \geqslant 0$ since $y^D-y \in D^*$, $y^D \in D$). There is $\varepsilon_1 > 0$ such that for all $\varepsilon \in (0, \varepsilon_1)$ the inequality $-\langle y^D,y^D-y\rangle + \varepsilon\|y^D\|^2 < 0$ holds. Hence $\|(1-\varepsilon)y^D - y\|^2 < \|y^D-y\|^2$; since D is a cone, $(1-\varepsilon)y^D \in D$ for sufficiently small $\varepsilon > 0$ and y^D cannot satisfy (7).

Lemma 3. If $\|y_D\| = \min_{d \in D^*+y} \|d\|$, then $y_D = y^D$.

Proof. Since y^D satisfies (8) and ((9), we have $\|d\|^2 = \|d-y^D\|^2 + \|y^D\|^2 + 2\langle d-y^D,y^D\rangle = \|d-y^D\|^2 + \|y^D\|^2 + 2\langle d-y,y^D\rangle \geqslant \|y^D\|^2$ for all $d-y \in D^*$. Since y_D is unique, $y_D = y^D$.

Lemma 4. The projection y^D on a cone in Hilbert space has the following properties:

(i) $\|y^D\| \leqslant \|y\|$

(ii) $\|y_1^D - y_2^D\| \leqslant \|y_1 - y_2\|$

(iii) If $y = y_1-y_2$ and $y_2 \in D^*$, then $\|y^D\| \leqslant \|y_1\|$

(iv) If $y=y_1-y_2$, $y_1 \in D$, $y_2 \in D^*$ and $\langle y_1,y_2\rangle = 0$ then $y^D=y_1$.

Proof. (i) Relation (9) implies $\|y^D\|^2 = \langle y^D,y\rangle \leqslant \|y^D\| \|y\|$.
(ii) We have $\|y_1^D-y_2^D\|^2 = \langle y_1^D-y_2^D, y_1-y_2-(y_1-y_1^D)+(y_2-y_2^D)\rangle =$

$= \langle y_1^D - y_2^D, y_1 - y_2 \rangle - \langle y_1^D, y_2^D - y_2 \rangle - \langle y_2^D, y_1^D - y_1 \rangle \leqslant \langle y_1^D - y_2^D, y_1 - y_2 \rangle$
$\leqslant \|y_1^D - y_2^D\| \; \|y_1 - y_2\|$ as a consequence of (8) and (9).

(iii) $\|y^D\|^2 = \langle y^D, y_1 - y_2 \rangle \leqslant \langle y^D, y_1 \rangle \leqslant \|y^D\| \; \|y_1\|$ if $y_2 \in D^*$.

(iv) Since $\langle y^D, y_2 \rangle \geqslant 0$, $\langle y_1, y_2 \rangle = 0$ and $y = y_1 - y_2$, hence $\langle y^D - y, y_2 \rangle \geqslant \|y_2\|^2$. Since $\langle y^D - y, y_1 \rangle \geqslant 0$ and $\|y^D - y\|^2 = \langle y^D - y, -y \rangle$, hence $\langle y^D - y, y_2 \rangle \geqslant \|y^D - y\|^2$. Therefore, $\|y^D - y_1\|^2 = \|y^D - y - y_2\|^2 = \|y^D - y\|^2 + \|y_2\|^2 - 2\langle y^D - y, y_2 \rangle \leqslant 0$, and $y^D = y_1$.

Lemma 5. The functional $\hat{Q} : \mathcal{H} \to R^1$ defined by $\hat{Q}(y) = 0.5 \|y^D\|^2$ is differentiable and has the gradient $\hat{Q}^*_y(y) = y^D$.

Proof. Consider the optimization problem: minimize $Q(d) = 0.5\|d\|^2$ for $d \in D^* + y$. The unique solution of the problem is y^D - see Lemma 3. Since the inequality $0.5\|d + \eta\|^2 \geqslant 0.5\|y^D + \eta\|^2$ is satisfied for all $d \in \mathcal{H}$ if and only if $\eta = -y^D$ (otherwise $d = -\eta$ yields a contradiction), hence $\eta = -y^D$ is the unique element such that $0.5\|d\|^2 + \langle \eta, d - y \rangle \geqslant 0.5\|y^D\|^2 + \langle \eta, y^D - y \rangle$ for all $d \in \mathcal{H}$. Therefore, $\eta = -y^D$ is the unique, normal Lagrange multiplier for the problem (observe that the cone D may have an empty interior; consider the positive cone in $L^2[t_o, t_1]$). Moreover, the mapping $N: \mathcal{H} \to \mathcal{H}$ such that $\eta = N(y) = -y^D$ is Lipschitzian - see Lemma 4, property (ii). Corollary 1 yields the conclusion of the lemma.

3. Application to penalty functional techniques.

The penalty functional methods have been applied to the cases when the constraining operator P is finite-dimmensional (see e.g. [5] , [10]) and in special cases of infinite-dimmensional operators (so called ε-technique, see [11]). The notion of projection on a cone makes it possible to generalize the penalty techniques to arbitrary infinite-dimmensional operators.

Consider the problem: minimize $Q(y)$ on $Y_p = \{y \in E: p - P(y) \in D\}$ where E is a normed space, $P : E \to \mathcal{H}$, D is a self-conjugate cone ($D^* = D$) in $\mathcal{H}$. Observe that most of the typical positive cones are self-conjugate; however, the last assumption is made in order to simplify the analysis, and the results can be generalised for the case when D is not self-conjugate.

Define the increased penalty functional $\Phi : E \times R^1 \to R^1$ by

$$\Phi(y, \varsigma) = Q(y) + 0.5\varsigma \|(P(y) - p)^D\|^2 \qquad (10)$$

Theorem 3. Let $\{\varsigma_n\}_1^\infty$ tend monotonically towards infinity. Suppose for each ς_n there is a y_n minimizing $\Phi(y, \varsigma_n)$ for $y \in E$. Then

(i) $\Phi(y_{n+1}, \varsigma_{n+1}) \geqslant \Phi(y_n, \varsigma_n)$

(ii) $\inf_{y\in Y_p} Q(y) \geq \Phi(y_n, \varsigma_n)$

(iii) $\lim_{n\to\infty} \varsigma_n \|(P(y_n) - p)^D\|^2 = 0$

(iv) If there exists $\hat{y} = \lim_{n\to\infty} y_n$ and Q, P are continuous, then $Q(\hat{y}) \leq Q(y)$ for all $y \in Y_p$.

(v) Denote $p_n = (P(y_n)-p)^D + p$. Then $\lim_{n\to\infty} p_n = p$ and each y_n minimizes $Q(y)$ over $Y_{pn} = \{y \in E: p_n - P(y) \in D\}$.

(vi) Denote $\eta_n = \varsigma_n (P(y_n)-p)^D$. If Q,P are differentiable, then η_n is the normal Lagrange multiplier for the problem of minimizing $Q(y)$ over $Y_{pn} =$ $= \{y \in E: p_n - P(y) \in D\}$ with $p_n = (P(y_n)-p)^D + p$.

Proof. The points (i)...(iv) are an easy generalization of the theorems presented in [5] . To prove (v) observe that $p_n - P(y_n) \in D^* = D$, according to lemma 2; hence $y_n \in Y_{pn}$. Moreover, $Q(y_n) + 0.5\varsigma_n \cdot \|(P(y_n) - p)^D\|^2 \leq Q(y) + 0.5\varsigma_n \|(P(y) - p)^D\|^2$ for all $y \in E$. But for all $y \in Y_{pn}$ we have $p_2 = (P(y_n) - p)^D - (P(y) - p) \in D = D^*$. Denote $p_1 = (P(y_n) - p)^D \in D$; hence $P(y) - p = p_1 - p_2$ and, according to lemma 4 p. (iii), we have $\|(P(y)-p)^D\| \leq \|(P(y_n) - p)^D\|$ for all $y \in Y_{pn}$. Therefore, $Q(y_n) \leq Q(y)$ for all $y \in Y_{pn}$.

To prove point (vi) observe that Φ is differentiable according to lemma 5 and $\Phi^*_y(y_n, \varsigma_n) = Q^*_y(y_n) + \varsigma_n P^*_y(y_n)(P(y_n) - p)^D = \theta$. Hence $\eta_n = \varsigma_n (P(y_n)-p)^D$ satisfies the Lagrange condition. Moreover, $\eta_n \in D^*$ and $\langle \eta_n, P(y_n) - p_n \rangle = \varsigma_n \langle (P(y_n) - p)^D, P(y_n) - p - (P(y_n)-p)^D \rangle = 0$ according to lemma 2. Therefore, η_n has the properties of a Lagrange multiplier - see Corollary 2, point (i).

The Lagrange functional corresponding to the penalty technique has a normal form - though no normality assumptions has been made. If the original problem is not normal, then the sequence $\{\eta_n\}_1^\infty$ does not converge. But the sequence $\{p_n\}_1^\infty$ converges to p. Hence the penalty functional technique approximates optimization problems by normal optimization problems.

Corollary 3. If the assumptions of Theorem 3, point (vi) are satisfied for all $p \in \mathcal{H}$, then the set of $p \in \mathcal{H}$ such that the optimization problem is normal is dense in $\mathcal{H}$.

The corollary explains why the Balakrishnan's ε-technique leads to a normal formulation of the maximum principle: singular optimal control problems are approximated by normal ones when applying penalty functionals. However, the increased penalty technique has one sincere drowback: the method becomes ineffectives computationally when ς increases. Thw computational effort necessary to solve the unconstrained problem of minimization of Φ increases rapidly with ς ; this is due to "steep valley" effects, well known in computational optimization.

To overcome this difficulty, another technique has been proposed [12].

Define the shifted penalty functional $\Psi : E \times \mathcal{H} \times R^1 \to R^1$ by

$$\Psi(y,v,\varsigma) = Q(y) + 0.5\,\varsigma\, \| (P(y) + v)^D \|^2 \qquad (11)$$

where $v \in \mathcal{H}$ is not necessarity equal-p. The theorem 3, points (v), (vi), can be restated in this case:

Theorem 4. Given $v \in \mathcal{H}$ and $\varsigma > 0$ suppose there exists y_v minimizing Ψ over E. Then:

(i) Denote $p_v = (P(y_v) + v)^D - v \in \mathcal{H}$. The element y_v minimizes $Q(y)$ over $Y_{pv} = \{ y \in E: p_v - P(y) \in D \}$.

(ii) Denote $\eta_v = \varsigma (P(y_v) + v)^D$. If P,Q are differentiable, then η_v is the normal Lagrange multiplier for the problem of minimization of Q over Y_{pv}.

The penalty shifting method consists of a suitable algorithm of changing v in order to achieve $p_v \to p$. An efficient algorithm was proposed first by Powell [13] for equality constraints in R^n and later generalized in [12] for inequality constraints. Stated in terms of projection on cones, the algorithm has the form:

$$v_{n+1} = (v_n + P(y_{vn}))^D - p \; ; \quad v_1 = -p \qquad (12)$$

Theorem 5. Suppose there exists a solution y_o and a normal Lagrange multiplier η_o for the problem: minimize $Q(y)$ over $Y_p = \{ y \in E: p - P(y) \in D \}$, where E is a normed space, $\mathcal{H}$ is Hilbert, D is a selfconjugate cone in $\mathcal{H}$, $Q: E \to R^1$ and $P : E \to \mathcal{H}$ are Frechet differentiable, $p = p_o \in \mathcal{H}$ is given. Suppose there is a neighbourhood $U(p_o)$ such that the optimization problem has solutions for all $p \in U(p_o)$ and the mapping $N: U(p_o) \to \mathcal{H}$, $\eta = N(p)$, is well-defined and Lipschitzian, $\| N(p_1) - N(p_2) \| \leq R_\eta \| p_1 - p_2 \|$ for all $p_1, p_2 \in U(p_o)$. Suppose the shifted penalty functional $\Psi(y,v,\varsigma)$ has a minimum with respect to $y \in E$ for each $\varsigma > \varsigma'$ and each v in a neighbourhood $U(v_1)$, $v_1 = -p_o$. Suppose v is changed iteratively: $v_1 = -p_o$, $v_{n+1} = (P(y_n) + v_n)^D - p_o$, where y_n are minimal points of $\Psi(y,v_n,\varsigma)$. Then:

(i) There exists $\varsigma'' \geq \varsigma''$ such that for all $\varsigma \geq \varsigma''$ the sequence $\{v_n\}_1^\infty$ converges to $v_o = \frac{1}{\varsigma} \eta_o - p_o$ and $\{v_n\}_1^\infty \subset U(v_1)$ whereas the sequence $\{p_n\}_1^\infty$ defined by $p_n = (P(y_n) + v_n)^D - v_n$ converges to p_o and $\{p_n\}_1^\infty \subset U(p_o)$.

(ii) Given any $\alpha > 0$ there exists $\varsigma_\alpha \geq \varsigma''$, $\varsigma_\alpha \geq \frac{1+\alpha}{\alpha} R_\eta$, such that $\varsigma \geq \varsigma_\alpha$ implies $\| p_{n+1} - p_o \| \leq \alpha \| p_n - p_o \|$, $\| v_{n+1} - v_o \| \leq \alpha \| v_n - v_o \|$. Thus the convergence is at least geometrical and an arbitrary convergence rate can be achived.

Proof. Since $v_o = v_1 + \frac{1}{\varsigma} \eta_o$, $v_o \in U(v_1)$ for sufficiently large ς and there is a neighbourhood $U(v_o) \subset U(v_1)$ such that $v_1 \in U(v_o)$. Since $p_1 = p_o + (P(y_1) - p_o)^D$, hence - by the Theorem 3 -

there is a sufficiently large ϱ such that $p_1 \in U(p_o)$.

Suppose $v_n \in U(v_o)$ and $p_n \in U(p_o)$. By the Theorem 4, $\eta_n = \varrho (P(y_n) + v_n)^{D^o}$ is the normal Lagrange multiplier for the problem of minimizing $Q(y)$ over Y_{pn}. Thus we have:

$$v_n = \frac{1}{\varrho}\eta_n - p_n \ ; \quad v_{n+1} = \frac{1}{\varrho}\eta_n - p_o : \quad v_o = \frac{1}{\varrho}\eta_o - p_o$$

and

$$v_n - v_o = \frac{1}{\varrho}(\eta_n - \eta_o) - (p_n - p_o) \ ; \quad v_{n+1} - v_o = \frac{1}{\varrho}(\eta_n - \eta_o)$$

Since $\eta_n = N(p_n)$ is Lipschitzian,

$$\|p_n - p_o\| \leq \frac{R_\eta}{\varrho}\|p_n - p_o\| + \|v_n - v_o\|$$

and, if $\varrho > R_\eta$

$$\|p_n - p_o\| \leq \frac{\varrho}{\varrho - R_\eta}\|v_n - v_o\|$$

On the other hand

$$\|v_{n+1} - v_o\| \leq \frac{R_\eta}{\varrho}\|p_n - p_o\|$$

hence

$$\|v_{n+1} - v_o\| \leq \frac{R_\eta}{\varrho - R_\eta}\|v_n - v_o\|$$

If ϱ is sufficiently large, $v_{n+1} \in U(v_o)$ and y_{n+1}, p_{n+1} exist. Moreower

$$\|p_{n+1} - p_o\| \leq \frac{R_\eta}{\varrho - R_\eta}\|p_n - p_o\|$$

and $p_{n+1} \in U(p_o)$. By induction, $\{v_n\}_1^\infty \subset U(v_o) \subset U(v_1)$ and $\{p_n\}_1^\infty \subset U(p_o)$. The convergence rate $\alpha = \frac{R_\eta}{\varrho - R_\eta}$ can be made arbitrarily small.

The assumptions of Theorem 5 can be made more explicit by investigating the conditions of the Lipschitz-continuity of $\eta = N(p)$, the existence of minimal y_n etc. However, these problems shall not be pursued in this paper. It should only be stressed that the assumption of Lipschitz-continuity of N is essential, what can be shown by simple examples even in R^1 - see [12]. Although an arbitrary convergence rate α can be achived, it is not practical to require too small α and too large ϱ, since the computational effort of solving the problem of unconstrained minimization of $\Psi(y,v,\varrho)$ becomes rather large in that case.

4. Application to gradient projection and reduction techniques.

Consider once more the problem of minimizing $Q(y)$ over $Y_p = \{y \in \mathcal{H}_y : p - P(y) \in D\}$ where $Q : \mathcal{H}_y \to R^1$, $P : \mathcal{H}_y \to \mathcal{H}_p$ are differentiable, $\mathcal{H}_y$, $\mathcal{H}_p$ are Hilbert, D is a positive cone in $\mathcal{H}_p$. Assume there is a $y_1 \in Y_p$ given and construct a cone $K(y_1) = \{\delta y \in \mathcal{H}_y : \exists_{\varepsilon_1 > 0} \forall_{\varepsilon \in (0,\varepsilon_1)} \ p - P(y_1) - \varepsilon P_y(y_1)\delta y \in D\}$. Assume the cone is external - see theorem 2 - to the set Y_p at y_1 (to make the assumption explicit, several regularity assumptions could be made). Assume the direction $d = -Q_y^*(y_1)$ does not belong to $K(y_1)$. However, the projection d^k of d on $K(y_1)$ provides for a good approximation of a permissible direction of improvement.

Assume now that $\{\delta y \in \mathcal{H}_y : P_y(y_1)\,\delta y \in D\}^* = P_y^*(y_1)\,D^*$ (again, to make the assumption explicit, we should use Farkas lemma - $-P_y^*(y_1)D^*$ being weakly closed - or Lyusternik theorem - $D = \{\theta\}$, $P_y(y_1)$ a surjection). This is the basic normality assumption which makes it possible to introduce a normal Lagrange multiplier η at a nonoptimal point y.

Since $K(y_1) = \{\delta y \in \mathcal{H}_y : -P_y(y_1)\,\delta y \in D_1\}$, $D_1 = D + \alpha(p-P(y_1)), \alpha \in R^1$ and $D_1^* = \{\eta \in D^* : \langle \eta, p-P(y_1)\rangle = 0\}$, hence $K^*(y_1) = \{-P_y^*(y_1)\eta : \eta \in D^*, \langle \eta, P(y_1) - p\rangle = 0\}$. When projecting d on $K(y_1)$, we obtain $d^k - d \in K^*(y_1)$ and $\langle d^k - d, d^k\rangle = 0$. If $d = -Q_y^*(y_1)$, we have:

$$
\begin{aligned}
&(i) \quad \eta \in D^*; \quad \langle \eta, p-P(y_1)\rangle = 0\\
&(ii) \quad d^k = -Q_y^*(y_1) - P_y^*(y_1)\eta\\
&(iii) \quad \langle P_y^*(y_1)\eta \,, Q_y^*(y_1) + P_y^*(y_1)\eta\rangle = 0
\end{aligned}
\qquad (13)
$$

Hence we have a normal Lagrange multiplier η for a nonoptimal point $y_1 \in Y_p$. It coincides with the optimal η for an optimal y_1, since $d^k = \theta$ if y_1 is optimal. The multiplier satisfies the usual conditions (i) ; but there is also an additional condition (iii) - trivial at an optimal point - which helps to determine η for a nonoptimal point.

In a general case it is not easy to make a constructive use of the set of conditions (13) (i), (iii). However, these conditions generalise the known notions of the Rosen gradient projection [14] or, in a special but very important case, of the Wolfe reduced gradient [15] .

Actually, assume $D = \{\theta\}$, $P_y(y_1)$ being onto; then $K(y_1)$ is the linear subspace tangent to Y_p at y_1 (the nullspace of $P_y(y_1)$). The conditions (i) are trivial, the condition (ii) amounts to an orthogonal gradient projection on the tangent subspace, and the condition (iii) results in an explicite value of the Lagrange multiplier and the Rosen gradient projection

$$
\begin{aligned}
\eta &= -\left[P_y(y_1)\,P_y^*(y_1)\right]^{-1} P_y(y_1)\,Q_y^*(y_1)\\
d^k &= -\left[I - P_y^*(y_1)\left[P_y(y_1)\,P_y^*(y_1)\right]^{-1} P_y(y_1)\right] Q_y^*(y_1)
\end{aligned}
\qquad (14)
$$

The notion of Wolfe's reduced gradient applies to several optimal control problems when $\mathcal{H}_y = \mathcal{H}_x \times \mathcal{H}_u$ and $\mathcal{H}_p = \mathcal{H}_x$. The most important feature of the class of problems is that the state $x \in \mathcal{H}_x$ can be explicitly determined (analytically or numerically) in terms of the control $u \in \mathcal{H}_u$. The statement of such a problem is: minimize $Q(x,u)$ for $(x,u) \in Y_o = \{(x,u) \in \mathcal{H}_x \times \mathcal{H}_u : P(x,u) = \theta \in \mathcal{H}_x\}$ where

$Q : \mathcal{H}_x \times \mathcal{H}_u \to R^1$ and $P : \mathcal{H}_x \times \mathcal{H}_u \to \mathcal{H}_x$ are Frechet differentiable and $P_x(x,u)$ is onto for all $(x,u) \in Y_o$. The last assumption corresponds to the requirement that x could be determined in terms of u.

If we take $y = (x,u)$, $P_y(y) = (P_x(x,u), P_u(x,u))$, $Q^*_y(y) = (Q^*_x(x,u), Q^*_u(x,u))$ and apply the Rosen's projection on the subspace $K = \{(\delta x, \delta u) \in \mathcal{H}_x \times \mathcal{H}_u : P_x(x,u)\delta x + P_u(x,u)\delta u = \Theta \in \mathcal{H}_x\}$, we obtain

$$d_x^k = - Q_x^*(x,u) - P_x^*(x,u)\eta$$
$$d_u^k = - Q_u^*(x,u) - P_u^*(x,u)\eta$$
$$\eta = -(P_x(x,u)P_x^*(x,u)+P_u(x,u)P_u^*(x,u))^{-1}((P_x(x,u)Q_x^*(x,u)+P_u(x,u)Q_u^*(x,u))$$

However, this is not the most useful way of introducing a Lagrange multiplier in that case. Since $P_x^{-1}(x,u)$ exists for $(x,u) \in Y_o$, we can change the original variables $y = (x,y)$ to $\tilde{y} = (y_1, y_2)$ where $\delta y_1 = \delta x + P_x^{-1}(x,u)P_u(x,u)\delta u$, $\delta y_2 = \delta u$. Now $P_{\tilde{y}}(y) = (P_x(x,u), 0)$ and $Q^*_{\tilde{y}}(y) = (Q^*_x(x,u), Q^*_u(x,u) - P_u(x,u)P_x^{-1}(x,u) Q^*_x(x,u))$. The tangent subspace becomes $K = \{\Theta\} \times \mathcal{H}_u$ and the projection is particularly simple. The new Lagrange multiplier (we keep to the original denotation) is

$$\eta = - P_x^{*-1}(x,u)\, Q_x^*(x,u) \tag{16}$$

and the new projection of the negative gradient is

$$\text{(i)} \quad - b_x^k = - Q_x^*(x,u) - P_x^*(x,u)\eta = \Theta$$
$$\text{(ii)} \quad - b_u^k = -Q_u^*(x,u) - P_u^*(x,u)\eta \tag{17}$$

The relation (i) has a simple interpretation: we choose the new Lagrange multiplier in such a way that the Lagrange functional $Q(x,u) + \langle \eta, P(x,u)\rangle$ doeas not depend in a linear approximation on δx. The Lagrange multiplier determined by)(16) differs obviously from the multiplier (15); however, at an optimal solution the multipliers are identical. The projection (15), (ii) is a generalization of the Wolfe's reduced gradient; in the original space it is a non-orthogonal projection (but it is a projection, since the variable transformation is one-to-one). The resulting technique has been in fact widely employed in many approaches of the calculus of variation and provides for many computational methods of optimization.

5. Reduced gradient and the basic variational equations

If the condition $P(x,u) = \Theta$ determines implicitly a mapping $S : \mathcal{H}_u \to \mathcal{H}_x$, $x = S(u)$, then instead of minimizing $Q(x,u)$ over Y_o we have to minimize $J(u) = Q(S(u),u)$ in $\mathcal{H}_u$. However, the determination of S might be cumbersone, and it is often useful to express the derivatives of J in terms of the derivatives of an equivalent Lagrange functional. This technique amounts actually to the gradient reduction technique. Assume Q,S are twice differentiable ; so is J. Denote $J(u+\delta u)=J(u)+ \langle b(u),\delta u\rangle + 0.5 \langle \delta u, A(u)\delta u\rangle + o(\|\delta u\|^2)$ (18)

where $b(u)$ is the gradient and $A(u)$ the hessian operator of the functional J. On the other hand

$$J(u) = L(\eta, x, u) = Q(x,u) + \langle \eta, P(x,u) \rangle \tag{19}$$

if

$$L^*_\eta(\eta, x, u) = P(x,u) = \Theta \tag{20}$$

which can be interpreted as the basic state equation of the optimal control problem. By choosing η according to (16) we obtain

$$L^*_x(\eta, x, u) = Q^*_x(x,u) + P^*_x(x,u)\eta = \Theta \tag{21}$$

which can be interpreted as the basic adjoint equation of the problem. Hence the gradient of the functional J (the reduced gradient of Q) is

$$L^*_u(\eta, x, u) = Q^*_u(x,u) + P^*_u(x,u)\eta = b(u) \tag{22}$$

The linear approximation of the state equation is

$$\delta x' = - P_x^{-1}(x,u)P_u(x,u)\,\delta u; \quad \delta x = \delta x' + o(\|\delta u\|) \tag{23}$$

Expanding $L(\eta, x + \delta x, u + \delta u)$ into second-order terms of δx, δu and applying (23) results in the hessian operator of the functional J

$$A(u) = L_{u\eta}L^{-1}_{x\eta}\, L_{xx}L^{-1}_{\eta x}L_{\eta u} - L_{u\eta}L^{-1}_{x\eta}L_{xu} - L_{ux}L^{-1}_{x\eta}L_{\eta u} + L_{uu} \tag{24}$$

with the denotation shortened in an obvious way; observe that $L_{\eta x} = P_x(x,u)$. However, the explicit expression for the hessian is not the most useful computationally. There is an alternate way: to expand $L(\eta + \delta\eta, x + \delta x, u + \delta u)$ into second-order terms and choose an appropriate $\delta\eta$. Hence the hessian operator can be determined by the set of equations

$$\begin{aligned}
&(i) \quad A(u)\,\delta u = L_{u\eta}\delta\eta + L_{ux}\,\delta x' + L_{uu}\,\delta u \\
&(ii) \quad \Theta = L_{x\eta}\delta\eta + L_{xx}\,\delta x' + L_{xu}\,\delta u \\
&(iii) \quad \Theta = L_{\eta x}\,\delta x' + L_{\eta u}\,\delta u
\end{aligned} \tag{25}$$

The equations (ii), (iii) - where (iii) is equivalent to (23) - are the linearisation of the adjoint and state equation and are called basic variational equations.

In some computational approaches an important problem is the inversion of the hessian operator, in order to determine the Newton's direction of improvement $d = -A^{-1}(u)b(u)$. Setting $\delta u = d$, $A(u)\delta u = -b(u)$ in (26) and assuming L^{-1}_{uu} exists, we get

$$\begin{aligned}
&(i) \quad d = - L^{-1}_{uu}(L_{u\eta}\delta\tilde{\eta} + L_{ux}\,\delta\tilde{x} + b(u)) \\
&(ii) \quad \Theta = (L_{x\eta} - L_{xu}L^{-1}_{uu}L_{u\eta})\delta\tilde{\eta} + (L_{xx} - L_{xu}L^{-1}_{uu}L_{ux})\delta\tilde{x} - L_{xu}L^{-1}_{uu}b(u) \\
&(iii) \quad \Theta = -L_{\eta u}L^{-1}_{uu}L_{u\eta}\delta\tilde{\eta} + (L_{\eta x} - L_{\eta u}L_{uu}L_{ux})\,\delta\tilde{x} - L_{\eta u}L^{-1}_{uu}\,b(u)
\end{aligned} \tag{26}$$

where (ii), (iii) are called canonical variational equations. Their solution is more difficult than the solution of basic variational equations; usually, the canonical equations represent a nontrivial two-point boundary problem, and the typical method of their solution is the reduction to a Riccati-type nonlinear equation.

II. Application to optimal control problems with delays.

6. Optimal control problem with delays: the gradient and the hessian.

Consider the problem: minimise the performance functional

$$Q(x,u) = \int_{t_0}^{t_1} f_0(x(t),x(t-\tau),u(t),u(t-\tau),t)dt + h(x(t_1)) \tag{27}$$

where x satisfies the process equation

$$P(x,u) = \Theta \Leftrightarrow \dot{x}(t) = f(x(t),x(t-\tau),u(t),u(t-\tau),t);$$
$$x(t) = \varphi_1(t) \quad \text{for} \quad t \in [t_0-\tau, t_0]$$
$$u(t) = \varphi_2(t) \quad \text{for} \quad t \in [t_0-\tau, t_0) \tag{28}$$

The analysis of the relations between this particular problem and the general Hilbert space problem leads to the known, following results.

Denote $\tilde{H} = -f_0 + \eta^* f$ and

$$\eta(t+\tau)=\psi(t);\ x(t+\tau)=z(t);\ x((t-\tau)=y(t);\ u(t+\tau)=w(t);\ u(t-\tau)=v(t) \tag{29}$$

Define the shifted hamiltonian function H by

$$H(\eta,\psi,z,y,w,u,v,t) = \tilde{H}\Big|_t + \mu\tilde{H}\Big|_{t+\tau};\ \mu = \begin{cases} 1 & \text{for } t \in [t_0, t_1-\tau] \\ 0 & \text{for } t \in (t_1-\tau, t_1] \end{cases} \tag{30}$$

where $\tilde{H}\Big|_{t+\tau}$ denotes the function $\tilde{H}$ with arguments evaluated at $t+\tau$. Then the process equation (29) can be rewritten in the form

$$\dot{x} = H_\eta^* \ ; \quad \varphi_1 \ ; \quad \varphi_2 \tag{31}$$

the adjoint equation - under appropriate differentiability assumptions-takes the form

$$\dot{\eta} = -H_x^* \ ; \quad \eta(t_1) = -h_x^*(x(t_1)) \tag{32}$$

and the gradient equation is

$$b = -H_u^* \tag{33}$$

These results apply for the case without any additional constraints save (29). If, for example, a final state-function $x(t) = \varphi_3(t)$ for $t \in [t_1-\tau, t_1]$ is given, the penalty functional approach is most useful. The basic variational equations take the form

$$\delta\dot{x}=H_{\eta x}\delta x+H_{\eta y}\delta y + H_{\eta u}\delta u + H_{\eta v}\delta v;\quad x(t) = 0 \ \text{for} \ t \in [t_0-\tau, t_0] \ ;$$
$$\delta\dot{\eta} = -H_{x\eta}\delta\eta - H_{x\psi}\delta\psi - H_{xy}\delta y - H_{xx}\delta x - H_{xz}\delta z - H_{xv}\delta v - H_{xu}\delta u - H_{xw}\delta w \ ;$$
$$\delta\eta(t_1) = -h_{xx}\delta x(t_1);$$
$$A\delta u= -H_{u\eta}\delta\eta - H_{u\psi}\delta\psi - H_{uy}\delta y - H_{ux}\delta x - H_{uz}\delta z - H_{uv}\delta v - H_{uu}\delta u - H_{uw}\delta w \tag{34}$$

The canonical variational equations have a complicated form

$$\begin{aligned} \delta\dot{\tilde{x}} &= A_1\delta\tilde{x} + B_1\delta\tilde{y} + C_1\delta\tilde{z} + A_2\delta\tilde{\eta} + B_2\delta\tilde{\xi} + C_2\delta\tilde{\psi} + \alpha_x \\ \delta\dot{\tilde{\eta}} &= A_3\delta\tilde{x} + B_3\delta\tilde{y} + C_3\delta\tilde{z} + A_4\delta\tilde{\eta} + B_4\delta\tilde{\xi} + C_4\delta\tilde{\psi} + \alpha_\eta \end{aligned} \tag{35}$$

where $\delta\tilde{\xi}(t)= \delta\tilde{\eta}(t-\tau), \delta\tilde{x}(t)=0$ for $t \in [t_0-\tau, t_0], \delta\tilde{\eta}(t_1) = -h_{xx}\delta\tilde{x}(t_1)$, A_j, B_j, C_j are matrices determined by second-order derivatives of the hamiltonian function H (assumed a matrix related to H_{uu}, H_{uv}, H_{uw} is invertible) and α_x, α_η are determined by the gradient b.

A numerical solution of the equations (34) presents no particular difficulties, since we can solve the advanced-type adjoint variational equation backwards after solving the retarded-type state variational equation. A numerical solution of the equations (35) presents a major computational problem, since they are of neutral type with two-point boundary conditions.

Several techniques have been proposed in order to solve the equations (35) or, equivalently, to invert the hessian operator of an optimal control problem with delays. Most of the techniques consist of a Riccati setting of an integral operator type. However, such a setting results in an integral or partial differential equation of Riccati type, which is also rather difficult to solve computationally. Recently, Chan and Perkins [19] proposed a simple iterative technique for solving equations (35), which is quite effective for quadratic problems. However, the Newton's method (or "second variation method") is iterative itself in non-quadratic cases. Therefore, it is rather reasonable to ommit the inversion of the hessian for non-quadratic optimal control problems with delays.

7. Universal conjugate direction and variable operator methods. A computational example.

The general formulation of an optimal control problem in Hilbert space is a natural basis for the application of several universal computational methods, well known for nonlinear programming problems in R^n and recently generalised for Hilbert space problems. There are two families of such methods.

The first family, called conjugate directions methods, is based on the concept of conjugacy or A-orthogonality.

The second family, called variable operator or variable matric methods, is based on the notion of the outer product in $\mathcal{H}$. A variable operator V_i is an aproximation of A^{-1} constructed iteratively by $V_{i+1} = V_i + \Delta V_i$ where ΔV_i is determined with help of outer products in $\mathcal{H}$. A discussion of general properties of these two families of optimization methods is given in [18]. The existing computational experience seems to indicate that these methods are more effective than the Newton's if the inversion of the hessian is difficult.

One of the conjugate gradient methods was applied as a subprocedure in solving the following computaional example - constructed by M. Jacobs and T.J. Kao in an unpublished paper and computed by St. Kurcyusz. The example illustrates the effectiveness of penalty functional techniques. The problem consist in achieving a given final complete state (a function of time) $x(t) = 0$ for $t \in [1,2]$ in a time delay system:
$\dot{x}(t) = -x(t-1) + u(t)$, $t \in [0,2]$; $x(t) = 1$, $t \in [-1,0]$, while minimizing the functional $Q = \frac{1}{2}\int_0^2 u^2\, dt$. The problem has an analytical

solution, hence provides for a good test of computational methods. The results achived by a penalty shifting technique applied to the final complete state are following:

No of iterations of penalty shift	Final state constraint violation	No of computed functional values	Penalty functional	Performance functional
0 (beginning)	1.0	1	51.0	-
1	9.4×10^{-3}	107	0.168	0.167
2	1.4×10^{-3}	88	0.171	0.169
3	4.0×10^{-4}	17	0.171	0.169

The number of computed functional values per iteration decreases (which is typical for penalty shifting methods) instead of increasing (typical for penalty increase methods).

Bibliography

[1] Dubovitski A.J., Milyutin A.A.: Extremal problems with constraints. Journal of Computational Mathematics and Mathematical Physics (Russian), Vol V, No 3, p.395-453, 1965.

[2] Neustadt L.W.: An abstract variational theory with applications to a broad class of optimization problems. SIAM Journal on Control, Vol. V, No 1, p. 90-137, 1967.

[3] Goldshtein J.G.: Duality theory in mathematical programming (Russian), Nauka, Moscow 1971.

[4] Pshenitshny B.N.: Necessary conditions of optimality (Russian), Nauka, Moscow 1969.

[5] Luenberger D.G.: Optimization by vector space methods. J. Wiley, N. York 1969.

[6] Neustadt L.W. A general theory of extremals. Journal on Computer and System Science, Vol. III, p.57-91, 1969.

[7] Girsanov I.W.: Lectures on mathematical theory of extremal problems. University of Moscow, 1970.

[8] Wierzbicki A.P.: Maximum principle for semiconvex performance functionals. SIAM Journal on Control, V.X No 3 p.444-459, 1972.

[9] Galperin A.M.: Towards the theory of permissible directions (Russian), Kibernetika No 2, p. 51-59, 1972.

[10] Fiacco, A.V., Mc Cormick G.P.: The sequential unconstrained minimization technique for nonlinear programming. Management Science. Vo.X, No 2, p. 360-366, 1964.

[11] Balakrishnan, A.V. A computational approach to the maximum principle. Journal of Computer and System Science, Vol.V, 1971.

[12] Wierzbicki A.P.: A penalty function shifting method in constrained static optimization and its convergence properties. Archiwum Automatyki i Telemechaniki, Vol. XVI, No 4, p. 395-416, 1971.

[13] Powell, M.J.D.: A method for nonlinear constraints in minimisation problems. In R. Fletcher: Optimization, Academic Press, N. York 1969.

[14] Rosen, J.B.: The gradient projection method for nonlinear programming. Part I, II. Journal of SIAM, Vol. VIII, p. 181-217, 1960, Vol. IX, p. 514-532, 1962.

[15] Wolfe, P. Methods of nonlinear programming. In J. Abadie: Nonlinear Programming, Interscience, J. Wiley, N. York, 1967.

[16] Horwitz L.B., Sarachik P.E.: Davidon's method in Hilbert Space. SIAM J. on Appl. Math., Vol. XVI, No 4, p. 676-695, 1968.

[17] Wierzbicki A.P.: Coordination and sensitivity analysis of a large scale problem with performance iteration. Proc. of V-th Congress of IFAC, Paris 1972.

[18] Wierzbicki A.P.: Methods of mathematical programming in Hilbert space. Polish-Italian Meeting on Control Theory and Applications, Cracow 1972.

[19] Chan H.C., Perkins W.R. Optimization of time delay systems using parameter imbedding. Proc. of V-th Congress of IFAC, Paris 1972.

SUFFICIENT CONDITIONS OF OPTIMALITY FOR CONTINGENT EQUATIONS

V.I. Blagodatskih

Mathematical Institute of USSR Academy of Sciences, Moscow, USSR

1. Statement of the problem

In this paper we prove sufficient conditions of optimality in the form of maximum principle for controllable processes which behaviour is described by contingent equation.

Let E^n be Euclidean n space of the state $x = (x_1, \dots, x_n)$ with norm $\|x\| = \sqrt{\sum_{i=1}^{n} x_i^2}$ and $\Omega(E^n)$ be space of all nonempty compact subsets of E^n with Hausdorff metric

$$\rho(F,G) = \inf \{d: F \subset S_d(G), \quad G \subset S_d(F)\}$$

where $S_d(M)$ denotes d neighborhood of set M in space E^n.

Let's consider controllable processes which behaviour is described by contingent equation

$$\dot{x} \in F(x), \tag{1}$$

or the differential inclusion as it is also called. Here $F : E^n \to \Omega(E^n)$ is certain given mapping. The absolutely continuous function $x(t)$ is the <u>solution</u> of the equation (1) on the interval $[0, T]$ if the inclusion $\dot{x}(t) \in F(x(t))$ is valed almost every-where on this interval.

Let M_0, M_1 be nonempty closed subsets of E^n. These subsets may be non-convex and non-compact. The solution $x(t)$ given on interval $[0, T]$ <u>does the transfer from the set</u> M_0 <u>into the set</u> M_1 <u>for the time</u> T if the conditions $x(0) \in M_0$, $x(T) \in M_1$ are satisfied. <u>The time-optimal control problem</u> is to define the solution of the equation (1), doing the transfer from the set M_0 into the set M_1 for minimal time.

Let G be an arbitrary nonempty closed subset of E^n. The function

$$C(\psi) = \max_{f \in G} (f, \psi) \tag{2}$$

of the vector $\psi \in E^n$ is called <u>the support function</u> of the set G.If the maximum in the expression (2) for the given vector ψ_0 is reached at the vector $f_0 \in G$, that is, $C(\psi_0) = (f_0, \psi_0)$, then the hyperplane $(x,\psi_0) = C(\psi_0)$ is called <u>the support hyperplane</u> for the set G at the point f_0 and the vector ψ_0 is called <u>the support vector</u> at the point f_0. In this case the inequality $(f - f_0, \psi_0) \leq 0$ is valed for any vector $f \in G$. As one can see from the condition (2) the support function $C(\psi)$ is single-valued function of the set G. On the contrary, if we know the support function $C(\psi)$, we can incite only convex hull of set G, that is,

$$\text{conv } G = \bigcap_{\psi \in E^n} \{g : (g, \psi) \leq C(\psi)\}.$$

If $F : E^n \to \Omega(E^n)$ is an arbitrary mapping, we can consider the support function of the set $F(x)$ for any $x \in E^n$; we shall denote this function by $C(x, \psi)$ and shall call by <u>the support function of the mapping</u> F:

$$C(x, \psi) = \max_{f \in F(x)} (f, \psi).$$

Next lemma follows directly from the definitions of the support function $C(x, \psi)$ and Hausdorff metric.

<u>Lemma 1</u>. <u>If the mapping</u> $F : E^n \to \Omega(E^n)$ <u>satisfies Lipshitz's condition (is continuous), then the support function</u> $C(x, \psi)$ <u>satisfies Lipshitz's condition (is continuous) in</u> x <u>for any fixed vector</u> $\psi \in E^n$. <u>On the contrary, if the support function</u> $C(x, \psi)$ <u>satisfies Lipshitz's condition (is continuous) in</u> x, <u>then the respective mapping</u> $\text{conv } F : x \to \text{conv } F(x)$ <u>satisfies Lipshitz's condition (is continuous)</u>.

Together with the inclusion (1) let's consider the differential inclusion

$$\dot{x} \in \text{conv } F(x). \tag{3}$$

<u>Lemma 2</u>. <u>If absolutely continuous function</u> $x(t)$ <u>is the solution of the equation (1) on the interval</u> $[0, T]$, <u>then the inequality</u>

$$(\dot{x}(t), \psi) \leq C(x(t), \psi) \tag{4}$$

<u>is valed almost everywhere on this interval for any vector</u> $\psi \in E^n$. <u>On the contrary, if the condition (4) is valed for absolutely continuous function</u> $x(t)$ <u>almost everywhere on interval</u> $[0, T]$ <u>for any vector</u> $\psi \in E^n$, <u>then this function is the solution of the equation (3) on the interval</u> $[0, T]$.

Proof of lemma 2 follows directly from the definitions of the support function $C(x, \psi)$ and of the solutions of differential inclusions.

Let $C_0(\psi)$ and $C_1(\psi)$ be support functions of the sets M_0 and M_1, respectively.

Maximum principle. Assume that the support function $C(x, \psi)$ is continuously differentiable in x and the solution $x(t)$ does the transfer from the set M_0 into the set M_1 on interval $[0,T]$. We shall say that the solution $x(t)$ satisfies the maximum principle on interval $[0, T]$ if there exists nontrivial solution $\psi(t)$ of adjoint system

$$(5) \qquad \dot{\psi} = - \frac{C(x(t), \psi)}{\partial x}$$

such that following conditions are valed:

A) the maximum condition

$$(6) \qquad (\dot{x}(t), \psi(t)) = C(x(t), \psi(t))$$

is valed almost everywhere on interval $[0, T]$;

B) transversality condition on the set M_0: vector $\psi(0)$ is the support vector for the set M_0 at the point $x(0)$, that is,

$$(7) \qquad C_0(\psi(0)) = (x(0), \psi(0)) ;$$

C) transversality condition on the set M_1 : vector $-\psi(T)$ is the support vector for the set M_1 at the point $x(T)$, that is,

$$(8) \qquad C_1(-\psi(T)) = (x(T), -\psi(T)).$$

Sufficient conditions of optimality in the form of maximum principle can be received in the following section.

2. The main result

The region of reachahility Y_T for the equation (1) is the set of all points $x_0 \in E^n$ from which we can do the transfer into the set M_1 for the time not exceeding T. The set M_1 is strongly stable if set M_1 lies interily in Y_τ for any $\tau > 0$. In particular, if set M_1 consists of the single point, then the definition of the strong stability coincides with the definition of the local controllability in small of the equation (1) at this point (1).

Theorem 1. Assume that the set M_1 is strongly stable and the support function $C(x, \psi)$ satisfies Lipshitz's condition in x for any fixed vector ψ, then the inclusion $Y_{\tau_1} \subset \text{int } Y_{\tau_2}$ is valed for any τ_1, τ_2, $0 \leqslant \tau_1 < \tau_2$.

Proof. Let $x_0 \in Y_{\tau_1}$, and $x(t)$ be the solution of the equation (1), doing the transfer from the point x_0 into the set M_1 for the time $\tau \leqslant \tau_1$, that is, $x(0) = x_0$, $x(\tau) \in M_1$. Since the set M_1 is strongly stable, $M_1 \subset \text{int } Y_{\tau_2 - \tau}$ and $x(\tau) \in \text{int } Y_{\tau_2 - \tau}$. Thus, there exists $\varepsilon > 0$ that $S_\varepsilon(x(\tau)) \subset Y_{\tau_2 - \tau}$. Since the support function $C(x, \psi)$ satisfies Lipshitz's condition in x, by lemma 1 mapping conv $F(x)$ also satisfies Lipshitz's condition. Thus, the theorem of the continuous dependence of the solution on the initial conditions is valed (2) for the inclution (3). It follows that there exists such neighbourhood $U(x_0)$ that there exists the solution $y(t)$ of the inclusion (3) for any point $y_0 \in U(x_0)$, and this solution does the transfer from the point y_0 into the set $S_{\frac{\varepsilon}{2}}(x(\tau))$ for the time τ, that is, $y(\tau) \in S_{\frac{\varepsilon}{2}}(x(\tau))$. Any solution of the inclusion (3) can be approximated with any accuraty by the solutions of the equation (1) (see theorem 2.2 in paper (3)), therefore there exists the solution $x^*(t)$ of the equation (1) with initial condition $x^*(0) = y_0$ that $\| x^*(t) - y(t) \| \leqslant \frac{\varepsilon}{2}$. Thus, $x^*(\tau) \in S_\varepsilon(x(\tau))$ and we can do the transfer from the point $x^*(\tau)$ into the set M_1 for the time $\leqslant \tau_2 - \tau$, that is, $y_0 \in Y_{\tau_2}$. Therefore, $U(x_0) \subset Y_{\tau_2}$, that is, $x_0 \in \text{int } Y_{\tau_2}$. Q.E.D.

The support function $C(x, \psi)$ for the equation (1) is concave in x if for any vector $\psi \in E^n$, for any points $x_1, x_2 \in E^n$ and for any numbers $\alpha, \beta \geqslant 0$, $\alpha + \beta = 1$, condition

$$(9) \qquad \alpha C(x_1, \psi) + \beta C(x_2, \psi) \leqslant C(\alpha x_1 + \beta x_2, \psi)$$

is valed. This condition is equivalent to that of concavity of the multivaled mapping $F(x)$, that is equivalent to condition

$$\alpha F(x_1) + \beta F(x_2) \subset F(\alpha x_1 + \beta x_2)$$

for any $x_1, x_2 \in E^n$.

If the support function $C(x, \psi)$ is continuously differentiable in x, then the condition (9) is equivalent (4) to the condition

$$\left(\frac{\partial C(x_1, \psi)}{\partial x}, x_2 - x_1\right) \geqslant C(x_2, \psi) - C(x_1, \psi)$$

for any vector $\psi \in E^n$ and for any points $x_1, x_2 \in E^n$. Let's define a weaker condition on the support function $C(x, \psi)$. Let's say that the support function $C(x, \psi)$ is concave in x at the point x_0 in the direction ψ_0, if the condition

$$(10) \quad \left(\frac{\partial C(x_0, \psi_0)}{\partial x}, x - x_0\right) \geqslant C(x, \psi_0) - C(x_0, \psi_0)$$

is valed for any $x \in E^n$.

Let $x(t)$ be solution of equation (1) on interval $[0, T]$, $\psi(t)$ be respective solution of adjoint system (5). Let's say that the solution $x(t)$ satisfies strong transversality condition on the set M_1, if the condition

$$(11) \quad C_1(-\psi(t)) < (x(t), -\psi(t))$$

is valed for any $0 \leqslant t < T$.

Note, that if the set M_1 is strongly stable, solution $x(t)$ satisfies maximum principle on interval $[0, T]$ and the support function $C(x, \psi)$ is concave in x at point $x(t)$ in the direction $\psi(t)$, $0 \leqslant t \leqslant T$, then the condition (II) is valed. Indeed, as it will be shown in the proof of theorem 2 under the given conditions vector $-\psi(t)$ is the support vector for the region of reachability Y_{T-t} at the point $x(t)$. And since $M_1 \subset \operatorname{int} Y_{T-t}$ for any $0 \leqslant t < T$ condition (II) is valed.

The main result of this paper is

Theorem 2. Assume that M_0, M_1 are nonempty closed subsets of E^n, the solution $x(t)$ of the equation (1) does the transfer from the set M_0 into the set M_1 on interval $[0, T]$ and it satisfies maximum principle on that interval and $\psi(t)$ is the solution of the adjoint system. Assume that the support function $C(x, \psi)$ is concave in x at the point $x(t)$ in the direction $\psi(t)$ for any $t \in [0, T]$

<u>and the solution</u> $x(t)$ <u>satisfies strong transversality condition on the set</u> M_1. <u>Then the solution</u> $x(t)$ <u>is optimal</u>.

<u>Proof</u>. Let $y(t)$ be an arbitrary solution of the equation (1) defined on the interval $[0, T]$. Inequality

$$(12)\qquad \frac{d}{dt}(y(t) - x(t), \psi(t)) \leq 0$$

is valed almost everywhere on this interval. Indeed, using lemma 2 and conditions (6) and (10), we get

$$\frac{d}{dt}(y(t)-x(t), \psi(t)) = (\dot{y}(t), \psi(t)) - (\dot{x}(t), \psi(t)) +$$

$$+ (y(t)-x(t), \dot{\psi}(t)) \leq C(Y(t), \psi(t)) - C(x(t), \psi(t)) -$$

$$- \left(\frac{\partial C(x(t), \psi(t))}{\partial x}, y(t) - x(t) \right) \leq 0.$$

Let Γ_τ, $0 \leq \tau \leq T$, be hyperplane passing through point $x(\tau)$ and orthogonal to vector $\psi(\tau)$. It is impossible to do the transfer from the hyperplane Γ_τ into the set M_1 for the time $\theta < T - \tau$. Indeed, let $y(t)$ be arbitrary solution of the equation (1) with the condition $y(\tau) \in \Gamma_\tau$. Integrating an inequality (12) on the interval $[\tau, \tau + \theta]$ we receive

$$(y(\tau + \theta) - x(\tau + \theta), \psi(\tau + \theta)) \leq 0.$$

From the strong transversality condition (11) it follows that

$$C_1(-\psi(\tau + \theta)) < (x(\tau + \theta), -\psi(\tau + \theta)) =$$

$$= (y(\tau + \theta) - x(\tau + \theta), \psi(\tau + \theta)) - (y(\tau + \theta),$$

$$\psi(\tau + \theta)) \leq (y(\tau + \theta), -\psi(\tau + \theta)),$$

that is, $C_1(-\psi(\tau + \theta)) < (y(\tau + \theta), -\psi(\tau + \theta))$. It means that point $y(\tau + \theta)$ does not belong to set M_1.

If point $y(\tau)$ satisfies the condition $(y(\tau)-x(\tau), \psi(\tau)) < 0$, then it is impossible to do the transfer from this point into the set M_1 for the time $T - \tau$. Indeed, integrating an inequality (12) on the interval $[\tau, T]$ we can receive $(y(T)-x(T), \psi(T)) < 0$.

It contradicts the transversality condition (8) on the set M_1, Thus, the hyperplane Γ_τ is the support for the region reachability $Y_{T-\tau}$ at the point $x(\tau)$ with the support vector $-\psi(\tau)$ for all $\tau \in [0, T]$.

From the transversality condition (7) on the set M_0 it follows that $M_0 \cap Y_T \subset \Gamma_0$. Thus, it is impossible to do the transfer from the set M_0 into the set M_1 for the time $< T$, that is the solution $x(t)$ is optimal.

Q.E.D.

Remark. Since the solution $x(t)$ of the equation (1) is also the solution of the equation (3), then by the theorem 2 the solution $x(t)$ is also optimal for the equation (3).

Corrollary. Assume that the set M_1 is strongly stable, the support function $C(x, \psi)$ is concave in x and the solution $x(t)$ does the transfer from the set M_0 into the set M_1 on interval $[0, T]$ and it satisfies maximum principle on this interval. Then the solution $x(t)$ is optimal.

Proof of this corrollary coincides with that of the theorem 2 but we have to use the result of theorem 1 instead of strong transversality condition (11).

In case sets M_0 and M_1 are the points and $F(x)$ is convex for all $x \in E^n$, this corrollary was proved in the author's paper (5).

Thus, to solve the time-optimal control problem, in case set M_1 is strongly stable and the support function $C(x, \psi)$ is concave in x, it is sufficient to find at least one solution of the equation (1) which satisfies maximum principle. Note, the solution may not be the single one. In case, we have some solution $x(t)$, $0 \leq t \leq T$, and want to know wether it is optimal or not, it is sufficient to verify all the conditions of theorem 2. Note, we have to verify the condition of concavity of the support function $C(x, \psi)$ only at the points $x(t)$ in the directions $\psi(t)$.

3. Examples

Now we consider "classical control process". It behaviour is described by the system of differential equations $x = f(x,u)$, $u \in U$. Then the support function is

$$C(x, \psi) = \max_{u \in U} (f(x,u), \psi).$$

The sufficient condition of optimality similar to that in the above corrollary for the liner control processes

$$\dot{x} = Ax + v, \quad v \in V$$

was proved in paper (6) taking into the account the additional assumption of convexity of sets M_0, M_1 and V. In paper (7) the condition of strong stability of the set M_1 was loosed up to strong transversality condition on the set M_1.

Example 1. Assume that the behaviour of the control system is described by differential equation of order n

$$(13) \qquad x^{(n)} = f(x, \dot{x}, \ldots, x^{(n-1)}, u),$$

where vector u belongs to the set $U(y)$ in space E , depending on $y = (x, \dot{x}, \ldots, x^{(n-1)})$. The set M_1 consists of one point x = 0. Suppose, that following conditions are satisfied:

1) functions $f_1(y) = \min_{u \in U(y)} f(y,u)$ and $f_2(y) = \max_{u \in U(y)} f(y,u)$

are continuously differentiable;

2) function $f_1(y)$ is convex and function $f_2(y)$ is concave;

3) point x = 0 is the interior point of the set $f(0,U(0))$.

Then the support function

$$C(y, \psi) = y_2\psi_1 + \ldots + y_n \psi_{n-1} + f_1(y) \frac{\psi_n - |\psi_n|}{2} + f_2(y) \frac{\psi_n + |\psi_n|}{2}$$

is concave in y . It was shown in paper (1) that the proccess (13) is locally controllable in small at the point x = 0 in the assumption 3). Thus, maximum principle is sufficient condition of optimality for proccess (13) in the assumption 1) - 3).

The oscillating objects which behaviour is described by differential equation of order 2 were considered in paper (8). Some conditions were received by the method of regular synthesis under which maximum principle is sufficient condition of optimality. It is easy to varify that oscillating objects satisfy the above assumption

1) - 3).

Thus, optimality of trojectories can be received directly from the above corrollary without regular synthesis. We can received more general results by the given method. For example, the condition of trajectories reaching switching lines under non-zero angles can be omitted. The support function $C(x,\psi)$ was concave in x in the example 1 and we made use of corrollary to show that the solutions, satisfying maximum principle, are optimal. The support function is not concave in the following example and sets M_0, M_1 and $F(x)$ are not convex but the given solutions satisfy all conditions of theorem 2, and thus, are optimal.

Example 2. Consider the control system

$$(14)\qquad \begin{aligned} \dot{x}_1 &= x_2 \\ \dot{x}_2 &= \frac{3 - x_1 e^{-x_1^2}}{2} + \frac{3 + x_1 e^{-x_1^2}}{2}\, u \\ u &= \pm 1 \end{aligned}$$

Set M_1 consists of two points $(\frac{5}{2}, 3)$ and $(\frac{7}{2}, 3)$, and set M_0 consists of two sets $\{x_2 = -3,\ x_1 \geqslant 0\}$ and $\{x_1 = 0,\ x_2 \leqslant -3\}$. Sets M_0, M_1 and $F(x)$ are not convex and set M_1 is not strongly stable for the given system. The support function

$$C(x,\psi) = x_2\psi_1 + 3\,\frac{\psi_2 + |\psi_2|}{2} - x_1 e^{-x_1^2}\,\frac{\psi_2 - |\psi_2|}{2}$$

is continuously differentiable in x, but is not concave in case of definition (9). The adjoint system is

$$\begin{aligned} \dot{\psi}_1 &= (1 - 2x_1^2)\, e^{-x_1^2}\, \frac{\psi_2 - |\psi_2|}{2} \\ \dot{\psi}_2 &= -\psi_1 \end{aligned}$$

Two solutions $x^1(t) = (\frac{3}{2}t^2 - 3t + \frac{5}{2},\ 3t - 3)$, $x^2(t) = (\frac{3}{2}t^2 - 3t + \frac{7}{2},\ 3t - 3)$ and the solution $\psi(t) \equiv 1$ of the adjoint system satisfy all conditions of the theorem 2 and both of them are optimal.

References

1. Blagodatskih, V.I., On local controllability of differential inclution (Russian), Differenc. Uravn. 9, No. 2, (1973) 361-362.
2. Filippov, A.F., Classical solutions of differential equations with multivalued right-hand sides (English trans.), SIAM Control, 5 (1967), 609-621.
3. Hermes, H., The generalized differential equation $\dot{x} \in R(t,x)$, Adv. in Math., 4, No. 2, (1970) 149-169.
4. Ponstein, J., Seven kinds of convexity, SIAM Review, 9, No. 1, (1967) 115-119.
5. Blagodatskih, V.I., Sufficient condition of optimality (Russian), Differens. Uravn., 9, No. 3, (1973),416-422.
6. Boltyanskii, V.G., Linear problem of optimal control (Russian), Differenc. Uravn., 5, No. 3, (1969) 783-799.
7. Dajovich, S., On optimal control theory in linear systems (Russian), Differenc. Uravn., 8, No. 9, (1972) 1687-1690.
8. Boltyanskii, V.G., Mathematical methods of optimal control, Moscow, (1969).

VARIATIONAL APPROXIMATIONS OF SOME OPTIMAL CONTROL PROBLEMS

TULLIO ZOLEZZI

Centro di Matematica e di Fisica Teorica del C.N.R.-Genova

1. Necessary and sufficient conditions are investigated such that the optimal states and controls, and the value of a general optimal control problem depend in a continuous way on the data.

Let us consider the sequence of problems $P_n, n \geq 0$, given by (u control, x state)

$$\min \left\{ \int_{t_{1n}}^{t_{2n}} f_n(t,x,u)\,dt + h_n[x(t_{1n}), x(t_{2n})] \right\},$$

$$\dot{x} = g_n(t,x,u) \text{ a.e. in } (t_{1n}, t_{2n}),$$

with constraints

$$(t_{1n}, x(t_{1n}), t_{2n}, x(t_{2n})) \in B_n, \quad (t, x(t)) \in A_n \text{ if } t \in [t_{1n}, t_{2n}],$$
$$u(t) \in V_n(t, x(t)) \text{ a.e. in } (t_{1n}, t_{2n}), \quad \|u\|_{L^p} \leq e_n.$$

P_o is given, and P_n, $n \geq 1$, is to be considered as "variational perturbation" of P_o. Assume that there exist optimal u_n, x_n for $P_n, n \geq 1$. <u>Variational convergence</u> of $\{P_n\}$ to P_o means the following : there exist optimal u_o, x_o for P_o such that (perhaps for some subsequence)

$$(u_n, x_n) \longrightarrow (u_o, x_o) \quad \text{(in some sense)},$$
$$\min P_n \longrightarrow \min P_o .$$

This means (a) existence of optimal controls for P_o ;
(b) "variational stability" of P_o if $P_n \to P_o$ variationally for "many" sequences P_n (depending obviously on convergence of $\{(u_n, x_n)\}$).

2. <u>Sufficients conditions for variational convergence</u>.

In the above generality, we get

$$\min P_n \longrightarrow \min P_o, \quad x_n \longrightarrow x_o \text{ uniformly}$$

(and, generally speaking, no "usual" convergence is obtained about u_n) under general conditions on A_n, B_n, V_n, e_n, and not very strong assumptions about f_n, g_n.

Moreover

g_n linear in u implies $u_n \rightharpoonup u_o$ in L^p.

Such convergence can fail for variable end time problems (simple examples show $\min P_n \not\to \min P_o$ for time - optimal problems with uniform convergence of data). The general case is considered in Zolezzi (3).

Assume now that t_{1n}, t_{2n} are fixed (for every n), and

$g_n(t,x,u) = a_n(t)x + b_n(t)\, u + c_n(t)$,

$f_n = 0$, h_n depending on $x(t_{2n})$ only,

$V_n(t,x)$ a compact polytope independent on (t,x) .

Moreover suppose that $\liminf h_n(y_n) \geq h_o(y_o)$ if $y_n \to y_o$, and for every z_o there exists z_n such that $\limsup h_n(z_n) \leq h_o(z_o)$. Then

$(a_n, c_n) \rightharpoonup (a_o, c_o)$, $b_n \to b_o$ in L^1 implies (for some optimal $u_n, n \geq 0$)

$\min P_n \to \min P_o$, $x_n \to x_o$ uniformly, $u_n \to u_o$ in every L^p, $p < \infty$.

The same conclusions hold with $b_n \rightharpoonup b_o$ in L^1 only, if either

(a) u is scalar, b_n is piecewise continuous and $b_n(t) = 0$ for no more than r points (r independent on n) ; also if g is non-linear in u, with a monotonicity assumption on $g(t,x,\cdot)$,

or

(b) $u \in R^s, s \geq 1$, and the following regularity assumption holds: for every $p,q \in \operatorname{extr} V_n$, any orthonormal basis $(y_1,\dots,y_m)$, $\phi_n^{-1}(t)b_n(t)(p-q)y_j = 0$ for no more than r points or intervals, r independent on n (ϕ_n beeing the principal matrix of $\dot{x} = a_n\, x$).

Moreover u_o is piecewise constant, and $u_n \to u_o$ uniformly on continuity intervals of u_o.

Same results hold if we minimize either

$$\int_{t_{1n}}^{t_{2n}} \left[a_n^* \, x + b_n^*(t,u(t)) \right] dt \quad ,$$

or
$$\int_{t_{1n}}^{t_{2n}} (f^*(t,x) + \alpha |u|^p) dt, \quad p > 1, \quad \alpha > 0 .$$

If P_o has uniqueness, same results hold when we minimize

$$\int_{t_{1n}}^{t_{2n}} f_n(t,x,u)\, dt$$

assuming that convergence of $\{f_n\}$ is "coercive", that is

$\liminf f_n(t,x,u) \geq f_o(t,x,u) + \varphi(|u-v|^p)$, $p > 1$,

φ convex and strictly increasing.

Clearly applications can be made to variational stability of classical problems of calculus of variations.

About the above results, see Zolezzi (4).

3. Necessary conditions for variational convergence.

Take

$$g_n(t,x,u) = a(t)x + b_n(t)u + c(t) ,$$

so perturbing now only b, and minimize

$$\min |y - x(T)| , t_{2n} = T \text{ fixed} , y \text{ given}$$

(minimum final distance problem). Then

$b_n \to b_o$ in L^1 is a necessary condition for strong convergence of optimal controls for y in some restricted region, when either P_o is (completely) controllable, or the regularity assumption holds and uniform convergence of optimal controls $u_n \to u_o$ does not destroy optimality of u_o.

Moreover

$b_n \to b_o$ in L^1 is a necessary condition to strong convergence of optimal controls minimizing

$$|y - x(T)| + z\, x(T) , \quad y \text{ and } z \text{ given} .$$

About such results see Zolezzi (5) .

4. Among the (few) result on such problems (applications of which can be found in many fields connected with optimization, for example perturbation, sensitivity problems) see results of Cullum (1), Kirillova (2).

Such known results are generalized and substantially extended in this work.

All the above mentioned results can be shown to be a by-product of a general method, called "variational convergence" by the present author, generalizing the classical direct method of the calculus of variations, and useful to obtain, for general minimum problems, both existence and "stability" under perturbations (from a variational point of view). See Zolezzi (6) about some abstract results on this subject.

References

(1) CULLUM, J. Perturbations and approximations of continuous optimal control problems.
Math.Theory of Control, edited by Balakrishnqn-Neustadt.
Academic Press, 1967.

(2) KIRILLOVA, F.M. On the correctness of the formulation of optimal control problems.
S.I.A.M.J. Control 1, 36-58(1963).

(3) ZOLEZZI, T. Su alcuni problemi debolmente ben posti di controllo ottimo.
RIC. di MAT. 21, 184-203(1972).

(4) ZOLEZZI, T. Su alcuni problemi fortemente ben posti di controllo ottimo.
To appear in ANN.MAT.PURA APPL.

(5) ZOLEZZI, T. Condizioni necessarie di stabilità variazionale per il problema lineare del minimo scarto finale.
To appear in B.U.M.I.

(6) ZOLEZZI, T. On convergence of minima.
To appear in B.U.M.I.

NORM PERTURBATION OF SUPREMUM PROBLEMS (*)

J. BARANGER , Institut de Mathématiques Appliquées, B.P. 53
38041 GRENOBLE Cédex FRANCE.

ABSTRACT

Let E be a normed linear space, S a closed bounded subset of E and J an u.s.c. (for the norm topology) and bounded above mapping of S into $\mathbb{R}$.

It is well known that in general there exists no $\bar{s} \in S$ such that

$$J(\bar{s}) = \operatorname*{Sup}_{s \in S} J(s)$$

(even if S is weakly compact).

For $J(s) = \|x-s\|$ (with x given in E), Edelstein, Asplund and Zisler have shown, under various hypotheses on E and S, that the set

$$(s) = \{x \in E \mid \exists\ \bar{s} \in S \text{ such that } \|\bar{s}-x\| = \operatorname*{Sup}_{x \in S} \|s-x\|\}$$

is dense in E.

Here we give analogous results for the problem

$$\operatorname*{Sup}_{s \in S} (J(s)+\|s-x\|)$$

These results generalize those of Asplund and Zisler and allow us to obtain existence theorems for perturbed problems in optimal control.

1. THE PROBLEM.

Let E be a normed linear space, S a closed and bounded subset of E and J an u.s.c. (for the norm topology) and bounded above mapping of S into $\mathbb{R}$.

We are looking for an $\bar{s} \in S$ such that

(1) $$J(\bar{s}) = \operatorname*{Sup}_{s \in S} J(s)$$

A particular (and famous) case of problem 1 is the problem of farthest points (i.e. $J(s) = \|x-s\|$, where x is given in E).

(*) This work is part of a thesis submitted at Université de Grenoble in 1973.

1.1. Problem (1) has no solution in general (even with S weakly compact and $J(s) = \|s-x\|$ with x given in E). Here is a counter example :

Let E be a separable Hilbert space with basis e_i, $i \in \mathbb{N}$

$S = \{e_i, i \in \mathbb{N}\} \cup \{0\}$ is weakly compact.

For any $x \in E$, we have :

$$\|x-e_i\|^2 = 1 + \|x\|^2 - 2(x,e_i)$$

Now suppose that $(x,e_i) > 0$, $\forall i \in \mathbb{N}$; then, we have :

$$\sup_{s \in S} \|x-s\| = \sqrt{1+\|x\|^2}$$, and this supremum is never attained.

1.2. Existence results for the problem of farthest points.

As we have just seen in 1.1., this problem has no solution in general ; however, Asplund [2], generalizing a result of Edelstein [1] — who himself generalized a result of Asplund [1] — has obtained the following :

Theorem (Asplund)

Let E be a reflexive locally uniformly rotund Banach space and S a bounded and norm closed subset of E. Then the subset of the $x \in E$ having farthest points in S is fat (*) in E.

2. THE PERTURBED PROBLEM.

As it is impossible to assert that problem (1) has a solution, we consider the perturbed problem.

Does there exist a $\bar{s} \in S$ such that

$$J(\bar{s}) + \|x-\bar{s}\| = \sup_{s \in S} (J(s) + \|x-s\|) \tag{2}$$

where

(3) S is a bounded and (norm) closed subset of the normed linear space E, J is an u.s.c. and bounded above mapping of S into $\mathbb{R}$, and x is given in E.

We shall call an $\bar{s} \in S$ verifying (2) a J farthest point (in short a JFP).

We have the following generalization of Asplund's result :

Theorem 1

Let E be a locally uniformly rotund and reflexive Banach space ; then under hypothesis (3) the subset of the $x \in E$ admitting a JFP in S is a fat subset of E.

Proof : The function $r(x) = \sup_{s \in S} (J(s)+\|x-s\|)$ is convex, lipschitzian with constant 1 and satisfies :

(*) A fat subset in E is a subset of E which contains the intersection of a countable family of open and dense subsets of E. By the Baire category theorem such a set is itself dense in E.

$$\sup_{x \in B(y,b)\ (*)} r(x) = \sup_x \sup_s (J(s)+\|x-s\|) = \sup_s [J(s) + \sup_x \|x-s\|]$$
$$= r(y)+b$$

Then Corollary of lemma 3 of Asplund [2] asserts that there exists a fat subset G of E such that for every $y \in G$ all $p \in \partial r(y)$ (**) have a norm equal to one. Take such a $p \in \partial r(y)$, $\|p\| = 1$. We have : $r(x) \geq r(y) + \langle p,x-y \rangle$, $\forall x \in E$. Therefore

(4) $$r(2y-x) \geq r(y) + \langle p,y-x \rangle, \ \forall x \in E.$$

E being reflexive there exists an $x \in B(y,r(y))$ (we can always suppose $r(y) > 0$) such that :

$$\langle p,x-y \rangle = -r(y).$$

Then (4) implies $r(2y-x) \geq 2r(y)$. The converse is trivial ; so we have

$$r(2y-x) = 2r(y).$$

Hence, for every $n \in \mathbb{N}$, there exists $s_n \in S$ such that :

(5) $$\|2y-x-s_n\| + J(s_n) \geq 2r(y) - \delta\left(\frac{1}{n}, \frac{y-x}{r(y)}\right)$$

where $\delta(\varepsilon,t)$ is the modulus of local uniform rotundity (***)

Set $$u_n = \frac{s_n+x-2y}{\|s_n+x-2y\|} \quad \text{if } \|s_n+x-2y\| \neq 0$$

$$= \frac{x-y}{\|x-y\|} \quad \text{elsewhere}$$

and

(6) $$t_n = s_n+u_n J(s_n) .$$

We have $t_n+x-2y = s_n+x-2y+u_n J(s_n)$

$$= u_n(\|s_n+x-2y\| + J(s_n))$$

so first

$$\|t_n+x-2y\| = |J(s_n) + \|s_n+x-2y\||$$

$= J(s_n) + \|s_n+x-2y\|$ this quantity being positive for n sufficiently large by (5),

and second

$$u_n = \frac{t_n+x-2y}{\|t_n-x-2y\|}$$

(*) $B(y,b)$ is the ball $\{x, \|x-y\| \leq b\}$

(**) $\partial r(y)$ is the sub-differential of r it y

(***) for $\|t\| = 1$ $\delta(\varepsilon,t) = \text{Inf}\ \{1 - \frac{1}{2}\|t+u\|, \|u\| = 1, \|u-t\| \geq \varepsilon\}$

Hence (5) gives

$$\|2y-x-t_n\| \geq 2r(y) - \delta(\tfrac{1}{n}, \tfrac{y-x}{r(y)})$$

and this implies

$$\|t_n - x\| \leq \tfrac{1}{n} r(y).$$

Thus, t_n converges towards x and u_n towards $\frac{x-y}{\|x-y\|} = u$.

Finally there exists a sequence s_{n_q} such that $\overline{\lim_n} J(s_n) = \lim_q J(s_{n_q}) = \theta$.

Taking the limit in q in (6) we see that s_{n_q} converge (for the norm topology) towards $s = x - u\theta$. Then

$$\begin{aligned} \|s-y\| + J(s) &= \|x-u\theta-y\| + J(s) \\ &= |\|x-y\|-\theta| + J(s) : \\ &\geq \|x-y\| + J(s) - \theta \\ &\geq \|x-y\| = r(y), \text{ because of the u.s.c. of J.} \end{aligned}$$

3. APPLICATION IN OPTIMAL CONTROL THEORY .

We shall limit ourselves to just one example.

The state equation is :

$$(7) \quad \begin{cases} -\nabla(u\nabla y) = f , & f \in L^2(\Omega) \\ y \in H_0^1(\Omega) \end{cases}$$

where Ω is open in $\mathbb{R}^n$.

A z_d being given in $L^2(\Omega)$ (or $H_0^1(\Omega)$), we are looking for a $\tilde{u} \in \mathcal{U}_{ad}$ such that

$$(8) \quad \|y(\tilde{u})-z_d\|_{L^2(\text{or } H_0^1)} = \operatorname*{Inf}_{u \in \mathcal{U}_{ad}} \|y(u)-z_d\|$$

where $\mathcal{U}_{ad}$ is a closed subset of

$$= \{u \in L^p(\Omega) \; ; \; 0 < \alpha \leq u(x) \leq \beta \text{ a.e}\}$$

We take $0 < p < 1$ in order to ensure the local uniform rotundity of $L^p(\Omega)$.
Notice that neither $\mathcal{U}_{ad}$ nor the mapping $u \to y(u)$ are convex.
It is impossible to apply the theorem of Asplund to problem (8), the hypothesis being too weak. But we can apply theorem 1 to obtain :

Theorem 2

For every $\varepsilon > 0$ the subset of the $w \in L^p(\Omega)$ such that there exists $\tilde{u} \in \mathcal{U}_{ad}$ satisfying

$$\|y(\tilde{u})-z_d\|_{L^2(\text{or } H_0^1)} - \varepsilon\|\tilde{u}-w\|_{L^p} = \operatorname{Inf} \|y(u)-z_d\|_{L^2(\text{or } H_0^1)} - \varepsilon\|u-w\|_{L^p}$$

is fat in $L^p(\Omega)$.

Proof : We apply theorem 1 with

$$J(u) = -\tfrac{1}{\varepsilon}\|y(u)-z_d\|$$

It remains only to show that J is u.s.c. In fact we have :

Lemma 1

$u \to y(u)$ is a continuous mapping from $L^p(\Omega)$ into $H_0^1(\Omega)$ (these two spaces being endowed with their norm topology).

Lemma 1 is a consequence of :

Lemma 2

$u \to y(u)$ is a continuous mapping from $L^p(\Omega)$ with its norm topology into $H_0^1(\Omega)$ with the weak topology.

Proof of lemma 2 : Let $u_m \in L^p$ converging towards $u \in L^p$ in norm. Put $y_m = y(u_m)$. The variational form of (7) gives

$$\int_\Omega u_m(\nabla y_m)^2 = \int_\Omega f\, y_m$$

Hence, by Poincarré's inequality

$$\alpha \int (\nabla y_m)^2 \leq \|f\|_{L^2} \|y_m\|_{H_0^1} \quad .$$

Therefore there exists a subsequence y_{m_j} which converges weakly in H_0^1 towards a y.

Let us now consider the variational form of (7)

$$\text{(9)} \qquad \int u_{m_j} \nabla y_{m_j} \nabla z = \int fz \qquad \forall z \in H_0^1$$

There exists a subsequence $u_{m_{j_k}}$ which converges towards u a.e. and

$$|u_{m_{j_k}}(x)-u(x)||\nabla z(x)| \leq 2\beta |\nabla z(x)| \in L^2(\Omega) \; .$$

Then, Lebesgue's theorem implies that $u_{m_{j_k}} \nabla z$ converges (for the norm topology in L^2) towards $u\nabla z$. We can now take the limit in (9) and obtain :

$$\int u \nabla y \nabla z = \int fz \qquad \forall z \in H_0^1 \quad ,$$

so $y = y(u)$. It is then trivial to obtain that the whole sequence y_m converges to y.

Proof of lemma 1 : Consider :

$$X_m = \int u_m(\nabla y_m - \nabla y)^2 \geq \alpha \int (\nabla y_m - \nabla y)^2 \geq 0 \; .$$

We shall show that X_m converges towards zero

$$X_m = \int u_m(\nabla y_m)^2 - 2\int u_m \nabla y_m \nabla z + \int u_m (\nabla y)^2 \; .$$

As the canonical injection from H_0^1 into L^2 is compact

$$\int u_m(\nabla y_m)^2 = \int f\, y_m \text{ converges towards } \int fy = \int u(\nabla y)^2 \quad .$$

$-2\int u_m \nabla y_m \nabla z$ converges towards $-2\int u \nabla y \nabla z$ (as we have seen in the proof of lemma 1).

Another application of Lebesgue's theorem shows that $\int u_m(\nabla y)^2$ converges towards $\int u(\nabla y)^2$.

4. OTHER RESULTS.

Using Asplund's techniques, Zisler [1] has obtained three theorems which can be generalized as follows.

Theorem 3.

Let E be a Banach space whose dual is a locally uniformly rotund and strongly differentiable space (SDS) (*), S a closed and bounded subset of E^* and J an u.s.c. and bounded above mapping of S into $\mathbb{R}$.

Then the subset of the $x \in E^*$ having a JFP in S is fat in E^*.

Theorem 4.

Let E be a weakly uniformly rotund (‡) Banach space, S a weakly compact subset of E and J as in theorem 3. Then the subset of the $x \in E$ having a JFP in S is fat in E.

Theorem 5.

Let E be a reflexive, Frechet differentiable Banach space, S a weakly compact subset of E′ and J as in theorem 3. The same conclusion as in theorem 3 is valid.

Proof : there is no difficulty to adapt the proofs given by Zisler using the same device as in theorem 2. We give here a proof of theorem 5 based on a different method than Zisler's.

X being reflexive there exists an x such that

$$\|x-y\| = r(y) \text{ and } r(2y-x) = 2r(y).$$

Hence there exists $s_n \in S$ such that :

$$\|2y-x-s_n\| \geq 2r(y) - \varepsilon_n \qquad (\varepsilon_n \to 0 \text{ when } n \to \infty)$$

There also exists $f_n \in X^*$ such that $\|f_n\| = 1$ and $f_n(2y-x-s_n) = \|2y-x-s_n\|$

then

$$r(y) \geq f_n(y-x) = f_n(2y-x-s_n) - f_n(y-s_n)$$

$$\geq 2r(y) - \varepsilon_n - \|y-s_n\| \geq r(y) - \varepsilon_n$$

So $f_n(\frac{y-x}{r(y)})$ converges towards $1 = \frac{\|y-x\|}{r(y)}$.

(*) An SDS is a Banach space in which every convex continuous function is strongly differentiable on a G_δ dense in is domain. (See [3] for more details).

(‡) A Banach space is weakly uniformly rotund if $\|x_n\| \to 1$, $\|y_n\| \to 1$, $\|x_n+y_n\| \to 2$ imply x_n-y_n tends weakly to zero.

A theorem of Šmulian [1] now states that f_n converges (for the norm topology in X^*) towards an f with $\|f\| = 1$. Moreover

$$r(y) \geq f(y-s_n) = f(2y-x-s_n) - f(y-x) \text{ so } \lim_{n\to\infty} f(y-s_n) = r(y).$$

S being weakly compact there exists a subsequence of s_n converging towards an $s \in S$. Such an s satisfies

$$\|s-y\| \geq f(y-s) = \lim_{n} f(y-s_n) = r(y)$$

the theorem is then proved for $J = 0$. The device used in theorem 2 gives now the general case.

Remark.

One may look for other perturbations than $\|x-s\|$; the case $\|x-s\|^2$ hase been solved by Asplund [4] when E is a Hilbert space. We have obtained, in collaboration with Temam [1], results for perturbation of the form $\omega(\|x-s\|)$ where ω is a positive, convex, increasing function such that $\lim_{u\to\infty} \omega(u) = \infty$. (E is supposed to be a reflexive Banach space having the property :

(H) If a sequence x_n converges weakly towards x and $\|x_n\|$ converges towards $\|x\|$, then $\|x_n-x\| \to 0$).

BIBLIOGRAPHY

ASPLUND, E. [1]. The potential of projections in Hilbert space (quoted in Edelstein [1]).

ASPLUND, E. [2]. Farthest point of sets in reflexive locally uniformly rotund Banach space. Israel J. of Maths 4 (1966) p 213-216.

ASPLUND, E. [3]. Frechet differentiability of convex functions. Acta Math 421 (1968) p 31-47.

ASPLUND, E. [4]. Topics in the theory of convex functions. Proceedings of NATO, Venice, june 1968. Aldo Ghizetti editor, Edizioni Oderisi.

BARANGER, J. [1]. Existence de solution pour des problèmes d'optimisation non convexe. C.R.A.S. t 274 p 307.

BARANGER, J. [2]. Quelques résultats en optimisation non convexe. Deuxième partie : Théorèmes d'existence en densité et application au contrôle. Thèse Grenoble 1973.

BARANGER, J., and TEMAM, R. [1]. Non convex optimization problems depending on a parameter. A paraître.

EDELSTEIN, M. [1]. Farthest points of sets in uniformly convex Banach spaces. Israel J. of Math 4 (1966) p 171-176.

SMULIAN, V.L. [1]. Sur la dérivabilité de la norme dans l'espace de Banach. Dokl. Akad. Nauk SSSR (N.S) 27 (1940), p 643-648.

ZISLER, V. [1]. On some extremal problems in Banach spaces.

ON TWO CONJECTURES ABOUT THE CLOSED-LOOP TIME-OPTIMAL CONTROL

Pavol Brunovský
Mathematical Institute Slovak Academy of Sciences
Bratislava, Czechoslovakia

Consider the linear control system

$$\text{(L)} \qquad \dot{x} = Ax + u$$

($x \in R^n$, A constant), with control constraints $u \in U$, where U is a convex compact polytope of dimension $m \leq n$ imbedded in R^n, containing the origin in its relative interior, and the problem of steering the system from a given initial position to the origin in minimum time.

While the theory of the time-optimal control problem for the systems (L) has been satisfactorily developed as far as the structure of the open-loop optimal controls is concerned, this is not the case of their synthesis - the closed-loop time-optimal control.

To synthesize the open-loop controls into a closed-loop controller is formally allways possible as soon as the optimal controls are unique. There are various reasons which make a synthesis desirable - the most important perheaps being that if a system which is under the action of a closed-loop optimal controller is deviated from its optimal trajectory by an instantaneous perturbation, the rest of its trajectory will again be optimal for the new initial position.

If the system (L) is normal, it is well known from the basic optimal control theory that the set $\mathcal{R}$ of initial points $x \in R^n$ for which the optimal control $u_x(t)$, $t \in [0, T(x)]$ ($T(x)$ being the optimal time of steering x to 0) exists is open in R^n, the optimal controls are unique, and they are piecewise constant with values at the vertices of U. Its synthesis v is obtained by $v(x) = u_x(0)$.

It is generally believed that the system under the action of the closed-loop controller v,

$$\text{(CL)} \qquad \dot{x} = Ax + v(x),$$

exhibits the following properties :

(i) its behavior is indeed optimal

(ii) its behavior will not be severely affected by small perturbations.

Formulating conjecture (i) more precisely, it means that the solutions of (CL) coincide with the optimal trajectories of (L); the sense in which (ii) can be understood, follows from Theorems 2 and 3 below.

Due to discontinuity of v, care has to be taken with the definition of solution of (CL). Numerous studies of discontinuous differential equations in the fifties lead to the conclusion that the classical (Carathéodory) solution may not represent well the behavior of the system modeled by such an equation. In particular, it does not characterize the so called sliding (chattering) which may occur along the surfaces of discontinuity. The necessity to modify the definition of solution is clearly seen if one tries to investigate the conjecture (ii).

The best and most universal definition of solution of a discontinuous differential equation is due to Filippov. This, applied to (CL), defines a solution of (CL) as a solution of the multivalued differential equation

$$(CL_0) \qquad \dot{x} \in Ax + V(x)$$

in the usual sense, where

$$V(x) = \bigcap_{\delta > 0} \bigcap_{\mu(N)=0} \text{co cl } v\,(B(x,\delta) \setminus N) ,$$

μ is the Lebesgue measure in R^n and $B(x,\delta)$ is the ball with center x and radius δ.

Thus, in order to justify (i), we have to prove that the Filippov solutions of (CL), i.e. the solutions of (CL_0), are optimal trajectories of (L).

Unfortunately, it turns out that this is not true in general and the following theorem, which settles completely the case n = 2, shows that the systems for which (i) is not true, are not exceptional.

We shall say that the closed-loop control v is "good", if every solution of (CL_0) is an optimal trajectory of (L) and, vice versa, every optimal trajectory of (L) is a solution of (CL_0).

<u>Theorem 1</u>. <u>Let</u> n = 2 <u>and let</u> (L) <u>be normal</u>. <u>Then</u>, v <u>is good if and only if the following conditions are not met</u> :

<u>A has two distinct real eigenvalues and there exists a vertex</u> w <u>of U such that</u> $\{\psi \mid \langle \psi, w\rangle = \max_{u \in U} \langle \psi, u\rangle\}$ <u>contains the eigenvector of</u> $-A'$ <u>corresponding to its larger eigenvalue but not the other</u>

eigenvector of $-A$.

In particular, w is allways good if $m = 1$.

A typical example of bad synthesis is given at the end of the paper : $v(x) = w_1$ on the negative x_1 - semiaxis, which has measure zero, and therefore this value is supressed in the Filippov definition: the solution of (CL_0) from a point $x = (x_1, 0)$ instead of being a solution of $\dot{x} = Ax + w_1$, will slide along the line of discontinuity $x_2 = 0$ with speed $\dot{x}_1 = - x_1 = co\{Ax + w_2, Ax + w_4\} \cap \{x_2 = 0\}$.

Let us now turn to the stability conjecture (ii). We shall model the pertur-bations as measurable functions p(t) satisfying the estimate $|p(t)| \leqslant \varepsilon$ and we shall be interested in statements, valid for any perturbation satisfying the given bound. Those can be conveniently expressed in terms of solutions of the multivalued differential equation

$$(CL_\varepsilon) \qquad \dot{x} \in Ax + V(x) + B(0, \varepsilon)$$

Namely, $\varphi(t)$ is a solution of (CL_ε) if and only if there exists a measurable function $p(t)$, $|p(t)| \leqslant \varepsilon$ such that $\varphi(t)$ is a solution of the equation

$$\dot{x} \in Ax + V(x) + p(t)$$

Of course, one cannot expect positive results in case the synthesis is not good for the unperturbed system. Therefore, we restrict ourselves to $n = 2$ and we shall assume that the system is normal and no vertex of U satisfies the conditions of Theorem 1. Under these assumptions the following is true :

Theorem 2. Let $m = 1$. Then, for any compact $K \subset \mathcal{R}$ and any $\eta > 0$ there exists an $\varepsilon > 0$ such that all solutions of (CL_ε) starting at points $x \in K$ reach the origin in a time not exceeding $T(x) + \eta$.

Theorem 3. Let $m = 1$. Then, given any compact $K \subset \mathcal{R}$ and any $\rho > 0$, there exists an $\varepsilon > 0$ such that any solution of (CL_ε) starting at a point $x \in K$ reaches $B(0, \rho)$ at time not exceeding $T(x)$ and stays in $B(0, \rho)$ afterwards

As far as higher dimensions are concerned, no definite results have yet be obtained. However, there are some reasons to believe that no synthesis is good if $n > 2$ and $m > 1$, while the case $n > 2$, $m = 1$ is unclear.

For the detailed version of the proofs of Theorems 1 - 3 cf. Brunovský.

Example.

$$\dot{x}_1 = -x_1 + u_1$$

$$\dot{x}_2 = x_2 + u_2 ,$$

$U = \{(u_1, u_2) \mid |u_1| + |u_2| \leqslant 1\} = \mathrm{co}\{w_1, w_2, w_3, w_4\}$, where

$w_1 = (1,0)$, $w_2 = (0, 1)$, $w_3 = (-1, 0)$, $w_4 = (0, -1)$,

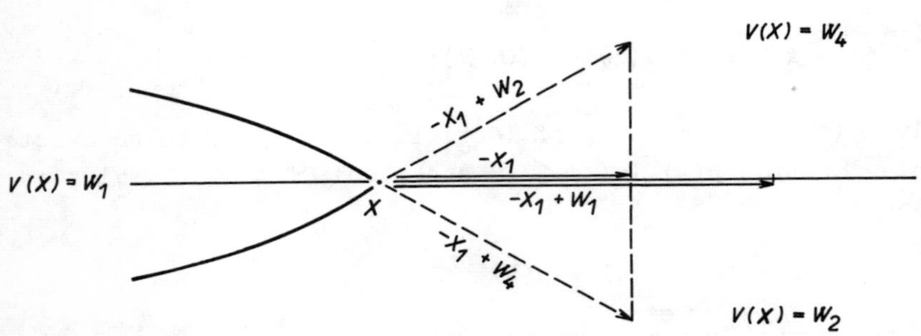

References :

Brunovský, P., The closed-loop time-optimal control, submitted for publication to SIAM Journal on Control

Filippov, A.F. Matematičeskij sbornik 51, 99-128 (1960)

COUPLING OF STATE VARIABLES IN THE OPTIMAL LOW THRUST ORBITAL TRANSFER PROBLEM

Romain HENRION
Aspirant F.N.R.S.
Aerospace Laboratory
University of Liège
BELGIUM

1. INTRODUCTION

The high specific cost of orbiting a satellite explains the importance of the low thrust orbital transfer problem. Indeed, mass and size of such an engine are much smaller than for an impulsive one, but its fuel expenditure can be as close as wanted, if final time is left open. However, the numerous studies undertaken have stressed two main difficulties :

- the possible existence of singular arcs (intermediate thrust arcs, which do not proceed directly from the PONTRYAGIN maximum principle), the optimality of which is not ensured;
- the high sensitivity of optimal trajectories with respect to the unknown initial conditions, a consequence of the two-point-boundary-value-problem introduced by the maximum principle.

We shall show here how these difficulties are reduced to a great extent by decoupling the state variables with repect to the thrust amplitude control.

2. PROBLEM FORMULATION

Using non-dimensional polar variables, the plane trajectory of a low thrust engine in a central inverse-square-force field may be described by the following equations :

$$
\begin{aligned}
\dot{r} &= u_r \\
\dot{\theta} &= \frac{u_\theta}{r} \\
\dot{u}_r &= \frac{u_\theta^2}{r} - \frac{1}{r^2} + \frac{\xi}{\mu} \sin \psi \\
\dot{u}_\theta &= - \frac{u_r u_\theta}{r} + \frac{\xi}{\mu} \cos \psi \\
\dot{\mu} &= - \frac{\xi}{c}
\end{aligned}
\tag{1}
$$

where r , θ , u_r , u_θ are position and velocity variables, μ the instantaneous mass and c the ejection velocity. Denoting by a the acceleration factor, we have the constraint

$$0 \leq \xi \leq a$$

The aim of optimization is to choose the control variables ψ and ξ so as to minimize some cost function, e.g. fuel consumption.

3. PONTRYAGIN MAXIMUM PRINCIPLE

Denoting by λ the adjoint state vector, the Hamiltonian takes the form

$$H = H_o + \xi \, H_1$$

where $H = \lambda' \cdot \dot{q} \, (q \, , \, \xi \, , \, \psi)$

As H is autonomous, we have the well-known equation

$$\dot{H} = H_t = 0$$

which shows that the numerical value of H is a trajectory constant.
The adjoint state variables are governed by the equation

$$\dot{\lambda} = - H_q$$

or explicitly

$$\dot{\lambda}_r = \frac{u_\theta}{r^2} \lambda_\theta + \left(\frac{u_\theta^2}{r^2} - \frac{2}{r^3}\right) \lambda_{u_r} - \frac{u_r \, u_\theta}{r^2} \lambda_{u_\theta}$$

$$\dot{\lambda}_\theta = 0$$

$$\dot{\lambda}_{u_r} = - \lambda_r + \frac{u_\theta}{r} \lambda_{u_\theta}$$

$$\dot{\lambda}_{u_\theta} = - \frac{\lambda_\theta}{r} - 2 \, \frac{u_\theta}{r} \lambda_{u_r} + \frac{u_r}{r} \lambda_{u_\theta}$$

$$\dot{\lambda}_\mu = \frac{\xi}{\mu^2} \left[\sin \psi \cdot \lambda_{u_r} + \cos \psi \cdot \lambda_{u_\theta} \right]$$

Maximizing H according to the maximum principle, we get :

- for ψ $\left.\begin{aligned} \sin \psi &= \lambda_{u_r} / \lambda \\ \cos \psi &= \lambda_{u_\theta} / \lambda \end{aligned}\right\}$ where $\lambda = (\lambda_{u_r}^2 + \lambda_{u_\theta}^2)^{1/2}$

- for ξ $\quad \xi = 0 \quad$ if $H_1 < 0$

$\xi = a \quad$ if $H_1 > 0$

However, if H_1 vanishes for some finite period, the principle no longer fixes ξ for H is no longer an explicit function of ξ . This case corresponds to a so-called singular or intermediate thrust arc. Now, by deriving the condition $H_1 \doteq 0$ several times, it is possible to determine precisely that special value ξ which keeps $H_1 \doteq 0$. But, its optimality is by no means guaranteed. We shall show that by decoupling the present state variables with respect to ξ , we shall be led to a reduced system, governed non-linearly by a new control. Applying then the LEGENDRE-CLEBSCH condition, we easily establish all the known optimality conditions for singular arcs.

4. PRACTICAL IMPORTANCE OF THE DECOUPLING OPERATION

The practical, i.e. numerical, importance of the decoupling operation is shown on figure 1. We represented the flow-diagram starting with ξ , of the polar variables. One easily sees that the coupling of this system is extremely strong, as all the variables influence one another directly. Indeed, there are only two stages.

It is clear that this is very prejudicial to numerical precision and stability : some error in any variable has an immediate consequence on all the other ones. But, the larger the number of stages between two given variables, the lower the speed and magnitude of the influence of an error on the first upon the value of the second. Therefore, our aim is to increase this number of stages. We shall see that by making use of canonical transformations, it can be brought up to four. Results show also a serious decrease in sensitivity to errors in the unknown initial values.

5. CANONICAL TRANSFORMATIONS

Because of the hamiltonian formulation of the problem, the decoupling is best achieved by making use of canonical transformations. They are described by the equation

$$\lambda' \, d q - H \, dt = \Lambda' \, d Q - K \, d T + d F \qquad (2)$$

which tranforms the set (q , t , λ , H) into the set (Q , T , Λ , K).F stands for the corresponding generating function. Except from the last one, all the transformations we shall use are of the MATHIEU type where $F \equiv 0$.

5.1. 1st Transformation

Initial set : r , θ , u_r , u_θ , μ

$$\lambda_r \, , \, \lambda_\theta \, , \, \lambda_{u_r} , \, \lambda_{u_\theta} \, , \, \lambda_\mu$$

Associated differential form : $\lambda_r \, dr + \lambda_\theta \, d\theta + \lambda_{u_r} \, du_r + \lambda_{u_\theta} \, du_\theta + \lambda_\mu \, d\mu - H \, dt$ (3)

The first transformation we shall do, has been given by FRAEIJS de VEUBEKE [1], and is described by the relations optimizing ψ . ψ is now a new state variable :

$$\lambda_{u_r} = \lambda \sin \psi$$
$$\lambda_{u_\theta} = \lambda \cos \psi \tag{4}$$

New differential form : $\lambda_r \, dr + \lambda_\theta \, d\theta + \lambda d \, u_\lambda + \lambda_\psi \, d\psi + \lambda_\mu \, d\mu - H \, dt$ (5)

Substituting (4) into (3) and identifying to (5), we get

$$u_r = u_\lambda \sin \psi - \frac{\lambda_\psi}{\lambda} \cos \psi$$
$$u_\theta = u_\lambda \cos \psi + \frac{\lambda_\psi}{\lambda} \sin \psi \tag{6}$$

Making use of (4) and (6), the new expression of the Hamiltonian is readily derived :

$$H = H_o + \xi \, H_1$$

$$H_o = u_\lambda \left(\sin \psi \, . \, \lambda_r + \cos \psi \frac{\lambda_\theta + \lambda_\psi}{r}\right) + \frac{\lambda_\psi}{\lambda} \left(\frac{\lambda_\theta + \lambda_\psi}{r} \sin \psi - \cos \psi \, \lambda_r\right) - \frac{\sin \psi}{r^2} \lambda$$

$$H_1 = \frac{1}{\mu} \lambda - \frac{1}{c} \lambda_\mu$$

The new equations of state q and costate λ can then be computed from

$$\dot{q} = H_\lambda$$

$$\dot{\lambda} = - H_q$$

(7) now shows that at present, there are left only three variables directly governed by ξ , namely u_λ , μ and λ_μ .
Their equations have the following forms :

$$\dot{u}_\lambda = f(q,t,\lambda) + \frac{\xi}{\mu} \quad ; \quad \dot{\mu} = -\frac{\xi}{c} \quad ; \quad \dot{\lambda}_\mu = \frac{\xi\lambda}{\mu^2}$$

5.2. 2nd Transformation

The aim of this transformation is to eliminate u_λ 's dependancy on ξ . Applying the method inspired from hydrodynamics and used by FRAEIJS de VEUBEKE[2] and KELLEY, KOPP and MOYER[3], we solve the equation

$$\frac{du_\lambda}{\frac{1}{\mu}} = \frac{d\mu}{-\frac{1}{c}}$$

which produces $\quad u_\lambda = - c \ln \mu + w \quad$ (8)

The integration constant w will be used instead of u_λ as a new state variable. The associated differential form is :

$$\lambda_r \, dr + \lambda_\theta \, d\theta + \lambda_w \, dw + \lambda_\psi \, d\psi + \sigma \, d\mu - H \, dt \tag{9}$$

Substituting (8) into (5) and identifying to (9), we get

$$\lambda = \lambda_w$$
$$\lambda_\mu = \sigma + \frac{c}{\mu} \lambda_w \tag{10}$$

The expression of the switching function H_1 is now :

$$H_1 = - \frac{1}{c} \sigma \tag{11}$$

It shows that a present, μ is the only variable still directly controlled by ξ . It may therefore be considered as a new control variable replacing ξ . However, trying to apply the corresponding LEGENDRE-CLEBSCH condition, we see that it is trivially satisfied. We shall therefore continue our decoupling operations by decoupling now the system with respect to μ .

5.3. 3rd Transformation

As for the second transformation, the equations

$$\frac{dr}{\sin\psi} = \frac{rd\theta}{\cos\psi} = \frac{rd\psi}{\cos\psi}$$

lead to the change of variables

$$\begin{aligned} g &= r\cos\psi \\ h &= r\sin\psi \\ \omega &= \theta - \psi \end{aligned} \qquad (12)$$

which implies

$$\lambda_g = \cos\psi \cdot \lambda_r - \frac{\sin\psi}{r}(\lambda_\psi + \lambda_\theta)$$

$$\lambda_h = \sin\psi \cdot \lambda_r + \frac{\cos\psi}{r}(\lambda_\psi + \lambda_\theta)$$

$$\lambda_\omega = \lambda_\theta$$

Figure 2 gives the corresponding flow-diagram. We can draw now the following clues :

1) ξ now governs only μ , whereas h is the only non-ignorable variable governed by μ .

   Moreover h appears non-linearly in all the equations. The application of the LEGENDRE-CLEBSCH condition to h, considered as a new control, will produce a useful condition. This point will be developed in the next paragraph;

2) The coupling of the present system is much weaker than for the polar variables. Indeed, it seems that there are now four stages. However, for numerical integrations, the first stage is eliminated, as μ has an analytical solution. In paragraph 8, we show how the number of four stages can be restored.

6. SINGULAR ARCS

A singular arc being characterized by the vanishing of the switching function H_1 , we get by multiple derivation :

$$H_1 \doteq 0 \qquad \sigma \doteq 0 \qquad (13)$$

$$\dot{H}_1 \doteq 0 \qquad \lambda_h \doteq 0 \qquad (14)$$

$$\ddot{H}_1 = 0 \qquad \frac{\lambda_g^2}{\lambda_w} - \frac{\lambda_w (g^2 - 2h^2)}{(h^2 + g^2)^{5/2}} = 0 \qquad (15)$$

As FRAEIJS de VEUBEKE[1] has shown that $\lambda_w = \lambda \neq 0$, this implies

$$g^2 \geqslant 2\,h^2 \tag{16}$$

which is equal to the well-known relation

$$-\frac{1}{\sqrt{3}} \leqslant \sin\psi \leqslant \frac{1}{\sqrt{3}} \tag{17}$$

(13) and (14) now reduce the Hamiltonian to

$$H = H_o = \frac{\lambda_g \cdot \lambda_\omega}{\lambda_w} - \frac{3\,h^3\,\lambda_w}{(h^2 + g^2)^{5/2}} \tag{18}$$

Figure 2 shows that ξ influences the variables appearing in H only through h. So, as already indicated above, h may be considered as a new control. Indeed, (15) expresses the stationarity of H with respect to h.

$$0 = \frac{\partial H}{\partial h} = \frac{\lambda g^2}{\lambda_w} - \frac{\lambda_w\,(g^2 - 2\,h^2)}{(h^2 + g^2)^{5/2}} \tag{19}$$

Applying the LEGENDRE-CLEBSCH condition :

$$0 > \frac{\partial^2 H}{\partial h^2} = \frac{3\,\lambda_w \cdot h}{(h^2 + g^2)^{7/2}}\left[\,3\,g^2 - 2\,h^2\,\right] \tag{20}$$

By (16), this leads to

$$h \leqslant 0 \tag{21}$$

and in conjunction with (17) :

$$-\frac{1}{\sqrt{3}} \leqslant \sin\psi \leqslant 0 \quad . \tag{22}$$

Now

$$\dddot{H}_1 = 0 \qquad -6\,g\,h\,(g^2 + h^2)\,\lambda_g + \eta\,h\,(3\,g^2 - 2\,h^2)\,\lambda_w + g\,(g^2 - 4\,h^2)\,\lambda_\omega \doteq 0 \tag{23}$$

$$\ddddot{H}_1 = 0 \qquad \frac{\xi}{\mu}\,h\,(3\,g^2 - 2\,h^2)\,\lambda_w^2 - 3\,h\,(3\,g^2 + 2\,h^2)\,\lambda_w \cdot H + 3\,\eta^2\,(g^2 - 2\,h^2)\,\lambda_w^2 + \eta\,g\,(6\,h^2\,\lambda_g - 12\,g^2\,\lambda_g - 2\,h\,\lambda_\omega)\,\lambda_w$$

$$+ (3\,g^2 - 4\,h^2)\,\lambda_\omega^2 + h\,(13\,g^2 - 8\,h^2)\,\lambda_g\,\lambda_\omega$$

$$+ 3(4\,g^4 - 2\,h^4 + 3\,g^2\,h^2)\,\lambda_g^2 = 0 \tag{24}$$

where $\eta \equiv -\,c\,\ell n\,\mu + w + \dfrac{g\,\lambda_g + h\,\lambda_h}{\lambda_w}$

Equation (24) gives the value of ξ . The corresponding arc is physically feasable if $0 \leqslant \xi \leqslant a$.

Special cases

H being autonomous and $\dot{\lambda}_\omega = 0$, the numerical values of H and λ_ω are constant on the whole trajectory.

a) $H = \lambda_\omega = 0$: this case corresponds to free final time and polar angle. One easily shows[1] that necessarily $\xi \doteq 0$. So, in this case there are no singular, intermediate thrust arcs.

b) $H = 0$; $\lambda_\omega \neq 0$ (LAWDEN's spiral)
Eliminating λ_g , λ_ω and w from (24) by (15), (18) and (23), we get

$$\xi = 3\mu\,\frac{h\,(g^2 - 2\,h^2)(9\,g^4 + 4\,h^4 - 7\,g^2\,h^2)}{(g^2 + h^2)(3\,g^2 - 2\,h^2)^3}$$

(16) and (24) then imply $\xi \leqslant 0$, which means that LAWDEN's intermediate thrust arc is not optimal. It therefore follows that optimal arcs can only exist for $H \neq 0$.

With this result, one easily shows then, that (21) and (22) actually reduce to

$$h < 0$$

$$-\frac{1}{\sqrt{3}} \leqslant \sin\psi < 0$$

7. NEW CONSTANT OF MOTION (SINGULAR ARCS)

Making use of (12), (13) and (14), it is straightforward to show that the following expression is constant on a singular arc :

$$C = 3\,H\,t + w\,\lambda_w - 2\,g\,\lambda_g$$

Clearly, it may prove very useful for numerical integrations of singular arcs.

8. TWO ADDITIONAL CANONICAL TRANSFORMATIONS

In order to improve decoupling, two additional transformations are useful.

8.1. 4th Transformation

First, we shall substitute to g, h and w the new variables α , β and v according to the relations

$$\lambda_\alpha = g\,\lambda_w$$

$$\lambda_\beta = h\,\lambda_w$$

$$v = w - \alpha\, g - \beta\, h$$

The associated differential form

$$\lambda_\alpha\, d\alpha + \lambda_\beta\, d\beta + \lambda_v\, dv + \lambda_\omega\, d\omega + \sigma\, d\mu - H\, dt$$

leads to

$$\alpha = -\lambda_g/\lambda_w$$

$$\beta = -\lambda_h/\lambda_w$$

$$\lambda_v = \lambda_w$$

8.2. 5th Transformation

With $\qquad -c\,\ell n\,\mu + v = \eta$

the differential form

$$\lambda_\alpha\, d\alpha + \lambda_\beta\, d\beta + \lambda_\eta\, d\eta + \lambda_\omega\, d\omega + \kappa\, d\mu - H\, dt$$

implies by (10) :

$$\lambda_\eta = \lambda_v$$

$$\kappa = \sigma + \frac{c}{\mu}\lambda_v = \lambda_\mu$$

Putting now

$$\lambda_\eta = e^\rho$$

We are led to $\qquad -\eta\, e^\rho = \lambda_\rho$

and the corresponding differential form with a non-vanishing generating function

$$\lambda_\alpha\, d\alpha + \lambda_\beta\, d\beta + \lambda_\omega\, d\omega + \lambda_\mu\, d\mu + \lambda_\rho\, d\rho - H\, dt - d\lambda_\rho$$

Hamiltonian :

$$H = H_o + \xi H_1$$

$$H_o = \beta \lambda_\rho - 2 \alpha \beta \lambda_\alpha + (\alpha^2 - \beta^2) \lambda_\beta - \alpha \lambda_\omega$$

$$- \lambda_\beta \exp(3\rho)(\lambda_\alpha^2 + \lambda_\beta^2)^{-3/2} \tag{25}$$

$$H_1 = \exp(\rho)/\mu - \lambda_\mu / c$$

State equations :

$$\dot{\alpha} = - 2 \alpha\beta + 3 \exp(3\rho) \lambda_\alpha \lambda_\beta (\lambda_\alpha^2 + \lambda_\beta^2)^{-5/2}$$

$$\dot{\beta} = \alpha^2 - \beta^2 + \exp(3\rho)(2 \lambda_\beta^2 - \lambda_\alpha^2)(\lambda_\alpha^2 + \lambda_\beta^2)^{-5/2}$$

$$\dot{\rho} = \beta$$

$$\dot{\omega} = - \alpha \tag{26}$$

$$\dot{\mu} = - \frac{\xi}{c}$$

$$\dot{\lambda}_\alpha = 2 \beta \lambda_\alpha - 2 \alpha \lambda_\beta + \lambda_\omega$$

$$\dot{\lambda}_\beta = - \lambda_\rho + 2 \alpha \lambda_\alpha + 2 \beta \lambda_\beta$$

$$\dot{\lambda}_\rho = 3 \exp(3\rho) \lambda_\beta (\lambda_\alpha^2 + \lambda_\beta^2)^{-3/2} - \xi \exp(\rho)/\mu$$

$$\dot{\lambda}_\omega = 0$$

$$\dot{\lambda}_\mu = \xi \exp(\rho)/\mu^2$$

Figure 3 displays the corresponding flow-diagram. Even if μ is eliminated, there are now four stages. So, the decoupling of this system is very strong and it should lead to a serious improvement in numerical precision and sensitivity. Especially the last variable ρ should be rather insensitive to a sudden change, i.e. a switch of ξ , as it is separated by four stages from ξ . The following application confirms this conjecture.

9. APPLICATION

In order to check the quality of the system (26), we have used it to describe a fuel-optimal transfer trajectory, with open final time and angle. We have $H = \lambda_\omega = 0$ and consequently, there are no singular arcs, but only thrusting ($\xi = a$) and coasting ($\xi = 0$) arcs. Now, there are only five equations which must necessarily be integrated. Indeed :

- the "orbital transfer of variables", introduced by FRAEIJS de VEUBEKE[1] allows to jump the coasting arcs;
- $\lambda_\omega = 0$;
- ω is ignorable;
- λ_μ is ignorable, provided the switching function H_1 is replaced by $- H_o$;
- adopting $t_o = - \frac{c}{a} \mu_o$, μ may be replaced by $- \frac{a}{c} t$.

At present, with

$$\lambda_1 = e^{-\rho} \lambda_\alpha$$

$$\lambda_2 = e^{-\rho} \lambda_\beta$$

$$\lambda_3 = e^{-\rho} \lambda_\rho$$

it is easily seen that ρ is also ignorable, and that only the five following equations are left :

$$\dot{\alpha} = - 2 \alpha\beta + 3 \lambda_1 \lambda_2 (\lambda_1^2 + \lambda_2^2)^{- 5/2}$$

$$\dot{\beta} = \alpha^2 - \beta^2 + (2 \lambda_2^2 - \lambda_1^2)(\lambda_1^2 + \lambda_2^2)^{- 5/2}$$

$$\dot{\lambda}_1 = \beta \lambda_1 - 2 \alpha \lambda_2$$

$$\dot{\lambda}_2 = \beta \lambda_2 + 2 \alpha \lambda_1 - \lambda_3$$

$$\dot{\lambda}_3 = - \beta \lambda_3 + 3 \lambda_2 (\lambda_1^2 + \lambda_2^2)^{-3/2} + \frac{c}{t}$$

As the two following equations are ignorable, their integration has no effect on precision :

$$\dot{\rho} = \beta$$

$$\dot{\omega} = - \alpha$$

The corresponding flow-diagram is similar to figure 3, if μ and λ_μ are suppressed and if λ_1 , λ_2 and λ_3 are substituted to λ_α , λ_β and λ_ρ : there are four stages.

Numerical results

We adopted $a = 0.03$

$c = 0.3228$

which corresponds to rather a low thrust level. The numerical stability and precision difficulties of this problem are well known :

- Figure 4 displays the behaviour of ρ on one coasting and two thrusting arcs. According to the theoretical forecast , the join of the curves at the junction points where ξ switches, is very smooth and neat. This confirms the high degree of decoupling announced by the flow-diagram;
- By integrating several trajectories forward and then backward, the new system of variables exhibited on increase of precision of at least 2 significant digits with respect to the polar variables;
- The sensitivity with respect to unknown initial values was reduced by 1 and mostly 2 significant digits. As a consequence, convergence was achieved at points where the polar variables did diverge.

Moreover, these improvements increased with the length of the trajectory and the number of switchings.

10. CONCLUSION

The decoupling operation presented in this paper leads to two important results :

- an easy, direct theoretical examination of singular arcs;
- a considerable increase in numerical precision and decrease in sensitivity with respect to initial conditions.

REFERENCES

1. FRAEIJS de VEUBEKE, B.
"Canonical Transformations and the Thrust-Coast-Thrust Optimal Transfer Problem"
Astronautica Acta, 4, 12 , 323-328, (1966)

2. FRAEIJS de VEUBEKE, B.
"Une généralisation du principe du maximum pour les systèmes bang-bang avec limitation du nombre de commutations"
Centre Belge de Recherches Mathématiques. Colloque sur la Théorie Mathématique du Contrôle Optimal. Vander, 55-67, Louvain (1970)

3. KELLEY, H.J., KOPP, R.E. and MOYER, H.G.
"Singular Extremals" in "Topics in Optimization"
(ed. G. LEITMANN), Ac. Press, chap. 3, 63-101, New York (1967)

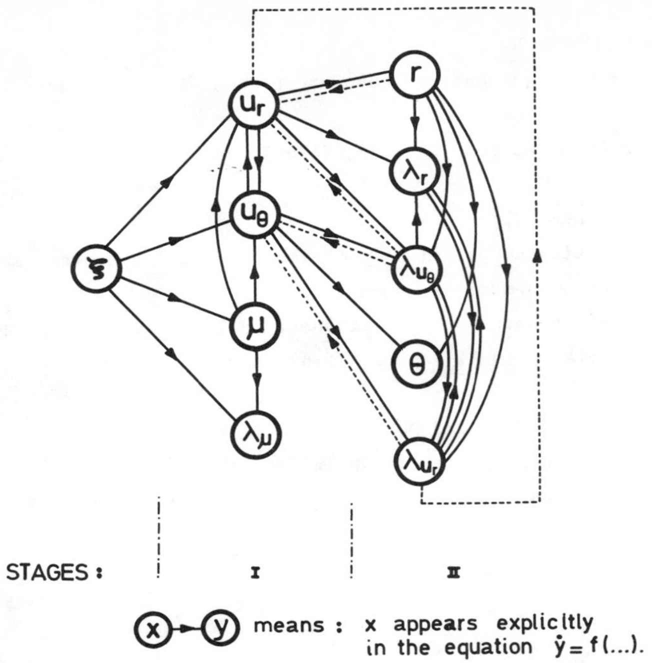

FIGURE 1

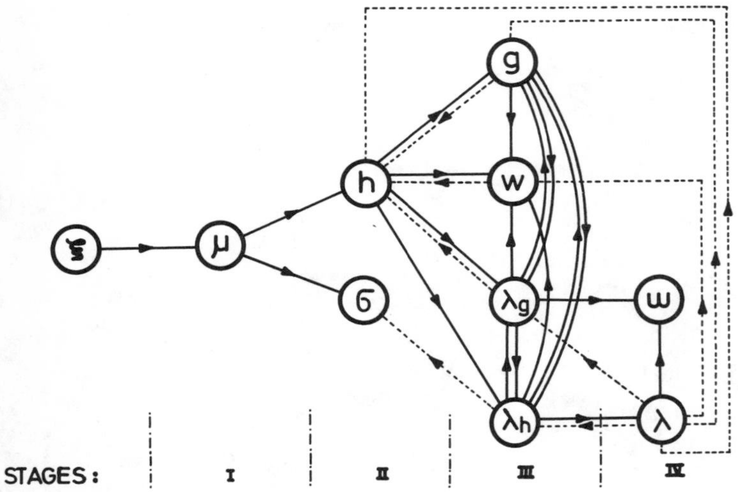

FIGURE 2

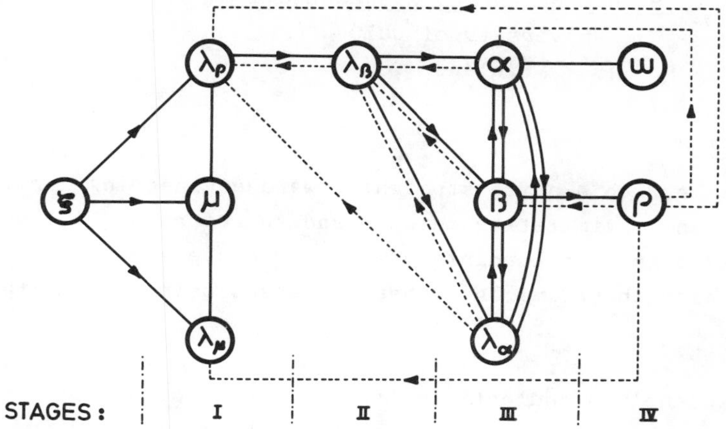

FIGURE 3

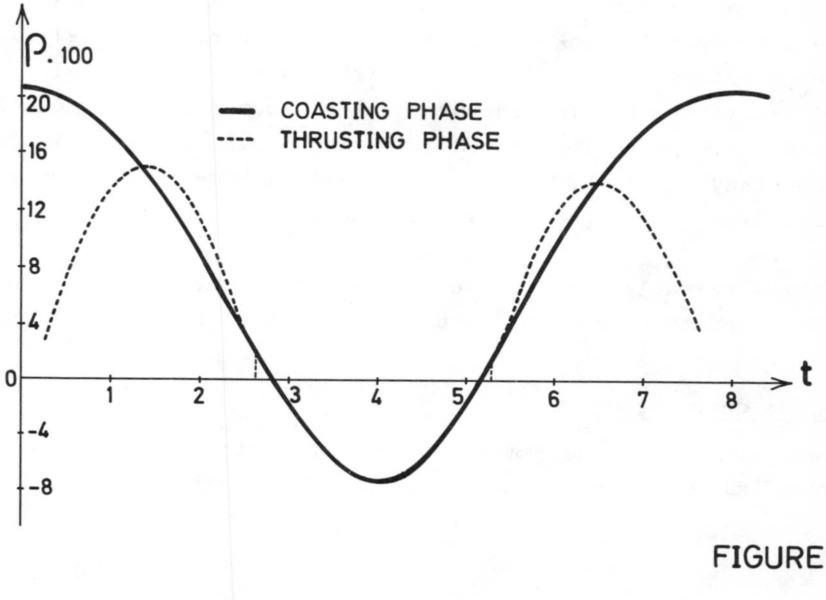

FIGURE 4

OPTIMIZATION OF THE AMMONIA OXIDATION PROCESS USED IN THE MANUFACTURE OF NITRIC ACID

P. Uronen and E. Kiukaanniemi
University of Oulu
Finland

1. Introduction

This work deals with a quite straightforward engineering application of modelling and optimization methods and therefore we will concentrate on the practical points of view, because similar methods may be used also in optimizing other processes using ageing type of catalysts.

To be able to analyse and optimize the operation of a plant, one must first have a mathematical model describing the real behaviour of the process. The modelling of a plant can be done in many different ways; one can use the physical and chemical relationships the process is based on, or one can try to find the model experimentally. So we can have different models for the same process.

The models based on physical and chemical relationships have a meaningful general structure, ie. the right control and state variables will be automatically taken into account. However, the model will include generally so many nonlinear differential and partial differential equations that the use of model without many approximations is impossible. The empirical models normally can be quite simple in form, but they do not have the same technically meaningful properties as models derived using the physical and chemical laws.

Therefore a suitable combination of these both methods may be a good compromise and this semi-empirical method has been used in this work.

2. Process description

The plant studied is the ammonia oxidation process used in the manufacture of nitric acid. Figure 1 shows schematically the whole pro-

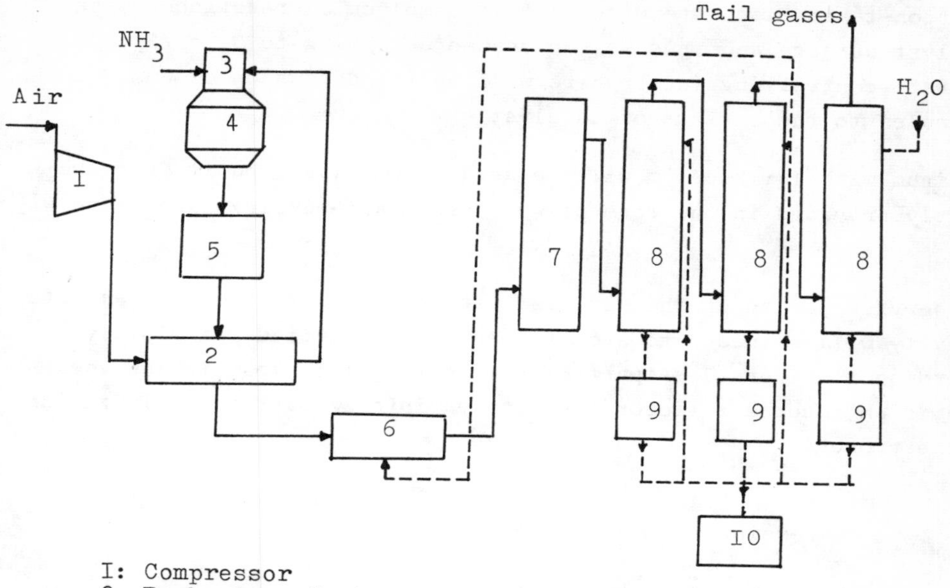

I: Compressor
2: Preheater of air
3: Mixer
4: Burner
5: Waste heat boiler
6: Cooler
7: Oxidation tower
8: Acid coolers
9: Resulting acid

Fig. I. Schematic representation of a nitric acid plant.

cess. The raw materials are air and gaseous ammonia which are fed after mixing (about 10-11 Vol % ammonia in the mixture) to a reactor, where ammonia is catalytically burned to nitric oxide and water.

The next steps are cooling of the reaction products, oxidation of nitric oxide to nitric dioxide and then absorbing NO_2 into water in several countercurrent towers to form nitric acid (55-60%).

The first step, the oxidation of ammonia, is the most critical for the whole process. Various investigators /3/ have shown that the conversion in the oxidation of NO and in absorbtion is very high (97-99%) and stable. Therefore the optimization of the ammonia oxidation process will mainly determine the optimum of the whole unit.

The details of the chemical and kinetic phenomena included in the catalytic oxidation of ammonia are not yet fully known. Probably the reaction takes place stepwise as fast bimolecular reactions on the catalyst surface and thus we can assume according to Oele /2/ that the rate controlling factor here will be the diffusion of ammonia molecules to the surface of catalyst.

Platinum with 10% rhodium will be used as catalyst and it is conveniently provided in the form of wowen sieves (4-6 sieves) of fine wires.

The service period of the catalyst varies from 3 to 5 months and its activity will decrease towards the end of the period principally according to Figure 2. A part of platinum will be lost in the use and this is an important factor to be taken into account in optimization of the process.

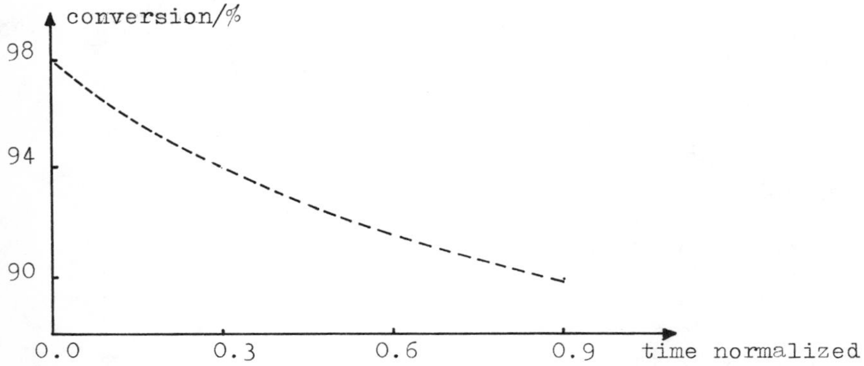

Fig. 2. A decreasing type activity curve of the platinum catalyst.

3. Mathematical model

Based on the physical and chemical phenomena involved in the above process we can qualitatively conclude that the most important factors affecting on the conversion of the oxidation process are: ammonia to air ratio in feed, feed temperature, total gas charge on sieves, plant pressure, number of sieves and sieve dimensions.

In following we will assume that the pressure will remain constant. As mentioned above we can assume that the rate of reaction will be controlled by the chemisorbtion of ammonia moles to the surface of the catalyst.

Oele /2/ has derived the following empirical formula for the mass transfer in turbulent flow

$$(1) \qquad \alpha = \frac{0.924 \cdot \lambda \cdot Re^{0.33}}{P\, c_p\, d}, \text{ where}$$

- α is the mass transfer coefficient
- λ is the thermal conductivity of the gas mixture
- Re is the Reynolds number
- P is the overall pressure
- c_p is the molar spesific heat of the mixture
- d is the diameter of the catalyst wire.

Using now well-known formula for driving force in mass transfer, mass balance and the definition of conversion and Reynolds number we get equation (2):

$$(2) \qquad \ln(1-x) = \frac{\lambda n f(11.82\, m_o - 28.6)}{c_p \mu^{0.33} d^{0.67} G^{0.67}}, \text{ where}$$

- x is the conversion of NH_3 to NO
- f is characteristic factor for sieves representing wire area - sieve area ratio
- n is number of sieves
- μ is the dynamic viscosity of the mixture
- G is the total gas charge per sieve area
- m_o is the ammonia content in feed

Curievici and Ungureanu /1/ have shown that if we know m_o and T_o, the temperature of feed, we can with a reasonable accuracy calculate the mean temperature at the sieves and thus also the values of λ, thermal conductivity, μ, dynamic viscosity and c_p, molar spesific heat as linear functions of m_o and T_o in the normal operation range. So the following model can be derived

$$(3) \qquad \ln(1-x) = \frac{T_o \cdot 10^{-3}(94m_o - 228.7) + 320m_o^2 - 748.1m_o - 70.67}{T_o \cdot 10^{-3}(1.43 + 3.84m_o) + 13.06m_o^2 + 5.03m_o + 6.51} \times$$

$$\frac{n \cdot f}{(27.45 \cdot 10^{-3} \cdot T_o + 93.44m_o + 13.61)^{0.33} (d \cdot 10^6)^{0.67} G^{0.67}}$$

This model can be adapted to a real plant using equation (3) multiplied by a correction factor k which can be determined using the nominal operating point values of the plant.

When this is done for example at three conversion levels representing three points on the ageing curve of the catalyst, we can get the time dependence for the correction factor k. So the ageing of the catalyst can be expressed as a polynomial of the normalized time t (normalized so that 1.0 represents 100 days).

An other possibility for matching the model to plant data would be to keep all numerical coefficients in equation (3) as unknown and if a necessary amount of conversion measurements could be made we could formulate a quadratic error function and the values for the coefficients could then be found by minimizing this error function.

The difficulty here is the reliable measurement of the conversion, so that the ageing curve of the catalyst could be estimated.

4. Simulation

To test the model an extensive examplewise digital simulation study was done. In this study the effect of small changes in process conditions (m_o, G, T_o, n, f) on the conversion was investigated. This was carried out by a program written in FORTRAN IV. Figures 3, 4, 5 and Tables I and II represent the results. The simulation was carried out at one conversion level which corresponds to one age of the catalyst i.e. one point on the ageing curve of the catalyst.

The results of the simulation show good conformity with operational experiences and with previously published results concerning the

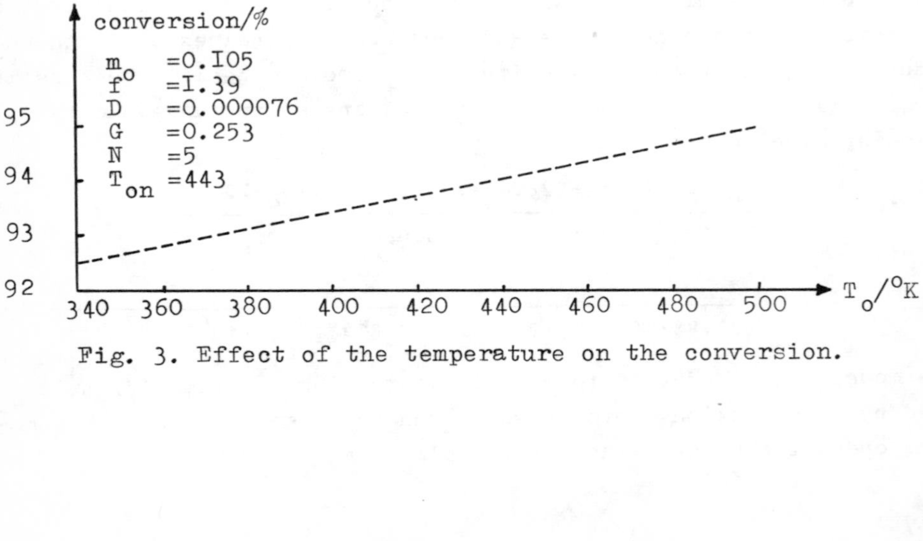

Fig. 3. Effect of the temperature on the conversion.

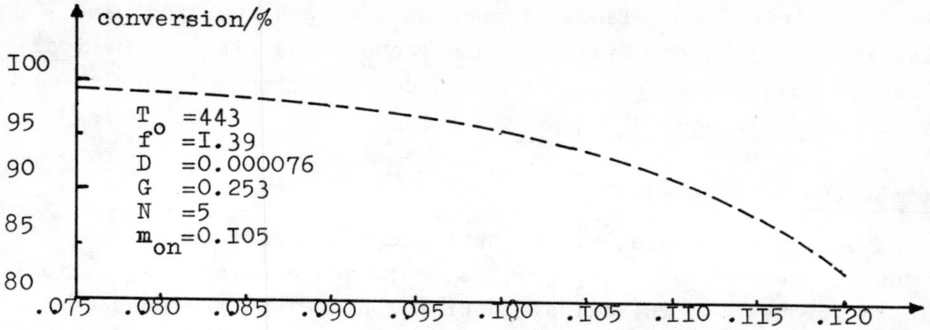

Fig. 4. Effect of the mixture strength in feed on the conversion.

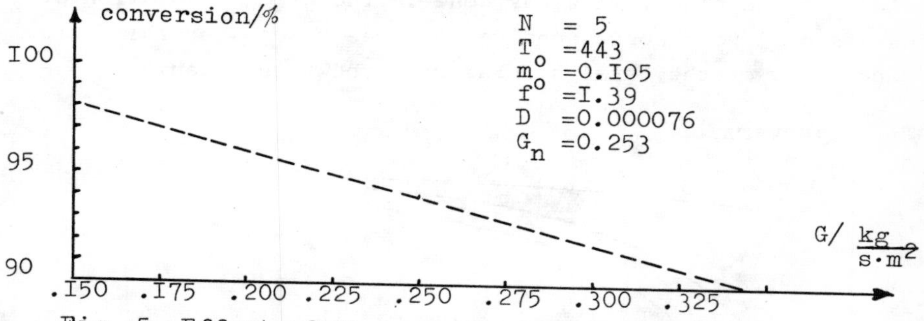

Fig. 5. Effect of the gas charge on the conversion.

TABLE I:

The effect of number of sieves on the conversion

Constants:

T = 443
m$_o$ = 0.105
f^o = 1.39
D = 0.000076
G = 0.253

Conversion	number of sieves
0.815120	3
0.894678	4
0.940000	5
0.965819	6
0.980528	7
0.988907	8
0.993680	9
0.996400	10
0.997949	11
0.998831	12

TABLE II:

The effect of sieve dimensions on the conversion

Constants:

T = 443
m$_o$ = 0.105
G^o = 0.253
N = 5

Conversion	D/f
0.940000	0.000076/1.39
0.951976	0.000076/1.5
0.941904	0.000060/1.2
0.990603	0.000040/1.5

behaviour of nitric acid plants. For example we can see, that increasing total gas charge will decrease conversion, Fig. 4. Physically, increase in gas charge means shorter contact time and thus the result can be explained.

5. Optimization

The model presented can be used to optimize the operation of the plant during one period of service. As mentioned earlier the critical part in the process is the oxidation of ammonia. Thus the conversion of the other parts of the process can be approximated as constants.

The conversion of the oxidation process can be calculated with aid of the model as a function of the air feed around the nominal operating point values of the plant. This was done in the example solution at three conversion levels corresponding to three values for the factor k i.e. three ages of the catalyst. Figure 6 shows the result.

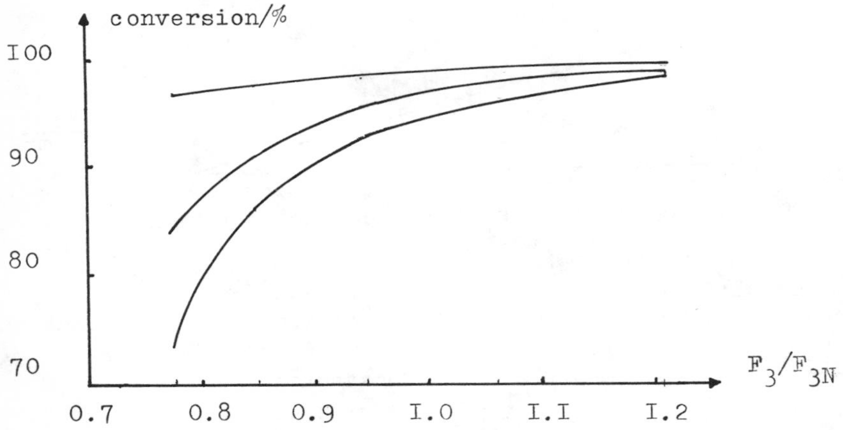

Fig. 6. Effect of air flow on conversion.

These results can be combined by matching time polynomials to these curves and thus the final result for conversion will be function of the normalized time t and the air feed.

(4) $x = x(t_1F_3) = a_o(t)+a_1(t)F_3+a_2(t)F_3{}^2$

The objective function for optimization purposes can now be formed. The total variable costs minus the price of produced nitric acid and export steam integrated over the service period will be used for minimization:

$$(5)\quad J = \int_{t_o}^{t_f} \left\{ K_1 \frac{F_2}{F_{2N}} + (K_2+K_3)\frac{F_3}{F_{3N}} - K_4\frac{T}{T_N} - K_5\frac{F_2}{F_{2N}} - x(t_1F_3) \right\} dt$$

In this equation $K_{1,\ldots,5}$ are weighted cost coefficients which must be individually evaluated for each plant. F_2 means the flow rate of ammonia and T is the temperature rise in the burner. Subscript N denotes for the nominal values. The first three terms in equation (5) mean the cost of ammonia, the cost of energy needed for compression of air and the cost of platinum losses which as a first approximation depend on the flow rate. The fourth term means the value of export steam and the temperature rise can be shown to be easily calculated when the values of F_3, the air feed and m_o, the ammonia concentration in feed are known /4/. The last term includes the conversion and expresses the value of produced nitric acid.

The linear relationships have been assumed here between the costs and relative changes from nominal values and this is reasonable, if the variations are not very large.

The exact values for coefficients K_i should be determined from real cost curves of the plant.

As constraints for this optimization problem will be:

1) $m_o \leq 0.13$, the ammonia contents in feed is less than 13% which is the lower explosion limit of ammonia-air mixture
2) $0.7\ F_{3N} \leq F_3 \leq 1.1\ F_{3N}$, the capacity of air compressor.

The solution of this optimization task can be evalueated by using regressive dynamic programming. Using some preliminary numerical values the solution was computed. Figures 7 and 8 represent the results graphically. From these curves we can conclude that the optimal strategy means a small air flow and high concentration of ammonia at the beginning of the service period and air feed F_3 will increase and ammonia concentration will decrease towards the end of the service period. Normal strategy is that ammonia concentration will be held constant over the whole period.

The preliminary calculations also show that the optimal control strategy derived here will give about 2% better value for the objective function in comparison with the conventional method. The realization of the proposed method would be quite easy also with conventional

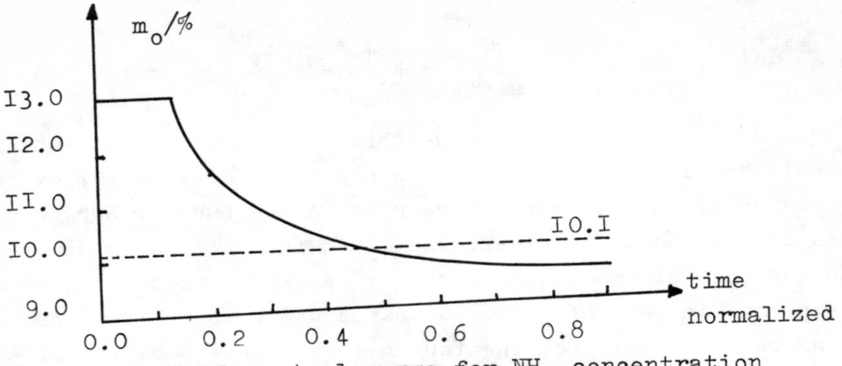

Fig. 7. Optimal control curve for NH_3-concentration in feed.

—— dynamic programming

---- constant feed

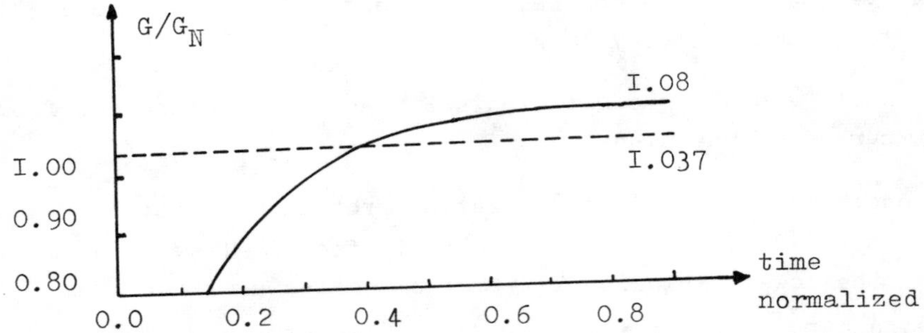

Fig. 8. Optimal control curve for the plant capacity.

—— dynamic programming

---- constant feed

instruments providing that reliability and repeatability of the conversion measurement methods are remarkably improved.

6. Summary

A semiempirical mathematical model for the industrial ammonia oxidation process has been derived. The model takes also the ageing of the catalyst into account. The model is used to simulate the behaviour of the plant and the results show a good conformity with the operational experiences and published qualitative results.

Based on this model an examplewise optimization study of the whole nitric acid plant has been carried out. The objective function used is the total variable costs minus total revenue value of production integrated over one service period of the catalyst.

This minimization problem can be solved by using dynamic programming and preliminary numerical calculations give about 2% better result than the conventional strategy in plant operation.

7. References

/1/ Curievici, I., Ungureanu, S.T., An analogical model of the reactor for ammonia catalytic oxidation, Buletinul Institutului Politehnic Din Iasi, Serie Noua, Tomul, XIV (XVIII), Fasc. 1-2, 227, 1968.

/2/ Oele, A.P., Technological aspects of the catalytic combustion of ammonia with platinum gauze elements, Chem. Eng. Sci. 3, 1-2, 146, 1958.

/3/ Roudier, Houssier and Tracez, An Examination of nitric acid process yields, Information Chimie, Vol. 12, 6, 27.

/4/ Uronen, P., Kiukaanniemi, E., Optimization of a nitric acid plant, Process Technology International, December 1972, Vol. 17, No. 12.

STOCHASTIC CONTROL WITH AT MOST DENUMERABLE NUMBER OF CORRECTIONS

J. Zabczyk

Institute of Mathematics PAN

Warsaw, POLAND

Let us consider Markov processes $X^1,\ldots,X^k$ defined on a state space E. By Markov process X^i we shall mean an object $(\Omega, x_t, \mathcal{F}_t, \mathcal{F}, \Theta_t, P_x^i)$ where Ω is a sample space with elements ω being mappings from $[0,+\infty)$ into E, x_t are random variables such that $x_t(\omega) = \omega(t)$ for all $t \geqslant 0$, $\omega \in \Omega$; $\mathcal{F}_t, \mathcal{F}$ are σ-algebras: $\mathcal{F}_t = \sigma\{x_s;\ s \leqslant t\}$, $\mathcal{F} = \sigma\{x_s;\ s < +\infty\}$; Θ_t are the shift operators: $\Theta_t : \Omega \to \Omega$, $\Theta_t\omega(s) = \omega(t+s)$ and probability measures P_x^i are defined on $\mathcal{F}$. The measures P_x^i describe the motion of the process X^i starting from $x \in E$.

A strategy π is a sequence $\pi = ((\tau_i, d_i))$ where $0 = \tau_1 \leqslant \tau_2 \leqslant \ldots$ are stopping times with respect to $\{\mathcal{F}_t\}$ and $d_1, d_2, \ldots$ are functions mapping Ω into $\{1,2,\ldots,k\}$, measurable with respect to $\mathcal{F}_{\tau_1}, \mathcal{F}_{\tau_2}, \ldots$ respectively, (the stopping times $\tau_1, \tau_2, \ldots$ may be interpreted as moments of corrections and the functions $d_1, d_2, \ldots$ indicate the processes chosen at the moments $\tau_1, \tau_2, \ldots$).

To simplify the next definition we shall suppose that the controler chooses the moment τ_{i+1} knowing the past of the process beginning from τ_i. That means $\tau_{i+1} = \Theta_{\tau_i}(\tau'_{i+1})$ and $d_{i+1} = \Theta_{\tau_i}(d'_{i+1})$ where τ'_{i+1} is any stopping time and d'_{i+1} any $\mathcal{F}$-measurable function.

Every strategy π defines new measures $P_x^{N,\pi}$, P_x^{π}, $N=1\ldots,$

$x \in E$ which satisfy the conditions

$$P_x^{1,\pi}(A) = P_x^{d_1(x)}(A) \quad , \qquad A \in \mathcal{F}$$

$$P_x^{N,\pi}(A \cap \theta_{\tau_N}(B)) = E_x^{N-1,\pi}(P_{x_{\tau_N}}^{d_N}(B) \; ; A) \quad , A \in \mathcal{F}_{\tau_N}, B \in \mathcal{F},$$

$$P_x^{\pi}(A) = P_x^{N,\pi}(A) \; , \quad A \in \mathcal{F}_{\tau_N} \; , \quad B \in \mathcal{F} \quad , \quad N = 2,3,\ldots$$

Let $c_1, \ldots, c_k, g$ be some non-negative Borel functions defined on E. Let us put

$$v_N^{\pi}(x) = E_x^{\pi}\left(\int_0^{\tau_{N+1}} g(x_s)\, ds - \sum_{i=1}^{N} c_{d_i}(x_{\tau_i}) \right)$$

$$v_\infty^{\pi}(x) = \begin{cases} \lim_N v_N^{\pi}(x) \; , & \text{if the limit exists} \\ -\infty \; , & \text{otherwise} \end{cases}$$

$$v_N(x) = \sup_{\pi} v_N^{\pi}(x) \quad , \quad v_\infty(x) = \sup_{\pi} v_\infty^{\pi}(x) \; , \qquad x \in E.$$

There holds the following theorem (see [4])

Theorem 1. If the functions $c_1, \ldots, c_k$ $G^1 g, \ldots, G^k g$, where $G^i g(x) = E_x^i\left(\int_0^{+\infty} g(x_s) ds \right)$, are bounded and (finely) continuous then

1) v_N are bounded, Borel functions also, $v_N \uparrow v_\infty$,

2) $v_{N+1}(x) = \sup_{\tau, i} \left[E_x^i\left(\int_0^{\tau} g(x_s) ds + v_N(x_\tau) \right) - c_i(x) \right]$

3) $v_\infty(x) = \sup_{\tau, i} \left[E_x^i\left(\int_0^{\tau} g(x_s) ds + v_\infty(x_\tau) \right) - c_i(x) \right]$.

Let us consider an example

Example. Let E be an interval $[-\alpha, \alpha]$ and

$$dX_t^1 = -dt + dw_t \; , \qquad dX_t^2 = dt + dw_t \; .$$

Let $-\alpha, \alpha$ be stopping points for both processes. If the cost functions c_1, c_2 are equal to a number $c > 0$ and the reward is equal to the exit time from the open interval $(-\alpha, \alpha)$ then (see [4]) there exist two numbers $\alpha_0, \gamma_0 > 0$ such that the best strategy, in the case of $N = +\infty$, is the following: if $\alpha > \alpha_0$ and the starting point x is in $(-\alpha, 0]$ choose the process X^2 and wait until it reaches the point γ_0 and at that point choose the process X^1 and wait until it reaches the point $-\gamma_0$ and so on; if $\alpha \leq \alpha_0$ and the starting point x is in $(-\alpha, 0]$ choose the process X^2 and $\tau_2 = +\infty$. The number γ_0 satisfies the equation

$$\text{th} 2\gamma = 2\gamma - c .$$

It is rather surprising that the correction points $\gamma_0, -\gamma_0$ do not depend on α.

Prof. E. Dynkin posed the problem to find the best strategy for the model considered in the example if, at any moment $t \geq 0$ the controller knows the states of the process only at moments $\tau_i \leq t$. As far as we know the exact solution is not found yet. In virtue of the solution of the example it seems reasonable to use the following strategy (for simplicity let us suppose that $\alpha > \alpha_0$ and $x \in (-\alpha, 0]$): choose the process X^2 and as the moment of the next correction the moment $-x + \gamma_0$.

Now we give "excessive characterization" of the function v_∞. A Borel function ψ is said to belong to $\mathcal{E}^i$ if it is finaly continuous (see R. Blumenthal, R. Getoor [1]), $\psi \geq -c_i$ and for all $x \in E$ and all stopping times τ

$$\psi(x) + c_i(x) \geq E^i_x(\psi(x_\tau))$$

<u>Theorem 2. Under the same assumptions as in Theorem 1, the function v_∞ is the least function v which satisfies the conditions</u> $v - G^i g \in \mathcal{E}^i$ for $i = 1, 2, \ldots, k$.

Proof. Since, for every $i = 1,2,\dots,k$

$$v_\infty(x) \geq E_x^i\left(\int_0^\tau g(x_s)ds + v_\infty(x_\tau)\right) - c_i(x)$$

and $$E_x^i\left(\int_0^\tau g(x_s)ds\right) = E_x^i\left(\int_0^{+\infty} g(x_s)ds\right) - E_x^i\left(\int_\tau^{+\infty} g(x_s)ds\right) =$$

$$= G^i g(x) - E_x^i(G^i g(x_\tau))$$

therefore $c_i(x) + (v_\infty - G^i g)(x) \geq E_x^i(v_\infty - G^i g)(x_\tau)$.

Now, let $v - G^i g \in \mathcal{E}^i$ for $i = 1,2,\dots,k$ then $v \geq v_N$ for $N = 1,2,\dots$. Indeed $v \geq E_x^i\left(\int_0^\tau g(x_s)ds + v(x_\tau)\right) - c_i(x)$ so $v \geq G^i g - c_i$, and therefore $v \geq v_1$. If $v \geq v_{N-1}$ then $v(x) \geq E_x^i\left(\int_0^\tau g(x_s)ds + v_{N-1}(x_\tau)\right) - c_i(x) = v_N(x)$ for suitable i.

This completes the proof.

Analogous theorems can be proved if we consider the stopping time problem. In that case we define

$$w_N^\pi = \max_{2 \leq i \leq N} E_x^\pi\left[\varphi(x_{\tau_i}) - \sum_{j<i} d_j(x_{\tau_j})\right]$$

$$w_\infty^\pi = \sup_N w_N^\pi$$

$$w_N = \sup_\pi w_N^\pi \quad \text{and} \quad w_\infty = \sup_\pi w_\infty^\pi$$

where φ is a bounded, Borel measurable and finely continuous function defined on E.

The equations 2) and 3) from Theorem 1 have now the form

2´) $$w_{N+1}(x) = \sup_{\tau,i}\left[E_x^i(\varphi \cup w_N)(x_\tau) - c_i(x)\right], \quad \varphi \cup w_N = \max(\varphi, w_N)$$

3´) $$w_\infty(x) = \sup_{\tau,i}\left[E_x^i(\varphi \cup w_\infty)(x_\tau) - c_i(x)\right], \quad \varphi \cup w_\infty = \max(\varphi, w_\infty)$$

An example of that kind $(N = 2)$ was considered in [3].

An excessive characterization of w_∞ in the case of $c_i \equiv 0$

$i = 1,2,\ldots,k$ was given by Griegelionis, Shiryaev [2].

References

[1] R.M. Blumenthal, R.K. Getoor, Markov processes and Potential Theory, Academic Press, New York - London, 1968

[2] B.I. Griegelionis, A.N. Shiryaev, On controlled Markov processes and the Stefan problem, Problemy Peredachi Informacii, 4(1968), pp. 60-72

[3] J. Zabczyk, A mathematical correction problem, Kybernetika, 8(1972), pp.317-322

[4] J. Zabczyk, Optimal control by means of switchings, Studia Mathematica 45(1973), pp.161-171

DESIGN OF OPTIMAL INCOMPLETE STATE FEEDBACK CONTROLLERS FOR LARGE LINEAR CONSTANT SYSTEMS

W.J. Naeije, P. Valk, O.H. Bosgra
Delft University of Technology, Delft,
The Netherlands.

SUMMARY

In this paper the theory of linear optimal output feedback control is investigated in relation to its applicability in the design of high-dimensional linear multivariable control systems. A method is presented which gives information about the relative importance of the inclusion of a state vector element in the output feedback. The necessary conditions of the optimization problem are shown to be a set of linear/quadratic algebraic matrix equations. Numerical algorithms are presented which take account of this linear/quadratic character.

I. INTRODUCTION

In the design of feedback controllers for linear, time-invariant systems of high dimension, implementation restrictions may result in the additional constraint that the control is a function of only a limited set of elements of the state vector. Optimization theory can then still be used if the structure of the controller, as specified in advance, is used as an additional constraint relation. In this paper the controller will be assumed to be a time-invariant matrix of feedback gains. As many servomechanism- and tracking problems can be reduced to regulator problems with a time-invariant feedback matrix, the optimal output regulator problem as discussed here can be viewed as the basis of a wide class of controller design problems.

In existing literature on this subject, the necessary conditions of the mathematical optimization problem are derived for a deterministic problem setting. As questions regarding existence and uniqueness of solutions are easiest solved in the case of optimization over a finite time interval, this problem has drawn attention first [1]. Extension of the results to the infinite time interval showed the dependence of the time-invariant feedback matrices upon the initial conditions of the problem. Moreover, only necessary conditions could be given, and existence and uniqueness of the solutions could not be quaranteed [2,3].

The corresponding stochastic output regulator problem [4,5] leads to essentially similar necessary conditions. The necessity to choose noise intensity matrices in this case is equivalent to choosing initial conditions in the deterministic case. So the widely used technique to assume initial conditions, uniformly distributed on the surface of the n-dimensional unit sphere [1] is equivalent to assuming a unity noise intensity matrix. However, in the stochastic problem such a trick can be replaced by the proper choice of a noise intensity matrix, based on knowledge of or assumptions on the physical background of the system. This motivates the use of a stochastic problem setting here.

In using the optimal output regulator theory in controller design, two main problems arise for which no sufficient solution exists in literature. At first, the designer needs information about the relative importance of the inclusion of a state vector element in the output feedback, to be able to make a suitable compromise between implementation costs and improved control behaviour. Secondly, numerical solution of the matrix equations which constitute the necessary conditions of the optimization problem must be performed using a suitable algorithm. This paper treats these two problems after a short recapitulation of the optimization problem.

II. NECESSARY CONDITIONS

Consider the time-invariant linear system

$$\dot{x}(t) = A\,x(t) + B\,u(t) + H\,w(t) \qquad E\{x(t_o)\} = 0 \tag{1}$$

$$y(t) = C\,x(t) \tag{2}$$

with state-vector $x(t)$, input vector $u(t)$, noise vector $w(t)$ and output vector $y(t)$ of dimensions n, m, r, k respectively. $w(t)$ is Gaussian white noise, characterized by $E\{w(t)\} = 0$, $E\{w(t)\;w^T(\tau)\} = \Phi\;\delta(t-\tau)$ in which $\Phi \geqslant 0$ is a time-invariant positive semidefinite intensity matrix.
Constraining the feedback to

$$u(t) = F\,y(t) = F\,C\,x(t),\; F\,C = L \tag{3}$$

leads to the optimization problem: choose F as to minimize the quadratic performance index

$$J = E\{x^T(t)\;Q\;x(t) + u^T(t)\;R\;u(t)\} \tag{4}$$

with $x(t)$, $u(t)$ and F satisfying (1), (2) and (3).
If $S = E\{x(t)\;x^T(t)\}$ is the solution of the variance equation for the closed-loop system:

$$\Gamma(S,F) = (A + BFC)S + S(A + BFC)^T + H\;\Phi\;H^T = 0 \tag{5}$$

the performance index can be written:

$$J = tr\{(Q + C^TF^TB^TRBFC)S\} \tag{6}$$

Using the matrix-minimum principle [6] by introducing a Hamiltonian function

$$H(S,F,P) = J + tr\{\Gamma(S,F)P^T\} \tag{7}$$

in which P is the adjoint matrix, the necessary conditions follow from:

$$\begin{aligned} -\left.\frac{\delta H(S,F,P)}{\delta S}\right|_{*} &= \dot{P} = 0 \\ \left.\frac{\delta H(S,F,P)}{\delta P}\right|_{*} &= \Gamma(S,F) \\ \left.\frac{\delta H(S,F,P)}{\delta F}\right|_{*} &= 0 \end{aligned} \tag{8}$$

The feedback matrix F, resulting from the necessary conditions, is:

$$F = -R^{-1}B^TPSC^T(CSC^T)^{-1} \tag{9}$$

in which the adjoint matrix $P = P^T > 0$ is the solution of

$$P(A + BFC) + (A + BFC)^TP + Q + C^TF^TRFC = 0 \tag{10}$$

and the variance matrix $S = S^T \geqslant 0$ is the solution of (5). The resulting performance criterion is

$$J^* = tr\{PH\Phi H^T\} \tag{11}$$

A necessary and sufficient condition for the existence of a solution F which minimizes (6) is that the system (1), (2) be output feedback stabilizable. Necessary and sufficient conditions for output feedback stabilizability are presently unknown; a sufficient condition is given in [7]. Evidently, a necessary condition is that (A,B) be stabilizable and (D^T,A) be detectable with $Q = DD^T$ [8]. If an optimizing solution F exists, uniqueness cannnot be proved generally. So the practical use of the necessary conditions consists of finding a numerical solution to the equations (5), (9) and (10) in which $P > 0$ and $S \geqslant 0$, implying that the closed loop system is stable, and investigating the nature of this solution, e.g. by comparing the performance index belonging to it with the performance index obtained with optimal complete state feedback.

From the structure of the feedback matrix F follows, that the optimal output feedback forms a minimum variance estimate $\hat{x}(t)$ of $x(t)$, given the observation $y(t) = C\,x(t)$. This linear estimate $\hat{x}(t)$ is determined by the projection: [9, p. 88]

$$\hat{x}(t) = SC^T(CSC^T)^{-1}\ C\ x(t) \tag{12}$$

and this vector is used in the same feedback structure as is encountered in optimal complete state feedback:

$$u(t) = -R^{-1}B^TP\ \hat{x}(t) \tag{13}$$

In the sequel, with no loss of generality the output matrix C will be assumed to be $C = [\,C_1 \mid 0\,]$ with C_1 non-singular. In that case, (12) can be written as

$$\hat{x}(t) = \left[\begin{array}{c|c} I & 0 \\ \hline S_{12}^T S_{11}^{-1} & 0 \end{array}\right] x(t),\quad S = \left[\begin{array}{c|c} S_{11} & S_{12} \\ \hline S_{12}^T & S_{22} \end{array}\right] \tag{14}$$

with partitioning consistent with the partitioning of C.

III. SELECTION OF OUTPUT VARIABLES

For a practical application of the theory of optimal output feedback, information is required about the importance of inclusion of each element of the state vector in the output feedback. Comparing all possible choices by computing the resulting performance indices seems unrealistic from a computational point of view. However, any computationally less involved method of comparison must be of an approximative nature. Here a method is developed based on a comparison of trajectories of an ideal system (e.g. using optimal feedback of the complete state vector) with trajectories of the same system under conditions of output feedback.

Let $x^o(t)$ be the trajectory of the system under consideration and L^o the optimal feedback matrix using complete state feedback:

$$\begin{aligned} \dot{x}^o(t) &= A\ x^o(t) + B\ u^o(t) + H\ w(t) \\ u^o(t) &= L^o\ x^o(t) \end{aligned} \tag{15}$$

Let x(t) be the trajectory of the system with output feedback matrix L:

$$\begin{aligned} \dot{x}(t) &= A\ x(t) + B\ u(t) + H\ w(t) \\ u(t) &= L\ x(t) = FC\ x(t) \end{aligned} \tag{16}$$

The difference between both trajectories, $e(t) = x(t) - x^o(t)$, is governed by the equation [2]

$$\dot{e}(t) = (A + BL)\ e(t) + B\ (L-L^o)\ x^o(t). \tag{17}$$

Define the source term vector q(t) in (17) as

$$q(t) = B\ (L-L^o)\ x^o(t) \tag{18}$$

Assuming that A + BL is stable, the difference between both trajectories will be minimum in some sense if q(t) is minimal. Introduce as a source term objective function

$$I = E\{q^T(t)\, q(t)\} = E\{tr(q(t)\, q^T(t))\} \tag{19}$$

Inserting (18) in (19) and interchange of the expectation and trace operators leads to

$$I = tr\left[(L - L^o)\, S^o (L - L^o)^T\right] \tag{20}$$

where S^o is the variance matrix of the optimal system using complete state feedback, which is the symmetric positive (semi)definite solution of the Lyapunov equation

$$(A + BL^o)S^o + S^o(A + BL^o)^T + H\Phi H^T = 0 \tag{21}$$

The necessary condition for minimizing I over all possible choices L within the constraint L = FC is

$$\left.\frac{\partial I}{\partial F}\right|_{*} = 0 \tag{22}$$

Elaboration of (22) leads to

$$L^* = L^o S^o C^T (C S^o C^T)^{-1} C \tag{23}$$

assuming the inverse exists. Note that L^*, the output feedback matrix minimizing the source term in (17), is obtained by applying a projector operating upon the optimal complete state feedback matrix L^o. [10, Ch. XI] . The corresponding extreme value of the source term objective function is

$$I^* = tr\left[(L^* - L^o)\, S^o (L^* - L^o)^T\right] \tag{24}$$

Using the adopted structure of the output matrix C and the corresponding partitioning of S^o and L^o,

$$S^o = \left[\begin{array}{c|c} S^o_{11} & S^o_{12} \\ \hline S^{To}_{12} & S^o_{22} \end{array}\right] \qquad L^o = \left[\begin{array}{c|c} L^o_1 & L^o_2 \end{array}\right] \tag{25}$$

and combining (24) and (23) leads to

$$I^* = tr\left[L^o_2 (S^o_{22} - S^{To}_{12} S^{o\,-1}_{11} S^o_{12}) L^{oT}_2\right] \tag{26}$$

As a computational alternative, (26) can be converted into [11]

$$I^* = tr\left[L^o_2 \tilde{S}^{-1}_{22} L^{oT}_2\right] \tag{27}$$

with

$$\tilde{S} = (S^o)^{-1} = \left[\begin{array}{c|c} \tilde{S}_{11} & \tilde{S}_{12} \\ \hline \tilde{S}^T_{12} & \tilde{S}_{22} \end{array}\right] \tag{28}$$

and partitioning consistent with (25).

Equation (26) and (27) provide a computationally feasible means of estimating the relative importance of inclusion of each state vector element in the output feedback structure. Assume the optimal complete state feedback parameters L^o and S^o are known. Assume further that a set of l state vector elements U_l are selected on technological grounds as possible measured output elements ($l \leqslant n$). From the set U_l, $k \leqslant l$ elements are to be selected to form the output vector. This selection may be performed as follows.
Compute all I_i^* by omitting the i-th element from U_l, $i = 1,2...l$ and determine $I_j^* = \min_i I_i^*$. Form U_{l-1} by omitting the j-th element from U_l. This process can be repeated up to U_k. It should be noted, that I^* in eq. (24) forms the square of a matrix norm of the difference of L^o and a projection of L^o, and that this projection, L^*, is not used as a feedback matrix in the selection procedure. This implies that a selection procedure, based on equation (24), can still be useful in those cases in which a feedback matrix L^* does not stabilize the system in eq. (17). Based on eq. (24), other selection procedures are possible and further engineering constraints can be included in the decision process. Also the optimal complete state feedback matrices L^o and S^o can be replaced by corresponding other matrices which render "ideal" system behaviour.

IV. SOLUTION OF NECESSARY CONDITIONS

For a numerical solution of the optimization problem two basic approaches are:

a. Use of numerical function minimisation algorithms based on the gradient of the performance index with respect to the parameter matrix L;
b. Numerical solution of the matrix equations e.g. by iterative procedures.

The gradient of the performance index with respect to the parameter matrix L can be analytically derived, using a suggestion by Kwakernaak for a similar problem [12, Ch. 5.7]. This derivation (see appendix I) gives as a result:

$$\frac{\partial J}{\partial L_1} = 2\left[RL_1S_{11} + B^TP\begin{bmatrix} S_{11} \\ \hline S_{12}^T \end{bmatrix} \right] \tag{29}$$

in which the output feedback matrix L has the structure

$$L = \left[L_1 \mid 0 \right] \tag{30}$$

and L and S are partitioned consistent with the adopted structure of the output matrix C. S and P follow from

$$(A + BL)S + S(A + BL)^T + H\Phi H^T = 0 \tag{31}$$

$$P(A + BL) + (A + BL)^T P + Q + L^T RL = 0 \tag{32}$$

Numerical algorithms that use (29) need the solution of (31) and (32) for each step where gradient evaluation is required. As the number of gradient evaluations generally is large, either in a pure gradient method with a small step size or in quadratically convergent search methods, this class of numerical techniques does not seem to be easily applicable for high-dimensional systems. Moreover, no use is made of insight into the analytical properties of the equations nor the analytically known solution of the feedback matrix L, equation (9).

For high-dimensional systems, solution of the matrix equations (5,9,10) by successive approximation techniques seems useful. At first, properties of the matrix equations must be investigated.
Equations (10,9) can be written as

$$\begin{aligned} A^T P + PA + Q - PBR^{-1}B^T P + \left[I - C^T(CSC^T)^{-1}C \right] . \\ . \, PBR^{-1}B^T P \left[I - SC^T(CSC^T)^{-1}C \right] = 0 \end{aligned} \tag{33}$$

Using the adopted structure for C and the partitioning of S (14), the projector in (33) can be written:

$$\left[I - C^T(CSC^T)^{-1}CS \right] = \left[\begin{array}{c|c} 0 & -S_{11}^{-1}S_{12} \\ \hline 0 & I \end{array} \right] \tag{34}$$

thus (33) becomes:

$$\begin{aligned} F(P,S) = A^T P + PA + Q - PBR^{-1}B^T P + \left[\begin{array}{c|c} 0 & -S_{11}^{-1}S_{12} \\ \hline 0 & I \end{array} \right] . \\ . \, PBR^{-1}B^T P \left[\begin{array}{c|c} 0 & 0 \\ \hline -S_{12}^T S_{11}^{-1} & I \end{array} \right] = 0 \end{aligned} \tag{35}$$

If consistent partitionings are made:

$$P = \left[\begin{array}{c|c} P_{11} & P_{12} \\ \hline P_{12}^T & P_{22} \end{array} \right] \qquad F(P,S) = \left[\begin{array}{c|c} F_{11} & F_{12} \\ \hline F_{12}^T & F_{22} \end{array} \right] \qquad A = \left[\begin{array}{c|c} A_{11} & A_{12} \\ \hline A_{21} & A_{22} \end{array} \right] \tag{36}$$

then F_{22} turns out to be linear in P_{22} and independent of S:

$$F_{22} = P_{22}A_{22} + A_{22}^T P_{22} + P_{12}^T A_{12} + A_{12}^T P_{12} + Q_{22} = 0 \tag{37}$$

while F_{11} and F_{12} are quadratic in P_{11} and P_{12}. So F(P,S) is a mixed linear/quadratic matrix equation in P.

It can be noted that the appearance of S-terms in (35) can be eliminated by applying a similarity transformation on the state vector:

$$x^+ = \left[\begin{array}{c|c} I & 0 \\ \hline -S_{12}^T S_{11}^{-1} & I \end{array}\right] x \tag{38}$$

Although this transformation can not be computed in advance and so has no practical significance, it does not affect the linear/quadratic character of (35) and it shows that the role of S in equation (35) is only limited to a transformation of state space.

As a conclusion, a computational algorithm primarily must meet the requirements set by the linear/quadratic character of F(P,S) = 0.

At present, two important algorithms for equations (5,9,10) are the Axsäter algorithm [13, p. 313], based on [5] and later adapted to the time-invariant case by Levine and Athans [14], and a simpler algorithm suggested by Anderson and Moore [13, p. 314]. Main disadvantages of these algorithms are:

a. The algorithms do not take advantage of the properties of the matrix equations as mentioned before;
b. The algorithms require a stabilizing initial output feedback matrix which can be difficult to determine;
c. The main drawback of both algorithms is the necessary condition that the closed-loop system matrix remains stable in the course of the iteration. The algorithms provide no guarantee for this stability, and practical applications show that in all but the simplest cases both algorithms fail for this reason. Also the fact that Axsäter's algorithm converges in the performance index $J = tr(PH\Phi H^T)$ [5,14] is only valid as long as the closed-loop system remains stable, and so has no significance as a proof of convergence.

Better algorithms might be developed if the mentioned properties of the equations (5,9,10) are taken into account:

a. The presence of the S-terms in eq. (10) results in a transformation of the state space;

b. The equation (10) has a mixed linear/quadratic character; as a result of the adopted structure of the output matrix C, the linear and the quadratic parts appear in separate partitions of the matrix equation.

If a Newton-Raphson algorithm could be analytically derived for the equations (5,9,10), these properties would be incorporated in the algorithm. However, the complexity of the equations prohibits such an approach. Thus, a linear converging algorithm is the only realizable proposition. As the role of S in eq. (10) is limited, Newton-Raphson applied to eq. (10) separately, assuming P as the only variable, may provide a basis for an algorithm. The result is: (see appendix II).

$$P_{k+1}(A - BR^{-1}B^TP_k) + (A - BR^{-1}B^TP_k)^TP_{k+1} + \alpha_k P_{k+1}BR^{-1}B^TP_k\alpha_k^T$$
$$+ \alpha_k P_k BR^{-1}B^TP_{k+1}\alpha_k^T + Q + P_kBR^{-1}B^TP_k - \alpha_k P_k BR^{-1}B^TP_k\alpha_k^T = 0 \quad (39)$$

$$\alpha_k = \left[\begin{array}{c|c} 0 & -S_{11}^{-1}S_{12} \\ \hline 0 & I \end{array}\right]_K$$

Eq. (39) is a linear matrix equation in P_{k+1}. Numerical solution requires the use of a Kronecker-product and this is inefficient or impossible for high-dimensional systems. As Lyapunov matrix equations can be efficiently solved, even for high-dimensional systems [15,16], an adaptation of (39) to the Lyapunov structure is desired. Three algorithms that perform this step will be suggested. All are based on the replacement in eq. (39) of the term

$\alpha_k\left[P_{k+1}BR^{-1}B^TP_k + P_kBR^{-1}B^TP_{k+1} - P_kBR^{-1}B^TP_k\right]\alpha_k^T$ by a term ΔQ_k:

$$P_{k+1}(A - BR^{-1}B^TP_k) + (A - BR^{-1}B^TP_k)^TP_{k+1} + Q + P_kBR^{-1}B^TP_k + \Delta Q_k = 0 \quad (40)$$

$$\text{(Algorithm I)} \quad \Delta Q_k = \alpha_k P_k BR^{-1}B^TP_k\alpha_k^T \quad (41)$$

$$\text{(Algorithm II)} \quad \Delta Q_k = \alpha_k \left[\begin{array}{c|c} 0 & 0 \\ \hline P_{12}^T & PR \end{array}\right]_K BR^{-1}B^T \left[\begin{array}{c|c} 0 & P_{12} \\ \hline 0 & PR \end{array}\right]_K \alpha_k^T \quad (42)$$

$$\text{(Algorithm III)} \quad \Delta Q_k = \alpha_k\left[\left[\begin{array}{c|c} 0 & 0 \\ \hline P_{12}^T & PR \end{array}\right]_K BR^{-1}B^TP_k + P_kBR^{-1}B^T\left[\begin{array}{c|c} 0 & P_{12} \\ \hline 0 & PR \end{array}\right]_K - P_kBR^{-1}B^TP_k\right]\alpha_k^T \quad (43)$$

PR in (42,43) is a predicted value for P_{22} and is determined by an equation representing (37), the linear partition of (35):

$$PR_kA_{22} + A_{22}^TPR_k + (P_{12})_kA_{12} + A_{12}^T(P_{12})_k + Q_{22} = 0 \quad (44)$$

Due to the appearance of the closed-loop system matrix $(A - BR^{-1}B^TP_k)$ in (40), the iteration in S, based on (5) can be chosen as

$$S_k(A - BR^{-1}B^TP_k)^T + (A - BR^{-1}B^TP_k)S_k + H\Phi H^T \tag{45}$$

$$+ BR^{-1}B^TP_k \left[\begin{array}{c|c} 0 & 0 \\ \hline 0 & S_{22}-S_{12}^TS_{11}^{-1}S_{12} \end{array}\right]_{k-1} + \left[\begin{array}{c|c} 0 & 0 \\ \hline 0 & S_{22}-S_{12}^TS_{11}^{-1}S_{12} \end{array}\right]_{k-1} P_kBR^{-1}B^T = 0$$

So the algorithms consist of iteratively solving (40,45) with ΔQ_k given by (41), (42) or (43). In the latter two cases, also (44) must be solved at each step. These algorithms have the following properties:

1. If $(A - BR^{-1}B^TP_o)$ is stable, then $(A - BR^{-1}B^TP_k)$ is stable and $P_k > 0$, $k = 1,2...$ (algorithms I, II; see appendix III).
2. The initial stabilizing feedback matrix $-BR^{-1}B^TP_o$ is a state feedback matrix, allowing the algorithm to start on the stable optimal state feedback matrix. Known algorithms require initial stabilizing output feedback.
3. Comparing with known algorithms, the range of convergence is significantly increased due to the guarantee of stability of the closed-loop system matrix and because the linear and quadratic partitions of the matrix equation (10) are treated separately in the algorithms. However, the conditions for convergence can not explicitly be given due to lack of knowledge about existence and uniqueness of the solutions to eq. (5,9,10).
4. In algorithms II and III, A_{22} must be a stable matrix for (44) to be efficiently solvable [15,16]. Under this condition, $PR = P_{22}$ in a stationary solution of the algorithms as can easily be proven by regarding the fact that P_{22} is a positive definite solution of a quadratic matrix equation and hence is unique.

IV. APPLICATIONS

In the application to technological systems, the suggested output selection procedure has provided satisfactory results. No numerical problems were encountered. An application to a 14-dimensional open-loop stable boiler system [17] is shown in fig. 1. The application of the resulting ordering of state vector elements in optimal output feedback with a subsequent decreasing number of output elements yields subsequent increasing values of the performance index, fig. 2. The fact that this sequence is flat over a considerable range can be interpreted as a satisfactory result of the selection algorithm. The results of fig. 2 were obtained using algorithm II. Comparing the convergence properties of the proposed algorithms with the algorithms of Axsäter and of Anderson/ Moore as applied to the same system, showed as a result that the Anderson/Moore algorithm failed when using 7 or less output variables and the Axsäter algorithm when using 6 of less. Both failures were due to instability of the closed-loop system matrix. The proposed algorithms I, II, III showed convergence up to 3, 2 and 1 output variables respectively. The speed of convergence for 8 output variables is shown in fig. 3. The general experience is that convergence slows down with decreasing number of output variables, fig. 4. It should be mentioned that the algorithms II and III exhibited practically identical behaviour. For all Lyapunov equation computations, the accelerated series method of R.A. Smith [15] was used.

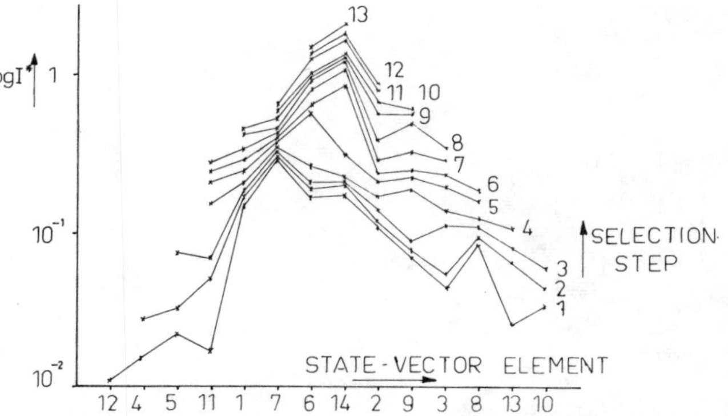

fig.1 Selection procedure applied to 14-dim. boiler system.

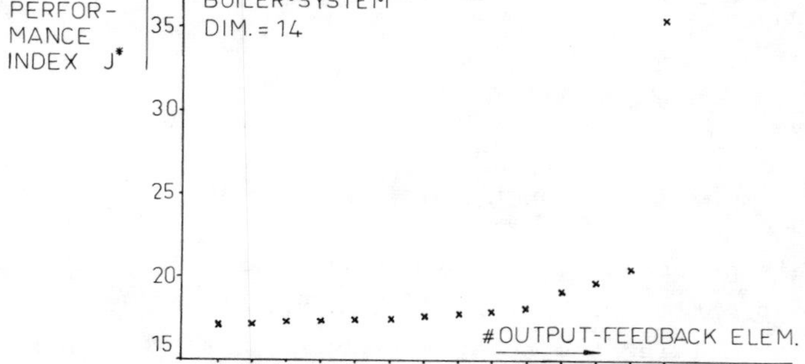

fig.2 Performance of optimal output-feedback dependent upon number of output elements.

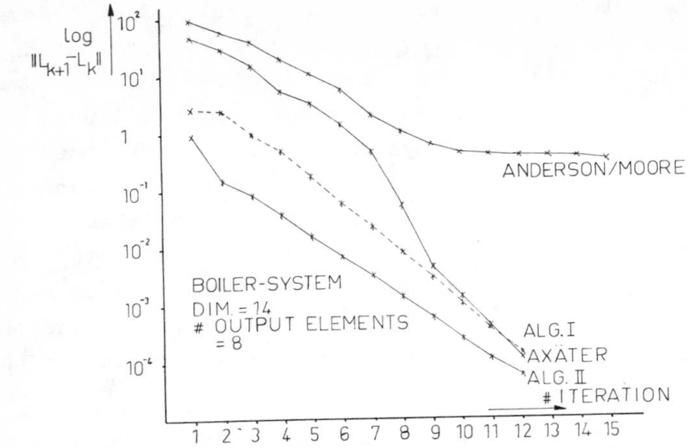

fig.3 Convergence behaviour of numerical algorithms.

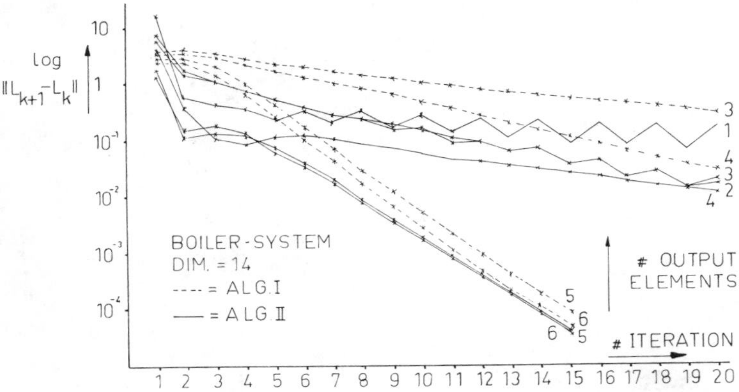

fig.4 Convergence of proposed algorithms dependent upon number of output elements.

V. CONCLUSIONS

The results of this paper involve an algorithm for selection of output variables in linear optimal output feedback control and improved numerical algorithms for solving the necessary conditions of optimal output feedback; these algorithms take into account analytical properties of the relevant linear/quadratic matrix equations. With these new results, the range of applicability of optimal output theory in linear output controller design certainly can be increased, as only very few [18,19] numerical applications have appeared that use existing algorithms. However, further improvements in the use of the suggested algorithms may be expected if questions regarding necessary and sufficient conditions for existence and uniqueness of solutions to the relevant matrix equations are better understood.

REFERENCES

1. D.L. Kleinman, M. Athans, The design of suboptimal linear time-varying systems. IEEE Trans. Aut. Contr. 13(1968), 150-160.
2. R.L. Kosut, Suboptimal control of linear time-invariant multivariable systems subject to control structure constraints. Ph.D. diss. Univ. Pennsylvania, 1969; also IEEE Trans. Aut. Contr. 15(1970), 557-563.
3. W.S. Levine, T.L. Johnson, M. Athans, Optimal limited state variable feedback controllers for linear systems. IEEE Trans. Aut. Contr. 16(1971), 785-793.
4. P.J. McLane, Linear optimal stochastic control using instantaneous output feedback. Int. J. Contr. 13(1971), 383-396.
5. S. Axsäter, Sub-optimal time-variable feedback control of linear dynamic systems with random inputs. Int. J. Contr. 4(1966), 549-566.
6. M. Athaus, The matrix minimum principle. Inf. Contr. 11(1968), 592-606.
7. M.T. Li, On output feedback stabilizability of linear system. IEEE Trans. Aut. Contr. 17(1972), 408-410.
8. A.K. Nandi, J.H. Herzog, Comments on "Design of single-input system for specified roots using output feedback". IEEE Trans. Aut. Contr. 16(1971), 384-385.
9. D.G. Luenberger, Optimization by vector space methods. Wiley, N.Y. 1969
10. M.C. Pease, Methods of matrix algebra. Academic Press, New York 1965.
11. T.E. Fortmann, A matrix inversion identity. IEEE Trans. Aut. Contr. 15(1970), 599.
12. H. Kwakernaak, R. Sivan, Linear optimal control systems. Wiley-Interscience, New York 1972.
13. B.D.O. Anderson, J.B. Moore, Linear optimal control. Prentice-Hall, Englewood Cliffs, N.J., 1971.
14. W.S. Levine, M. Athans, On the determination of the optimal constant output feedback gains for linear multivariable systems. IEEE Trans. Aut. Contr. 15(1970), 44-48.
15. R.A. Smith, Matrix equation $XA + BX = C$. SIAM J. Appl. Math. 16(1968), 198-201.
16. P.G. Smith, Numerical solution of the matrix equation $AX + XA^T + B = 0$. IEEE Trans. Aut. Contr. 16(1971), 278-279.
17. O.H. Bosgra, Application of optimal output control theory to a model of external power station boiler dynamic behaviour. Report N-95, Lab. Meas. Contr., Delft Univ. Techn., Stevinweg 1, Delft, The Netherlands, 1973.
18. E.J. Davison, N.S. Rau, The optimal output feedback control of a synchronous machine. IEEE Trans. Pow. App. Syst. 90(1971), 2123-2134.
19. M. Ramamoorty, M. Arumugam, Design of optimal constant-output feedback controllers for a synchronous machine, Proc. IEE 119(1972), 257-259.

APPENDIX I

Let λ be an element of L. Partial differentiation of eq. (31) gives:

$$\frac{\partial}{\partial\lambda}(A+BL).S+(A+BL)\frac{\partial S}{\partial\lambda}+\frac{\partial S}{\partial\lambda}(A+BL)^T+S.\frac{\partial}{\partial\lambda}(A+BL)^T = 0 \quad \text{(A1)}$$

$$\frac{\partial J}{\partial\lambda}=\frac{\partial}{\partial\lambda}\,\mathrm{tr}\left[(Q+L^TRL)S\right] \quad \text{using (6)}$$

$$= \mathrm{tr}\left[(Q+L^TRL)\frac{\partial S}{\partial\lambda}+\frac{\partial}{\partial\lambda}(L^TRL).S\right]$$

$$= \mathrm{tr}\left[-\{P(A+BL)+(A+BL)^TP\}\frac{\partial S}{\partial\lambda}+2\frac{\partial L}{\partial\lambda}.SL^TR\right] \quad \text{using (32)}$$

$$= 2\mathrm{tr}\left[\frac{\partial L}{\partial\lambda}SPB+\frac{\partial L}{\partial\lambda}SL^TR\right] \quad \text{using (A1)}$$

This last expression is equivalent to equation (29).

APPENDIX II

Equation (10), written in the form (35) as F(P) = 0, can be differentiated, using α as defined in (39):

$$dF=-dP.A-A^TdP+dPBR^{-1}B^TP+PBR^{-1}B^TdP-\alpha dPBR^{-1}B^TP\alpha-\alpha PBR^{-1}B^TdP\alpha= 0 \quad \text{(A2)}$$

Writing F and P as properly ordered vectors $\underline{F}$ and $\underline{P}$, (A2) becomes:

$$d\underline{F}=-(A\otimes I)d\underline{P}-(I\otimes A^T)d\underline{P}+(BR^{-1}B^TP\otimes I)d\underline{P}+(I\otimes PBR^{-1}B^T)d\underline{P}-(BR^{-1}B^TP\alpha\otimes\alpha)d\underline{P}$$
$$-(\alpha\otimes\alpha PBR^{-1}B^T)d\underline{P} = 0 \quad \text{(A3)}$$

In (A3), the derivative of $\underline{F}$ with respect to $\underline{P}$ is explicitly given. Inserting this derivative in the Newton-Raphson expression

$$\left[\frac{d\underline{F}}{d\underline{P}}\right]_K(\underline{P}_{k+1}-\underline{P}_k) = -F(\underline{P}_k) \quad \text{(4)}$$

and making the conversion from vectors back to matrices directly leads to (39).

APPENDIX III

Proof of property 1 for algorithms I, II: If $(A - BR^{-1}B^TP_k)$ is asymptotically stable, then by Lyapunov's theory $P_{k+1} > 0$, because in (40) $Q + P_kBR^{-1}B^TP_k + \Delta Q_k > 0$ and the pair $(A,Q^{\frac{1}{2}})$ is assumed to be detectable As (40) can be written as:

$$P_{k+1}(A-BR^{-1}B^TP_{k+1})+(A-BR^{-1}B^TP_{k+1})^TP_{k+1}+Q+P_{k+1}BR^{-1}B^TP_{k+1}+\Delta Q_k+$$
$$+(P_{k+1}-P_k)BR^{-1}B^T(P_{k+1}-P_k) = 0 \quad \text{(A5)}$$

and the solution P_{k+1} of (A5) is positive definite, by Lyapunov's theory $(A - BR^{-1}B^TP_{k+1})$ also is asymptotically stable.

CONTROL OF A NON LINEAR STOCHASTIC BOUNDARY VALUE PROBLEM

J.P. KERNEVEZ
Faculté des Sciences
6, Boulevard Gabriel
21000 - DIJON

J.P. QUADRAT, M. VIOT
I.R.I.A.
78 - ROCQUENCOURT

I - POSITION OF THE PROBLEM

The aim of this paper is to describe an optimal feedback control of a biochemical system described by partial differential equations and submitted to a random environment. Such biochemical systems have been described and studied in the deterministic case in J.P. KERNEVEZ [3] . In this section we give a short presentation of these membraneous systems, leading us to the stochastic model studied in the following sections. In section 2 are given some indications about existence and unicity of a solution for the state equations. In section 3 is given a way to approach an optimal feedback control of the system. In section 4 the particular case of a linear feedback is considered and numerical results are given. An artificial membrane separates 2 compartments 1 and 2. The membrane is made of inactive protein corretieulated with enzyme. In the compartments are some substrate S and some inhibitor I which are diffusing in the membrane.

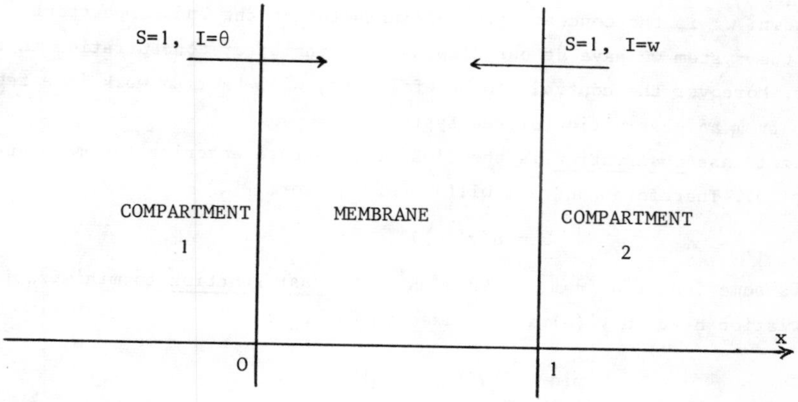

S is reacting in the membrane because of enzyme which is a catalyst of a biological reaction. In this paper we are interested only by the stationnary case. The evolution case will be treated in a paper to be published (J.P. QUADRAT, M. VIOT [7]). Let us call

$y(x)$ = substrate concentration at point x in the membrane $(0 < x < 1)$

$i(x)$ = inhibitor concentration.

The stationnary case equations are

$$(1.1)\quad \begin{cases} y''(x) = \dfrac{\sigma\, y(x)}{1+i(x)+|y(x)|} \; ; \; y(0) = y(1) = 1 \\ i(x) = wx + (1-x)\theta \end{cases}$$

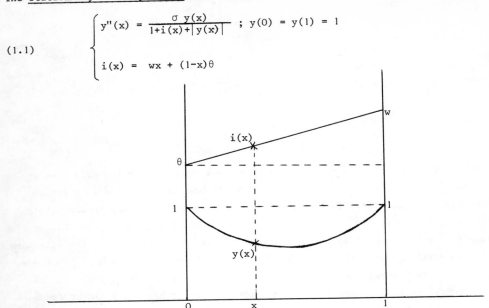

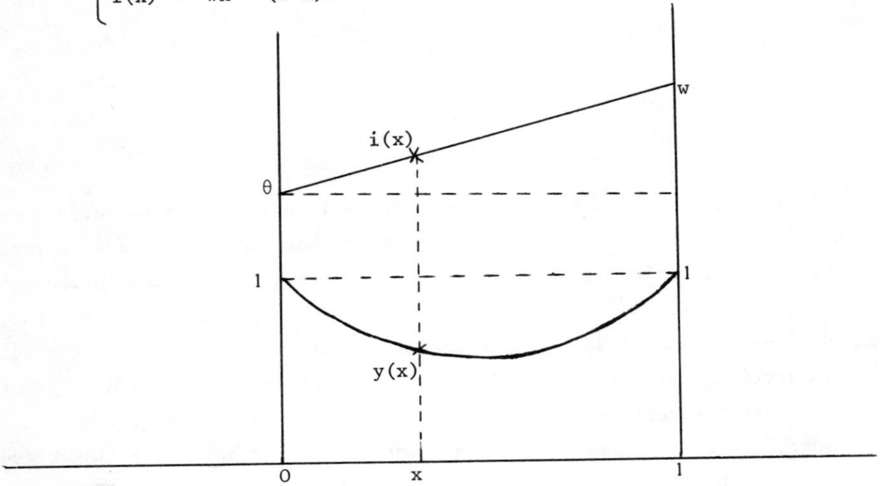

In (1.1) 2 parameters may be random : σ and w .

σ depends upon how much activator is in the system and this quantity of activator is not well-known. w is the concentration of inhibitor in the 2nd compartment.

To control the system we have at our disposal θ , inhibitor concentration in the 1rst compartment. Moreover the control, to be efficient, will have to work in a feedback closed loop from an observation of the system.

In the present case observation is the flux of substrate entering the membrane at x=0, that is $-y'(0)$. Therefore controls will be of the form

$$(1.2)\qquad \theta = u(y'(0))$$

where u is some function from $\mathbb{R}$ into $\mathbb{R}^+$. The cost function to minimize is the average deviation between $y'(0)$ and a fixed value z_d :

$$(1.3)\qquad \min_{u \in \mathcal{U}_{ad}} E|y'(0) - z_d|^2$$

where $\mathcal{U}_{ad}$ is some fixed subspace of functions with values in $\mathbb{R}^+$. For this apparently very simple unidimensionnal problem we are however faced with 2 main difficulties :

- using feedbacks (1.2) leads to a boundary value problem of the type

$$(1.4)\qquad y''(x) = F(y(x),y'(0),x), \quad y(0) \text{ and } y(1) \text{ given}$$

Existence and unicity of a solution for (1.4) are not standard. In section 2 we shall see that we must, for instance, impose to the feedback law u(.) to be monotone decreasing. This condition has a physical meaning (see remark 2.3) and led us to the notion of regulatory feedback.

. Another difficulty is the stochastic aspect of the control problem. We can no more use a variationnal approach, as in BENSOUSSAN (1) . Here we used an algorithm, called of "independant simulations", which is an extension of the strong law of large numbers to stochastic control problems. This method was already tested in the framework of stochastic dynamic programming (J.P. QUADRAT (5) , J.P. QUADRAT, M. VIOT (6)) and finds here a new field of applications (see section 3). In computations we looked for an optimum of (1.3) in a class of linear feedbacks :

(1.5)
$$\begin{cases} u(y'(0)) = \left(\alpha\,(y'(0)-z_d) + \beta\right)^+ \\ \alpha \leqslant 0 \quad \text{and bounded} \\ \beta \geqslant 0 \quad \text{and bounded} \end{cases}$$

This paper gives only a presentation of the ideas and of the main results. A more detailed study including proofs can be found in (7) .

II - EXISTENCE AND UNICITY OF A SOLUTION FOR THE STATE EQUATIONS

Let us call

(2.1)
$$u : \mathbb{R} \to \mathbb{R}^+ \quad \text{a continuous application}$$
$$\Omega = \mathbb{R}^2_+ = \{ (\sigma, w) \mid \sigma \geqslant 0 \ , \ w \geqslant 0 \}$$
$$\mu = \text{some measure of probability on } \Omega$$

Let us assume that

(2.2)
$$\int_\Omega (\sigma^2 + w^2)\, d\mu < \infty$$

We wish to solve the stochastic system

(2.3)
$$y'' = \frac{\sigma\, y}{1+i+|y|} \quad ; \quad \text{p.p.}x \in \left]0,1\right[\quad ; \text{a.s.}\omega$$
$$y(0) = y(1) \qquad \text{(p.s.)}$$

(2.4)
$$i(x,\omega) = w\, x + (1-x)\ \theta(\omega)$$

(2.5)
$$\theta(\omega) = u(y'(0,\omega)) \qquad \text{(a.s.)}$$

For a given ω , one can prove existence (but not unicity) of a solution for problem (2.3), (2.4), (2.5) by compacity methods (J.L. LIONS (4)). Then one finds a solution to the stochastic problem using a "measurable sections" theorem. (For a similar situation see A. BENSOUSSAN, R. TEMAM (2)). Therefore we can state the following result :

Theorem 2.1

Under hypothesis (2.1),(2.2), the stochastic system (2.3)-(2.5) admits a solution y such that

$$(2.6) \qquad y \in L^2(\Omega,\mu;H^2(0,1)) \; ; \quad 0 \leqslant y \leqslant 1 \quad \text{(a.s.)}$$

Remark 2.1

Without any feedback (2.5), the system would be

$$(2.7) \qquad y'' = \frac{\sigma \; y}{1+wx+(1-x)\theta+|y|} \quad ; \quad y(0) = y(1) = 1$$

The second member being monotone increasing with respect to y, one gets easily unicity of the solution of (2.7) for σ,w,θ given and positive.
Unfortunately, in our problem we lose this monotonicity because of (2.5).
Therefore the unicity problem must be approached in a different way. ■
For $\sigma > 0$, $w \geqslant 0$ given and θ varying in $\mathbb{R}^+$, let us call y_θ the solution of (2.7). Then we can prove the

Lemma 2.1

The function $v : \theta \to y'_\theta(0)$ is continuous, strictly increasing, with

for $\theta = 0$ $\qquad -\infty < y'_o(0) < 0$
for $\theta \to \infty$ $\qquad y'_\theta(0) \to 0$

Remark 2.2

For $\sigma > 0$ and $w \geqslant 0$ given, let us draw in the plane $(y'(0),\theta)$ the graphs of the function $v : \theta \to y'_\theta(0)$ and of a feedback $u:y'(0) \to \theta$.

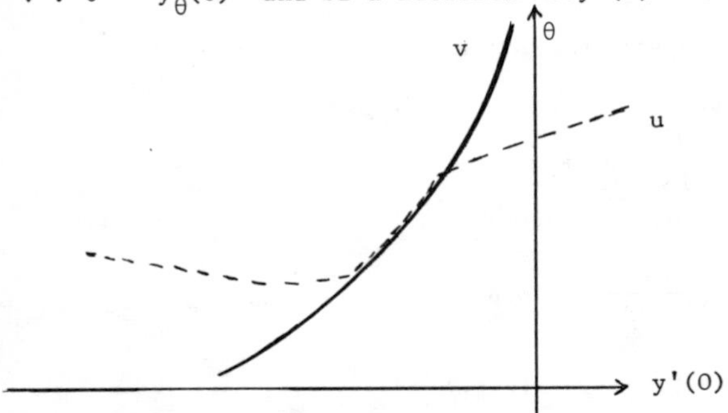

Every point of their intersection is such that

$$(2.8) \qquad \theta = u(y'_\theta(0))$$

i.e. the constraint (2.5) is satisfied. Therefore, in general, with any feedback law u, there is not a unique way for the system to work. For instance, if $u=v^{-1}$, it is clear that every solution y_θ, $\theta \geqslant 0$, of (2.7) will verify (2.8). But if u is monotone decreasing, v being strictly increasing, the 2 graphs have only one point of intersection, and this for every $\omega = (\sigma,w)$, $\sigma > 0$, $w \geqslant 0$, the case $\sigma = 0$ being trivial.
Therefore we have shown the

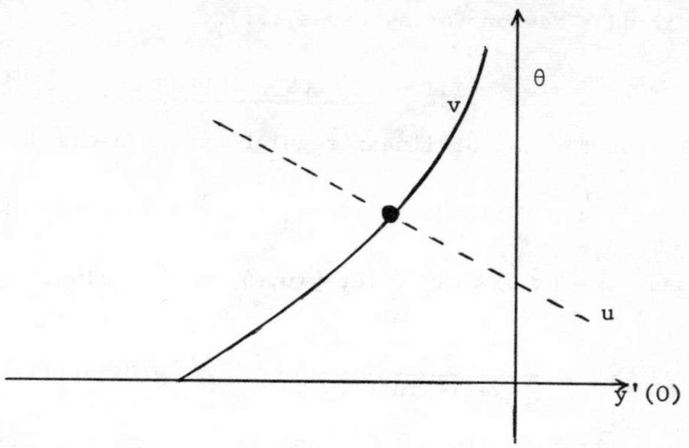

Theorem 2.2

If u is continuous, monotone decreasing from $\mathbb{R}$ into $\mathbb{R}^+$, the stochastic system (2.3-2.5) admits a unique solution.

Remark 2.3

The choice of a monotone decreasing feedback law u implies that when the flux of substrate entering the membrane, $-y'(0)$, is becoming less intense, the feedback u regulates it by lessening the inhibitor concentration in the 1rst compartment, so that the transformation of substrate in product increases, y decreases in the membrane and $-y'(0)$ increases. One can check the same regulatory effect of the feedback law u when $-y'(0)$ is increasing. Therefore one gets a stable steady state. This is what is expressed by the preceding result of unicity. We shall call regulatory the monotone decreasing feedbacks.

In the following sections we shall work mainly with linear regulatory feedbacks of the form

$$u(y'(0)) = \left(\alpha(y'(0)-z_d) + \beta\right)^+ \quad (\alpha \leqslant 0) \tag{2.9}$$

Numerical integration for ω fixed.

For a given $\omega = (\sigma, w)$ and a given feedback law (2.9) the system (2.3), (2.4), (2.9) is solved using the following under-relaxation method : let L_ε be the operator from $H^2(0,1)$ into $H^2(0,1)$ defined by

$$L_\varepsilon(y) = \varepsilon\left(1 - x \int_0^1 (1-s)F(y,y'(0))ds + \int_0^x (x-s)F(y,y'(0))ds\right) + (1-\varepsilon)y \tag{2.10}$$

$$F(y,y'(0)) = \frac{\sigma y}{1+wx+(1-x)\left(\alpha(y'(0)-z_d)+\beta\right)^+ + y} \tag{2.11}$$

Then the following sequence is defined

$$y_o = 1 \quad ; \quad y_{n+1} = L_\varepsilon(y_n) \tag{2.12}$$

For ε small enoug there is convergence of (2.12). (The integrals are approximated by a classical method : Newton-Cotes, Gauss,...).

III - THE CONTROL PROBLEM

Let $\mathcal{U}_{ad}$ be the set of linear regulatory feedbacks defined by

(3.1) $$u(y'(0)) = \left(\alpha\,(y'(0)-z_d)+\beta\right)^+ \;;\; -M_1 \leq \alpha \leq 0 \;;\quad 0 \leq \beta \leq M_2.$$

For $u \in \mathcal{U}_{ad}$ and $\omega = (\sigma, w) \in \mathbb{R}_+^2$, let $y(u,\omega)$ be the solution of equation

(3.2) $$y'' = \frac{\sigma\, y}{1+wx+(1-x)u(y'(0))+y} \;;\quad y(0) = y(1) = 1.$$

Let μ be a law of probability on $\mathbb{R}_+^2$ and z_d a number ; the cost function to minimize on $\mathcal{U}_{ad}$ is given by

(3.3) $$J(u) = \int_{\mathbb{R}_+^2} |y'(u,\omega)(0) - z_d|^2 \, d\mu(\omega)$$

It can be shown, using hypothesis (2.2), that the cost function J(.) is continuous with respect to the parameters α and β defining the linear feedback u. So that <u>the control problem</u> (3.1), (3.2), (3.3) <u>admits an optimal solution</u>, for every measure verifying (2.2).

<u>Remark</u> 3.1

When the measure μ is a discrete measure of the form

(3.4) $$\mu = \frac{1}{r} \sum_{j=1}^{r} \delta_{\omega_j}$$

$$\omega_1, \ldots, \omega_r \in \mathbb{R}_+^2 \;,\; \delta_{\omega_j} \text{ Dirac measure at point } \omega_j \,,$$

the problem (3.2), (3.3) can be written

(3.5) $$y_j'' = \frac{\sigma_j y_j}{1+w_j x+(1-x)u(y'(0))+y_j} \;;\quad y_j(0) = y_j(1) = 1 \;;\; j=1,\ldots,r$$

(3.6) $$\min_{u \in \mathcal{U}_{ad}} \quad J(u) = \frac{1}{r} \sum_{j=1}^{r} |y_j'(0) - z_d|^2$$

For any measure μ, the idea is to discretize it in a sequence μ_r

(3.7) $$\mu_r = \frac{1}{r} \sum_{j=1}^{r} \delta_{\omega_j}$$

(3.8) $(\omega_1, \ldots, \omega_j, \ldots)$ = sequence of independant simulations of ω according to the law μ .

Then we must solve the problem (3.5), (3.6); this is possible by purely deterministic methods (see section 4). This procedure is justified by the following result of convergence ; let ϕ_r (resp ϕ) the minimum cost associated to the measure (3.7) (resp the initial measure μ) ; let $\tilde{u}_r = (\tilde{\alpha}_r, \tilde{\beta}_r)$ an optimal linear feedback for μ_r, then

Theorem 3.1

For almost every sequence of independant simulations,

(3.9) $$\lim_r \phi_r = \phi$$

and every convergent subsequence of the sequence $(\tilde{\alpha}_r, \tilde{\beta}_r)$ converges towards an $(\tilde{\alpha}, \tilde{\beta})$ which is optimal for μ.

Remark 3.2

It is clear that theorem 3.1 is an extension of the strong law of large numbers. It can be also expressed in the following abstract form : Let $(X_\alpha(\omega))_{\alpha \in A}$ be a family of integrable random variables and for every $\alpha \in A$, let $(X^j_\alpha(\omega))_{j \geqslant 1}$ be a sequence of independant random variables following the same law that X_α. Then under some hypothesis of continuity of X in α and of compacity of the set A, one proves that :

(3.10) $$\min_{\alpha \in A} \frac{1}{r} \sum_{j=1}^{r} X^j_\alpha(\omega) \text{ converges almost surely towards } \min_{\alpha \in A} E(X_\alpha).$$

Moreover, for ω fixed, let $\tilde{\alpha}_r(\omega)$ such that

(3.11) $$\frac{1}{r} \sum_{j=1}^{r} X^j_{\tilde{\alpha}_r}(\omega) = \min_{\alpha \in A} \frac{1}{r} \sum_{j=1}^{r} X^j_\alpha(\omega).$$

Then every convergent subsequence of the sequence $(\tilde{\alpha}_r(\omega))$ converges towards an $\tilde{\alpha}$ such that

(3.12) $$E(X_{\tilde{\alpha}}) = \min_{\alpha \in A} E(X_\alpha)$$

In our case $\alpha = u$, $X_\alpha(\omega) = |y'(u,\omega)(0) - z_d|^2$. So that the linearity of feedbacks is not essential in the conclusion of theorem 3.1.

IV - OPTIMAL LINEAR FEEDBACK CONTROL

Let $\mathcal{U}_{ad}$ be given by (3.1) and let $(\omega_1, \ldots, \omega_r)$ be r independant simulations of the random parameters $\omega = (\sigma, w)$. From (3.5) and (3.6) the cost function to minimize becomes

(4.1) $$J^r(\alpha,\beta) = \frac{1}{r} \sum_{j=1}^{r} |y'_j(0) - z_d|^2 \; ; \; -M_1 \leqslant \alpha \leqslant 0 \; ; \; 0 \leqslant \beta \leqslant M_2 .$$

We use a gradient method with respect to α and β. Let us assume that,

(4.2) $$\alpha(y'_j(0) - z_d) + \beta > 0 \qquad \forall j = 1 \ldots r,$$

then the partial derivatives $\frac{\partial J^r}{\partial \alpha}$, $\frac{\partial J^r}{\partial \beta}$ can be obtained by the following

Theorem 4.1

The gradient of J^r is given by the relations

(4.3) $$\frac{\partial J^r}{\partial \alpha} = -\frac{1}{r} \sum_{j=1}^{r} \int_0^1 \lambda_j \frac{\partial F_j}{\partial \alpha} dx$$

(4.4) $$\frac{\partial J^r}{\partial \beta} = -\frac{1}{r}\sum_{j=1}^{r}\int_0^1 \lambda_j \frac{\partial F_j}{\partial \beta}\, dx$$

(4.5) $$F_j(y,z,\alpha,\beta) = \frac{\sigma_j y}{1+w_j x+(1-x)\left(\alpha(z-z_d)+\beta\right)+y}$$

the λ_j , $j=1,\dots,r$ being obtained by integration of the primal and dual systems :

(4.6) $$y_j'' = F_j(y_j, y_j'(0),\alpha,\beta) \quad ; \quad y_j(0) = y_j(1) = 1$$

(4.7) $$\begin{cases} \lambda_j'' = \dfrac{\partial F_j}{\partial y}(y_j, y_j'(0),\alpha,\beta)\,\lambda_j \\ \lambda_j(0) = 2(y_j'(0)-z_d) - \displaystyle\int_0^1 \frac{\partial F_j}{\partial z}\lambda_j dx \\ \lambda_j(1) = 0 \end{cases}$$

Remark 4.1

The integration of (4.6) was made by under-relaxation using the operator

(4.8) $$\begin{cases} M_\varepsilon(\lambda) = \varepsilon\{(1-x)\left(2(y'(0)-z_d) - \displaystyle\int_0^1 \frac{\partial F}{\partial z}\lambda\, dx\right) - x\displaystyle\int_0^1 (1-s)\frac{\partial F}{\partial s}\lambda\, ds + \\ + \displaystyle\int_0^x (x-s)\frac{\partial F}{\partial y}\lambda\, ds\} + (1-\varepsilon)\,\lambda \;. \end{cases}$$

The sequence $\lambda_0 = 0, \lambda_{n+1} = M_\varepsilon(\lambda_n)$ is then converging towards a solution of (4.6) if ε is taken small enough.

Remark 4.2

The optimal open loop control problem is included in the preceding one, by taking $\alpha = 0$, β varying in $(0,M_2)$. In the following we can compare the performances of these 2 types of control and verify the improvement given by the feedback part.

NUMERICAL RESULTS

In the following 3 pictures we see respectively :

. In the 1rst one σ is random on $(30;40)$ with a uniform distribution, w is fixed at $w=6$ and we have looked for an optimal open loop control to approach $z_d = -2$. All the curves are between the 2 extreme ones corresponding to $\sigma=30$ and $\sigma=40$

. In the 2nd one σ is still random and w fixed as in the 1rst picture, but now we have looked for an optimal linear feedback control and the values of $y'(0)$ for all the curves between $\sigma = 30$ and $\sigma = 40$ fit very closely to $z_d=-2$.

. In the 3rd picture σ is fixed at 36 and w is a random variable equally distributed between 3.75 and 8.25. This time again an optimal linear feedback control gives a good minimization of the deviation between $y'(0)$ for all the curves and $z_d=-2$.

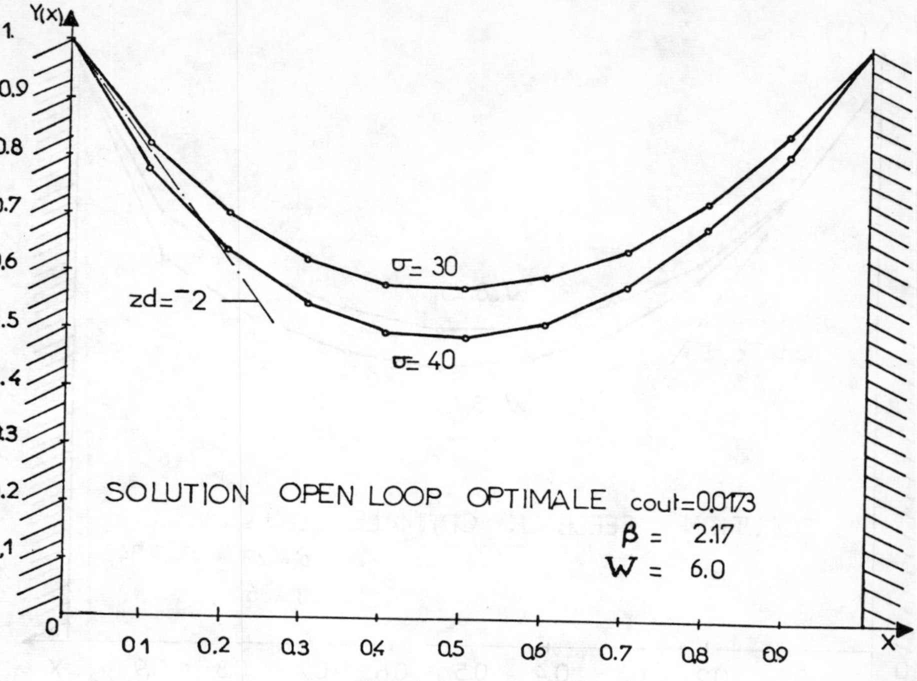
Y(X)
1.
0.9
0.8
0.7
0.6
0.5
0.4
0.3
0.2
0,1
0
σ= 30
σ= 40
zd=⁻2
SOLUTION OPEN LOOP OPTIMALE
cout=0.0173
β = 2.17
W = 6.0
0.1
0.2
0.3
0.4
0.5
0.6
0.7
0.8
0.9
X

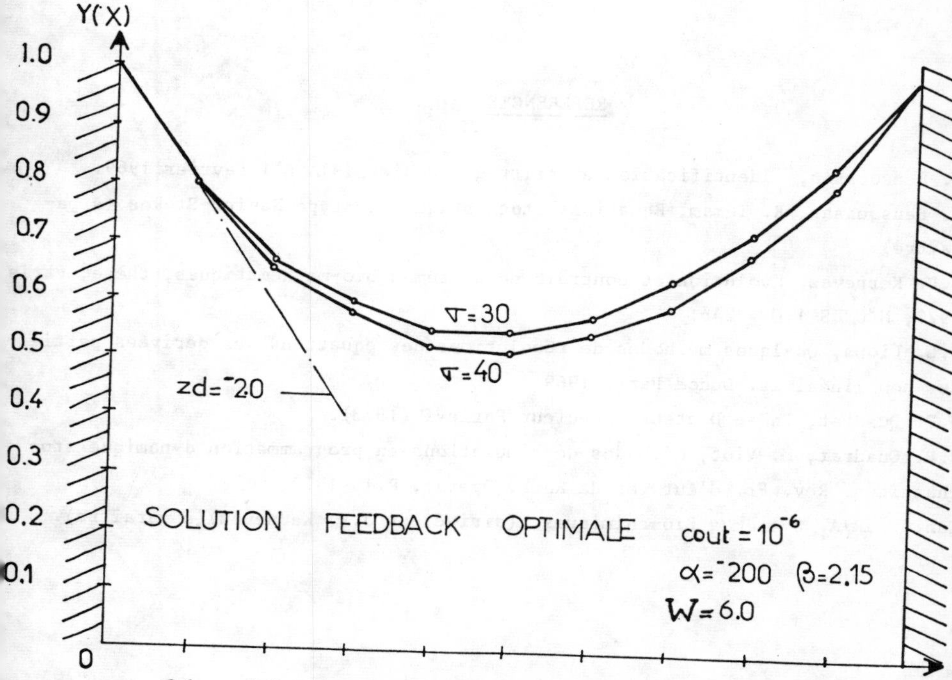
Y(X)
1.0
0.9
0.8
0.7
0.6
0.5
0.4
0.3
0.2
0.1
0
σ=30
σ=40
zd=⁻20
SOLUTION FEEDBACK OPTIMALE
cout = 10⁻⁶
α=⁻200 β=2.15
W=6.0
0.1
0.2
0.3
0.4
0.5
0.6
0.7
0.8
0.9
X

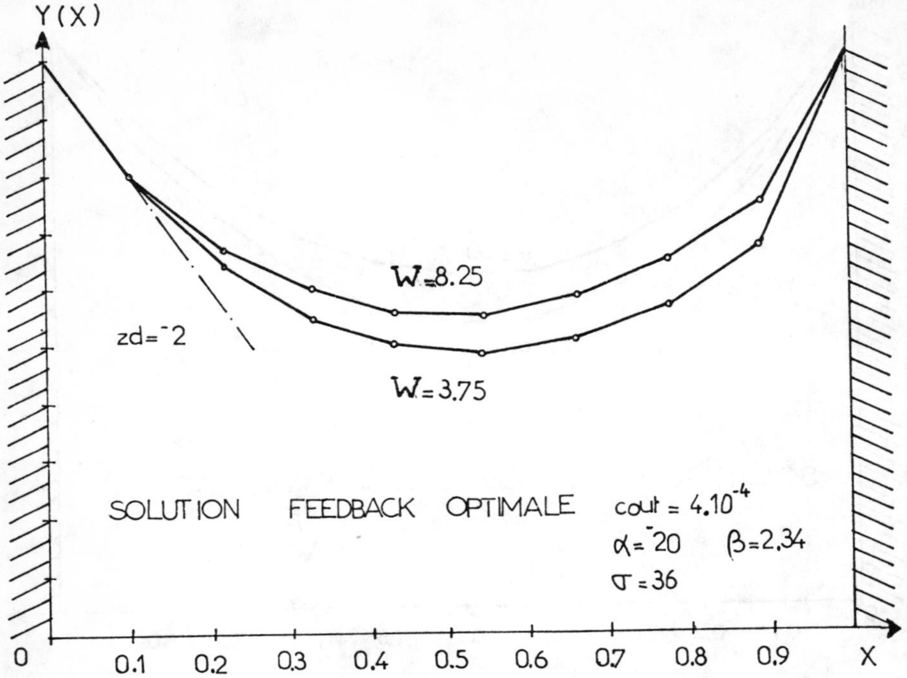

REFERENCES

(1) A. Bensoussan, Identification et filtrage, Cahier IRIA n°1 Février 1969.

(2) A. Bensoussan, R. Temam, Equations stochastiques du type Navier-Stokes (à paraître).

(3) J.P. Kernevez, Evolution et contrôle de systèmes bio-mathématiques, thèse, Paris 1972, N°CNRS 1.O. 7246.

(4) J.L. Lions, Quelques méthodes de résolutions des équations aux dérivées partielles non linéaires. Dunod Paris 1969.

(5) J.P. Quadrat, Thèse Docteur-Ingénieur ParisVI (1973).

(6) J.P. Quadrat, M. Viot, Méthodes de simulations en programmation dynamique stochastique. Rev. Fr. d'Aut. et de Rech. Operat. R.1. 1973.

(7) Cahier IRIA. Systèmes Bio-chimiques (dirigé par J.P. Kernevez)(à paraître).

AN ALGORITHM TO ESTIMATE SUB-OPTIMAL PRESENT VALUES FOR UNICHAIN MARKOV PROCESSES WITH ALTERNATIVE REWARD STRUCTURES

S. Das Gupta
Electrical Engineering Department
Jadavpur University
Calcutta 700029, INDIA

1. INTRODUCTION

Howard's algorithm (3), henceforth to be mentioned as H-algorithm, determines in a closed form, the optimal decision for a class of discrete Markov processes in the infinite horizon and associated alternative reward structures. His approach is reasonably general and is applicable, among others, to problems for discount factor β between 0 and 1 and as well as, with some modification, for $\beta = 1$. The algorithm however becomes rather uneconomic for

(a) large-scale systems

and or (b) discount factors close to unity

Finkbeiner and Runggaldier (2) proposed an algorithm, henceforth to be mentioned as FR-algorithm, which is essentially a sub-optimal algorithm that approached the optimal values by changing from one policy decision to another better one and improving upon the present values, often still further, by some additional iterations to any pre-assigned degree of accuracy. since the process of optimization exhibits contraction properties (4,2), the authors provided a formula to estimate, at the end of each iteration, the number of iterations necessary to bring the estimated present value vector within a pre-assigned neighbourhood of the optimal present value vector. When, in particular, this number is reduced to zero or negative, the last decision and the present value vector are taken as the respective sub-optimal values sought. The advantage of this method lies in the fact that the computation stops when a desired accuracy is reached. However, it also runs into difficulty for cases when, in particular, the discount factor is close to unity.

2. A SUB-OPTIMAL ALGORITHM

For large-scale systems, the approximate apriori estimate of certain quantities help the computational algorithm considerably. Both H and FR algorithms require starting values to initiate the computation. If these are chosen on the basis of the quick initial estimates a large amount of iterative computation may be reduced. This led to the derivation of estimates of steady-state probability and gain in ref.(1). Using the notations of Howard (3), we define the first order estimate of the steady state probability distribution as $\pi^{(1)}$, given by

$$\pi^{(1)} = \frac{1}{N} e^T P \qquad (2.1)$$

and in general, the m th. order estimate of π will be

$$\pi^{(m)} = \frac{1}{N} e^T P^m = \pi^{(m-1)} P \qquad (2.2)$$

where e is a column vector, each element of which is unity and where P is the transition probability matrix. Consequently the m th. order estimate of gain, $g^{(m)}$, will be defined as

$$g^{(m)} = \pi^{(m)} q \qquad (2.3)$$

where q is the immediate reward vector. Evidently, both $\pi^{(m)}$ and $g^{(m)}$ converge respectively to the steady-state probability π and average gain g, as m approaches infinity.

For discounted Markov processes, the estimate of g plays an important role in the present algorithm. Here we start with an arbitrary policy to estimate v_i corresponding to the present value vector v given by

$$v = (I - \beta P)^{-1} q \qquad (2.4)$$

from the approximate expression

$$v_i = (I + \beta P) q \qquad (2.5)$$

where I is an identity matrix of proper order. Then v_i is fed into the policy improvement routine (3) to find a new policy. If the new policy does not match the old policy, a new set of v_i is to be determined by the value determination algorithm according to eqn. (2.5). A little consideration will show that this is already available from the results of the preceding policy iteration algorithm.

Eventually, when there is a match between two consecutive policy decisions, the last policy is taken as the sub-optimal policy and the last estimate of the present value, v_i, is modified as follows to give the corrected estimate of the present value v_{est}

$$v_{est} = v_i + \frac{\beta^2}{1 - \beta} eg \tag{2.6}$$

It is shown in the appendix that a sufficient condition to guarrantee that the value v_{est} is a better estimate of v than the value v_i will be when

$$\pi^{(2)} > \frac{1}{2}\pi \tag{2.7}$$

which is generally not too difficult to satisfy in actual processes.

The estimate v_{est} may then be run for further iterations of FR-algorithm with pre-set error level or used as the starting value of v in the H-algorithm.

3. DISCUSSION OF THE RESULTS

Severalproblems were solved on the computer with various values of discount factor. Howard's Taxi-Cab problem was, for instance, one of them. The results on computing time were compared with those corresponding to the pure FR-algorithm run of the same problems for the same desired accuracy levels. A saving of about 10 to 20% was frequent.

When compared with the corresponding H-algorithm, the sub-optimal policies almost always coincided with the optimal policy*. Thus only one iteration cycle of H-algorithm was necessary

---

*One exception cited in ref. (1) for $\beta = 1$

in such cases.

It is not necessary to use the exact value of g in eqn.(2.6). Estimates of g may be used. Use of $g^{(1)}$ in eqn. (2.6) and of $g^{(4)}$ has shown any hardly noticeable difference in the computer time saved. In all the cases considered, $g^{(4)}$ came within $\frac{1}{2}$% of the value of g.

4. CONCLUSION

A new algorithm to obtain a sub-optimal policy and estimate of present values for a class of discounted discrete Markov processes having alternative reward structure in an infinite horizon have been discussed. Based on initial estimates of steady-state probability and gain, this algorithm determines a policy and estimates the present value vectors which could either be used as it is or in conjunction with FR-algorithm or H-algorithm depending upon the accuracy requirement. In both the later cases it generally accelerates the process of computation.

5. REFERENCE

1. Das Gupta, S., Int. J. Control, Vol 14, No.6,1031-40 (1971)
2. Finkbeiner, B., and Runggaldier,W., Computing methods in Optimization Problem-Vol 2, ed. by Zadeh, L. A., and Balakrishnan, A. V., (1965)
3. Howard, R. A., Dynamic Programming and Markov Processes, Technology Press and Wiley, (1965)
4. Liusternik, L. A., and Sobolev, V. J., Elements of Functional Analysis, Ungar,Rinehart and Winston, New York, (1961)

APPENDIX I

It will be shown here that eqn. (2.6) will give a better estimate of present value provided that the inequality (2.7) is satisfied. We assume that the transition probability matrix , $P = [p_{ij}]$ has distinct eigenvalues with

$$\sum_{j=1}^{N} p_{ij} = 1 \qquad \text{(I-1)}$$

where N is the order of the matrix P. Since the largest eigenvalue of P is 1, we can write the matrix P in the following form (1)

$$P = S + \sum_{i=1}^{N-1} \lambda_i T_i \qquad \text{(I-2)}$$

where S is the constituent matrix of P corresponding to the eigenvalue 1, and T_i are the other constituent matrices with respect to the other (N-1) eigenvalues λ_i. Obviously S is given by,

$$\lim_{n \to \infty} P^m = S = e\pi \qquad \text{(I-3)}$$

where e is a column vector each element of which is unity. In view of (I-3) and eqn. (2.6) we may express the correction term in eqn. (2.6) as

$$\frac{\beta^2}{1-\beta} eg = \sum_{m=2}^{\infty} \beta^m S q \qquad \text{(I-4)}$$

since

$$\pi q = g \qquad \text{(I-5)}$$

The actual present value v, according to eqn. (2.4) is

$$v = (I - \beta P)^{-1} q = (I + \beta P)q + \sum_{n=2}^{\infty} \beta^n P^n q$$
$$= v_i + a \qquad \text{(I-6)}$$

with

$$a = \sum_{n=2}^{\infty} \beta^n P^n q \qquad \text{(I-7)}$$

Similarly

$$v_{est} = v_i + \sum_{m=2}^{\infty} \beta^m S q = v_i + b \qquad \text{(I-8)}$$

with

$$b = \sum_{m=2}^{\infty} \beta^m S q \qquad \text{(I-9)}$$

To find condition for which the distance

$$\| v - v_{est} \|_E \leq \| v - v_i \|_E \qquad \text{(I-10)}$$

we have only to find the condition when

$$a \cdot b - \tfrac{1}{2} \| b \|_E^2 \geq 0 \qquad \text{(I-11)}$$

where a dot between two vectors signify inner product. This implies that

$$\sum_{m=2}^{\infty} \sum_{n=2}^{\infty} \beta^{m+n} q^T (\pi^T \pi^{(n)} - \tfrac{1}{2} \pi^T \pi) q \geq 0 \qquad \text{(I-12)}$$

in view of relations (I-7), (I-9) and (2.2).

Now a sufficient condition that the R.H.S. of eqn (I-12) be positive or zero will be when $\pi^T \pi^{(n)} - \frac{1}{2} \pi^T \pi$ is positive semi-definite for all values of n between 2 and ∞. This in turn requires that

$$\pi^{(n)} - \tfrac{1}{2} \pi > 0 \qquad \text{(I-13)}$$

for all n between 2 and ∞, where, in general, we mean by $x > y$, that each element of x is greater than the corresponding element of y.

According to eqn (2.2),

$$\pi^{(n+1)} = \pi^{(n)}P \tag{I-14}$$

also as

$$\pi = \pi P \tag{I-15}$$

we have

$$(\pi^{(n+1)} - \tfrac{1}{2}\pi) = (\pi^{(n)} - \tfrac{1}{2}\pi)P \tag{I-16}$$

Now if it is known apriori that

$$\pi^{(n)} - \tfrac{1}{2}\pi > 0 \tag{I-17}$$

and thatnone of the columns of P can have all zero entries

$$\pi^{(n+1)} - \tfrac{1}{2}\pi > 0 \tag{I-18}$$

Thus we have only to check if the inequality (2.7) is satisfied to show that v_{est} is a better estimate than v_i

SOME RECENT DEVELOPMENTS IN NONLINEAR PROGRAMMING

by G. Zoutendijk, University of Leyden, Netherlands

I. INTRODUCTION

The general nonlinear programming problem will be defined as

$$\max \{ f(x) \mid x \in R \subset E^n \}, \tag{1}$$

with

$$R = \{ x \in E^n \mid f_i(x) \leq b_i,\ i \in I_1;\ f_i(x) = b_i,\ i \in I_2;\ x \in L \}, \tag{2}$$

in which I_1 and I_2 are finite index sets and $L = \{x \mid Ax \leq c\}$ a convex polyhedron; R is supposed to be connected and to satisfy some regularity conditions (like being the closure of its interior); the function $f(x)$ is continuous. Usually differentiability of the functions f and f_i and existence of second partial derivatives is also assumed.

Three special cases can be distinguished :

1. All constraints are linear, to be subdivided into
   a. linear programming if $f(x)$ is linear;
   b. quadratic programming if $f(x)$ is quadratic and
   c. (general) linearly constrained nonlinear programming.

   From a computational point of view two important subclasses of the last class may be considered, the nearly linear problems (few nonlinearities in the objective function of a relatively simple nature and many linear constraints) and the highly nonlinear problems (few variables and constraints and a highly nonlinear objective function).
2. There are no constraints: unconstrained optimization.
3. There are also nonlinear constraints.

   Again it makes sense to consider the subclasses of nearly linear and highly nonlinear problems. Some methods will only work for convex programs (f concave, R convex).

II. LINEAR PROGRAMMING AND UNCONSTRAINED OPTIMIZATION

In linear programming the problems are usually large and structured; there are relatively few non-zero elements in the coefficients matrix. The productform algorithm has been successfully applied to the solution of these large problems. Re-inversion techniques have gradually become more sophisticated in that they better succeed in representing the inverse of the basis by means of a minimum number of non-zero elements. For reasons of numerical stability special decomposition methods are being applied for the inverse. For this the reader is referred to Bartels and Golub (1969) as well as to Forrest and Tomlin (1972). Many special methods have been developed for special structures. Much success has been obtained with the so-called generalized upper bound technique (see Dantzig and Van Slyke, 1967).

In unconstrained optimization (max $f(x)$) most of the methods are hillclimbing methods. Most widely used is the variable metric method. Writing $g^k = \nabla f(x^k)$ and $\Delta g^k = g^{k+1} - g^k$, the formulae are :

x^0 arbitrary,

$H_0 = I$ (or any other n by n positive definite and symmetric matrix),

$s^k = H_k g^k \; (k = 0, 1, 2 \ldots)$,

$$H_{k+1} = H_k - \lambda_k \frac{s^k (s^k)^T}{(\Delta g^k)^T s^k} - \frac{H_k \Delta g^k (\Delta g^k)^T H_k}{(\Delta g^k)^T H_k \Delta g^k},$$

$x^{k+1} = x^k + \lambda_k s^k$ with

λ_k determined by solving the one dimensional problem max $f(x^k + \lambda s^k)$.

This method like most unconstrained optimization methods has the quadratic termination property, i.e. it is finite for $f(x)$ quadratic, $f(x) = p^T x - \frac{1}{2} x^T C x$. In that case it can be easily shown that the following relations hold :

1. $H_n = C^{-1}$,
2. $(s^i)^T C s^j = 0, i \neq j$ or equivalently $(\Delta g^i)^T s^j = 0 \; (j > i)$,

the directions s^i are mutually conjugate.

This variable metric method, suggested by Davidon and further developed by Fletcher and Powell (1963), is a member of a class of methods (see Broyden, 1967).

Writing $q^k = \frac{H_k \Delta g^k}{(\Delta g^k)^T H_k \Delta g^k} - \frac{s^k}{(\Delta g^k)^T s^k}$ and H_{k+1}^{VM} for the matrix updated according to the variable metric method the update formula for a general member of the family reads :

$$H_{k+1} = H_{k+1}^{VM} + \alpha q^k (q^k)^T, \quad \alpha \text{ arbitrary.}$$

Recently Dixon (1972) has shown that all members of the family generate identical points and directions (provided the line search is carried out in a perfect way). This does not mean that all methods are the same in practice when line searches are not carried out in a perfect way and numerical stability of the H-matrix becomes of importance. In practice it has sometimes be worthwhile to reset the H-matrix to the unit matrix periodically. Research is going on to find methods in which no line searches are required.

Alternatively for the unconstrained maximization problem one could use a method of conjugate (feasible) directions.

These methods work along the following lines :

1. x^0 arbitrary,
2. for $k = 0, 1, \ldots, n-1$:

   in x^k require for s^k : $(\Delta g^h)^T s^k = 0, \; h = 0, 1, \ldots, k-1,$
   $(g^k)^T s^k > 0;$

   $x^{k+1} = x^k + \lambda_k s^k$ with λ_k maximizing $f(x^k + \lambda s^k)$;
3. after n steps :
   a. either start afresh with $x^0(\text{new}) = x^n(\text{old})$;
   b. or use a moving tableau, i.e. require $(\Delta g^h)^T s^k = 0$ for
   $h = k-n+1, \; k-n+2, \ldots, k-1.$

Depending on additional requirements to fix the directions s^k, i.e. depending on the direction generator chosen another method will result. Variant b. in step 3 is usually better from the computational point of view. This family of methods of conjugate directions has been first proposed by Zoutendijk (1960, 1970[a], 1973). Recently the computational aspects and convergence properties of some of these methods have been worked out in detail. All these methods have the quadratic termination property.

A special method from this class is the following :

1. x^0 arbitrary, $P_0 = I$, calculate g^0, $l = 0$;

2. for $k = 0, 1, \dots n-1$:

   a. $s^k := \max \{(g^k)^T s \mid (\Delta g^h)^T s = 0,\ h = 0, 1, \dots, k-1;\ s^T P_l s \leq 1\}$;

   b. $\lambda_k := \max f(x^k + \lambda s^k)$;

   c. $x^{k+1} = x^k + \lambda_k s^k$, calculate $g^{k+1} = \nabla f(x^{k+1})$;

3. near-optimality test; if not passed :

4. $P_{l+1}^{-1} = \sum_{h=0}^{n-1} - \lambda_h \frac{s^h (s^h)^T}{(\Delta g^h)^T s^h}$, $x^0(\text{new}) = x^n(\text{old})$; $l := l+1$, go to 2 .

To solve the direction problem 2 a. we need P_l^{-1} rather than P_l , so that no inversion is necessary after step 4. It is even possible to give an explicit formula for s^k :

$$s^k = Q_k g^k;\quad Q_0 = P_l^{-1};\quad Q_{k+1} = Q_k - \frac{Q_k \Delta g^k (\Delta g^k)^T Q_k}{(\Delta g^k)^T Q_k \Delta g^k}.$$

Since for a quadratic function $P_1^{-1} = C^{-1}$ the matrix P_l^{-1} in the general case can be considered to be an approximation to the inverse Hessian, so that the first step of each cycle of n steps is a quasi Newton step. For a quadratic function the method will terminate after at most n steps. If the steplengths are arbitrarily chosen during the first n steps and in step $n+1$ $\lambda = 1$ is chosen, then the method will also terminate in the maximum of the quadratic function. For a general function we may therefore expect that this metricized norm method is less crucially dependent on the accuracy of line searches which might be an important advantage. The method has been developed by Hestenes (1969) and, independently, by Zoutendijk (1970^b).

III. LINEARLY CONSTRAINED NONLINEAR PROGRAMMING.

The linearly constrained nonlinear programming problem

$$\max \{ f(x) \mid Ax \leq b \} \qquad (3)$$

can be solved by applying one of the methods of conjugate feasible directions.

1. $x^0 \in \{x \mid Ax \leqq b\}$, $s^0 \in \{s \mid a_{i.}^T s \leqq 0 \text{ if } i \in I(x^0)\} \cap \{s \mid (g^0)^T s > 0\}$

   (here $a_{i.}^T$ are the rows of A and $I(x^0) = \{i \mid a_{i.}^T x^0 = b_i\}$);

2. $\lambda_k = \min(\lambda_k', \lambda_k'')$ with $\lambda_k' := \max f(x^k + \lambda s^k)$ and $\lambda_k'' = \max\{\lambda \mid A(x^k + \lambda s^k) \leqq b\}$;

3. if $\lambda_k = \lambda_k'$ add $(\Delta g^k)^T s = 0$ to the direction problem;

   if $\lambda_k = \lambda_k''$ add $a_{i.}^T s = 0$ to the direction problem for the hyperplanes just hit and omit the conjugacy relations.

4. s^k has to satisfy all the relations added to the direction problem during previous steps as well as $(g^k)^T s^k > 0$. If no s^k can be found either the oldest one of the relations $(\Delta g^k)^T s = 0$ has to be omitted or one of the relations $a_{i.}^T s = 0$ has to be replaced by $a_{i.}^T s \leqq 0$ (the one with the most negative dual variable should be taken).

   If, for some i, $a_{i.}^T s^k < 0$, then remove this relation from the next direction problem.

5. $x^{k+1} = x^k + \lambda_k s^k$.

Again any direction generator may be chosen, so that we have outlined a whole class of methods.

It is also possible to adapt a variable metric method to problems involving non-linear constraints. One of the possibilities is :

At x^0 : find s^0 by solving through complimentary pivoting (see Zoutendijk, 1973)

$$\max\{(g^0)^T s \mid a_{i.} s \leqq 0,\ i \in I(x^0);\ s^T s \leqq 1\}.$$

The solution can be written in the form $s^0 = H_0 g^0$ with $H_0 = I - A(x^0)^T\{A(x^0)A(x^0)^T\}^{-1} A(x^0)$, $A(x^0)$ consisting of those $a_{i.}$, $i \in I(x^0)$ for which $a_{i.}^T s^0 = 0$ and the dual variable > 0 (essential constraints in x^0).

At x^k, $k = 0, 1, 2, \ldots$:

$$s^k = H_k g^k,\quad \lambda_k = \min(\lambda_k', \lambda_k''),\quad x^{k+1} = x^k + \lambda_k s^k;$$

if $\lambda_k = \lambda_k'$ update H_k according to a variable metric formula;

if $\lambda_k = \lambda_k''$ then $H_{k+1} = H_k - \dfrac{H_k a_{r.} a_{r.}^T H_k}{a_{r.}^T H_k a_{r.}}$ ($a_{r.}^T x = b_r$ being the hyperplane just hit);

If p is the number of rows of $A(x^0)$, then start afresh after $n-p$ steps with x^0 (new) $= x^{n-p}$ (old).

IV. GENERAL NONLINEAR PROGRAMMING

In general nonlinear programming we can distinguish :

1. direct methods;
2. barrier function and penalty function methods;
3. primal-dual methods;
4. special methods for special problems.

As far as the special methods are concerned we may mention :

a. Separable programming (objective and constraint functions separable, see Miller (1973));

b. Geometric programming for problems of the type

$$\min \{ h_0(x) \mid h_i(x) \leq 1, \quad i=1,\dots,m; \quad x > 0 \},$$

in which the $h_i(x)$ are posynomials, i.e. functions of the type

$$h(x) = \sum_{k=1}^{p} c_k \left\{ \prod_{j=1}^{n} x_j^{d_{kj}} \right\}, \; c_k > 0, \; x_j > 0.$$

It can be shown that the dual of a geometric program is a linearly constrained nonlinear programming problem (see Zangwill (1969)).

c. Convex programming ($f(x)$ concave, R convex);

for these problems we have the cutting plane method, developed by Cheney and Goldstein (1959) and - independently - by Kelley (1960) as well as a decomposition method developed by Wolfe (1967). These methods are dual to each other.

The direct methods are especially suited for the nearly linear problems (large, structured, few nonlinearities). There are three different approaches :

a. Direct extension of the methods of feasible directions; however, instead of requiring $a_i^T s \leq 0$ like in the case of linear constraints we require

$\nabla f_i(\tilde{x})^T s < 0$ when $f_i(\tilde{x}) = b_i$ (in practice when $f_i(\tilde{x}) \geqq b_i - \varepsilon$). Hence we restrict the search to the interior of the cone of feasible directions. One way of doing this has been outlined by Zoutendijk (1960).

b. Interior point methods like the modified feasible directions method (Zoutendijk, 1966). Having an interior point $\tilde{x}$ (interior with respect to the nonlinear constraints) and a linearization $L(\tilde{x}) \supset R$ we maximize $\nabla f(\tilde{x})^T x$ within $L(\tilde{x})$ leading to a solution $\tilde{y}$; we then maximize $f(\tilde{x} + \lambda(\tilde{y} - \tilde{x}))$ as a function of λ within R which may either result in an interior maximum x' or in a boundary point $\tilde{z}$. In the latter case the linear relation $\nabla f_i(\tilde{z})^T (x - \tilde{z}) \leqq 0$ with i denoting the constraint just hit will be added to the linearized constraints set while a new interior point will be chosen on the line connecting $\tilde{x}$ and $\tilde{z}$. If R is convex no feasible point will be cut off. In the former case the conjugacy relation $\{\nabla f(x') - \nabla f(\tilde{x})\}^T (x - x') = 0$ can be added to the linearized constraints set while x' can be taken as new interior point. This procedure can be adapted to non-convex regions; nonlinear equalities are difficult to handle, however. Conjugacy relations have to be omitted if no progress can otherwise be made.

c. Hemstitching methods where in $\tilde{x} \in R$ we are allowed to make a small step in a direction tangent to the cone of feasible directions, so that in the case of a nonlinear constraint we will leave the feasible region. By projecting the point so obtained onto the intersection of the nonlinear surfaces concerned we obtain a new and better feasible point. This approach is being taken successfully in the Generalized Reduced Gradient Method (Abadie and Carpentier, 1969). Nonlinear equalities can also be handled by this method.

Barrier function and penalty function methods are well-known and widely used to solve nonlinear programs of not too large a size. Let the problem be defined by (1) and (2). Then an example of a mixed method is :

solve max $\{g(x, \rho_k) \mid x \in L\}$ for $\rho_0 > \rho_1 > \rho_2 > \dots \downarrow 0$,

$$g(x, \rho) = f(x) + \rho \sum_{i \in I_1} \log \{b_i - f_i(x)\} - \frac{1}{\rho} \sum_{i \in I_2} \{b_i - f_i(x)\}^2,$$

The starting point x^0 should satisfy the relations $f_i(x^0) < b_i$, $i \in I_1$. The same will then hold for the subproblem solution $\hat{x}^0$ which will be starting point for the next subproblem, etc. See further Fiacco and Mc Cormick (1968) and Lootsma (1970).

In primal-dual methods we try to improve the primal and the dual variables more or less simultaneously. This could for instance be done by using a generalized Lagrangean function (Roode, 1968), i.e. a function such that

1. $\max_{x \in L} \min_{u \in D} \psi(x,u) = \min_{u \in D} \max_{x \in L} \psi(x,u)$

   with D being a convex subset of a space of a certain dimension p.

   The two problems will be called the primal and the dual problem, respectively.

2. The primal problem is equivalent to the original problem.

3. The dual problem can be solved relatively easily.

In that case it makes sense to solve the dual problem instead. Writing

$$\varphi(u) = \max_{x \in L} \psi(x,u)$$

we have to solve the problem

$$\min \{\varphi(u) \mid u \in D\}$$

which is unconstrained if $D = E^p$ and linearly constrained $D = \{u \in E^p \mid u \geq 0\}$. Each function evaluation $\varphi(u)$ entails the solution of a linearly constrained maximization problem in x.

From this it follows that for reasons of computational efficiency the number of function evaluations should be as small as possible. Buys (1972) has developed a method of this type for the function

$$\psi(x,u) = f(x) - \sum_{i \in I} h\{f_i(x), u_i\},$$

$$\begin{aligned} h\{f_i(x), u_i\} &= u_i\{f_i(x) - b_i\} + \tfrac{1}{2}\beta\{f_i(x) - b_i\}^2 \quad \text{if } i \in I_2 \text{ or } i \in I_1 \text{ and } f_i(x) - b_i \geq -\frac{u_i}{\beta}; \\ &= -\frac{u_i^2}{2\beta} \quad (\beta > 0) \qquad \text{if } i \in I_1 \text{ and } f_i(x) - b_i \leq -\frac{u_i}{\beta} \end{aligned}$$

Recently Robinson (1973) has reported another primal-dual method which looks quite promising. Expanding $f_i(x)$ in a Taylor serious with respect to x^k:

$f_i(x) = f_i(x^k) + \nabla f_i(x^k)^T (x - x^k) + \ldots$ and writing

$$l_i^{(k)}(x) = f_i(x^k) + \nabla f_i(x^k)^T (x - x^k),$$

assuming that at step k we have available vectors x^k, u^k we solve the linearly constrained problem :

$$\max \left\{ f(x) - \sum u_i^k \left(f_i(x) - l_i^{(k)}(x)\right) \,\middle|\, l_i^{(k)}(x) \leq b_i \,,\ i \in I_1 ;\ l_i^{(k)}(x) = b_i \,,\ i \in I_2 ;\ x \in L \right\}$$

which results in a new point x^{k+1} and a new dual solution u^{k+1} (dual variables of the linearized constraints). Convergence to the original nonlinear programming problem can be proved.

APPLICATIONS

Although nonlinear programming methods have been and are being applied to many different problems - makroeconomic planning, refinery scheduling/plant optimization, design and control engineering, economic growth models, pollution abatement models, approximation under constraints and resource conservation models - the number of applications is still limited. There are several reasons for this. First the is highly nonlinear, non-convex, so that local optima cannot be avoided. Secondly the word nonlinear is negative by definition. Often there is little theoretical knowledge about the process other than that the relation between certain variables is not linear; empirical relations without theoretical foundation might be dangerous to use since they might not hold anymore after some time.

Then there is the problem of data organization and model updating which is already tremendous in the linear programming case. Finally there are few computer codes available and those available are not very sophisticated; up to now there is little commercial motivation for computer manufacturers and software houses to supply these codes. This might change in the future, however, when the need increases to obtain better solutions to some of the nonlinear problems we have to face.

REFERENCES

Abadie, J. and Carpentier, J.	1969	Generalization of the Wolfe reduced gradient method to the case of nonlinear constraints, pp. 37-47 of R. Fletcher (ed.), Optimization, Academic Press.
Bartels, R.H. and Golub, G.H.	1969	The simplex method of linear programming using LU decomposition, Comm. ACM 12, 266-268.
Broyden, C.G.	1967	Quasi-Newton methods and their application to function minimization, Math. of Computation 21, 368-381.
Buys, J.D.	1972	Dual algorithms for unconstrained optimization problems, thesis, Univ. of Leyden.
Cheney, E.W. and Goldstein A.A.	1959	Newton's method for convex programming and Tchebycheff approximation, Num. Math. 1, 253-268.
Dantzig, G.B. and Van Slyke, R.	1967	Generalized upper bounding techniques, J. Comp. Systems Sci. 1, 213-226.
Dixon, L.C.W.	1972	Quasi-Newton algorithms generate identical points, Math. Progr. 2, 383-387.
Fiacco, A.V. and Mc Cormick,G.P.	1968	Nonlinear programming: sequential unconstrained minimization techniques, Wiley.
Fletcher, R. and Powell, M.J.D.	1963	A rapidly converging descent method for minimization, Computer Journal 6, 163-168.
Forrest, J.J.H. and Tomlin, J.A.	1972	Updated triangular factors of the basis to maintain sparsity in the product form simplex method, Math. Progr. 2, 263-278.
Hestenes, M.R.	1969	Multiplier and gradient methods, Journal Optimization Theory and Appl. 4, 303-320.

Kelly, J.E. 1960 The cutting plane method for solving convex programs, J. Soc. Industr. Appl. Math. 8, 703-712.

Lootsma, F.A. 1970 Boundary properties of penalty functions for unconstrained minimization, thesis, Eindhoven Techn. University.

Miller, C. 1963 The simplex method for local separable programming, pp. 89-100 of R.L. Graves and P. Wolfe (eds), Recent Advances in Math. Programming, Mc Graw-Hill.

Robinson, S.M. 1972 A quadratically-convergent algorithm for general nonlinear programming problems, Math. Progr. 3, 145-156.

Roode, J.D. 1968 Generalized Lagrangean Functions, thesis, Univ. of Leyden.

Wolfe, P. 1967 Methods of nonlinear programming, pp. 97-131 of J. Abadie (ed.), Nonlinear Programming, North-Holland.

Zangwill, W.I. 1969 Nonlinear programming, Prentice-Hall.

G. Zoutendijk 1960 Methods of Feasible Directions, Elsevier.

1966 Nonlinear programming: a numerical survey, J. Soc.Industr. and Appl.Math.Control 4, 194-210.

1970[a] Nonlinear programming, computational methods, pp. 37-85, in J. Abadie (ed.), Nonlinear and integer programming, North-Holland.

1970[b] Some algorithms based on the principle of feasible directions, pp. 93-122 of J.B. Rosen, O.L. Mangasarian and K. Ritter, (eds.), Nonlinear programming, Academic Press.

1973 On linearly constrained nonlinear programming, in: A.R. Goncalvez (ed.), Proceedings of the Figueira da Foz Nato Summerschool on Integer and Nonlinear Programming.

PENALTY METHODS AND AUGMENTED LAGRANGIANS IN NONLINEAR PROGRAMMING

R. Tyrrell Rockafellar
Dept. of Mathematics, University of Washington
Seattle, Washington 98195 U.S.A.

The usual penalty methods for solving nonlinear programming problems are subject to numerical instabilities, because the derivatives of the penalty functions increase without bound near the solution as computation proceeds. In recent years, the idea has arisen that such instabilities might be circumvented by an approach involving a Lagrangian function containing additional, penalty-like terms. Most of the work in this direction has been for problems with equality constraints. Here some new results of the author for the inequality case are described, along with references to the current literature. The proofs of these results will appear elsewhere.

Equality Constraints

Let $f_o, f_1, \ldots, f_m$ be real-valued functions on a subset X of a linear topological space, and consider the problem

(1) $\quad$ minimize $f_o(x)$ over $\{x \in X \mid f_i(x) = 0 \text{ for } i=1,\ldots,m\}$.

The augmented Lagrangian for this problem, as first introduced in 1958 by Arrow and Solow [2], is

$$L(x,y,r) = f_o(x) + \sum_{i=1}^{m} [rf_i(x)^2 + y_i f_i(x)], \tag{2}$$

where $r \geq 0$ is a penalty parameter and $y = (y_1, \ldots, y_m) \in R^m$. In fact, this is just the ordinary Lagrangian function for the altered problem in which the objective function f_o is replaced by $f_o + rf_1{}^2 + \ldots + rf_m{}^2$, with which it agrees for all points satisfying the constraints.

The motivation behind the introduction of the quadratic terms is that they may lead to a representation of a local optimal solution in terms of a local unconstrained minimum. If $\bar{x}$ is a local optimal solution to (1) with corresponding Lagrange multipliers $\bar{y}_i$, as furnished by classical theory, the function

$$L_o(x,\bar{y}) = f_o(x) + \sum_{i=1}^{m} \bar{y}_i f_i(x)$$

*This work was supported in part by the Air Force Office of Scientific Research under grant AF-AFOSR-72-2269.

has a stationary point at $\bar{x}$ which is a local minimum relative to the manifold of feasible soltuions. However, this stationary point need not be a local minimum in the unconstrained sense, and L may even have negative second derivatives at $\bar{x}$ in certain directions normal to the feasible manifold. The hope is that by adding the terms $rf_i(x)^2$, the latter possibility can be countered, at least for r large enough. It is not difficult to show this is true if $\bar{x}$ satisfies second-order sufficient conditions for optimality (cf. [1]).

The augmented Lagrangian gives rise to a basic class of algorithms having the following form:

$$(3)\quad \begin{cases} \text{Given } (y^k,r^k), \text{ minimize } L(x,y^k,r^k) \text{ (partially ?) in} \\ x \in X \text{ to get } x^k. \text{ Then, by some rule, modify } (y^k,r^k) \\ \text{to get } (y^{k+1},r^{k+1}). \end{cases}$$

Typical exterior penalty methods correspond to the case where $y^{k+1} = y^k = 0$ and $r^{k+1} = \alpha r^k$ (α = some factor > 1). In 1968, Hestenes [10] and Powell [19] independently drew attention to potential advantages of the case

$$(4)\qquad y^{k+1} = y^k + 2r^k \nabla_y L(x^k,y^k,r^k), \quad r^{k+1} \geq r^k.$$

The same type of algorithm was subsequently proposed also by Haarhoff and Buys [9] and investigated by Buys in his thesis [4]. Some discussion may also be found in the book of Luenberger [13]. Recently Bertsekas [3] has obtained definitive results in the case where an ε-bound on the gradient is used as the stopping criterion for the minimization at each stage. These results confirm that the convergence is essentially superlinear when $r^k \to \infty$. Various numerical experiments involving modifications of the Hestenes-Powell algorithm still in the pattern of (3) have been carried out by Miele and his associates [15], [16], [17], [18]; see also Tripathi and Narendra [26]. Some infinite-dimensional applications have been considered by Rupp [24], [25].

An algorithm of Fletcher [6] (see also [7], [8]) may, in one form, be considered also as a "continuous" version of (3) in which certain functions of x are substituted for y and r in $L(x,y,r)$; one then has a single function to be minimized. The original work of Arrow and Solow [2] also concerned, in effect, a "continuous" version of (3) in which x and y values were modified simultaneously in locating a saddle point of L.

Inequality Constraints.

For the inequality-constrained problem,

(P) minimize $f_o(x)$ over $\{x \in X | f_i(x) \leq 0, i = 1,\ldots,m\}$,

it is not immediately apparent what form the augmented Lagrangian should have, but the natural generalization turns out to be

$$(5) \qquad L(x,y,r) = f_o(x) + \sum_{i=1}^{m} \lambda(f_i(x),y_i,r),$$

where

$$(6) \qquad \lambda(f_i(x),y_i,r) = \begin{cases} rf_i(x)^2 + y_if_i(x) & \text{if } f_i(x) \geq -y_i/2r, \\ -y_i^2/4r & \text{if } f_i(x) \leq -y_i/2r. \end{cases}$$

In dealing with (5), the multipliers y_i are <u>not</u> constrained to be nonnegative, in contrast with the ordinary Kuhn-Tucker theory. This Lagrangian was introduced by the author in 1970 [20] and studied in a series of papers [21], [22], [23], the main results of which will be indicated below. It has also been treated by Buys [3] and Arrow, Gould and Howe [1]. Related approaches to the inequality-constrained problem may be found in papers of Wierzbicki [27], [28], [29], Fletcher [7], Kort and Bertsekas [11], Lill [12], and Mangasarian [14].

To relate the augmented Lagrangian to penalty approaches, it should be noted that by taking $y = 0$ one obtains the standard "quadratic" penalty function. Observe also that the classical Lagrangian for problems with inequalities can be viewed as a limiting case:

$$(7) \qquad \lim_{r \downarrow 0} L(x,y,r) = L_o(x,y) = \begin{cases} f_o(x) + \sum_{i=1}^{m} y_if_i(x) & \text{if } y \geq 0, \\ -\infty & \text{if } y \ngeq 0. \end{cases}$$

The following properties of (5)-(6) can be verified [21], [23]: $L(x,y,r)$ is always concave in (y,r), and it is continuously differentiable (once) in x if every f_i is differentiable. Furthermore, it is convex in x if (X and) every f_i is convex; the latter is referred to as the <u>convex case.</u> Higher-order differentiability is not inherited by L from the functions f_i along the "transition surfaces" corresponding to formula (6). However, as will be seen from Theorem 4 below, most of the interest in connection with algorithms and their convergence centers on the local properties of L in a neighborhood of a point $(\bar{x},\bar{y},\bar{r})$ such that $\bar{x}$ is a local optimal solution to (P), $\bar{y}$ is a corresponding multiplier vector in the classical sense of Kuhn and Tucker, and $\bar{r} > 0$. If the multipliers $\bar{y}_i$ satisfy the complementary slackness conditions, as usually has to be assumed in a close analysis of convergence, it is clear that none of the "transition surfaces" will pass through $(\bar{x},\bar{y},\bar{r})$, and hence L will be two or three times continuously differentiable in some neighborhood of $(\bar{x},\bar{y},\bar{r})$, if every f_i has this order of differentiability. (Certain related Lagrangians recently proposed by Mangasarian [14] inherit higher-order differentiability everywhere, but they are not concave in (y,r).)

The class of algorithms (3) described above for the equality case may also be studied in the inequality case. In particular, rule (4) gives an immediate generalization of the Hestenes-Powell algorithm. We have shown in [22] that in the finite-dimensional convex case, this algorithm always converges globally if, say, an optimal solution $\bar{x}$ exists along with a Kuhn-Tucker vector $\bar{y}$. This is true even if the minimization in obtaining x^k is only approximate in a certain sense. The multiplier vectors y^k converge to some particular Kuhn-Tucker vector $\bar{y}$, even though the problem may possess more than one such vector. For convex and nonconvex problems, results on local rates of convergence in the equality case are applicable if the multipliers at the locally optimal solution in question satisfy complementary slackness conditions.

Dual Problem.

The main theoretical properties of the augmented Lagrangian, fundamental to all applications, can be described in terms of a certain dual problem corresponding to the global saddle point problem for L. To shorten the presentation here, we henceforth make the simplifying assumption that X is compact and the functions f_i are continuous. It must be emphasized that this assumption is not required, and that the more general setting is in fact the one treated in [21], [22], [23]. It should also be clear that our focus on inequality constraints involves no real restriction. Mixtures of equations and inequalities can be handled in much the same way.

The dual problem which we associate with (P) in terms of the augmented Lagrangian L is

(D) maximize $g(y,r)$ over all $y \in R^m$ and $r > 0$, where

$$g(y,r) = \min_{x \in X} L(x,y,r) \qquad \text{(finite)}.$$

Note that constraint $y \geq 0$ is not present in this problem. Nor does the condition $r > 0$ represent a true constraint, since, as is easily seen, $g(y,r)$ is nondecreasing as a function of r for every y. Thus the dual problem is one of unconstrained maximization. Further, $g(y,r)$ is concave in (y,r), and in the convex case it is continuously differentiable, regardless of the differentiability of f_i [21].

THEOREM 1[23]. $\min(P) = \sup(D) = \lim g(y^k, r^k)$, <u>where</u> $(y^k, r^k)_{k=1}^{\infty}$ <u>denotes an arbitrary sequence with</u> y^k <u>bounded and</u> r^k .

THEOREM 2[23]. <u>Let</u> $(y^k, r^k)_{k=1}^{\infty}$ <u>denote any sequence with</u> $\sup(D) = \lim g(y^k, r^k)$ <u>and</u> y^k <u>bounded</u> (<u>but not necessarily with</u> $r^k \to \infty$). <u>Let</u> x^k <u>minimize</u> $L(x, y^k, r^k)$ <u>over</u> X <u>to within</u> ε^k, <u>where</u> $\varepsilon^k \downarrow 0$. <u>Then all cluster points of the sequence</u> x^k <u>are optimal solutions to</u> (P).

If $y^k \equiv 0$, Theorem 1 asserts the familiar fact in the theory of penalty functions that

$$(8) \qquad \min(P) = \lim_{r^k \to \infty} \min_{x \in X} [f_o(x) + r^k \sum_{i=1}^{m} \max^2\{0, f_i(x)\}].$$

More generally, it suggests a larger class of penalty-like methods in which still $r^k \to \infty$, but y^k is allowed to vary. Perhaps, through a good rule for choosing y^k, such a method could yield improved convergence and thereby reduce some of the numerical instabilities associated with having $r^k \to \infty$. Theorem 2 even holds out the attractive possibility of algorithms in which both y^k and r^k remain bounded. The fundamental question here is whether a bounded maximizing sequence (y^k, r^k) exists at all for (D). In other words, under what circumstances can it be said that the dual problem has an optimal solution $(\bar{y}, \bar{r})$?

It is elementary from Theorem 1 and the definition of the dual that a necessary and sufficient condition for $(\bar{y}, \bar{r})$ to be an optimal solution to (D) and $\bar{x}$ to be a (globally) optimal solution to (P) is that $(\bar{x}, \bar{y}, \bar{r})$ be a (global) saddle point of L. The following theorems on saddle points therefore show that our question about the existence of bounded maximizing sequences (y^k, r^k) has an affirmative answer for "most" problems.

THEOREM 3 [21]. <u>In the convex case</u>, $(\bar{x}, \bar{y}, \bar{r})$ <u>is a saddle point of</u> L <u>if and only if</u> $(\bar{x}, \bar{y})$ <u>is a saddle point of the classical Lagrangian</u> L_o <u>in</u> (7).

THEOREM 4 [23]. <u>Suppose that</u> $\bar{x} \in \text{int } X \subset R^n$, <u>and that each</u> f_i <u>is differentiable of class</u> C^2 <u>near</u> $\bar{x}$.

(a) <u>If</u> $(\bar{x}, \bar{y}, \bar{r})$ <u>is a global saddle point of</u> L, <u>then</u> $(\bar{x}, \bar{y})$ <u>satisfies the second-order necessary conditions</u> [5, p.25] <u>for local optimality in</u> (P), <u>and</u> $\bar{x}$ <u>is globally optimal</u>.

(b) *If* $(\bar{x},\bar{y})$ *satisfies the second-order sufficient conditions* [5,p.30] *for local optimality in* (P) *and* $\bar{x}$ *is uniquely globally optimal, then* $(\bar{x},\bar{y},\bar{r})$ *is a global saddle point of* L *for all* $\bar{r}$ *sufficiently large.*

Part (b) of Theorem 4 stengthens a local result of Arrow, Gould and Howe [1] involving assumptions of complementary slackness and the superfluous constraint $y \geq 0$. A corresponding local result has also been furnished by Mangasarian [14] for his different family of Lagrangians. It is shown in [23] that the existence of a dual optimal solution $(\bar{y},\bar{r})$ depends precisely on whether (P) has a second-order stability property with respect to the ordinary class of perturbations.

REFERENCES

1. K. J. Arrow, F. J. Gould and S. M. Howe, "A general saddle point result for constrained optimization", Institute of Statistics Mimeo Series No. 774, Univ. of N. Carolina (Chapel Hill), 1971.

2. K. J. Arrow and R. M. Solow, "Gradient methods for constrained maxima, with weakened assumptions", in *Studies in Linear and Nonlinear Programming*, K. Arrow, L. Hurwicz and H. Uzawa (editors), Stanford Univ. Press, 1958.

3. D. P. Bertsekas, "Combined primal-dual and penalty methods for constrained minimization", SIAM J. Control, to appear.

4. J. D. Buys, "Dual algorithms for constrained optimization", Thesis, Leiden, 1972.

5. A. V. Fiacco and G. P. McCormick, *Nonlinear Programming: Sequential Unconstrained Optimization Techniques*, Wiley, 1968.

6. R. Fletcher, "A class of methods for nonlinear programming with termination and convergence properties", in *Integer and Nonlinear Programming*, J. Abadie (editor), North-Holland, 1970.

7. R. Fletcher, "A class of methods for non-linear programming III: Rates of convergence", in *Numerical Methods for Non-linear Optimization*, F. A. Lootsma (editor), Academic Press, 1973.

8. R. Fletcher and S. Lill, "A class of methods for nonlinear programming, II: computational experience", in *Nonlinear Programming*, J. B. Rosen, O. L. Mangasarian and K. Ritter (editors), Academic Press, 1971.

9. P. C. Haarhoff and J. D. Buys, "A new method for the optimization of a nonlinear function subject to nonlinear constraints", Computer J. 13 (1970), 178-184.

10. M. R. Hestenes, "Multiplier and gradient methods", J. Opt. Theory Appl. 4(1969), 303-320.

11. B. W. Kort and D. P. Bertsekas, "A new penalty function method for constrained minimization", Proc. of IEEE Decision and Control

Conference, New Orleans, Dec. 1972.

12. S. A. Lill, "Generalization of an exact method for solving equality constrained problems to deal with inequality constraints", in Numerical Methods for Nonlinear Optimization, F. A. Lootsma (editor), Academic Press, 1973.

13. D. G. Luenberger, Introduction to linear and nonlinear programming, Addison-Wesley, 1973, 320-322.

14. O. L. Mangasarian, "Unconstrained Lagrangians in nonlinear programming", Computer Sciences Tech. Report #174, Univ. of Wisconsin, Madison, 1973.

15. A. Miele, E. E. Cragg, R. R. Iver and A. V. Levy, "Use of the augmented penalty function in mathematical programming, part I", J. Opt. Theory Appl. 8(1971), 115-130.

16. A. Miele, E. E. Cragg and A. V. Levy, "Use of the augmented penalty function in mathematical programming problems, part II", J. Opt. Theory Appl. 8(1971, 131-153.

17. A. Miele, P. E. Moseley and E. E. Cragg, "A modification of the method of multipliers for mathematical programming problems", in Techniques of Optimization, A. V. Balakrishnan (editor), Academic Press, 1972.

18. A. Miele, P. E. Moseley, A. V. Levy and G. M. Coggins, "On the method of multipliers for mathematical programming problems", J. Opt. Theory Appl. 10(1972), 1-33.

19. M. J. D. Powell, "A method for nonlinear optimiztion in minimization problems", in Optimization, R. Fletcher (editor), Academic Press, 1969.

20. R. T. Rockafellar, "New applications of duality in convex programming", written version of talk at 7th International Symposium on Math. Programming (the Hague, 1970) and elsewhere, published in the Proc. of the 4th Conference on Probability (Braşov, Romania, 1971).

21. R. T. Rockafellar, "A dual approach to solving nonlinear programming problems by unconstrained optimization", Math. Prog., to appear.

22. R. T. Rockafellar, "The multiplier method of Hestenes and Powell applied to convex programming", J. Opt. Theory Appl., to appear.

23. R. T. Rockafellar, "Augmented Lagrange multiplier functions and duality in nonconvex programming", SIAM J. Control, to appear.

24. R. D. Rupp, "A method for solving a quadratic optimal control problem", J. Opt. Theory Appl. 9(1972), 238-250.

25. R. D. Rupp, "Approximation of the classical isoperimetric problem", J. Opt. Theory Appl. 9(1972), 251-264.

26. S. S. Tripathi and K. S. Narendra, "Constrained optimization problems using multiplier methods", J. Opt. Theory Appl. 9(1972), 59-70.

27. A. P. Wierzbicki, "Convergence properties of a penalty shifting algorithm for nonlinear programming porblems with inequality constraints", Archiwum Automatiki i Telemechaniki (1970).

28. A. P. Wierzbicki, "A penalty function shifting method in constrained static optimization and its convergence properties", Archiwum Automatyki i Telemechaniki 16(1971), 395-416.

29. A. P. Wierzbicki and A. Hatko, "Computational methods in Hilbert space for optimal control problems with delays", these proceedings.

ON INF-COMPACT MATHEMATICAL PROGRAMS

by

Roger J.-B. Wets

timum is attained at a feasible point) and (ii) the problem is stable (there exist scalars - often called Lagrange multipliers - that can be associated to the constraints and allow us to replace the mathematical problem by an unconstrained problem whose optimal solution is identical to that of the original problem). For convex mathematical problems, stability coincides with solvability of the dual program. These properties correspond to some properties of the variational function of the problem. If our original problem is: Find $\inf f(x)$ subject to $G(x) \leq 0$, where G is a vector valued map; then one way to define the variational function P is as:

$$P(u) = \operatorname*{Inf}_{x}\{f(x) \mid G(x) \leq u\} .$$

It is easy to verify that if $P(0)$ is finite and P is locally Lipschitz at 0 on its effective domain $(\{u \mid P(u) < +\infty\})$ then the original problem is stable; if P is lower semicontinuous at 0 then the problem is assymptotically stable also calles dualizable, see e. g. [6,7,12,13]. In this paper one finds various theorems which allow us to conclude that the variational function possesses the appropriate properties when the objective and the constraints of the original problem satisfy compactness-type assumptions. In order to extend these results to certain classes of control problems and stochastic programs, we develop in section 4 some further properties for composition of inf-compact functions.

2. NOTATIONS AND TERMINOLOGY

The class of functions to be minimized is limited to those with domain a separable reflexive Banach space and range $]-\infty,+\infty]$ and who are not identically $+\infty$. Say f a member of this class, then its effective domain is $\operatorname{dom} f = \{x \mid f(x) < +\infty\}$. The $(\alpha-)$ level set of f is by definition $L_\alpha(f) = \{x \mid f(x) \leq \alpha\}$ with $\alpha \in \mathbb{R}$. It is well known that a function is lower semicontinuous (l.s.c.) if all its level sets are closed. f is inf-compact if $L_\alpha(f)$ is compact for all $\alpha \in \mathbb{R}$. This terminology was introduced by Moreau who developed with Rockafella and Valladier the basic properties of these functions, see e.g. [4,5,8, 11]. An inf-compact function always possesses a minimum. The epigraph of a function f is the subset of $]-\infty,+\infty[\times X$ such that epi f =

Supported by N.S.F. Grant No. GP-31551 May 1973

$\{(\alpha,x) \mid \alpha \geq f(x)\}$. The <u>indicator function</u> of a subset C is denoted by ψ_C and is defined by $\psi_C(x) = 0$ if $x \in C$ and $= +\infty$ otherwise. By κ we shall denote a set valued function from a space U into the subsets of a space X , i.e. for each given u , $\kappa(u)$ is a subset of X possibly the empty set. Typically $\kappa(u) = \{x \mid x$ satisfies some property which depends on $u\}$, more specifically $\kappa(u)$ could be $\{x \mid (x,u) \in D\}$ where D is a fixed subset of $X \times U$. The map κ is said to be <u>upper semicontinuous</u> if its graph is closed in $X \times U$. Such a map is said to be <u>Lipschitz</u> if the function $d(\kappa(u^1),\kappa(u^2))$ is Lipschitz where d denotes the Hausdorff distance between $\kappa(u^1)$ and $\kappa(u^2)$. We remind you that given P,Q two subsets of X then $d(P,Q) = \max(\delta(P,Q), \delta(Q,P))$ where.

$$\delta(P,Q) = \operatorname*{Inf}_{x \in P} \operatorname*{Sup.}_{y \in Q} \|x-y\|$$

Usually, one defines the Hausdorff distance only for compact sets and d then is real valued, in what follows we use the extended definition given here.

3. PROJECTION THEOREMS

π will denote the canonical projection of $X \times U$ onto U . Suppose $f : X \times U \to]-\infty,+\infty]$ and not $\not\equiv +\infty$. By πf we denote the <u>projected</u> function of f , i.e. the function such that $\pi f(u) = \operatorname{Inf} f(x,u)$. It is easy to verify that $\operatorname{cl} \pi \operatorname{epi} f = \operatorname{epi} \operatorname{cl} \pi f$. In particular we have that $\pi f(u) = +\infty$ if $\operatorname{dom} f \quad \{(x,u) \mid x \in X\} = \emptyset$. We can view πf as the variational function of an optimization problem where u represent the perturbation. Thus, whatever can be said about πf can also be translated into properties of the variational function.

Proposition 1. Suppose $f : X \times U \to]-\infty,+\infty]$, $f \not\equiv +\infty$ and f is convex then πf is convex on U .

<u>Proof:</u> This proposition is an immediate consequence of the definition since a function is convex if and only if its epigraph is convex and epi πf is then the (linear) projection of the convex set epi f .

Proposition 2. Suppose $f : X \times U \to]-\infty,+\infty]$, $f \not\equiv +\infty$, X and U both reflexive separable Banach spaces and f is weakly inf-compact. Then πf is weakly inf-compact.

<u>Proof:</u> It suffices to show that the sets $L_\alpha(\pi f)$ are weakly compact for all $\alpha \in \mathbb{R}$. But this follows from the fact that $L_\alpha(\pi f) = \pi L_\alpha(f)$,

Given a mathematical program - by which we mean a constrained optimization problem in finite or infinite dimensional spaces - it is or is not a "well-set" problem, i.e. (i) the problem is <u>solvable</u> (the opi.e. the $L_\alpha(\pi f)$ are simply (linear) projections of weakly compact sets and thus are weakly compact.

Proposition 3. Suppose $f : X \times U \to]-\infty,+\infty]$, $f \not\equiv +\infty$, X and U separable relexive Banach spaces and $f(x,u) = g(x) + \psi_{\kappa(u)}(x)$ with g weakly inf-compact and κ weakly upper semicontinuous. Then πf is weakly lower semicontinuous.

Proof: First observe that for fixed $u \in L_\alpha(\pi f)$ implies that there exists x such that $(x,u) \in L_\alpha(f)$, this follows from the inf-compactness of f in x for fixed u. Moreover if $(x,u) \in L_\alpha(f)$ then $x \in L_\alpha(g)$ since $f(x,u) = g(x)$ whenever $f(x,u) < +\infty$. Let us now consider the sequence $\{u^i\} \in L_\alpha(\pi f)$ converging weakly to u^o. By the above remarks there exists a corresponding $\{x^i\} \in L_\alpha(g)$. By weak compactness of $L_\alpha(g)$, $\{x^i\}$ contains a converging subsequence to a point in $L_\alpha(g)$, say x. The weak upper semicontinuity of κ implies that $x \in \kappa(u^o)$, i.e. $\psi_{\kappa(u^o)}(x) = 0$. Thus

$$\pi f(u^o) \leq f(x) \leq \alpha .$$

i.e. $u^o \in L_\alpha(\pi f)$. From this follows the lower semicontinuity of f.

Proposition 4. Suppose $f : X \times U \to]-\infty,+\infty]$ with X and U finite dimensional Euclidean spaces and $f(x,u) = g(x) + \psi_{\kappa(u)}(x)$ where g in inf-compact, convex, $\kappa(u) = \{x \mid (x,u) \in P\}$ and P is a nonempty convex polyhedral subset of $X \times U$. Then either πf is inf-compact or $L_{\alpha^o}(\pi f)$ - the set of minimum points of πf - is unbounded.

Proof: If P is bounded then it follows from proposition 2 that πf is inf-compact which completes the proof in this case. Thus, it suffices to consider the case when P is unbounded. From proposition 1 and 3 we know that f is convex and l.s.c. and we have always that $\alpha^o \geq \operatorname{Inf} \pi f \geq \operatorname{Inf} g > -\infty$ by inf-compactness of g. This implies that the infimum of πf is finite. If the infimum is not actually attained then from the properties of l.s.c. convex functions [10] it follows that there exists a ray, say $\{u^o + \lambda v, \lambda \geq 0\}$ in $\operatorname{dom} \pi f$ such that πf is strictly decreasing on this ray. For each fixed u, $f(x,u)$ is inf-compact in x, thus, for each λ there exists $x(\lambda)$ such that $f(u^o+\lambda v, x(\lambda)) = \pi f(u^o+\lambda v)$ with $x(\lambda) \in \kappa(u^o+\lambda v)$. Now, since P is a convex polyhedron, it can always be written as $\{(x,u) \mid Ax + Bu = p,\ x \geq 0,\ u \geq 0\}$ where A and B are matrices (Weyl-Minkowski). Since $x(\lambda) \in \kappa(u^o+\lambda v)$ it follows that for all $\lambda \geq 0$, $Ax(\lambda) + Bu^o + \lambda Bv = p$ with $x(\lambda) \geq 0$, $u^o + \lambda v \geq 0$. On the other hand for λ sufficiently large $x(\lambda) \in L_{\alpha^o+\varepsilon}(g)$ for ε arbitrarily small, thus the sequence $x(\lambda)$ contains a convergent subsequence to a point of $L_{\alpha^o}(g)$, say $\bar{x}$. From the above it follows that $A\bar{x} + Bu^o - p = Bv(-\lim \lambda)$ which now implies that $Bv = 0$. Thus $\bar{x} \in \kappa(u^o+\lambda v)$ for all $\lambda \geq 0$ and since by lower semicontinuity $g(\bar{x}) \leq \liminf g(x(\lambda))$, we can not have a strictly decreasing sequence on values for πf on the ray $\{u^o + \lambda v, \lambda \geq 0\}$.

The polyhedral restriction is by no means superfluous in the preceding proposition. To see this consider the following simple example:

$$\text{Find } \inf g(x) = x + \psi_{[0,1]}(x)$$

$$\text{subject to } x \geq 0 , \quad ux \geq 1 .$$

the corresponding variational function $\pi f = u^{-1}$ for $u \geq 1$ and $+\infty$ otherwise. The function g is clearly inf-compact and $\kappa(u) = \{x \mid x \geq 0, x \geq u^{-1}\}$ is an aupper semicontinuous set valued function. In this case $\operatorname{Inf} \pi f(u) = \alpha^o = 0$ but $L_{\alpha^o}(\pi f) = \emptyset$. In fact πf is convex and $L_\alpha(\pi f)$ is unbounded for all $\alpha > \alpha^o$.

Proposition 5. Suppose $f : X \times U \to]-\infty,+\infty]$ with X and U separable reflexive Banach spaces, and $f(x,u) = g(x) + \psi_{\kappa(u)}(x)$ where g is Lipschitz on X and $\kappa(u)$ is Lipschitz on its effective domain $\{u \mid \kappa(u) \neq \emptyset\}$. Then πf is Lipschitz on $\operatorname{dom} \pi f$.

<u>Proof:</u> It suffices to show that there exists a constant M such that $|\pi f(u) - \pi f(v)| \leq M \|u-v\|$ for all u,v dom f.

It is easy to verify that if $\pi f(u) = -\infty$ then the hypotheses imply that $\pi f(v) = -\infty$. Thus let us assume that $\pi f(u)$ and $\pi f(v)$ are finite and that $\pi f(u) \leq \pi f(v)$. Let us also assume that there exist x and y such that $\pi f(x,u) = \pi f(u)$ and $f(y,v) = \pi f(v)$, i.e. that the infimum of f is actually attained for u and v. Moreover, assume that there is a $\bar{y} \in \kappa(v)$ such that $\|\bar{y}-x\| = \operatorname{Inf} \{\|y-x\| \mid y \in \kappa(v)\}$, we have then

$$\begin{aligned} |\pi f(u) - \pi f(v)| &\leq |\pi f(u) - f(\bar{y},v)| \\ &\leq B \cdot \|x-\bar{y}\| \\ &\leq B \cdot K \cdot \|u-v\| \end{aligned}$$

where the first inequality follows from the fact that $\pi f(u) \leq \pi f(v) \leq f(\bar{y},v)$, the second inequality follows from the Lipschitz property of f on X (with constant B) and the last inequality follows from the Lipschitz property of κ (with constant K) since $\|\bar{y}-x\| \leq d(\kappa(u), \kappa(v)) \leq K\|u-v\|$. Now, if either the infimum of g on $\kappa(u)$ is not actually attained (or on $\kappa(v)$) or given a point in $\kappa(u)$ there is no closest point in $\kappa(v)$ the standard modification (using ε-optimal points) of the above arguments yields the proof.

4. A COMPOSITION THEOREM

In order to apply some of the above results to control problems and stochastic programming problems, we need to extend a result of Moreau [4], viz.: If f and g are inf-compact so is $f + g$. To prove the proposition 6 below:

Proposition 6. Suppose $f(x,\xi) : X \times \Xi \to]-\infty,+\infty]$ is a family of functions where X is a finite Euclidean space. Moreover, suppose that the $f(x,\xi)$ are inf-compact, convex in x for all and μ-measurable in ξ for all x . Let μ be a probability measure on Ξ. Then $F(x) = \int f(x,\xi)d\mu$ is a convex, inf-compact function.

We need to introduce some concepts which are standard in Convex Analysis but might still be somewhat unfamiliar to the mathematical programmer. We only need these concepts to prove proposition 6, they will not be used in further developments. Say $f \not\equiv +\infty$ is convex function with range $]-\infty,+\infty]$, then its <u>recession function</u> $f0^+$ is defined by

$$fo^+(y) = \lim_{\lambda\to+\infty} \lambda^{-1}|f(x+\lambda y) - f(x)| .$$

It can be shown that this limit is independent of x and that $f0^+$ is a positively homogeneous convex function. See |9] for a detailed description of the properties of recession functions. In particular it is easy to show that:

Lemma. Suppose $f : X \to]-\infty,+\infty]$ is convex and lower semicontinuous. Then f is inf-compact if and only if $f0^+(y) > 0$ for all $y \neq 0$.

Since in proposition 6 we allow for the possibility of infinite valued functions in the integrand, we have to give a meaning to the $\int$ sign. In fact all what is required is a slight modification of the standard Lebesgue-Stieltjes integral. As usual, we define the integral $\int \cdot d\mu$ as the sum of its positive part and of its negative part with the understanding (i) that either part may possibly be divergent, and (ii) that if $\mu\{\xi|f(x,\xi) = \pm\infty|\} \neq 0$ then the positive (negative) part is automatically defined to be $\pm\infty$; in addition we adopt the convention that the integral is $+\infty$ whenever the positive part is $+\infty$ whatever be the value of the negative part. If the integrand is almost everywhere finite, it corresponds to the Lebesgue-Stieltjes integral. This integral possesses essentially the same properties as the standard Lebesgue-Stieltjes integral except that subaddivity replaces the usual additivity property.

Proof of proposition 6: The proof of convexity can be found in [14] but can in fact be reconstructed quite easily if one observes that $F(x)$ is a convex combination of convex functions. Remains to establish inf-compactness, which in view of the lemma above is equivalent to showing that $FO^+(y) > 0$ for $y \neq 0$ since F is convex. From $fO^+(y,\xi) > 0$ for all $y \neq 0$ and for ξ , it follows readily from a weakend version of the dominated convergence theorem (we do not require here that the integral be convergent) that

$$FO^+(y) \geq \int fO^+(y,\xi)\,d\mu > 0 \ .$$

5. APPLICATIONS

A. Nonlinear programs. Let $f : \mathbb{R}^n \to]-\infty,+\infty]$, $f \not\equiv +\infty$ and let $G : \mathbb{R}^n \to \mathbb{R}^m$, i.e. G is a vector valued function with components G_i , $i=1,\dots,m$. A nonlinear program is then

$$\text{Find } \inf\ f(x)$$
$$\text{subject to } G(x) \leq 0$$

where by $\leq$ we denote componentwise ordering. The standard variational function associated with this problem, is given by

$$P(u) = \{\text{Inf } f(x) \mid G(x) \leq u\} \ .$$

It is obvious that all the projection theorems of section 3 provide sufficient conditions to establish dualizability of the nonlinear program (P is l.s.c. at 0) or stability (P is locally Lipschitz at 0). Moreover if the function G is continuous and f is inf-compact then the nonlinear program is solvable.

We examine somewhat further the case when the objective function as well as the contraint functions G_i , $i=1,\dots,m$ are convex functions. Since the G_i 's are continuous (they are convex-finite on $\mathbb{R}$) it follows that for all scalars u_i , the feasible region of the convex program: Find $\text{Inf } f(x)$ subject to $G(x) \leq u$, is a closed convex set.

For each i, the set $L_{u_i}(G_i)$ is closed and convex and the feasible region is given by $\bigcap_i L_{u_i}(G_i)$. One can also prove that.

Proposition 7. Let G_i, $i=1,\ldots,m$ be a class of lower semicontinuous convex functions. Suppose that for some u, the ray $\{x^o + \lambda y, \lambda \geq 0\}$ is contained in $C = \{x \mid G(x) \leq u\}$. Then for all v such that $D = \{x \mid G(x) \leq v\} \neq \emptyset$ and for any $x \in D$ we have that the ray $\{x + \lambda y$, $\lambda \geq 0$ is also contained in D.

<u>Proof:</u> Since C and D can be represendted as the intersection of level sets, it suffices to show $\{x^o + \lambda y, \lambda \geq 0\}$ $L_{u_i}(G_i)$ implies that for any $x \in L_{v_i}(G_i)$, $\{x + \lambda y, \lambda \geq 0\}$ is also contained in $L_{v_i}(G_i)$. One can prove that from basic properties of convex functions (see e.g. [10, p. 139]) or rely as in section 4 on the properties of the recession function and observe that under the above hypotheses $G_i 0^+(y)$ must be ≤ 0 which then automatically yields the conclusion.

As a corollary of the above proposition we have

Proposition 8. Suppose G is the constraint map of a convex program and $\{x \mid G(x) \leq 0\}$ is nonempty compact, then $\{x \mid G(x) \leq u\}$ is compact for all vectors u for which it is nonempty.

<u>Proof:</u> If the only ray contained in $\{x \mid G(x) \leq 0\}$ is the trivial ray, i.e. with direction $y = 0$, then by proposition 7 it is also the only ray contained in the closed convex set $\{x \mid G(x) \leq u\}$.

If $\{x \mid G(x) \leq 0\}$ is compact and we restrict the pertubations u of the original problem to a compact neighborhood of the origin (which is the only region of interest anyway) and if f is inf-compact, it follows not only that the original problem is solvable but we can then apply propositions 1 and 2 to obtain further properties of this problem. It is now easy to see how in a similar fashion one could also apply propositions 3, 4 and 5 to this class of problems.

B. Stochastic Programs. Provided the original problem satisfies some very weak assumptions, to each "convex" stochastic program correspond an equivalent deterministic convex program. However, it is not usually possible to find an explicit simple analytic representation of this deterministic problem and consequently it might be very difficult to establish if a given problem is or is not well set. It is not possible to develop here the full implications of the results of section 3 and 4. We only intend to indicate here some of the more elementary applications. Some partial results have already appeared in the litterature, see e.g. [13] for an application of a version of proposition 5 and [1] where a version of proposition 3 appeared first and also some applications are indicated. The following problem will be called here a <u>stochastic program:</u>

$$\text{Find} \quad \inf_x z = f(x) + E\,\{\operatorname{Inf}_y q(y,\xi) \mid H(x,y,\xi) \le 0\}$$

$$\text{subject to} \quad G(x) \le 0\ .$$

The functions $f : \mathbb{R}^n \to]-\infty,+\infty]$ and $G_i : \mathbb{R}^n \to \mathbb{R}$, $i=1,\ldots,m$ are convex functions in x ; $q : \mathbb{R}^n x \quad \to]-\infty,+\infty]$ and $H_j : \mathbb{R}^{\bar{n}} x \Xi \to \mathbb{R}$ are also convex in y and (x,y) respectively and measurable with respect to a probability measure μ that defines the distribution of the random variable ξ . Let

$$Q(x,\xi) = \operatorname{Inf}_y \{q(y,\xi) \mid H(x,y,\xi) \le 0\}$$

and

$$\mathcal{Q}(x) = E\{Q(x,\xi)\}$$

where E (expectation) is defined in the same manner than the integral $\int \cdot\, d\mu$ in section 4. The equivalent deterministic program is then written as

$$\text{Find} \quad \inf z = f(x) + \mathcal{Q}(x)$$

$$\text{subject to} \quad G(x) \le 0\ .$$

We shall assume that it can be shown that $Q(x,\xi)$ is measurable with respect to ξ (see e.g. [2] and [14] for a proof of the linear case). Then $\mathcal{Q}(x)$ is convex since from proposition 1 it follows that $Q(x,\xi)$ is convex in x . We shall assume that $q(x,\xi)$ is inf-compact in y and that the set valued function κ defined by $\kappa(x,\xi) = \{y \mid H(x,y,\xi) \le 0\}$ is upper simicontinuous in x . Both assumptions are quite natural. Since $q(y,\xi)$ represents a penalty to be payed for selecting a recourse action y and one might expect that the cost increases strictly with $\|y\|$. See for example the simple recourse model [15] where the penalty is a function of the difference between x and ξ . The upper semicontinuity of κ is more technical but it is hard to imagine a practical problem that would fail to satisfy this assumption. With these assumptions, it follows directly from proposition 3 that $Q(x,\xi)$ is lower semicontinuous in x . This property can also be proved for $\mathcal{Q}(x)$ if some weak integrability condition is satisfied. If now f is also l.s.c. and $\{x \mid G(x) \leq 0\}$ is compact, it follows then from proposition 3 again that the equivalent problem is solvable and at least asymptotically stable (dualizable).

If in addition one can show that the sets $\kappa(x,\xi)$ are bounded then we can use propositions 2 and 6 to show that $\mathcal{Q}$ is inf-compact. Proposition 2 can be used again to obtain the properties of the variational function since $f + \psi_{\{x \mid G(x) \leq 0\}}$ is then also inf-compact.

C. Control problems governed by partial differential equations.
Let us just mention here that the inf-compactness assumption can be used to obtain existence and regularity conditions rather than coercivity as used in [3]. It is obvious that all coercive functions are inf-compact, the only part we loose is the uniqueness of the solution.

References

[1] S. Gartska: "Regularity Conditions for a Class of Convex Programs", Manuscript (1972).

[2] P. Kall: "Das zweiteilige Problem der stochastischen linearen Programmierung", Z. Wahrscheinlichkeitstheorie verw. Gebiete 8 (1966), 101-112.

[3] J.-L. Lions: Optimal Control of Systems Governed by Partial Differential Equations, Springer-Verlag, Berlin (1971).

[4] J.-J. Moreau: "Fonctionelles Convexes", Seminaire sur les Equations aux derivées partielles, Collège des France,Paris (1966-1967).

[5] J.-J. Moreau: "Theorèmes inf-sup, C. R. Acad. Sci. Paris, 258 (1964), 2720-2722.

[6] R. T. Rockafellar: "Duality and Stability in Extremum Problems Involving Convex Functions", Pacific J. Math.,21(1967),167-187.

[7] R. T. Rockafellar: "Duality in Nonlinear Programming" in Mathematics of Decision Sciences, Lectures in Applied Mathematics, Providence, R.I., 11 (1968), 401-422.

[8] R. T. Rockafellar: "Level Sets and Continuity of Conjugate Convex Functions", Trans. Amer. Math. Soc., 123 (1966), 46-63.

[9] R. T. Rockafellar: Convex Analysis, Princeton University Press, Princeton, N.J. (1969).

[10] J. Stoer and C. Witzgall: Convexity and Optimization in Finite Dimension I, Springer-Verlag, Berlin (1970).

[11] M.Valadier: "Integration de Convexes Fermés notamment d'Epigragraphes. Inf.-Convolution Continue", R.I.R.O., R-2 (1970), 57-73.

[12] R. Van Slyke and R. Wets: "A Duality Theory for Abstract Mathematical Programs with Applications to Optimal Control Theory", J. Math. Anal. Applic. 22 (1968), 679-706.

[13] D. Walkup and R. Wets: "Some Practical Regularity Conditions for Nonlinear Programs", SIAM J. on Control, 7 (1969), 430-436.

[14] D. Walkup and R. Wets: "Stochastic Programs with Recourse", SIAM J. Appl. Math., 15 (1967), 1299-1314.

[15] W. Ziemba: "Stochastic Programs with Simple Recourse", Manuscript (1972).

Mathematisches Institut
der Universität zu Köln
5000 Köln 41
Weyertal 86-90

and

Department of Mathematics
University of Kentucky
Lexington, Kentucky

NONCONVEX QUADRATIC PROGRAMS, LINEAR COMPLEMENTARITY PROBLEMS, AND INTEGER LINEAR PROGRAMS (*)

F.Giannessi †   E.Tomasin ††

Abstract.

The problem of nonconvex quadratic programs is considered, and an algorithm is proposed to find the global minimum, solving the corresponding linear complementarity problem. An application to the general complementarity problem and to 0-1 integer programming problems, is shown.

1 - Introduction.

The aim of this paper is to study the general quadratic programming problem, i.e. the problem of finding the minimum of a quadratic function under linear constraints. Such a problem is often met in many fields of mathematics, mechanics, economics, and so on. If the objective function is convex, the problem is well known, both theoretically and computationally [2, 4, 6], while, when the objective function is nonconvex, the problem of finding the global minimum is still open, even if there are many methods to find a local minimum. Among the methods proposed to solve the general case [3, 5, 11, 13], two kinds of approaches can be distinguished: a) enumerative methods [3, 11], and b) cutting plane methods [5, 13]. While the former approach seems to be not so efficient as expected, the latter till now, gave rise to methods which are not always finite [16].

In this paper a method to solve the general quadratic programming problem is proposed, where the quadratic problem is substituted by an equivalent linear complementarity problem, and this is solved by a particular cutting plane method [8]. This way no method of pratically efficiency is produced but after a successive investigation of the method and of the respective properties, by modifying somewhere the initial bases of the method, the algorithm was implemented [14]. Thus, the vertices of a convex polyedron are found, which minimize a linear function and satisfy a given condition, for example complementarity (this way the problem of nonconvex quadratic programming is solved) or a 0-1 integer condition. Then the method can be used also to solve a general linear complementarity problem which can be met independently from the quadratic

(*) Research supported by National Groups of Functional Analysis and its Applications of Mathematical Commitee of C.N.R.

† Department of Operations Research and Statistical Sciences, Univ. of PISA, Via S.Giuseppe, 22, PISA, ITALY.

†† Mathematical Institute, Ca' Foscari University, VENICE, ITALY.

programming problem, or 0-1 programs.

2-The general quadratic programming problem.

The problem with which we are concerned is, without loss of generality, the following [1]:

$$P : \min \varphi(x)=c^T x + \frac{1}{2}x^T Dx, \quad x\in \mathcal{X} =\{x:Ax\geq b, \quad x\geq 0\}$$

where A is a matrix of order mxn and D is a matrix of order nxn. The following theorems hold:

THEOREM 2.1 (Kuhn-Tucker [1, 10]) *If* x *is a local minimum for* P, *there exist vectors* y, u, v, *such that:*

$$c + Dx - A^T y - u = 0 \tag{2.1a}$$

$$Ax - v = b \tag{2.1b}$$

$$x,\ y,\ u,\ v \geq 0 \tag{2.1c}$$

$$x^T u = y^T v = 0 \tag{2.1d}$$

THEOREM 2.2 ([4], p.146). *If* (x, y, u, v) *is a solution of* (2.1), *then the equality*

$$\varphi(x) = \frac{1}{2}(c^T x + b^T y) \tag{2.2}$$

holds

THEOREM 2.3 ([14]).

If i) $\mathcal{X} \neq \emptyset$

ii) $\mathcal{X} \neq \{0\}$

iii) $\mathcal{X}$ is compact

then:

$$\inf \{\frac{1}{2}\ (c^T x + b^T y)\} = -\infty.$$

Define

$$\hat{A} = \begin{pmatrix} D & -A^T \\ A & 0 \end{pmatrix};\ \hat{b}=\begin{pmatrix} -c \\ b \end{pmatrix};\ \hat{c}=\frac{1}{2}\begin{pmatrix} c \\ b \end{pmatrix};\ z'=\begin{pmatrix} x \\ y \end{pmatrix};\ z"=\begin{pmatrix} u \\ v \end{pmatrix};\ z=\begin{pmatrix} z' \\ z" \end{pmatrix}$$

THEOREM 2.4 *The linear complementarity problem:*

$$\min \hat{c}^T z', \quad z'\in\{z:\hat{A}z' - z" = \hat{b};\quad z'^T z" = 0,\quad z \geq 0\} \tag{2.3}$$

is equivalent [2] *to* P.

---

1 - The "T" as superscript and "min" denote transposition and global minimum respectively.

Theorem 2.4, which is a staightforward consequence of theorems 2.1, 2.2 shows that, to solve P, we have to solve a linear complementarity problem [7, 9]. Then, in the following sections, the linear complementarity problem, will be considered.

3 - The linear complementarity problem.

For sake of simplicity define now

$$x'^T=(x_i),\quad x''^T=(x_{m+i}),\quad A=(a_{ij}),\quad a^T=(a_{i0}),\quad c^T=(c_j),\quad i=1,..m;\ j=1,..n=2m,\quad x^T=(x'^T,\ x''^T)$$

and consider the problem:

$$Q : \min c^T x,\quad x \in X = \{x : Ax = a;\quad x'^T x''=0;\quad x \geq 0\}$$

Now the problem Q will be considered, instead of (2.3), which is a particular case of Q. Then, define X^* the convex closure of the points of X, which verify $x'^T x''=0$; the following theorems hold:

THEOREM 3.1 *The set of optimal solutions of Q is a face (in particular a vertex) of X.*

PROOF. A vector $x \in R$ must have at least n-m=m zero elements. Then either an optimal solution (shortly o.s.) of Q is a vertex of X, or it belongs to a face of X, whose points are o.s. of Q. The result follows.

THEOREM 3.2 *Q is equivalent* [3] *to the linear problem.*

$$\min c^T x\quad ;\quad x \in X^* \tag{3.1}$$

which is a consequence of theorem 3.1.
If X^* would be known, the problem (3.1) could be solved in place of Q. As X^* is unknown but it is sufficient to know only a subset of X^* containing an o.s. of Q, such a subset is determined to reduce Q to a linear programming problem.

4 - A cutting plane method to solve Q. The case of nondegeneracy.

Consider the problem:

$$Q_0 : \min c^T x\quad ;\quad x \in X$$

The method previously outlined, consists firstly in solving Q_0, as an o.s. of Q_0 satisfying

$$x'^T x'' = 0$$

is also an o.s. of Q and then of P.

---

2 - In the sense that the first n elements of an optimal solution of the latter one are an optimal solution of the former one.

3 - In the sense that they have the same o.s. and the same minimum.

But, by the theorem 2.3 Q_0 has no finite o.s.;then in section 6 a transformation is defined, such that a problem equivalent to Q_0, having finite o.s.,is obtained. Suppose then, that Q_0 has finite o.s. If an o.s. such that $x'^T x"=0$ terminate the algorithm. Otherwise, a linear inequality is determined, such that it is not satisfied by V_0 (where V_0 is the vertex corresponding to the current o.s.), but is satisfied by every other vertex of X.

This abviously can be realized in many ways; here the strongest inequality is generated. This happens when every vertex in X adjacent to V_0 strictly verifie the generated inequality. If there is no degeneracy, this can be easily performed; otherwise,it is more complicated. In this section the case of non degeneracy is considered.

Let the coordinates of V_0 be the basic solution (briefly, b.s.) of the reduced form:

$$x_i + \alpha_{i,m+1} x_{m+1} + \ldots + \alpha_{in} x_n = \alpha_{i0} \qquad i=1,\ldots, m. \tag{4.1}$$

of the system $Ax = a$, where $x_{m+1} = \ldots = x_n = 0$.

If the n-vector

$$(\alpha_{10},\ldots,\alpha_{m0}, \; 0, \; 0, \; \ldots \; 0) \tag{4.2}$$

satisfies $x'^T x" = 0$, then it is o.s. of Q too.

Otherwise, a convex polyhedron is determined, which has only the vertices of X, but V_0.

By the hypothesis of non degeneracy, it is:

$$\alpha_{i0} > 0 \qquad i = 1,\ldots, m. \tag{4.3}$$

Define

$$\bar{x}_j = \text{Sup}\{x_j \; : \; x_i = \alpha_{i0} - \alpha_{ij} x_j \geq 0, \;\; i=1,\ldots, m\}, \quad j=m+1,\ldots,n.$$

$$\alpha_{m+1,j} = \begin{cases} 0 & , \text{ if } \bar{x}_j = +\infty \\ 1/\bar{x} & , \text{ if } \bar{x}_j < +\infty \end{cases} \quad ; \; j = m+1,\ldots, n;$$

$$\alpha^T = (\alpha_{m+1,m+1},\ldots,\alpha_{m+1,n})$$

The inequality

$$\alpha^T x" \geq 1 \tag{4.4}$$

is said a *cut* for X. The intersection, $\bar{X}$, between X and the halfspace (4.4) is the required polyhedron, as it is shown by the following:

<u>THEOREM 4.1</u> *the inequality* (4.4), i) *is not verified by* V_0; ii) *is weakly verified by all the vertices of* X *which are adjacent to* V_0, iii) *is verified by every other vertex of* X; iiii) *every vertex of* X *is a vertex of* $\bar{X}$ *too.*

which is easily prooved [8]. The cut (4.4) has interesting properties [8].

Let Q_1 be the problem obtained by adding to Q_0 the constraints

(4.5) $\alpha^T x'' - x_{n+1} = 1 \; ; \quad x_{n+1} \geq 0$

As (4.2) does not verify $x'^T x'' = 0$, the addition of (4.5) to Q does not change its solutions. Q_1 is then associated to Q in place of Q_0. An o.s. of Q_1 weakly verifies (4.4), so that it is an element of $\mathcal{A}_X(V_0)$, i.e. of the set of vertices adjacent to V_0 in X. Let V_1 be o.s. of Q_1, and x_h, $h>m$, the nonbasic variable (shortly, n.v.) that must become basic to go from V_0 to V_1. The current system is then:

(4.6) $$\begin{cases} x_i + \alpha'_{i,m+1} x_{m+1} + \ldots + \alpha'_{i,h-1} x_{h-1} + \alpha'_{i,h+1} x_{h+1} + \ldots + \alpha'_{i,n+1} x_{n+1} = \alpha'_{i,0} & i=1,\ldots m \\ x_h + \alpha'_{m+1,m+1} x_{m+1} + \ldots + \alpha'_{m+1,h-1} x_{h-1} + \alpha'_{m+1,h+1} x_{h+1} + \ldots + \alpha'_{m+1,n+1} x_{n+1} = \alpha_{m+1,0} \end{cases}$$

If V_1 at $x_j=0$ $j=m+1,\ldots n+1$, $j \neq h$, verifies $x'^T x''=0$, then it is o.s. of Q too. Otherwise, the procedure is iterated and another inequality is generated which is not satisfied by V_1, but is satisfied by every other vertex of the feasible region of Q_1. This is realized in next section.

5 - <u>The cut in the case of degeneracy</u>.

Assume that V_1 does not satisfy $x'^T x''=0$. Then to determine a cut, like in section 4, which cuts off V_1, the definition of cut has now to be enlarged; in fact $\mathcal{A}_{\bar{X}}(V_1)$ may contain more than $n+1-(m+1)=m$ elements, so that it would be impossible to determine an inequality like (4.4), weakly verified by all the elements of $\mathcal{A}_{\bar{X}}(V_1)$. Remark that in (4.6) exists at least one $\bar{I} \in \{i=1\ldots m\}$ such that $\alpha'_{\bar{I}0} = 0$, so that to define a generic cut is to define a cut when degeneracy occurs, i.e. when (4.3) are not verified. Then consider the following linear programming problem:

(5.1a) $\min \; (c_1 x_1 + \ldots + c_N x_N)$

(5.1b) $x_i + \alpha_{i,M+1} x_{M+1} + \ldots + \alpha_{iN} x_N = \alpha_{i0}, \quad i=1,\ldots M \; ; \quad j=1,\ldots N$

(5.1c) $x_j \geq 0$

and assume that the vector:

(5.2) $(x_i = \alpha_{i0}, \quad i=1,\ldots,M; \quad x_i = 0, \quad i=M+1,\ldots,N)$

is an o.s. of (5.1), which does not verify $x'^T x''=0$.
If $M=m$, $N=n$ (5.1b) coincide with (4.1); after the first cut (4.5) has been determined, if $M=m+1$ and $N=n+1$, (5.1b) coincide with (4.6). Assume, without loss of generality [4]

(5.3) $\alpha_{i0} > 0, \quad i=1,\ldots,\bar{M}; \quad \alpha_{i0} = 0, \quad i=\bar{M}+1,\ldots,M \quad , \quad (0<\bar{M}<M).$

To define a cut which does not verify only (5.2) among the vertices of (5.1b,c) consider the latest $M-\bar{M}$ of (5.1b), which may be equivalently

4 - In fact, if $\alpha_{i0}=0$, $i=1,\ldots,M$, then (5.1b,c) has only one vertex which does not verify $x'^T x''=0$, and then Q has no o.s.; if $\alpha_{i0}>0$, $i=1,\ldots,M$, we are in the case of section 4.

written

$$(5.4) \qquad \alpha_{i,M+1}x_{M+1} + \ldots + \alpha_{iN}x_N \leq 0, \quad i=\bar{M}+1,\ldots,N$$

Define $C_0 = \{(x_{M+1},\ldots,x_N): x_{M+1} \geq 0; \ \ldots \ ; \ x_N \geq 0\}$ and C_r, $r=1,\ldots,M-\bar{M}$ as the intersection among C_0 and the first r halfspaces (5.4). To define a cut in this case firstly the convex polyhedral cone $C=C_{M-\bar{M}}$ must be determined. This is realized in a gradual way; in fact starting from C_0 whose edges are trivially known, the edges of C_{r+1} are obtained knowing the edges of C_r. This way $C_{M-\bar{M}}$, and then C, is determined [8]. Assume then to know the parametric equations

$$(5.5) \qquad x_j = \beta_{ij}t, \quad j=M+1,\ldots,N \ ; \ t \geq 0; \quad i=1,\ldots,k; \quad \beta_{ij} \geq 0$$

of the edges of C respectively indicated

$$H_i \ \ldots \ H_k .$$

The set $\mathcal{A}_X(V_0)$ of the vertices of the convex polyhedron X, adjacent to V_0 must now be determined [5]. Then put

$$(5.6) \qquad \bar{t}_s = \text{Sup}\,\{t:\alpha_{i,M+1}\,\beta_{s,M+1} + \ldots + \alpha_{iN}\beta_{sN} \leq \alpha_{i0}, \quad i=1,\ldots,\bar{M}\} \ s=1,\ldots,k$$

(5.6) can be assumed to be finite [6].
Then the elements of $\mathcal{A}_X(V_0)$ are the points [7]

$$(5.7) \qquad V_s = (\bar{x}_{sj} = \beta_{si}\,\bar{t}_s, \quad j=M+1,\ldots,N), \quad s=1,\ldots,k .$$

Now a cut can be defined in this case.
If $k \leq N-M$, using the method described in section 4, an inequality like (4.4) said here too cut, can be easily determined. If $k>N-M$ an inequality like (4.4) may not exist, then more then one inequality is generated. Define

$$I = \{1,\ldots,k\} \ ; \quad \bar{x}_j = \sum_{s\in I} \bar{x}_{sj}, \quad j=M+1,\ldots,N$$

$$V^T = (\bar{x}_{M+1},\ldots,\bar{x}_N) \ ; \quad \alpha^T = (\alpha_{M+1,M+1},\ldots,\alpha_{M+1,N})$$

and consider the problem

$$(5.8) \qquad \min V^T\alpha \ , \quad \alpha \in \{\alpha : V_s^T\,\alpha \geq 1, \quad s\in I\} .$$

The feasible region of (5.8) is given by:

$$(5.9) \qquad V_s^T\,\alpha - \beta = 1 \qquad s\in I; \quad \beta \geq 0$$

where $\beta^T = (\beta_1,\ldots,\beta_k)$ is a slack vector.

---

5 - The notations X, V_0, $\mathcal{A}_X(V_0)$ of section 3, 4 are used again.

6 - See the following section.

7 - V_s denote both the point and the (N-M)-uple.

The meaning of (5.8) is obvious: an o.s. of (5.8) $(\alpha^*_{M+1,j},\ j=M+1,..,N)$ gives the coefficients of the inequality

$$\alpha^*_{M+1,M+1}x_{M+1}+\ldots+\alpha^*_{M+1,N}x_N\geq 1 \tag{5.10}$$

which, among the ones verified by all the vertices of x but V_0 minimizes the sum of the differences between the two terms of (5.10) evaluated at every element of $\mathcal{A}_X(V_0)$. As (5.8) has finite o.s. (5.10) always exists. If all (5.7) weakly verify (5.10), it is said a *cut* for (5.1) and our aim is attained. Otherwise, another b.s. distinct from the optimal one is determined. Let it be the following

$$(\alpha^*_{M+2,j}\quad ,\quad j=M+1,N)\quad ,$$

to which by the definition of the system of constraints corresponds the inequality

$$\alpha^*_{M+2,M+1}x_{M+1}+\ldots+\alpha^*_{M+2,N}x_N\geq 1 \tag{5.11}$$

analogous to (5.10), which is considered together with it. If every (5.9) weakly verify at least one equation of the system (5.10), (5.11), then this is said *cutting system* or briefly *cut* for the problem (5.1). Otherwise, another b.s. of the problem (5.8) is determined and then another inequality like (5.11) is generated. As to every b.s. of the (5.8) corresponds a N-M-1 face of the polyhedron $\bar{X}$, iterating the procedure a finite number of times, a system of inequalities

$$\alpha^*_{M+1,M+1}x_{M+1}+\ldots+\alpha^*_{M+1,N}x_N\geq 1,\quad i=1,\ldots,\hat{M}-M \tag{5.12}$$

is determined which is said a *cutting system*, shortly a *cut* for the problem (5.1) and which is added to (5.1b). Then, an o.s. of the new problem is determined; this is of course an element of $\mathcal{A}_X(V_0)$. Iterating the procedure described above, in a finite number of steps an optimal solution of Q is reached. The finiteness of the method is justified by the following theorem:

<u>THEOREM 5.1</u> ([8]). *The polyhedra* $\bar{X}$ *and* X_0 *have the same vertices*

where X^0 is the union of the differences between X and the convex hull of its vertices, and the convex hull of the vertices of X, but V_0.

6 - Determination of an optimal solution of Q_0

By the theorem (2.3) Q_0 has no finite optimal solutions. Nevertheless with a transformation it is possible to obtain a problem equivalent to Q_0 having finite o.s. Let $\hat{Q}$ and $\hat{Q}_0$ denote respectively Q and Q_0, both with the additional constraints:

$$x_1+x_2+\ldots\ldots+x_n+x_{n+1}=Q_{00};\quad x_{n+1}\geq 0 \tag{6.1}$$

Then, by the following:

<u>THEOREM 6.1</u> ([8]). *Q has finite o.s., iff a real* Q_{00} *exists, such that* $\hat{Q}$ *has finite o.s. satisfying the inequality* $x_{n+1}>0$;

$\hat{Q}_0$ can be considered in place of Q_0, when the last has no finite o.s., if Q_{oo} is large enough. Theorems 3.1 and 3.2 hold for $\hat{Q}$ too. Remark that, after the above transformation, there are vertices of the feasible region of $\hat{Q}_0$ which are not vertices and then b.s. for Q_0. It is useful to eliminate such vertices, as the computation increases rapidly when the number of iterations, and then the degeneracy of the problems, Q_0, Q_1 ... Q_n, becomes larger. This is realized, as indicated in [14], by a transformation, here briefly described. The following parametric capacity constraint is added, in place of (4.1)

$$x_1+x_2+.....+x_n+x_{n+1}=\alpha, \text{ with } x_{n+1}\geq 0 \tag{6.2}$$

where α is a parameter.
Let $\hat{Q}_0(\alpha)$ be the problem $\hat{Q}_0$ with α in place of Q_{oo}, $V_0(\alpha)$ an optimal vertex of $\hat{Q}_0(\alpha)$ and $X(\alpha)$ the feasible region of $\hat{Q}_0(\alpha)$. All the vertices of $X(\alpha)$ adjacent to $V_0(\alpha)$, which are not vertices of X, are determined. Remark that every vertex of X, is such that

$$\lim_{\alpha\to+\infty} a_{i0}(\alpha) = +\infty \tag{6.3}$$

for at least one $i \in \{i=1,...m\}$. Then the set of vertices adjacent to every vertex satisfying (6.3) and such that they are vertices both in X and in $X(\alpha)$, is determined. Such a set is considered in place of $\mathcal{A}_X(V_0)$; it is said $\mathcal{A}_{X(\alpha)}(V_0(\alpha))$. The method descibed in section 5 is then applied to $\mathcal{A}_{X(\alpha)}(V_0(\alpha))$ and a system analogous to (5.12) is obtained. This procedure is justified remarking that:

$$\lim_{\alpha\to+\infty} X(\alpha) = X$$

and that the vertices of $X(\alpha)$ which are not vertices of X, verify (6.3). The following
Proposition. *The polyhedron defined by the constraints of* Q_0 *and by the system* (5.12) *is the convex hull of the vertices of* X;
holds.
In fact every vertex of X belongs to the polyhedron and every inequality of the system (5.12) defines a facet of the polyhedron itself.

7 - Connections between nonconvex and concave quadratic programming problems.

The idea previously described and the subsequent algorithm are not so efficient to be applied in the form outlined above. It is then necessary to implement the algorithm; to this aim, remark that the central problem of this theory, is the concave programming problem, to which it is always possible to restrict. In fact, it is known that for a given nonconvex quadratic programming problem, it is possible a decomposition in convex, and concave sub-problems [3]. Thus, given a nonconvex quadratic problem, it is enough to be able to solve convex and concave subproblems. The solution of the former one can be easily obtained by the known methods. The method of the previous sections can be used for the latter one. To this aim consider the problem P of section 2; the so called facial decomposition can be used. It consists in a tree like proce

dure, beginning with a single node corresponding to X itself; from this node there are branches leading to the (n-1) dimensional faces of X, let they be F_i,(i=1,...,k), where k can be easily determined. [17]
If $\varphi(x)$ is either convex, or concave on each of F_i, $\forall_i$=1,...,k the procedure can be stopped. Otherwise, from each of F_i, where $\varphi(x)$ is neither convex, nor noncave, there are branches going to the (n-2) dimensional faces of X. To these also the procedure described above is applied, until a set of sub-problems of P, eithers convex, or concave is found. Each convex sub-problem, can be solved using one of the efficient existing algorithms, while for the concave ones, the algorithm previously described can be implemented as is shown in the following section.

8 - A sufficient condition for the optimality in a concave quadratic programming problem.

Consider the problem P, assuming that $\varphi(x)$ is strctly concave, and the complementarity problem (2.3) equivalent to P. Let $(\bar{x}, \bar{y}, \bar{u}, \bar{v})$ be an o.s. of the linear programming problem associated to (2.3), i.e. a vertex of the polyhedron (2.1a,b,c). This always happens, when any problem Q_0, Q_1,...,Q_n, previously considered, has been solved. If the o.s. thus obtained, satisfies the complementarity condition, a solution of Q, and then of P, is at hand. Otherwise, using the algorithm described above, such a vertex is cut off, even if $(\bar{x},\bar{v})$ is o.s. of P. More precisely, let Q_i, i=0,1,...,r, be the problems which are to be solved, before obtaining a vertex satisfying the complementarity condition, and let (x^i, y^i, u^i, v^i) be the corresponding o.s.; (x^r,v^r) is o.s. of P, but it may be that (x^i, v^i) is o.s. of P with i<r. Since the complementarity condition is the only sufficient condition, which is at hand to decide if (x^i, v^i) is o.s. of P, all the problems Q_i i=0,1,...,r are to be solved. This happens frequently, and as iterating the procedure requires much computation, it is obvious to try to stop the algorithm if (x^i, v^i) is o.s. of P, even if (x^i, y^i, u^i, v^i) does not satisfy the complementarity condition. If P is strictly concave, an o.s. in necessarily a vertex of X. Then, when (x^i, v^i) is a vertex of X a well known condition [15] can be used, to decide if it is o.s. of P. Consider firstly the case of nondegeneracy.

I The case of nondegeneracy

Let $(x^{(0)}, v^{(0)}, y^{(0)}, u^{(0)})$ be an o.s. of the current i-th problem Q_1 and let $(x^{(0)}, v^{(0)})$ be a nondegenerate vertex of X. In this hypothesis, there are n vertices $(x^{(i)}, v^{(i)})$ i=1,...,n adjacent to it. Define t^0 the n-vector of the nonbasic variables, associated to $(x^{(0)}, v^{(0)})$ and t^i i=1,...,n, the vectors associated to $(x^{(i)}, v^{(i)})$ in the space having t^0 as origin.
In the n vertices adjacent to t^0 let it be:

$$\varphi(t^i) \geq \varphi(t^0), \forall\ i=1,\dots,n. \tag{8.1}$$

On the straight lines containing the edges originating in t^0, consider n points t^{*i}, i=1,...,n, such that:

$$\varphi(t^{*i}) = \varphi(t^0), \forall\ i=1,\dots,n, \tag{8.2}$$

which necessarily exist, as $\varphi(x)$ is concave.
Remark that

(8.3) $$t^{*i} = \lambda_i t^0 + (1-\lambda_i) t^i , \forall i=1,\dots,n ,$$

where
$$\lambda_i = - \frac{(t^0) - (t^i)}{t^0 D t^0 + t^i D t^i} ,$$

and then $$\lambda_i \leq 0 , \forall i=1,\dots,n ,$$

i.e. the points t^{*i} are nonconvex linear combination of t^0 and of t^i. Consider now the hyperplane passing through the n points t^{*i}; its equation is:

(8.4) $$g(t) = \sum_{i=1}^{i=n} t_i / (1-\lambda_i) t_i^i -1$$

Remark that $g(t^0) = -1$. Consider now the following problem:

(8.5a) $$\max g(t)$$

subject to

(8.5b) $$Ax-v = b$$

where (8.5b) have to be suitably expressed in the space having t^0 as origin. If the linear programming problem (8.5) has o.s. t', such that

(8.6) $$g(t') \leq 0$$

t^0 is an optimal basis for P, i.e. $(x^{(0)}, v^{(0)})$ is o.s. of P, even if $(x^{(0)}, v^{(0)}, y^{(0)}, u^{(0)})$ does not satisfy the complementarity condition; if

(8.7) $$g(t') > 0,$$

a vertex where the following inequality may hold

$$\varphi(t) < \varphi(t^0)$$

exists, so that the procedure of the algorithm of section 5 must be iterated.
Analogous considerations may be done, for the case of degeneracy of $(x^{(0)}, v^{(0)})$, see [14].

9 - An algorithm for nonconvex quadratic problems.

After the remarks of the preceding sections here an algorithm to solve nonconvex quadratic programming problems is briefly outlined:

Step 1 Let the problem P be given. Its feasible region can be decomposed, in such a way that P is replaced by a finite number of convex, and strictly concave problems. The first ones, can be solved by any algorithm for convex quadratic programming problems. To solve the second ones, use step 2.

Step 2 The concave programming problem is transformed into the complementarity problem. Determine Q_0. Put r=0 and go to step 3.

Step 3 The linear parametric problem of section 6 is solved and the set $\mathcal{A}_{X(\alpha)}(V_0^{(\alpha)})$ is determined. Go to step 4.

Step 4 A cut is determined; the constraints which define the present cut, are in addition to Q_r. Go to step 5.

Step 5 Q_r is solved (if r>0, an o.s. of Q_r is quickly available). If an o.s. of Q_r verifies the complementarity condition, an o.s. of P is obtained as a subvector of it; terminate the algorithm. Otherwise, go to step 6.

Step 6 If the subvector of the o.s. of Q_r, corresponding to a solution of P, is a vertex of P, go to step 7; otherwise, put r=r+1 and go to step 5.

Step 7 The problem (8.5) is determined and solved. If the o.s. of (8.5), verifies (8.6) terminate the algorithm; if it verifies (8.7), put r=r+1 and go to step 5.

REFERENCES

[1] - ABADIE J., *On the Khun-Tucker Theorem*. In "Nonlinear programming", J.Abadie (ed.), North-Holland Publ. Co., 1967, pp.19-36.

[2] - BEALE E.M.L., *Numerical Methods*. In "Nonlinear Programming", J.Abadie (ed.), North-Holland Publ. Co., 1967, pp.133-205.

[3] - BURDET C.A., *General Quadratic Programming*. Carnegie-Mellon Univ. Paper W.P. -41-71- 2, Nov. 1971.

[4] - COTTLE R. W., *The principal pivoting method of quadratic programming*. In "Mathematics of the decision sciences, Part I, eds. G.B.Dantzig and A.F.Veinott Jr. American Mathematical Society, Providence, 1968, pp.144-162.

[5] - COTTLE R.W. and W.C. MYLANDER, *Ritter's cutting plane method for nonconvex quadratic programming*. In "Integer and nonlinear programming", J;Abadie (ed.), North-Holland Publ. Co., 1970, pp.257-283.

[6] - DANTZIG C.B., *Linear Programming and Extension*. Princeton Univ. Press, 1963.

- DANTZIG G.B., A.F. VEINOTT, *Mathematics of the Decision Sciences*. American Mathematical Society, Providence, 1968.

[7] - EAVES B.C., *On the basic theorem of complementarity*. "Mathematical Programming", Vol.1, 1971, n.1, pp.68-75.

[8] - GIANNESSI F., *Nonconvex quadratic programming, linear complementarity problems, and integer linear programs*. Dept. of Operations Research and Statistical Sciences, Univ. of PISA, ITALY. Paper A/1, January 1973.

[9] - KARAMARDIAN S., *The complementarity problem*. "Mathematical Programming", Vol.2, 1972, n.1, pp.107-123.

[10] - KUHN H.W. and A.W. TUCKER, *Nonlinear programming*. In: "Second Berkeley Symp. Mathematical Statistics and Probability", ed. J.Neyman, Univ. of California Press, Berkeley, 1951, pp.481-492.

[11] - LEMKE C.E., *Bimatrix Equilibrium Points and Mathematical Programming*. "Management Science", Vol.11, 1965, pp.681-689.

[12] - RAGHAVACHARI M., *On connections between zero-one integer programming and concave programming under linear constraints*.

[13] - RITTER K., *A method for solving maximum problems with a nonconcave quadratic objective function*. Z.Wharscheinlichkeitstheorie, Vern. Geb.4, 1966, pp. 340-351.

[14] - TOMASIN E., *Global optimization in nonconvex quadratic programming and related fields*. Dept. of Operations Research and Statistical Sciences, Univ. of Pisa, September 1973.

[15] - TUI HOANG, *Concave programming under linear constraints*. Soviet Math., 1964, pp.1437-1440.

[16] - ZWART P.B. *Nonlinear programming: Counter-examples to global optimization algorithms proposed by Ritter and Tui.* Washington Univ., Dept. of Applied Mathematics and Computer Sciences School of Ingeeniring and Applied Science. Report No. Co -1493-32- 1972.

[17] - BURDET *The facial decomposition method.* Graduate School of Industrial Administration Carnegie Mellon Univ. Pittsbrgh, Penn. May 1972.

A WIDELY CONVERGENT MINIMIZATION ALGORITHM WITH QUADRATIC TERMINATION PROPERTY

by GIULIO TRECCANI, UNIVERSITÀ DI GENOVA

I. INTRODUCTION AND NOTATIONS

We shall consider methods for minimizing a real valued function of n real variables $\varphi: R^n \longrightarrow R$ of the following type :

1.1 $$d_k = x_{k+1} - x_k = \alpha_k p_k$$

were p_k, the search direction, is a vector in R^n and α_k, the scalar stepsize, is a suitable nonnegative real number.

Two properties of methods of this kind will be considered, convergence and quadratic termination.

Assume that $\varphi(x)$ has an unique absolute minimum point x^*; then a method 1.1 is said to be globally convergent for the function φ, if every solution x_i of 1.1, starting at any point $x_o \in R^n$, is convergent to x^*.

Assume now that $\varphi(x)$ is a convex quadratic function; then a method 1.1 is said to have the quadratic termination property if for every initial point x_o it minimizes $\varphi(x)$ in at most n interations.

In the following it will be assumed that

1.2 $\varphi(x)$ in continuously differentiable in R^n ;

1.3 $\varphi(x)$ is bounded from below ;

1.4 every level set of $\varphi(x)$ is a bounded set ;

1.5 there is one and only one point $x \in R^n$ such that grad $\varphi(x^*) = 0$.

Then it can be proved that x^* is the absolute minimum point of $\varphi(x)$ and that the level sets are connected. We remark however that $\varphi(x)$ has no convexity property.

We shall construct a modification of the well known Fletcher-Reeves conjugate gradient method, which will be proved to be convergent and to have the quadratic termination property.

Even though accurate line minimization is required in our method, it will be proved to converge without any convexity assumption and search direction restoration, while the quadratic termination property is conserved; this seems to be a somewhat new result respect to the classical conjugate gradient methods of Hestenes-Stiefel, Polak-Ribiere and Fletcher-Reeves. This algoritm has been deduced as an application of the theory of Discrete Semi Dynamical Systems (DSDS)(see [1]) ; a short summary of definitions and results of this theory which will be used for our purposes is in section 3.

1.6 Basic Notations.

$G(x)$ is the gradient of at the point x R^n.

$g_i = g(x_i)$,

$d_i = x_{i+1} - x_i$.

$|x|$ is the euclidean norm of x R^n.

R^+ is the set of nonnegative real numbers.

I^+ is the set of nonnegative integers

$\mathcal{X}$ is a metric space.

$F(X)$ is the set of compact non empty subsets of X

2. THE ALGORITHM

2.1 $x_{k+1} = x_k + d_k$

2.2 $d_k = \alpha_k \; p_k$

2.3 $A_k = \left\{ \alpha \in R^+ : \text{grad } \varphi (x_k + \alpha p_k)^T p_k = 0 \right\}$

2.4 $\alpha_k = 0$ if $A_k = \emptyset$, $\alpha_k = \text{Min } A_k$ otherwise

2.5 $\rho_k = \dfrac{g_k^T g_{k+1}}{|g_k||g_{k+1}|}$ if $g_{k+1} \neq 0$, $\rho_k = 0$ otherwise.

2.6 $\nu_k = \dfrac{d_k^T g_k}{|d_k||g_k|}$ if $g_k \neq 0$, $\nu_k = 1$ otherwise

2.7 $p_o = -g_o$

2.8 $p_{k+1} = -g_{k+1} + \dfrac{|g_{k+1}|^2}{|g_k|^2} \left\{ 1 - \rho_k^{\nu_k} \right\} p_k$.

3. COMMENTS AND IMPLEMENTATION OF THE ALGORITHM

3.1 Theorem.

The solution of the algorithm 2, starting at any point x_o R^n, is an infinite sequence $\{x_i\}$, such that if for some $k \in I^+$ we have $x_{k+1} = x_k$, then $x_i = x^*$ for $i \geq k$. In addition, if $x^* \notin \{x_i\}$, then $\varphi_i = \varphi(x_i)$ is a strictly decreasing sequence of real numbers.

Proof. Assume that $x_o, \ldots, x_k$ can be computed by algorithme 2 and that $x_i \neq x^*$ for $i = 0, \ldots, k$. Then the following properties hold for $0 \leq i \leq k$:

(i) $p_i \neq 0$, $p_i^T g_{i+1} = 0$

(ii) $d_i \neq 0$

(iii) $\nu_i > 0$

(iV) $\varphi(x_{i+1}) < \varphi(x_i)$.

These properties hold for $i = 0$; indeed $p_o = - g_o$ by 2.7 and $g_o \neq 0$ by assumption, hence $\alpha_o > 0$ and $d_o \neq 0$; on the other hand $\nu_o = 1$ and $\varphi(x_1) < \varphi(x_o)$. Assume that these properties are true for $i \leq k$, and prove them for i+1. Since $\nu_i > 0$ and $g_i \neq 0$, p_{i+1} can be computed by 2.8; then we have $p_{k+1}^T g_{i+1} =$

$= - |g_{i+1}|^2 > 0$, since $g_{i+1}^T p_i = 0$ and $g_{i+1} \neq 0$. It follows that

$p_{i+1} \neq 0$, x_{i+2} can be computed and $d_{i+1} \neq 0$, $\varphi(x_{i+2}) < \varphi(x_{i+1})$, $\nu_{i+1} > 0$

and the statement follows by induction.

Assume now that $x = x^*$ for some $k \in I$, while $x_j \neq x^*$ for $j < k$.
Then by 2.8 we have $p_k = 0$, which implies $d_k = 0$, $x_{k+1} = x_k$ and $p_i = d_i = 0$ for every $i \geq k$.
It is clear that algorithm 2 is an ideal algorithm not only because the line search 2.4-2.4 is assumed to be exact, but also because no stopping rule is given and in the computation of ν_k very small quantities can be involved in the denominator, even if g_k is not very small.
For these reasons,maintag the assumption of exact line searches, we propose an equivalent form of the algorithm 2.

3.2. ALGORITHM

1. Set $i = 0$, $p_o = - g_o$

2. Compute $\alpha_i = \text{Min } A_i = \text{Min}\left\{\alpha \in R^+ : \text{grad}\,\varphi(x_i + \alpha p_i)^T p_i = 0\right.$

3. Compute $d_i = \alpha_i p_i$.

4. Compute $x_{i+1} = x_i + d_i$

5. If $|g_{i+1}| < \varepsilon$ stop, otherwise go to 6.

6. $$\rho_i = \frac{|g_{i+1}^T g_i|}{|g_i|\,|g_{i+1}|}$$

7. $$\nu_i = \frac{|g_i|}{|p_i|}$$

8. $$p_{i+1} = -g_{i+1} + \frac{|g_{i+1}|^2}{|g_i|^2}\left\{1 - \rho_i^{\nu_i}\right\} p_i$$

9. Set $i = i+1$ and go to 2.

3.3 THEOREM

If $x_o \neq x^*$, then either there is $k \in I^+$ such that the algorithm stops at a point x_k such that $|g_k| < \varepsilon$, and $\{x_o, \ldots, x_k\}$ belongs to the solution of algorithms 2 starting at x_o, or the solution of 3.2 is an infinite sequence x_i such that $|g_i| \geq \varepsilon$ and $\varphi(x_{i+1}) < \varphi(x_i)$ for every $i \in I^+$, and $\{x_i\}$ is the solution of algorithm 2 starting at x_o.

Proof. We have only to prove that $\nu_k = \dfrac{|g_k|}{|p_k|} = \dfrac{|g_k^T d_k|}{|g_k|\,|d_k|}$.

From the proof of theorem 3.1 it follows that if $g_i \geq \varepsilon$ for $0 \leq i \leq k$, then $d_k \neq 0$ and $\alpha_k \neq 0$, while $p_k^T g_k = -|g_k|^2$, which implies that :

$$\frac{|g_k^T d_k|}{|g_k|\,|d_k|} = \frac{|g_k^T \alpha_k d_k|}{|g_k|\,|\alpha_k d_k|} = \frac{|g_k^T p_k|}{|g_k|\,|p_k|} = \frac{|g_k|^2}{|g_k|\,|p_k|} = \frac{|g_k|}{|p_k|}.$$

Now we shall prove that algorithms 2 and 3.2 have quadratic termination property.

3.4. THEOREM

If $\varphi(x)$ is a convex quadratic function of n real variables, then $\varphi(x)$ is minimized by 2 and 3.2 in at most n interations.

Proof. Indeed if φ is quadratic, $\rho_i = 0$ and 2 is the Fletcher-Reeves conjugate gradient method, which generates conjugate directions and minimizes $\varphi(x)$ in at most n iterations.

C.E.D.

4. DISCRETE SEMI DYNAMICAL SYSTEMS.

Assume that X is a local compact complete metric space. The norm generated by the metric is denoted by $|\ \ |$. F(X) is the set of non empty compact subsets of X. A map $\pi : X \longrightarrow F(X)$ is said to be upper semicontinuous at $x \in X$ if

$$\lim_{y \to x} \ \max_{z \in f(y)} \ \min_{v \in f(x)} |z-v| = 0 .$$

4.1 A DSDS is the triple (X, I^+, f) where $f : X \times I^+ \longrightarrow F(X)$ is such that :

(i) $f(x,0) = \{x\}$, $\forall\ x \in X$

(ii) $f\left[f(x,k),h\right] = f(x,k+h)$, $\forall\ x \in X,\ h,\ k \in I^+$

(iii) f is upper semicontinuous in $X \times I^+$.

4.2. A solution of the DSDS through $x \in X$ is a map $\sigma : \mathcal{J} \rightarrow X$ such that :

(i) $I^+ \subseteq \mathcal{J} \subseteq I$

(ii) $\sigma(0) = x$

(iii) $\sigma(j) \in f\left[\sigma(k),\ j-k\right]$, $\forall\ j,k \in \mathcal{J}$, $j \geq k$

(iv) there is no proper extension of σ which has the properties (i), (ii) and (iii).

4.3 A set $M \subseteq X$ is said to be weakly positively invariant if for every $x \in M$ there is a solution σ of the DSDS such that $\sigma(k) \in M$ for every $k \in I^+$.

4.4 The positive limit set of a solution σ of the DSDS in the set :

$$L^+(\sigma) = \left\{ y \in X \ : \exists \{k_n\} \subseteq I^+ ,\ k_n \longrightarrow +\infty ,\ \text{such that } \sigma(k_n) \rightarrow y \right\}$$

4.5 A Liapunov function for a solution σ is a real valued function $\varphi : X \longrightarrow R$ such that $\left\{\varphi\left[\sigma(i)\right]\right\}$ is a non increasing sequence of real numbers.

The following two properties hold :

4.6. Property.

$L^+(\sigma)$ is a weakly positively invariant set.

4.7 Property

A continuous Liapunov φ for a solution σ is constant on $L^+(\sigma)$.

Properties 4.6 and 4.7 will be crucial in our convergence proof.

5. CONSTRUCTION OF A DSDS WHICH IS RELATED TO THE ALGORITHM 2.

Let X be the following set :

$$X=\{(x_1,x_2) \in R^n \times R^n : \varphi(x_2) \leq \varphi(x_1)\}$$

where $\varphi : R^n \longrightarrow R$ satisfies 1.2 , 1.3, 1.4 and 1.5.

Then X is a closed subset of $R^n \times R^n$ and is a locally compact complete metric space with the euclidean metric.

Assume that :

5.1 $p : X \longrightarrow F(R^n)$ is upper semicontinuous in a nonempty subset $X' \subseteq X$.

5.2 $a : X \times R^n \longrightarrow F(R^+)$ is upper semicontinuous in $X \times R^n$.

5.3 $h : X \longrightarrow F(R^n)$ defined as $h(x_1,x_2) = \{ y \in R^n ; y = \alpha p ,$

$\alpha \in a(x_1,x_2,p), \; p \in p(x_1,x_2)\}$ is such that $\varphi(y) \leq \varphi(x_2)$ for every $y \in h(x_1,x_2)$ and is upper semicontinuous in $X \setminus X'$.

Then it is easy to see that h is upper semicontinuous in X and the map :

5.4 $f : X \longrightarrow F(X) \qquad f(x_1,x_2) = (x_2,x_2+h(x_1,x_2))$

is supper semicontinuous in X, hence the triple (X,I^+,f^k) is a DSDS.

Assume now that the set :

5.5. $A(x_1,x_2,p) = \{ \alpha \in R^+ : \text{grad}\, \varphi(x_2+\alpha p)^T p = 0, \varphi(x_2+\alpha p) \leq \varphi(x_2)$

is empty. Then $a(x_1,x_2,p) = \{0\}$. On the otherhand, if the set $A(x_1,x_2,p) \neq \emptyset$,

we set :

5.6 $a(x_1,x_2,p) = \left\{\alpha \in A(x_1,x_2,p) : \varphi(x_2+\alpha p) \leq \varphi(x_2+p \text{ Min } A)\right\}$.

The map $a : X \times R^n \longrightarrow F(R^+)$ is clearly upper semicontinuous.

We define a map $p : X \longrightarrow F(R^n)$ as follows.

5.7 if $x_2 = x^*$ $p(x_1,x_2) = 0$

5.8 if $x_1 = x_2 \neq x^*$ $p(x_1,x_2) = -g(x_2)$

5.9 if $x_1 \neq x_2 \neq x^*$, $g^T(x_1)g(x_2) = 0$ $p(x_1,x_2) = \{y \in R^n : |y| \leq 1$,

$$y = -\alpha_1 g(x_2) + \alpha_2(x_2-x_1) , \alpha_1 \geq 0, \alpha_2 \geq 0\} .$$

5.10 if $x_1 \neq x_2 \neq x^*$, $g^T(x_1)g(x_2) \neq 0$, $(x_1-x_2)^T g(x_1) = 0$

$$p(x_1,x_2) = -g(x_2) + \frac{|g(x_2)|^2}{|g(x_1)|^2} (\log \frac{|g^T(x_1)\, g(x_2)|}{|g(x_1)||g(x_2)|}) |g(x_1)| \frac{(x_1-x_2)}{|x_1-x_2|} .$$

5.11 if $x_1 \neq x_2 \neq x^*$, $g^T(x_1)g(x_2) \neq 0$, $g^T(x_1)(x_1-x_2) \neq 0$

$$p(x_1,x_2) = -g(x_2) + \frac{|g(x_2)|^2}{|g(x_1)|^2} \left\{1 - \frac{g^T(x_1)g(x_2)}{|g(x_1)||g(x_2)|}^{\frac{|(x_1-x_2)^T g(x_1)|}{|x_1-x_2||g(x_1)|}} \right\} .$$

$$\cdot \frac{|g(x_1)|^2}{|(x_1-x_2)^T g(x_1)|} (x_2-x_1).$$

5.12 Property

The map $p(x_1,x_2)$ is upper semicontinuous for $g^T(x_1)g(x_2) \neq 0$, while the map

$h(x_1,x_2)$ is upper semicontinuous for every $(x_1,x_2) \in X$.

From the definitions 5.5 - 5.6 of the map $a(x_1,x_2,p)$, it follows that if we set :

(5.13) $\Phi(x_1,x_2) = \varphi(x_1) + \varphi(x_2)$

then $\Phi : X \rightarrow R$ is a Liapunov function for every solution σ of the DSDS defi ned above.

5.14 Theorem

If σ is any solution of the DSDS defined by 5.5-5.12 and $k \in I^+$ is such that $\Phi[\sigma(k)] = \Phi[\sigma(k+1)] = \Phi[\sigma(k+2)]$, then $\sigma(i) = (x^*,x^*)$ for every $i \geq k$.

Proof. Let $\sigma(k) = (x_1,x_2)$, $\sigma(k+1) = (x_2,x_3)$ and $\sigma(k+2) = (x_3,x_4)$ then by hypothesis and by 5.13 $\varphi(x_1) = \varphi(x_2) = \varphi(x_3) = \varphi(x_4)$; indeed $\Phi[\sigma(k)] = \Phi[\sigma(k+1)]$ implies $\varphi(x_1) + \varphi(x_2) = \varphi(x_2) + \varphi(x_3)$, $\varphi(x_1) = \varphi(x_2) = \varphi(x_3)$.

Hence it is sufficient to prove that $x_1 = x^*$ for some $i \in (1,2,3,4)$.

Clearly, if $x_3 \neq x^*$, it must be $x_1 \neq x_2$, $g^T(x_2)(x_2-x_1) \geq 0$, and also $x_2 \neq x_3$, since $x_2 = x_3$ would imply $\varphi(x_4) < \varphi(x_3)$.

Then we have $x_1 \neq x_2 \neq x_3$, $g^T(x_2)(x_2-x_1) \geq 0$, $g^T(x_3)(x_3-x_2) = 0$.

Now if $g^T(x_3)g(x_2) \neq 0$, then by 5.10-5.11 $p[\sigma(k+1)] = p(x_2,x_3) \neq 0$ and $p^T(x_2,x_3)g(x_3) = -|g(x_3)|^2$ 0, which would imply $\varphi(x_4) < \varphi(x_3)$.

On the other hand if $g^T(x_3)g(x_2) = 0$ it would be $p_2^T(x_2,x_3)g(x_3) =$

$= \alpha_1\ g(x_3)\ ^2 + \alpha_2(x_3-x_2)^T g(x_3) = -\alpha_1 |g(x_3)|^2 < 0$, which would imply $\varphi(x_4) < \varphi(x_3)$.

In any case we have proved that $x_3 \neq x^*$ implies a contradiction, hence

$x_1 = x_2 = x_3 = x_4 = x^*$.

Q.E.D.

As a consequence of theorem 5.14 the following property holds.

5.15 Theorem.

For every solution σ of the DSDS defined by 5.5-5.11, $L^+(\sigma)$ is a nonempty set and $L^+(\sigma) = (x^*,x^*)$.

Proof. Since Φ is a Liapunov function for any solution, the level sets of $\varphi(x)$ are strongly positively invariant, and since by 1.4 they are bounded, $L^+(\sigma)$ must be nonempty.

As $\varphi(x)$ is continuous in R^n, Φ is continuous in X and by 4.7 is constant on $L^+(\sigma)$.
Since by 4.6 $L^+(\sigma)$ is weakly positively invariant, if $(x_1,x_2) \in L^+(\sigma)$ then $(x_1,x_2) = (x^*,x^*)$ by theorem 5.14.

C.E.D.

6. CONVERGENCE OF ALGORITHM 2.

Let x_1 be any solution of algorithm 2. Then if we set :

6.1 $\sigma(0) = (x_o,x_o)$, $\sigma(1) = (x_o,x_1)$, $\sigma(k) = (x_{k-1},x_k)$ then any maximal extension of σ is a solution of the DSDS.

Indeed only the cases 5.7, 5.8, 5.8 and 5.11 can be verified.
It follows that $L^+(\sigma) = (x^*,x^*)$ by theorem 5.15, which implies that $x_i \longrightarrow x^*$ for $i \longrightarrow +\infty$.

7. CONCLUSIONS.

We have constructed a DSDS whose properties imply the convergence of a conjugate gradient method, which is a modification of the Fletcher and Reeves method and has the quadratic termination property.
The convergence is global for functions continuously differentiable, bounded from below, having bounded level sets and one and only one critical point.
However these assumptions are not restrictive, since if the function is simply continuously differentiable, the algorithm converges to a local minimum point if the initial point is chosen in any bounded level set containing one and only critical point.

BIBLIOGRAPHY

1 G.P.Szego and G.Treccani : Semigruppi di Trasformazioni Multivoche in I^+. Symposia Mathematica, Vol.VI, Academic Press, London,New York, 1971, pp.287-307.

2 G.P.Szego and G.Treccani : Axiomatization of minimization algorithms and a new conjugate gradient method in G.Szego "Minimization Algorithms", 1972 , Academic Press,Inc. New York.

A HEURISTIC APPROACH TO COMBINATORIAL OPTIMIZATION PROBLEMS (o)

by

E. Biondi, P.C. Palermo
Istituto di Elettrotecnica ed Elettronica
Politecnico di Milano

1. Introduction

It is well known that the exact solution of large scale combinatorial problems (for instance scheduling, sequencing, delivery, plant-location, travelling salesman problem ....) implies hard or prohibitive difficulties concerning the computation time and the memory storage.

These problems are generally approached by heuristic techniques, frequently developed " ad hoc", which allow to determine efficiently near-optimal solutions.

This paper outlines a general heuristic approach, based on the conceptual framework of Dynamic Programming, which seems to be effective in a large class of combinatorial problems.

Some efficient algorithms, successfully tested in classical flow-shop scheduling and delivery problems, are discussed.

2. The approach

Denote

$f(x,d)$ a separable objective function,

x the discrete set of state variables,

d the discrete set of decision variables,

$$g(x) = \min_{d} f(x,d) \quad .$$

(o) This work was supported by C.N.R. (National Research Council) and by foundation Fiorentini-Mauro.

According to the principle of Dynamic Programming the overall optimization problem is decomposed into a sequence of linked sub-problems concerning sub-sets of decision variables.

Let

d^k be the decision variable (s) at the k-th stage, (k=1,...,N),

$D^k=\{d_1^k,\dots,d_{n_k}^k\}$ the feasible definition set of d^k,

x^k the state variables at the k-th stage,

f_i^k the pay-off of the decision d_i^k at the k-th stage,

x_i^{k+1} the state variables at the stage (k+1) after the decision d_i^k at the k-th stage,

$\bar{d}^k = \{d-d^1 - \dots - d^{k-1}\}$,

$\bar{D}^k$ the definition set of $\bar{d}^k$.

At the k-th stage the basic dynamic program involves the following computation :

determine $$g(x^k)= \min_{d^k} \{f_i^k + \min_{\bar{d}^k} f(x_i^{k+1}, \bar{d}^{k+1})\} =$$

$$= \min_{d^k} \{f_i^k + g(x_i^{k+1})\} . \qquad (1)$$

Let

$\lambda \cdot h(x_i^{k+1})$ be a parametric estimate of $g(x_i^{k+1})$,
where $h(x_i^{k+1})$ is a known (suitably defined) evaluation function of $g(x_i^{k+1})$
and $\lambda > 0$ is an adjustment factor of $h(x_i^{k+1})$,

λ_i^k be the (unknown) exact value of the adjustment factor of $h(x_i^{k+1})$,

i.e. $$g(x^k) = \min_{d^k} \{f_i^k + \lambda_i^k \ h(x_i^{k+1})\} \qquad (2)$$

In the following, problem (2) is dealt with.
Heuristic Search algorithms are developed in accordance with some assumptions about the adjustment factors λ_i^k .
Such assumptions allow to reduce the computational complexity of problem (1) by meaningful criteria.

2.1 Assumption 1

$$\lambda_i^k = \lambda \qquad (k=1,\ldots,N) \quad , \quad (\forall k,\ i=1,\ldots,n_k) \ .$$

The assumption is very strong and rather rough.
However it allows to generate a large set S of sub-optimal solutions by an efficient iterative algorithm .
At each iteration a different value is assigned to the parameter λ and a solution is determined by solving the sequence of sub-problems :

$$\min_{d^k} \ \{f_i^k + \lambda\ h(x_i^{k+1})\} \qquad (k=1,\ldots,N) \quad , \quad \lambda \text{ given} \ .$$

The minimum cost solution within the set S is selected.
The computational effort is limited because the cost of the solution computed by the algorithm is piecewise constant with respect to λ , i.e. the same solution may be determined in a suitable λ-range (see fig. 1 and section 3 below).

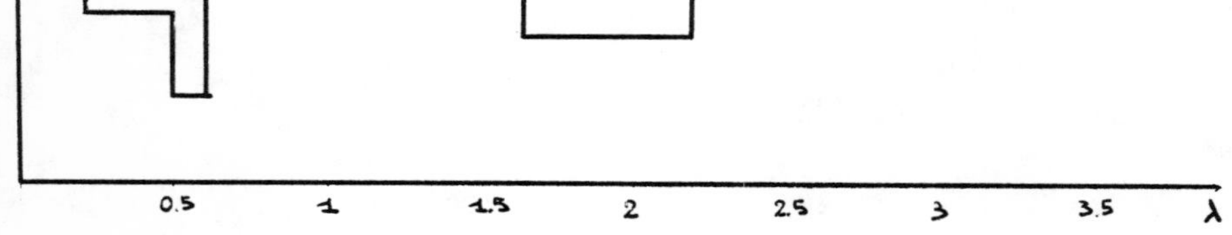

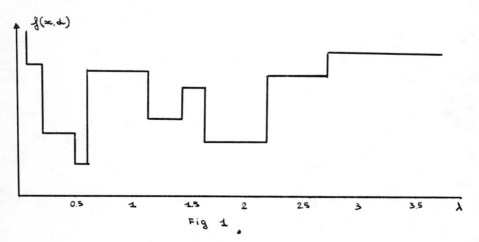

Fig 1

2.2 Assumption 2

$$\lambda_i^k = \lambda^k \qquad (i=1,\dots,n_k) \ , \ (k=1,\dots,N) \quad .$$

This assumption (more plausible than assumption 1) leads to a conspicuous cut of the definition set D^k .

In fact

d_j^k , $(j=1,\dots,n_k)$, is deleted from D^k

if, for every $\lambda^k>0$, there exists a decision $d_i^k \in D^k$ such that

$f_i^k + \lambda^k . h(x_i^{k+1}) \leq f_j^k + \lambda^k h(x_j^{k+1})$ (exclusion test I).

Let $R_1^k \subseteq D^k$ be the definition set of d^k after the application of the exclusion test I.

Efficient Branch-Search and Branch - and Bound algorithms may be developed by applying a suitable branching criterion on the set R_1^k .

An effective procedure is now outlined.

Let U^k be an upper bound of $g(x^k)$ determined by a classical heuristic method or by the iterative algorithm based on assumption 1. Assume that $\alpha U^k, (0<\alpha<1)$, is a close estimate of $g(x^k)$ (the value of the parameter α may turn out from empirical analyses of the quality of the heuristic method used for the generation of the upper bound).

Denote

$$\lambda_{Li}^k = \frac{\alpha U^k - f_i^k}{h(x_i^{k+1})} \qquad (d_i^k \in R_1^k) \quad .$$

The following Branching Criterion is now introduced:

at the k-th stage the decision $d_o^k \in R_1^k$ such that

$$\lambda_{Lo}^{k} = \max_{d_i^k \in R_1^k} \lambda_{Li}^{k}$$

is selected for branching.
The geometric interpretation of the exclusion test I and of the branching criterion is shown in fig. 2 .

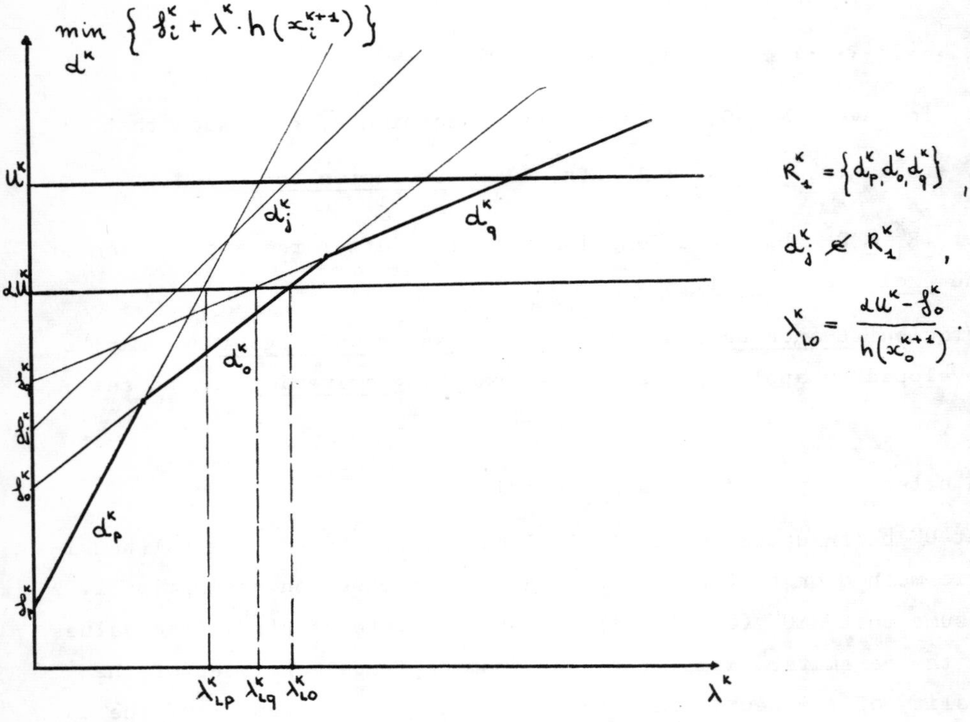

Fig. 2

2.3. <u>Assumption 3</u>

$$|\lambda_i^k - \lambda_j^k| \leq \beta_H \qquad (i,j) \text{ , } (i,j=1,\ldots,n_k) \text{ , } (k=1,\ldots,N)$$

This assumption is always true for a suitable value of the parameter β_H, which clearly depends on the assumed evaluation function $h(x_i^{k+1})$.

The closer $h(x_i^{k+1})$ is to $g(x_i^{k+1})$, the smaller is the λ-range containing the values $\lambda_i^k,(i=1,\ldots,n_k)$, and a small value β_H is likely.

According to assumption 3 some decisions may be deleted from D^k. In fact, consider the example shown in fig. 3 .

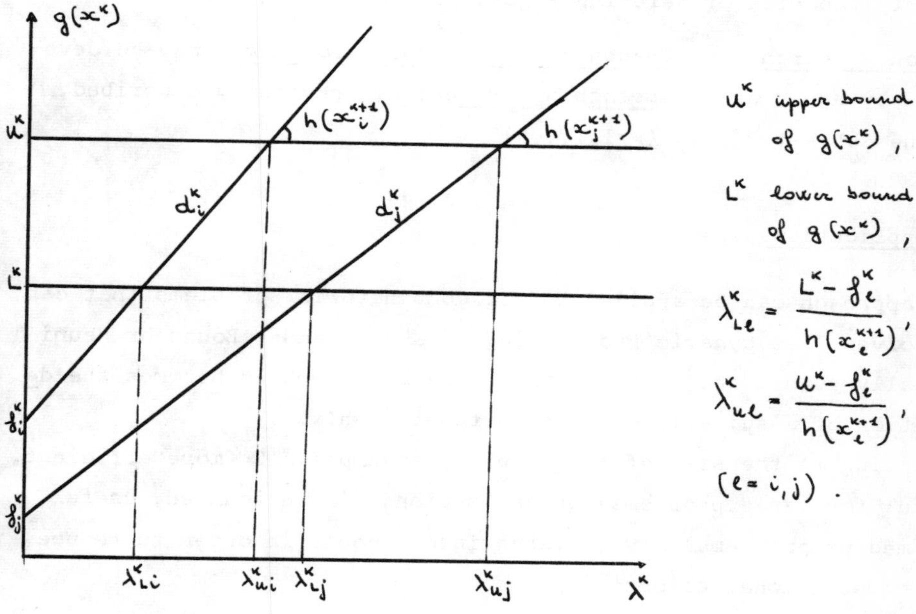

Fig. 3 .

It turns out that d_i^k is not preferred to d_j^k if

$$\min\{|\lambda_{uj}^k - \lambda_{ui}^k|, |\lambda_{Lj}^k - \lambda_{Li}^k|\} > |\lambda_j^k - \lambda_i^k| \quad .$$

Consequently the following test may be performed :

d_i^k is deleted from D^k according to assumption 3, if there exists a decision $d_j^k \in D^k$ such that

$$\min\{|\lambda_{uj}^k - \lambda_{ui}^k|, |\lambda_{Lj}^k - \lambda_{Li}^k|\} \geq \beta_H \qquad \text{(exclusion test II)}$$

Experimental values of the parameter β_H may be tested in order to delete those decisions from D^k, which have a small probability of belonging to an optimal solution.
The resulting subset $R_2^k \subseteq D^k$ is generally larger than the sub-set R_1^k found in accordance with assumption 2 (which corresponds to $\beta_H = 0$).
Clearly the risk of deleting a good choice is less.

Branch - Search and Branch - and - Bound algorithms may be developed by applying the Branching Criterion, previously described, on the sub-set R_2^k , $(k=1,\ldots,N)$.

3. Applications

The approach can be applied to all combinatorial problems that can be solved via Dynamic Programming, Branch - and - Bound or Heuristically guided search algorithms. The main problem becomes the definition of a suitable evaluation function $h(x)$.
According to the size of the problem, assumption 2 (more efficient, useful for large problems) or assumption 3 (more founded, useful for medium problems) may be taken into account in order to reduce the computational effort.

Experiments have been performed with reference to classical delivery and flow-shop scheduling problems, whose statement and formulation are now briefly summarized.

3.1 Consider the following delivery problem :

a set of customers, each with a known location and a known requirement for commodity, is to be supplied from a single warehouse by vehicles of known capacity.
The objective is to design the routes which minimize the delivery cost, subject to a capacity constraint for the vehicles and meeting the customers' requirements.

Let

$x = \{x_i | i=0,1,\dots,n\}$ be a set of elements representing the warehouse (i=0) and the customers (i=1,...,n),

$u=\{u_{ij}\}$, $(i,j=0,1,\dots,n)$, the set of links available for the transport,

$x^k \subseteq \{x-x_o\}$ the set of customers to be supplied after (K-1) stages,

d^k the feasible route to be selected at the k-th stage.

Assume

$$h(x_i^{k+1}) = \sum_{x_j \varepsilon x_i^{k+1}} u_{oj} .$$

The iterative algorithm [4] and a branch-search algorithm [13] according to the outlined branching criterion have been experimented in large problems.

The computational results show that the minimum cost solution attained by the iterative algorithm gives a satisfactory near-optimal solution of the problem (a typical diagram of the results is drawn in fig. 1).

The quality of the solution supplied by the branch-search algorithm is practically equivalent (or just better), while the computation time is considerably shorter.

The tests of both algorithms show remarkable advantages with respect to the exact and heuristic methods proposed for the same problem in the literature.

3.2 The classical three-machine flow-shop scheduling problem is now considered:

n jobs must be processed through three machines (A,B,C) in the same technological order. All jobs are processed once and only once on each machine. A job cannot be processed on a machine before it has been completed on the preceding machine in the technological order.

The objective is to schedule the flow-shop so that the completion time of the last job on the last machine ("makespan") is minimized.

Denote

a_i, b_i, c_i the processing times of job i , (i=1,...,n), on the machines A,B,C ;

d the job-set;

d^k the job to be processed at the k-th stage (k=1,..,n);

$\bar{d}^k = \{d-d^1-\ldots-d^{k-1}\}$;

D^k the job-set to be processed at the k-th stage;

$$B(x^k) = \max\{B(x^{k-1}), \sum_{\{d-\bar{d}^{k+1}\}} a_j\} + b_i \quad ;$$

$$C(x^k) = \max\{C(x^{k-1}), B(x^k)\} + c_i \quad ;$$

$B(x^1) = a+b$ where a,b,c are the processing times of the

$C(x^1) = B(x^1)+c$ first job of the scheduling ;

$$f_i^k = C(x^k) - C(x^{k-1}) \quad .$$

Assume

$$h(x^{k+1}) = \max \begin{cases} \sum_{\bar{d}^{k+1}} c_j \ , \\ \sum_{\bar{d}^{k+1}} b_j + \min_{\bar{d}^{k+1}} c_j - (C(x^k) - B(x^k)) \ , \\ \sum_{d^{k+1}} a_j + \min_{\bar{d}^{k+1}} (c_j + b_j) - (C(x^k) - \sum_{\{d-\bar{d}^{k+1}\}} a_j) \end{cases}$$

A Branch - and - Bound technique has been applied according to the <u>Exclusion Test II</u> and the <u>Branching Criterion.</u> The results show that the procedure allow to determine a near-optimal solution in large problems in a very effective way. It has been checked that such solution generally corresponds to the optimal one in small size problems.

REFERENCES

1 N. Agin "Optimum seeking with Branch and Bound", Mgmt Sci. 13, 176-186, (1966).

2 S. Ashour " An experimental investigation and comparative evaluation of flow-shop scheduling techniques" , Opns. Res. 18, 541-549, (1970).

3 E. Balas "A note on the Branch - and - Bound Principle", Opns. Res. 16, 442-445, (1968).

4 E. Biondi, P.C. Palermo, C. Pluchinotta "A heuristic method for a delivery problem", 7-th Mathematical Programming Symposium 1970 The Hague - The Netherlands.

5 H. Campbell, R. Dudek, M. Smith "A heuristic algorithm for the n job - m machine sequencing problem", Mgmt. Sci. 16, 630-637, (1970).

6 N. Christofides, S. Eilon "An algorithm for the vehicle dispatching problem", Opl. Res. Q. 20, 309-318 (1969).

7 R.J. Giglio, H.M. Wagner "Approximate Solutions to the three-Machine Scheduling Problem", Opns. Res., 305-319 (1964).

8 P. Hart, N.J. Nilsson, B. Raphael "A formal basis of the heuristic determination of minimum cost paths" IEEE Trans. SSC 4, 100-107 (1968).

9 M. Held, R.M. Karp "A Dynamic Programming Approach to sequencing problems", J. Soc. ind. appl. Math. 10, 196-208 (1962).

10 E. Ignall, L. Schrage "Application of the Branch - and - Bound Technique to some flow-shop scheduling problems", Opns. Res. 13, 400-412 (1965).

11 G.B. McMahon, P.P. Burton "Flow-shop scheduling with Branch - and - Bound method", Opns. Res. 15, 473-481 (1967).

12 L.G. Mitten "Branch - and - Bound Methods: general formulation and properties", Opns. Res. 16, 442-445, (1968).

13 P.C. Palermo, M. Tamaccio "A Branch-Search algorithm for a delivery problem" Working Paper LCA 71-4 (1971).

A NEW SOLUTION FOR THE GENERAL SET COVERING PROBLEM

László Béla KOVÁCS
Computer and Automation Institute
Hungarian Academy of Sciences

1. Introduction. The theory and applications of the set partitioning and general set covering problem is discussed in the present paper. After the problem formulation, three applications are shown: bus route planning, airline-crew scheduling and a switching circuit design method. After a short survey of existing methods a new algorithm is introduced in section 5. The exact algorithm is based on the branch-and-bound principle, but no linear programming is used in determining the bounds. A heuristic procedure is utilized instead, which often gives the optimal or near optimal solution of the problem. The optimality in certain cases guaranted by lemmata 1-3. The last lemma also provides a lower bound for the branch-and-bound procedure, which is close to the real minimum of the sub-problems after a few steps - as practice shows. A computer program is written for a CDC 3300 computer. The results are promising. The program is being further tested and developed.

2. Problem and terminology. Let us consider the following problem:

$$\min \underline{c}^T \underline{x}$$

$$(1) \qquad A\underline{x} \geqq \underline{e}$$

$$x_j = 0 \text{ or } 1 \quad (j = 1, \ldots, n),$$

where A is a given $m \times n$ matrix of 0 and 1 elements is an $\underline{c}$ vector of positive integer elements and $\underline{e}$ is an m-vector, each component of which is 1. Problem (1) is known as the set covering problem for the following reason. It is given m subsets of the set

$$\mathcal{J}_0 = \{1, 2, \dots, n\},$$

$\mathcal{J}_1, \mathcal{J}_2, \dots, \mathcal{J}_m$. A cost $c_j > 0$ is associated to each element $j \in \mathcal{J}_0$. At least one element of each set $\mathcal{J}_1, \dots, \mathcal{J}_m$ is to be chosen i.e. each set is to be covered at a minimal total cost. In other words a set $\mathcal{J}$ is to be determined:

$$\min \left\{ \sum_{j \in \mathcal{J}} c_j \;\middle|\; \mathcal{J} \subset \mathcal{J}_0 ,\ \mathcal{J} \cap \mathcal{J}_i \neq \emptyset \quad (i = 1, \dots, m) \right\}$$

Problem (1) is obtained, if matrix is defined as

$$a_{ij} = \begin{cases} 1 & \text{if } j \in \mathcal{J}_i \\ 0 & \text{otherwise.} \end{cases}$$

The set $\mathcal{J}$ is the subscripts of variables x_j having the value 1 in the optimal solution of problem (1) .

The general set covering problem

$$\min \underline{c}^T \underline{x}$$

$$(2) \qquad A \underline{x} \geqq \underline{b}$$

$$x_j = 0 \text{ or } 1 \quad (j = 1, \dots, n)$$

may be interpreted a similar way. The only difference is, that each set $\mathcal{J}_i$ should be covered $b_i > 0$ times instead of just ones. Thus $\underline{b}$ is a given m -vector of positive integer components.

The set partitioning problem

$$\min \underline{c}^T \underline{x}$$

$$(3) \qquad A \underline{x} = \underline{e}$$

$$x_j = 0 \text{ or } 1 \quad (j = 1, \dots, n)$$

is the same as problem (1) , except that the inequalities are substituted by equations. The same interpretation may be used also for problem (3) , the only difference is that each set should be covered exactly ones. Each solution $\underline{x}$ of problem (3) however defines a partitioning of the set $\mathcal{J}_0$. Let

$$K_j = \bigcup_{j \in \mathcal{J}_i} \mathcal{J}_i$$

if $x_j = 1$ and $K_j = 0$ otherwise. Then obviously

$$\bigcup_j K_j = \mathcal{J}_0$$

and

$$K_r \cap K_s = \emptyset \quad \text{if } r \neq s$$

Thus the sets $K_1, \ldots, K_n$ give a partitioning of set $\mathcal{J}_0$. Therefore problem (3) may be stated as to determine the partitioning of set $\mathcal{J}_0$ at a minimal cost. It should be noted, that apart from the trivial case of identical columns in matrix A any postitioning uniquely determines the corresponding solution $\underline{x}$, if there is any.

3. Applications.

3.1 Planning of bus routes. It is given n possible bus routes with an attached cost c_j . There are m bus stops and A is the incidence matrix, i.e.

$$a_{ij} = \begin{cases} 1 & \text{if route } j \text{ goes through bus stop } i \\ 0 & \text{otherwise.} \end{cases}$$

A bus network of minimal cost is to be determined in such a way, that at least one bus route should go through each bus stop. If variable x_j has the value 1 or 0 depending on whether bus route j is realised or not, then problem (1) is the mathematical model for the bus route planning.

3.2 Airline-crew scheduling. Exactly one crew should be assigned to each of the given m flights in such a way that each crew obtaines an acceptable full assignement /limited number of flights, not too long working time etc./ The objective is to minimize the number of crews actually used. To solve this problem let us determine a great number of acceptable crew assignements and calculate matrix A:

$$a_{ij} = \begin{cases} 1 & \text{if flight } i \text{ is included in assignement } j \\ 0 & \text{otherwise} \end{cases}$$

and the meaning of the variables

$$x_j = \begin{cases} 1 & \text{if assignement } j \text{ is accepted} \\ 0 & \text{otherwise.} \end{cases}$$

Then a set pastitioning problem (3) is obtained with $c_j = 1$ $(j = 1,\dots,n)$.

3.3 Switching circuit design. A function $F(u_1, u_2, \dots, u_N)$ is called truth function, if both the variables $u_1, \dots, u_N$ and the function F may take only the values 0 and 1. The value 0 is also referred to as the off position of the switch or the truth value false. Similarly the value 1 is also interpreted as position on or truth value true. The costs of AND gate and OR gate are given. The problem is to realize the truth function $F(u_1, \dots, u_N)$ at a minimal cost. Let us suppose, that the function is given either in a tableau form or in a disjunctive normal form. The latter is the disjunction of different terms, where each term is a conjunction of a subset of variables $u_1, \dots, u_N$ and their negated form $1-u_1, \dots, 1-u_N$.

Let $\tilde{u}_k$ denote either variable u_k or its negated form $1-u_k$.

Definition. The conjunction $Q = \tilde{u}_{k_1} \tilde{u}_{k_2} \cdots \tilde{u}_{k_r}$ is a prime implicant of function F if $\tilde{u}_{k_1} \tilde{u}_{k_2} \cdots \tilde{u}_{k_r} = 1$ implies, that $F(u_1, \dots, u_N) = 1$

and no part of Q has the same property.

Then our problem may be transformed into a set covering problem by the following steps:

(i) Determine all prime implicants $Q_1, Q_2, \ldots, Q_n$ of function F and their attached costs $c_1, c_2, \ldots c_n$ /The number of conjunction in the prime implicant times the price of the AND gate plus the price of the OR gate counted once, because these prime implicants will be connected by the sign of disjunction/

(ii) Let $\underline{u}^1, \underline{u}^2, \ldots, \underline{u}^m$ denote all the vectors for which $F(u^k)=1$, and define the matrix A :

$$a_{ij} = \begin{cases} 1, & \text{if } Q_j(u^i)=1 \\ 0 & \text{otherwise.} \end{cases}$$

Then the truth function F may be written as

$$F = \sum_{j=1}^{n} Q_j x_j$$

and the problem becomes the set covering problem (1).

If the function F takes the value 1 for more then half of the values of the argument, then the function $1-F$ may be determined instead. There are also other tricks to decrease the size of the problem.

4. Survey of methods. Practically a version of any integer programming method may be tried for the set covering problem. On the other hand a great number of papers are already devoted to the subject, thus only a partial list of publications is mentioned here in which

each direction is represented by one or two papers.

First of all a version of the Gomory [5] cutting plane method, by Martin [10] is reported to be effective to set covering problems. A paper of House, Nelson and Rado [7] is devoted to the development of a special algorithm only for the set covering problem. The substance of the method is the construction of additional rows to matrix A for the exclusion of solutions not better than the best one obtained so far during the algorithm. The paper of Bellmore and Ratliff [3] also falls to the category of cutting plane methods with a substancially different type of cutting method.

A typical example for the use of branch-and-bound method is the paper of Lemke, Salkin and Spielberg [9] . The main difference between their and our approaches/discussed in section 4/ is, that they are solving linear programming subproblems for obtaining bounds and in the method of the present paper no linear programming is used at all. Most probably the structure of branching is also different.

Heuristic methods play an important role in large problems for several reasons. The airline-crew scheduling is solved by a heuristic method by Arabeyre, et all [1]. Garfinkel and Nemhouser [4] apply other heuristics for the set partitioning problem. A rounding process and consecutive fixing is used to decrease the size of the problem. The smaller problems so obtained are solved by existing integer programming methods. The group theoretic approach of integer programming is used by Thiriez [11] having the special advantage, that in most cases the determinant of the optimal basis B is usually small, because of the 0-1 elements in matrix A.

<u>5. A new solution for the set covering problem.</u> The description of the method consists of two part. The first part gives an explanation

of a heuristic method. The second one describes an exact method using the branch-and-bound principle and also the heuristic method for calculating bounds.

5.1. A heuristic method. Let us define the set of uncovered rows, I_k and the set of subscripts of unused variables, J_k after executing iteration k. At the beginning all rows are uncovered and no columns are used:

$$I_0 = \{1, 2, \ldots, m\} \quad , \qquad J_0 = \{1, 2, \ldots, n\}$$

Let us introduce the column count at iteration k as

$$n_j^k = \sum_{i \in I_k} a_{ij}$$

and the evaluation of unused variable at iteration k

$$r_j^k = \begin{cases} \dfrac{c_j}{n_j^k} , & if \quad n_j^k > 0 \\ \infty , & if \quad n_j^k = 0 \end{cases} \qquad for \quad j \in J_k$$

r_j^k is nothing but the row covering cost for each individual row, thus it gives an evaluation of variable j at iteration k.

At iteration $k+1$ the variable $x_{j_{k+1}}$ with the best evaluation is chosen:

$$r_{j_{k+1}} = \min_{j \in J_k} r_j^k$$

The new sets are easily calculated

$$J_{k+1} = J_k - \{j_{k+1}\} \quad and \quad I_{k+1} = I_k - M_{k+1},$$

where M_{k+1} is the set of subscripts of the rows covered by variable $x_{j_{k+1}}$:

$$M_{k+1} = \{ i \mid i \in I_0 ,\ a_{i j_{k+1}} = 1 \}.$$

The procedure is continued until either I_t or J_t becomes empty. In the first case we have obtained a feasible solution of the problem. In the second one no feasible solution exists. Let us suppose that a solution is obtained, i.e. $I_t = \emptyset$,

$$\hat{x}_j = \begin{cases} 1 & \text{if} \quad j = j_1, j_2, \dots, j_t \\ 0 & \text{otherwise.} \end{cases}$$

Then the following lemmata show how good this solution may be:

<u>Lemma 1</u>. If

$$(4) \qquad r^k_{j_{k+1}} \leq r^0_j \quad \text{for any} \quad 0 < k < t \quad \text{and} \quad 0 < j \leq n \quad (j \neq j_1, \dots, j_t)$$

then $\hat{x}$ is an optimal solution of problem (1).

<u>Proof</u>. Consider any solution $\tilde{x}$ of problem (1) and introduce the following notation :

$$(5) \qquad \hat{J} = \{ j \mid \hat{x}_j = 1 \}, \quad \tilde{J} = \{ j \mid \tilde{x}_j = 1 \}, \quad J = \hat{J} \cap \tilde{J}$$

We can suppose, that $\tilde{J} - \hat{J} \neq \emptyset$. Furthermore let

$$r = \min \{ r^0_j \mid 0 < j \leq n, \ j \neq j_1, \dots j_t \}.$$

Then

$$(6) \qquad \underline{c}^T \hat{\underline{x}} = \sum_{j_s \in \hat{J}} n^{s-1}_{j_s} r^{s-1}_{j_s} = \sum_{j_s \in J} n^{s-1}_{j_s} r^{s-1}_{j_s} + \sum_{j_s \in \hat{J} - J} n^{s-1}_{j_s} r^{s-1}_{j_s} .$$

The two expression may be handled seperately:

$$(7) \qquad \sum_{j_s \in J} n^{s-1}_{j_s} r^{s-1}_{j_s} = \sum_{j \in J} c_j$$

by the definition of $n^{s-1}_{j_s}$ and $r^{s-1}_{j_s}$. On the other hand

$$(8)\quad \sum_{j_s\in \hat{J}-J} n_{j_s}^{s-1} r_{j_s}^{s-1} \leq \sum_{j_s\in\hat{J}-J} n_{j_s}^{s-1} r \leq \sum_{j\in\tilde{J}-J} n_j^{o} r_j^{o} = \sum_{j\in\tilde{J}-J} c_j$$

The first inequality holds because of the definition of the number r and the second one is also valid, because $\tilde{\underline{x}}$ is a solution of problem (1) . (6) -(8) results, that

$$\underline{c}^T\hat{\underline{x}} \leq \underline{c}^T\tilde{\underline{x}}$$

for any $\tilde{\underline{x}}$ feasible solution of problem (1) , thus the lemma is proved.

The following lemma gives a stronger result:

Lemma 2. Let us define the numbers u_i for each row:

$$(9)\quad u_i = r_{j_s}^{s-1}, \text{ if } i\in M_s, \text{ but } i\notin M_1,\dots,M_{s-1}.$$

If

$$(10)\quad \sum_{i=1}^{m} a_{ij} u_i \leq c_j \qquad \text{for any } 0<j\leq n,\ j\neq j_1,\dots,j_t,$$

then $\hat{\underline{x}}$ is an optimal solution of problem (1).

Proof. Consider any feasible solution $\tilde{\underline{x}}$ of problem 1 . Because of the definition of numbers u_i

$$(11)\qquad \underline{c}^T\underline{x} = \sum_{i=1}^{m} u_i .$$

On the other hand, using the definition (5), the supposition (10) and the fact that $\tilde{\underline{x}}$ is a feasible solution of problem (1), i.e.

$$\sum_{j=1}^{n} a_{ij}\tilde{x}_j \geqq 1 \qquad (i=1,\dots,m)$$

We obtain

$$(12) \qquad \sum_{i=1}^{m} u_i \leqq \sum_{i=1}^{m} u_i \sum_{j=1}^{n} a_{ij}\tilde{x}_j = \sum_{j=1}^{n}\sum_{i=1}^{m} a_{ij}\tilde{x}_j u_i =$$

$$= \sum_{j\in\tilde{J}}\sum_{i=1}^{m} a_{ij}u_i \leqq \sum_{j\in\tilde{J}} c_j = \underline{c}^T\underline{\tilde{x}}.$$

Then (11) and (12) together gives the desired result, that

$$\underline{c}^T\underline{\hat{x}} \leqq \underline{c}^T\underline{\tilde{x}}$$

which proves the lemma.

The following lemma provides a lower bound for the branch-and-bound procedure if lemmata 1 and 2 did not prove the optimality of the feasible solution $\underline{\hat{x}}$ obtained by the above heuristic algorithm. The optimality of $\underline{\hat{x}}$ may also be proven sometimes by this result, see the notes after the lemma.

<u>Lemma 3.</u> Denote the optimum of problem (1) by z^*. Then

$$(13) \qquad z^* \geqq \sum_{i=1}^{m} f_i = F \,,$$

where f_i is the "minimal covering fraction" of row i:

$$(14) \qquad f_i = \min\{r_j^0 \mid 0 < j \leqq n,\ a_{ij} = 1\} \quad (i=1,\dots,m).$$

<u>Proof</u>. Define the following new problem

$$\min \sum_{j=1}^{n} \sum_{k=1}^{m} r_j^{o} y_{jk}$$

(15) $$\sum_{j=1}^{n} \sum_{k=1}^{m} g_{ijk} y_{jk} \geq 1 \qquad (i=1,\dots,m)$$

$$y_{jk} = 0 \text{ or } 1,$$

where

$$g_{ijk} \begin{cases} a_{ij}, \text{ if } j=k \\ 0 \quad \text{otherwise.} \end{cases}$$

To any feasible solution $\underline{x}$ there exists a solution of problem (15), in such a way that the objective functions are the same, e.g.:

$$y_{jk} = x_j \qquad (j=1,\dots,n; \; k=1,\dots,m).$$

On the other hand the minimum of problem (15) is obviously F . This proves the statetement (13) of the lemma.

<u>Notes</u>: 1° If $\underline{c}^T\hat{\underline{x}} = F$, then $\hat{\underline{x}}$ is on optimal solution of problem (1) .

2°. If $\underline{c}^T\hat{\underline{x}} > F$, then F gives a lower bound of the optimum of problem (1) which may be used in a branch-and-bound procedure.

<u>5.2 A branch-and-bound procedure for solving problem 1 .</u>

Many papers are devoted to the discussion of branch-and-bound procedures. See e.g. the survey of Lawler and Wood [8] . On the other hand the computer program is now developed further, thus only some basic caracteristics of the <u>present</u> comptur program is given here:

1° LIFO /last in first out/ rule is used for memory saving

2° Fast termination of subproblems is guaranteed if no solution of the subproblem exists. /The subproblem s are of the same form then problem (1), only some of the variables are fixed either at 0 or at 1.

3° Fast recanstruction of the next subproblem to be considered is made in a special way.

4° Good solutions are usually given in case of abnormal termination.

6. The computer program is written for a CDC 3300 computer in FORTRAN. The program is being tested and further developed. It solves problems up to 100 variables in a few seconds. A version of the program is suitable of solving the general set covering problem (2) without transforming it to the form (1). A similar approach may be used to solve also problem (3), but most probably in this case a somewhat different evaluation of variables /or group of variables/ is more useful.

References

[1] Arabeyre, et all
The Airline-Crew Scheduling Problem: A Survay.
Transportation Science, 3 /1969/ No 2.

[2] M.L.Balinski, On Maximum Matching, Minimum Covering and their Connections. Proceedings of the Princeton Symposium on Mathematical programming . /1970/ 303-312

[3] M.Belmore and D.Ratliff, Set Covering and Involuntary Bases, Management Science 18 /1971/ 194-206

[4] R.S.Garfinkel and G.L.Nemhouser, The Set-Partitioning Problem:

Set Covering with Equality constraints. Operations Research 17/1969/ 848-856.

[5] R.E. Gomory, An Algorithm for Integer Solutions of Linear Programs, 269-302 in [6] .

[6] R.L.Graves and Ph.Wolfe /eds./ Recent Advances in Mathematical Programming, McGraw-Hill /1963/

[7] R.W.House, L.O.Nelson, and T.Rado, Computer Studies of a Certain Class of Linear Integer Problems. Recent Advances of Optimization Techniques. Lavi and Vogl /eds./ Wiley /1966/ pp. 241-280.

[8] E.L.Lawler and D.E. Wood, "Branch-and-Bound Methods: A Survey", Operations Research, 14 /1966/ 699-719.

[9] C.E. Lemke, H.Salkin and K.Spielberg, Set Covering By Single Branch Enumeration with Linear Programming Subproblems Operations Research /to appear/

[10] G.T.Martin, An Accelerated Enclidean Algorithm for Integer Linear Programs, 311-318 in [6]

[11] H.Thiriez, The Set Covering Problem: A Group Theoretic Approach Revue Trancaise de Recherche Operationelle V-3 /1971/ 83-104.

A THEORETICAL PREDICTION OF THE INPUT-OUTPUT TABLE

Emil KLAFSZKY
Computer and Automation Institute,
Hungarian Academy of Sciences

1. The problem. Denote $I_1, I_2, \ldots I_i, \ldots I_m$ sources and $J_1, J_2, \ldots J_j, \ldots J_n$ sinks. Let $\alpha_{ij} \geqq 0$ be the amount of the quantity going from I_i to J_j, in other words the quantity of the production of I_i used up by J_j. We shall denote the $m \times n$ matrix (α_{ij}) by A and call it input-output table. The sums $\sum_{j=1}^{n} \alpha_{ij}$ (the total output of I_i) and the sums $\sum_{i=1}^{m} \alpha_{ij}$ (the total input of J_j) as usual are called the input-output marginal values.

The fundamental problem treated in this paper is: what we can say about a new input-output matrix $X = (\xi_{ij})$ which has the prescribed marginal output values $b = (\beta_1, \beta_2, \ldots \beta_m) > 0$ and input values $c = (\gamma_1, \gamma_2, \ldots \gamma_n) > 0$ respectively, when we know the current matrix $A = (\alpha_{ij})$.

More exactly what additional hypothesis can be set by the aid of the matrix A to ensure a "sufficiently good" solution of the system of equalities

$$(1) \quad \begin{cases} \xi_{ij} \geqq 0, \quad (i = 1, 2, \ldots m;\ j = 1, 2, \ldots n), \\ \sum_{j=1}^{n} \xi_{ij} = \beta_i, \quad (i = 1, 2, \ldots m), \\ \sum_{i=1}^{m} \xi_{ij} = \gamma_j, \quad (j = 1, 2, \ldots n), \end{cases}$$

which evidently has many solutions in general.

<u>2. The method RAS</u>. A well-known method [3-7] which ensures an sufficiently good solution works with the following hypothesis: The variables ξ_{ij} are of the form

$$(2)\qquad \begin{cases} \xi_{ij} = \rho_i \alpha_{ij} \sigma_j, & (i=1,2,\dots m;\ j=1,2,\dots n), \\ \rho_i > 0, & (i=1,2,\dots m), \\ \sigma_j > 0, & (j=1,2,\dots n). \end{cases}$$

We shall denote the diagonal matrix consisting of the numbers ρ_i $(i=1,2,\dots m)$ and σ_j $(j=1,2,\dots n)$ by R and S respectively. So the assumption (2) can be written in the form:

$$X = \mathrm{RAS}.$$

Hence the name of the method. This method is sometimes called as <u>Selejkovskij</u>- or <u>Fratar method</u> /after the first users of it/ and also as the <u>Gravity method</u> /after the principle of gravitation which was used to explain the hypothesis/.

It is easy to see that if there is a solution of the system (1)-(2) then the system (1) has a solution with the following properties

$$(3)\qquad \xi_{ij} = 0, \quad \text{if} \quad \alpha_{ij} = 0 \quad \text{and} \quad \xi_{ij} > 0, \quad \text{if} \quad \alpha_{ij} > 0$$

Conversly the above condition is also sufficient for the solvability of (1)-(2) as the following well-known theorem asserts.

THEOREM 1. There is a solution - unique in ξ_{ij} - of the systems (1)-(2) iff there is a solution of (1) satisfying (3).

In the following paragraph we shall use another hypothesis based on some information theoretic considerations which also leads to the method RAS. This treatment yields the above theorem as an easy corollary.

If we denote the solution X ensured by theorem 1 by $\mathcal{R}(A,b,c)$ then because of the uniqueness we get easily the following property of the method RAS.

COROLLARY | Suppose that for (A,b,c) and for (A,b^*,c^*) the system (1) - (2) is solvable. Then the relation

$$\mathcal{R}(A,b,c) = \mathcal{R}(\mathcal{R}(A,b^*,c^*),b,c)$$

holds; i.e. the "step by step" prediction has the same result as the "one step" one.

3. Treatment of the problem using the geometric programming.

In the following for the sake of simplicity of the treatment we may suppose withouth the loss of generality that the equalities

$$\sum_{i=1}^{m}\sum_{j=1}^{n}\alpha_{ij}=1, \quad \sum_{i=1}^{m}\beta_i=1, \quad \sum_{j=1}^{n}\gamma_j=1$$

hold.

We shall use the hypothesis as follows:

The prediction $X=(\xi_{ij})$ is considered to be "good" if it satisfies the system (1) and the I-divergence /information divergence or information surplus by Shannon-Wiener/ of the table X related to table A is minimal; i.e. if the function

$$\varphi(X)=\sum_{i=1}^{m}\sum_{j=1}^{n}\xi_{ij}\ln\frac{\xi_{ij}}{\alpha_{ij}} \qquad (4)$$

has the minimal value on the set of solutions of system(1).

The continuous extension to the whole non-negativ orthant of the function $\xi_{ij} \ln \frac{\xi_{ij}}{\alpha_{ij}}$ is zero if $\xi_{ij} = 0$.

Let us introduce the following notations:

$$Q = \{ (i,j) \mid \alpha_{ij} > 0, \ (i = 1,2,\dots m; \ j = 1,2,\dots n) \}$$

and

$$Q_I = \{ i \mid \exists j \ (i,j) \in Q \}, \quad Q_J = \{ j \mid \exists i \ (i,j) \in Q \}.$$

Since the function (4) has finite infinum on the solution set of (1) only if $\alpha_{ij} = 0$ implies $\xi_{ij} = 0$, thus we have to choise $\xi_{ij} = 0$ if $(i,j) \notin Q$.

Using this restriction we get a mathematical programming problem: the dual problem of a geometric programming as follows.

Minimize the function

$$(5) \qquad \sum_{(i,j) \in Q} \xi_{ij} \ln \frac{\xi_{ij}}{\alpha_{ij}}$$

on the solution set of the system

$$(6) \qquad \begin{cases} \xi_{ij} \geqq 0, & (i,j) \in Q, \\ \sum\limits_{\substack{j=1 \\ (i,j) \in Q}}^{n} \xi_{ij} = \beta_i, & (i = 1,2,\dots m), \\ \sum\limits_{\substack{i=1 \\ (i,j) \in Q}}^{m} \xi_{ij} = \gamma_j, & (j = 1,2,\dots n). \end{cases}$$

We may write the primal of this dual geometric programm [2] :

Find the supremum of the objective function

$$\ln \left(\prod_{i \in Q_I} \rho_i^{\beta_i} \cdot \prod_{j \in Q_J} \sigma_j^{\gamma_j} \right)$$

on the solution set of the system

$$(7) \quad \begin{cases} \sum_{(i,j)\in Q} \rho_i \alpha_{ij} \sigma_j \leq 1, \\ \rho_i > 0, \quad i \in Q_I, \\ \sigma_j > 0, \quad j \in Q_J. \end{cases}$$

In what follows we have to distinguish two cases. First we shall investigate the problem (5) - (6) when it is canonical (A.,) and secondly when it is not canonical but consistent (B.,).

A., The canonical property of the problem (5) - (6) means that the system (6) has a solution of property $\xi_{ij} > 0, ((i,j) \in Q)$. However this assertion is equivalent to the (3).

Using some results of the geometric programming we get the following assertions:

(i) For the dual geometric programm is canonical (and so consistent) and the feasibility set (6) is bounded, so the objective function (5) attains its minimum. This minimum is unique because of the strict convexity of the objective function.

(ii) The canonical property of the problem and the boundeness of the dual objective function (i) imply that the primal problem has optimal solution and the values of the two optimums are the same ([2] pp. 169 Theorem 1.).

(iii) The pair of solutions $\rho_i, \sigma_j, \xi_{ij}$ are optimal iff the equalities

$$\rho_i \alpha_{ij} \sigma_j \sum_{(i,j)\in Q} \xi_{ij} = \xi_{ij} \qquad \text{for all } (i,j) \in Q$$

hold ([2] pp. 167 Lemma 1.).

As $\sum_{(i,j)\in Q} \xi_{ij} = 1$, therefore the above assertion is equivalent to the following system of equalities

(8) $\xi_{ij} = \rho_i \alpha_{ij} \sigma_j$ for all $(i,j) \in Q$.

Using the above observations we get the following

THEOREM 2. Suppose that the prediction problem (with the parameters A, b, c) fulfills the canonical condition (3). Then the hypothesis of RAS method and our hypothesis based on the minimization of I-divergence yield the same result.

__Proof:__ Let $\xi_{ij}, \rho_i, \sigma_j$ be a solution of (1) - (2) i.e. the solution of the problem by the method RAS. The numbers ξ_{ij} fulfill (6) and the numbers $\xi_{ij}, \rho_i, \sigma_j$ fulfill (8). It is easy to see that they also satisfy (7) because

$$\sum_{(i,j)\in Q} \rho_i \alpha_{ij} \sigma_j = \sum_{(i,j)\in Q} \xi_{ij} = 1 .$$

So $\xi_{ij}, \rho_i, \sigma_j$ are a pair of optimal solutions.

On the contrary let $\xi_{ij}, \rho_i, \sigma_j$ be now a solution of the optimization problem. So they satisfy (8) and this means that they are represented in the form RAS ■

__The Theorem 1. is now the easy consequence of the Theorem 2. and the assertion (i).__

B., Let us suppose that the feasibility set (6) is consistent. Denote $E = (e_i) = (e^{(j)}) = (\varepsilon_{ij})$ an $m \times n$ matrix, where

$$\varepsilon_{ij} = \begin{cases} 1 & \text{if} \quad \alpha_{ij} > 0, \\ 0 & \text{if} \quad \alpha_{ij} = 0. \end{cases}$$

Denote further $\|E\| = \sum_{i=1}^{m} \sum_{j=1}^{n} \varepsilon_{ij}$, $\|e_i\| = \sum_{j=1}^{n} \varepsilon_{ij}$, $\|e^{(j)}\| = \sum_{i=1}^{m} \varepsilon_{ij}$.

It is clear that $\|E\| = \sum_{i=1}^{m} \|e_i\| = \sum_{j=1}^{n} \|e^{(j)}\|$.

Let us consider for $\varepsilon \geqq 0$ the modified prediction problem with parameters $A, b_\varepsilon, c_\varepsilon$ where

$$(9) \qquad \beta_i^{(\varepsilon)} = \frac{\beta_i + \|e_i\|\varepsilon}{1 + \|E\|\varepsilon}, \qquad \gamma_j^{(\varepsilon)} = \frac{\gamma_j + \|e^{(j)}\|\varepsilon}{1 + \|E\|\varepsilon}$$

It is obvious that if $\varepsilon \to 0$ then $\beta_i^{(\varepsilon)} \to \beta_i^{(0)}$ and $\gamma_j^{(\varepsilon)} \to \gamma_j^{(0)}$, where $\beta_i^{(0)} = \beta_i$ and $\gamma_j^{(0)} = \gamma_j$

The modified problem is a canonical one for all $\varepsilon > 0$ and so its optimal solution $\xi_{ij}^{(\varepsilon)*}$ can be written in the form RAS /Theorem 2./

We shall show that the "small modification" of the problem yields "small modified" optimal solution, namely the theorem is true a follows:

THEOREM 3. | The optimal solution is a continuous function of ε ; i.e. if $\varepsilon \to 0$ then $X_\varepsilon^* \to X_0^*$

Proof: Indirectly, let us suppose that there is a sequence $\varepsilon_1, \varepsilon_2, \dots \varepsilon_k, \dots \to 0$ for which $X_{\varepsilon_k}^* \to \overline{X}$ and $\overline{X} \neq X_0^*$.

Then, as the optimal solution of the original problem $(\varepsilon = 0)$ is unique we get the relation

(10) $\varphi(X_o^*) < \varphi(\bar{X})$

Because of the continuity of the functional φ we have the limit relations

$$\varphi(X_o^* + \varepsilon_k E) \to \varphi(X_o^*), \quad \text{if} \quad \varepsilon_k \to 0$$

and

$$\varphi(X^*_{\varepsilon_k}) \to \varphi(\bar{X}), \quad \text{if} \quad \varepsilon_k \to 0.$$

This and (10) imply the existence of an index k_o for which the relation

$$\varphi(X_o^* + \varepsilon_{k_o} E) < \varphi(X^*_{\varepsilon_{k_o}})$$

holds.

This however contradict the optimality of $X^*_{\varepsilon_{k_o}}$ because the $X_o^* + \varepsilon_{k_o} E$ is a feasible solution of the problem modified by ε_{k_o} ∎

There is a well-known method - different from that of RAS - due to Deming-Stephan [1] which works with the hypothesis of the minimization of the square contingency:

> The prediction $X = (\xi_{ij})$ is considered to be "good" if it satisfies the system (1) and the "distance" of the table X from the table A is minimal; i.e. if the function
>
> $$\chi(X) = \sum_{i=1}^{m} \sum_{j=1}^{n} \frac{(\xi_{ij} - \alpha_{ij})^2}{\alpha_{ij}}$$
>
> has the minimal value on the set of solutions of system (1).

There is a close connection between the functions φ and χ. Namely if we take the first term of Taylor series of the function $\ln \frac{\xi_{ij}}{\alpha_{ij}}$ around of point $\frac{\xi_{ij}}{\alpha_{ij}}$ instead of the function $\ln \frac{\xi_{ij}}{\alpha_{ij}}$

into the function of φ then we get exactly the function χ ; i.e.

$$\varphi(X)=\sum_{i,j}\xi_{ij}\ln\frac{\xi_{ij}}{\alpha_{ij}}\approx\sum_{i,j}\xi_{ij}\left(\frac{\xi_{ij}}{\alpha_{ij}}-1\right)=\sum_{i,j}\frac{(\xi_{ij}-\alpha_{ij})^2}{\alpha_{ij}}=\chi(X),$$

which proves our observation.

References

[1] Deming, W.E and Stephan, F.F. "On a least squares adjustment of a sampled frequency table when the expected marginal totals are know" Ann.math.Statist., 11/1940/ pp. 427-444.

[2] Duffin, R.J. and Peterson, E.L. and Zener,C. Geometric Programming, John Wiley, New York, 1966.

[3] D'Esopo,D.A. and Lefkowitz, B. "An algorithm for computing intersonal transfers using the gravity model." Opns. Res., 1963, 11.No.6. pp. 901-907.

[4] Fratar,Thomas J. "Vehicular Trip Distribution by Successive Approximations" Traffic Quarterly pp. 53-65 /January 1954/

[5] Шелейховский, Г. В. Транспортные основания композиции городского плана. Гипрогор, Л., 1963

[6] Stone,R. and Bates, J and Bacharach, M: A programme for Growth Input-output relationsships 1954-66. University of Cambridge, 1963.

[7] Stone, R. and Brown, A.: "A long term growth modell for the Britisch Economy" /In the book Europes Future in Figures Ed.Geary, R.C./

AN IMPROVED ALGORITHM FOR PSEUDO-BOOLEAN PROGRAMMING

Stanisław Walukiewicz, Leon Słomiński, Marian Faner
Polish Academy of Sciences
Institute of Applied Cybernetics
Warsaw, Poland

1. INTRODUCTION

Problems of nonlinear integer programming have been developed recently as applications to facility allocation [13,15], capital budgeting [12], transportation system management [14] and networks design [10]. The methods for solving such problems have been considered more rarely than the methods for solving linear tasks, so today we have three prominent methods for handling the above mentioned problems. They are as follows: (i) the implicit (in particular case explicit) enumeration method proposed by Lawler and Bell in 1967 [11]; (ii) the the method of transformation into equivalent linear problem developed by Fortet in 1959 [5] and Watters in 1967 [19] and (iii) the pseudo--Boolean programming described by Hammer and Rudeanu in 1968 [8,9,18]. The efficiency of the first two methods has been proved in practice [11, 17]. Many authors [7, 17] have pointed out the following disadvantages of pseudo-Boolean programming:

(i) procedure is hardly suitable for automatic computations because of the low degree of the formalization;

(ii) test used to determine the families of feasible solutions are quite weak;

(iii) the method falls into two independent parts (constructing the set of feasible solutions, determining the optimal solution or solutions) and the amount of computations seems to increase directly with both the number of constraints and the number of variables.

The H-R algorithm for pseudo-Boolean programming, given in the paper, has not the first of these disadvantages. The remaining two are reduced. This algorithm may be considered as an application of the branch and bound principle to discrete programming. In sections 3 and 5 we give a short description of the H-R algorithm.

Bradley [2, 3] has shown that every bounded integer programming problem is equivalent to infinitely many other integer programming problems. Then we may ask question which of the equivalent problems is the most suitable for branch and bound algorithm. We will discuss this question in section 4, and now we note, that the answer contains also the estimation of the efficiency of the method (ii). A summary of the computational experience with the H-R algorithm is presented in section 6.

2. THE PROBLEM

Let Q_n be an n-dimensional unit cube, i.e. a set of points $X = (x_1, x_2, \ldots, x_n)$, $x_j \in (\overline{0,1})$, $j \in (\overline{1,n})$, and let R be a set of real numbers. By a pseudo-Boolean functions we shall mean any real valued function $f(x_1, x_2, \ldots, x_n)$ with bivalent (0,1) variables, i.e. $f : Q_n \longrightarrow R$. It is easy to see that such a function can be written as an polynomial linear in each variable [8].

We don't distinguish, by means of a special graphical symbol, the logical variables, which take the value true (1) and false (0) from, the numerical variables, which take only two values 1 and 0. It will be clear from the context that either the variables x, y are logical or numerical. In addition there exist well known isomorphism between logical and arithmetical operations:

$$x \vee y = x + y - xy, \tag{2.1}$$

$$x \wedge y = xy, \tag{2.2}$$

$$x = 1 - x. \tag{2.3}$$

As the pseudo-Boolean programming is a numerical program we suppose that in each given pseudo-Boolean function at least a substitution (2.1) was done.

Now, it is possible to formulate a pseudo-Boolean programming problem. It is as follows:

Find such $X^* = (x_1^*, x_2^*, \ldots, x_n^*) \in Q_n$ (one or all) that

$$f(X^*) = \min_{X \in S} f(X), \tag{2.4}$$

where

$$S = \{X : g_i(X) \geqslant 0, \quad i \in (\overline{1,m})\}. \tag{2.5}$$

In this formulation f and g_i, $i \in (\overline{1,m})$, are pseudo-Boolean functions, S - a set of feasible solutions. If $S = \emptyset$ then the system of constraints (2.5) is inconsistent. We call the pair (m,n) the size of (2.4), (2.5) and S^* - the set of optimal solutions.

3. THE H-R ALGORITHM

The problem (2.4), (2.5) may be solved by means of the explicit enumeration as follows:

Let f_{ak} be an incumbent (the best feasible solution yet fiund). For each $X \in Q_n$ we check, if X fulfils the system (2.5) and afterwords, if $f(X) \leqslant f_{ak}$. The branch and bound principle allows:

(i) to execute this enumeration on the level of subsets of Q_n;

(ii) to omit in this enumeration some of subsets of Q_n without loss of any X^*.

It is obvious that in each application of this proncíple we may distinguish two main operations: branching and computation of bounds. Generally speaking, the computational efficiency mainly depends on the number of branching and in turn this number depends on the accuracy of the bounds computing. On the other hand, the time devoted to the bounds computing shouldn't be too long, otherwise the applications of the branch and bound principle is not efficient.

The H-R algorithm presented below, is a result of both theoretical considerations and practical experiments.

3.1. The General Description of the H-R Algorithm

Let S_i' be a feasible solution set of the i-th constraint and S_{i-1} be a feasible solution set of the first (i-1) constraints, $i \in (\overline{1,m})$. For the sake of definitness, it is assumed that $S_0 = Q_n$. Then

$$S_i = S_{i-1} \cap S_i', \quad i \in (\overline{1,m}). \tag{3.1}$$

Joining the system (2.5) in the (m+1)-th constraint

$$f_{ak} - f(X) \geq 0 \tag{3.2}$$

where f_{ak} is a parameter, we have converted the pseudo-Boolean programming problem into the problem of solving the system of (m+1) pseudo-Boolean inequalities.

It should be noted that it is not necessary to determine whole the set S (S'). It is sufficient, if the following relation hold

$$S^* \subset S_i, \quad i \in (\overline{1,m}). \tag{3.3}$$

So furthermore by S_i (S_i') we shall mean the set which, may be, doesn't contain all the feasible solutions of (2.4), (2.5) but which contains all optimal solutions.

In other words, the H-R algorithm determines the feasible solutions set of $g_1(X) \geq 0$ at the first step, and after by checking which of the $X \in S_1$ are satisfying $g_2(X) \geq 0$, i.e. it determines S_2 and so on up to determining S_m. Solving (3.2) for $X \in S_m$ we obtain all optimal solutions of (2.4), (2.5).

3.2. The Branching Process

We can bring each of the constraint of (2.5) to the following form (we omit index i here)

$$a_0 + a_1C_1 + a_2C_2 + \dots + a_kC_k \geq 0, \tag{3.4}$$

where a_0, $a_j \in R$, C_j – a Boolean conjunction (product) of some of the variables $x_1, x_2, \ldots, x_n$, $j \in (\overline{1,k})$, and

$$|a_1| \geqslant |a_2| \geqslant \ldots \geqslant |a_k|. \tag{3.5}$$

The H-R algorithm requires all sets to be described by means of characteristic functions and each characteristic function must be the Boolean disjunction of the disjoint reduced conjunctions defined in the following way:

A logical expression

$$h(X) = K\, D_1 D_2 \ldots D_p \tag{3.6}$$

is called a reduced conjunction in Q_n, if:

(i) K is a product of some of letters $\tilde{x}_j$, i.e. $\tilde{x}_j \in \{x_j, x_j\}, j \in (\overline{1,n})$. We not exclude the case $K = 1$.

(ii) Each D_k is a negation of a product (disjunction) of some of letters $\tilde{x}_j$, $j \in (\overline{1,n})$, such as that the letters appear in K don't appear in any D_k, $k \in (\overline{1,p})$.

The variables appear in K will be called fixed variables (in K).

It results from (3.6) that, if $K \neq 0$, then h(X) describe in a univocal way, some set $G \subset Q_n$, therefore we will sometimes write h(G) instead h(X). A set G we will call a family (of solutions). For the sake of definitness, it is assumed that if $K = 0$, then the corresponding set is empty.

Let $H(S_i)$ be a characteristic function of the feasible solution set of the first i constraints of (2.4), (2.5), $i \in (\overline{1,m+1})$. Then

$$H(S_i) = h_1(X) \vee h_2(X) \vee \ldots \vee h_q(X), \tag{3.7}$$

where $h_i(X) \wedge h_j(X) = 0$, if $i \neq j$ and q - a number of disjoint families of solutions.

By branching a given family G by means of a conjunction C we mean computing the reduced conjunction for the following products:

$$h(G_1) = h(G)\,C, \tag{3.8}$$

$$h(G_2) = h(G)\,\overline{C}. \tag{3.9}$$

So we obtain two new disjoint subfamilies by branching the given one.

3.3. The Computation of Bounds

Let (3.4) be a (i+1)-th constraint, where for simplicity we omit index (i+1). To obtain S_{i+1} we branch the family corresponding to $h_1(X)$ in (3.7) by means of C_1 according to (3.8) and (3.9). Next we branch each of these subfamilies by means of C_2 and so on up to branching by means of C_k. We repeat that action for each family in (3.7).

The essence of the H-R algorithm consists in excluding some families from above mentioned process. For each family we compute bounds which play a role of an argument in various fathoming criteria. In linear tasks we may use linear programming to compute the bounds as exactly as possible, but in nonlinear problem the computation of such bounds creates certain difficulties. Below, we present some fathoming criteria, which efficiency was investigated.

3.3.1. The Logical Criterion LC

If, for example, in accordance (3.8)

$$\text{LC:}\quad h_1(G_1) = h(X) \wedge C = 0, \tag{3.10}$$

then we say that G_1 has been excluded by logical criterion. In such a case

$$h_2(G_2) = h(X) = h(G). \tag{3.11}$$

Therefore we may make an assumption, that each family is not empty.

3.3.2. The Global Numerical Criteria GNC

These criteria are computed recursively for given constraint $g_i(X) \geqslant 0$, $i \in (\overline{1,m})$, without using information of S_{i-1}. Let $l(G)$ $(u(G))$ be a lower (upper) bound of $g_i(X)$ over the family G. We consider (3.4) and at the begining we have

$$l(Q_n) = a_0 + \sum_{a_i < 0} a_i, \tag{3.12}$$

$$u(Q_n) = a_0 + \sum_{a_i > 0} a_i. \tag{3.13}$$

After branching Q_n by means of conjunction C_1 we obtain, according to (3.8) and (3.9), two families G_1 and G_2 for which

$$l(G_1) = \begin{cases} l(Q_n) + a_1, & \text{if } a_1 > 0 \\ l(Q_n), & \text{if } a_1 < 0 \end{cases} \tag{3.14}$$

$$u(G_1) = \begin{cases} u(Q_n), & \text{if } a_1 > 0 \\ u(Q_n) + a_1, & \text{if } a_1 < 0 \end{cases} \tag{3.15}$$

$$l(G_2) = \begin{cases} l(Q_n), & \text{if } a_1 > 0 \\ l(Q_n) - a_1, & \text{if } a_1 < 0 \end{cases} \tag{3.16}$$

$$u(G_2) = \begin{cases} u(Q_n) - a_1, & \text{if } a_1 > 0 \\ u(Q_n), & \text{if } a_1 < 0 \end{cases} \tag{3.17}$$

For each conjunction C_j, $j \in (\overline{2,k})$, and a family G we obtain similar formulae, i.e. in (3.14)-(3.17) we replace Q_n by G and a_1 by a_j.

In results from (3.14)-(3.17) that in the branching process a lower bound of $g_i(X) \geqslant 0$ doesn't decrease and an upper bound of it doesn't increase.

The GNC consists in checking of two conditions for each family G:

$$\text{GNC1:} \quad l(G) \geqslant 0, \tag{3.18}$$
$$\text{GNC2:} \quad u(G) < 0. \tag{3.19}$$

In GNC1 is satisfying, then all points of G are solutions of $g_i(X) \geqslant 0$, therefore we may exclude G from branching process.

In GNC2 is satisfying, then G doesn't contain any solution of $g_i(X) \geqslant 0$ and we may exclude it from branching process. The families which are neither satisfying GNC1 nor GNC2 are branched in above described way.

3.3.3. The Local Numerical Criteria LNC

These criteria answer the question if branching of given family G of S_i by means of the conjunction of $g_{i+1}(X) \geqslant 0$ is necessary or not.

Let G be a family of S_i and let (3.6) be its reduced conjunction. We introduce two sets of index

$$J_0 = \{j : C_j K = 0\}, \tag{3.20}$$
$$J_1 = \{j : C_j K = K\}. \tag{3.21}$$

Let L(G) (U(G)) be a local lower (upper) bound of $g_{i+1}(X)$ 0 over a family G then we have

$$L(G) = l(G) + \sum_{\substack{j \in J_1 \\ a_j > 0}} a_j - \sum_{\substack{j \in J_0 \\ a_j < 0}} a_j, \tag{3.22}$$

$$U(G) = u(G) + \sum_{\substack{j \in J_1 \\ a_j < 0}} a_j - \sum_{\substack{j \in J_0 \\ a_j > 0}} a_j, \tag{3.23}$$

where l(G), u(G) we compute according to (3.12) and (3.13). Now we construct criteria very similar to (3.18) and (3.19) by replacing l(G) (u(G)) by L(G) (U(G)). We will mark them by LNC1 and LNC2 respectively.

If LNC1 is satisfying, then G is obviously the set of solutions of $g_{i+1}(X) \geqslant 0$. If LNC2 is satisfying, then G cannot contain any solution of $g_{i+1}(X) \geqslant 0$.

3.3.4. The Incumbent Criterion IC

In practice we often know a relatively good estimation for f_{ak}. Let F(G) be a lower bound of f(X) over G. If

$$\text{IC:} \quad F(G) > f_{ak}, \tag{3.24}$$

then G may be excluded from further considerations.

4. THE EFFICIENCY OF THE H-R ALGORITHM

Let (3.7) be a characteristic function of a feasible solutions set of the i-th constraint, i.e. we're replacing now S_i by S_i', $i \in (\overline{1,m})$.

Since all coefficients of considering constraint are real number, then there exist infinitely many formulation of this constraintthat $H(S')$ is sill a corresponding characteristic function of it. Therefore there exist infinitely many formulations of a given pseudo--Boolean programming problem.

It is known [5, 19] that every pseudo-Boolean programming problem may be converted into an equivalent linear problem. This convertion consists in replacing each nonlinear component appearing in goal function or/and in constraints by new variable and joining two additional linear constraints. So the efficiency of such linearization depends on the number of nonlinear components. According to [7] we may solve every integer linear problem with about 70 variables. For instance, in example 8 from [9] solved without computer, we have $r = 18$ nonlinear components $m = 3$, $n = 7$. So an equivalent problem has $n + r = 25$ variables and $m + 2r = 39$ constraints. For comparison, the biggest problem solved in [17] by means of IBM 7040 ($m = 7$, $n = 30$, $r = 10$) has 40 variables and 27 constraints.

We may reformulate the question that we set in Introduction in the following way: what factor does the efficiency of the H-R algorithm depend on. Basing on the up-to-date results we may establish two such factors: an order of solving constraints and a decomposition of system of constraints.

4.1. The Order of Solving Constraints

We introduce the concept of the optimal order of solving constraints by means of so-called "strenght" of given constraint $g_i(X) \geqslant 0$, $i \in (\overline{1,m})$.

Let (3.4) be such a constraint. We define the strength of it as

$$W_i = \left(-a_0 - \sum_{a_j < 0} a_j\right) \Big/ \left(\sum_{j=1}^{k} |a_j|\right), \quad i \in (\overline{1,m}). \tag{4.1}$$

If $W_i < 0$, then $g_i(X) \geqslant 0$, is redundant, i.e. $g_i(X) \geqslant 0$ for each $X \in Q_n$. If $W_i > 1$, then $g_i(X) \geqslant 0$ is inconsistent and $S = \emptyset$. So we may consider only such constraints for which

$$0 < W_i < 1 \tag{4.2}$$

because for cases $W_i = 0$ and $W_i = 1$ the unique solutions are known.

We can see that (4.1) is an estimation of the number of families in S_i'. Since in the H-R alforithm the enumeration is executed on the level of families, therefore we try to choose as the first, such a constraint for which the number of families is as small as possible, i.e. we choose $g_i(X) \geqslant 0$ for which W_i is the nearest either 1 or 0. So we can give the following priority rule in pseudo-Boolean programming: we put constraints in order according to nonincreasing priority $P(g_i)$, and

$$P(g_i) = \begin{cases} b(W_i - 0.5) & \text{for } 0.5 \leqslant W_i < 1 \\ (0.5 - W_i) & \text{for } 0 < W_i \leqslant 0.5 \end{cases} \tag{4.3}$$

We introduced coefficient $b > 1$ in (4.3) because if $W_i \approx 1$ then each family has more fixed variables. In practical computation we assume $b = 2$. In section 6 we present the computational results which illustrate the usefullness of putting the constraints in optimal order.

4.2. The Decomposition of Systems of Constraints

The idea of decomposition of a system of constraints consists in separating it into subsystems in such a way that constraints in each subsystem contain as much as possible common variables. Such separation make possible fathoming the great number of families by LC.

Let q_{ij} be a number of components of $g_i(X) \geqslant 0$ in which the variable x_j appears, $i \in (\overline{1,m})$, $j \in (\overline{1,n})$, $q_{ij} \geqslant 0$. The correlation coefficient of the i-th and k-th constraints we compute as

$$T_{ik} = \frac{1}{n} \sum_{j=1}^{n} q_{ij} q_{kj}, \quad i,k \in (\overline{1,m}) \tag{4.4}$$

We separate (2.5) according to the following procedure. Let $g_s(X) \geqslant 0$ be the strongest constraint. We compute T_{si} for all $i \in (\overline{1,m})$, $i \neq s$, and create a subsystem from these for which T_{si} takes the largest value. Next we find strongest constraint among remaining ones and follow the above described way.

Let Z be a subsystem containing m_z constraints. The priority of the subsystem we define as

$$P(Z) = \overline{W}(m_z/m), \tag{4.5}$$

where $\overline{W}$ is a mean value of strength of constraints belonging to Z. We begin computations with subsystem for which P(Z) is the largest.

5. THE IMPLEMENTATION OF THE H-R ALGORITHM

The H-R algorithm was written i ALGOL 1204 language and implemented for the second generation, Polish computer ODRA 1204 (storage capacity 16 k, access time 6 µsec, addition time 16 µsec). For comparison IBM 7090 - 32 k, 2.2 µsec, 4.4 µsec.

The H-R program is an adaptive one with two parts MASTER and EXECUTOR. The first part obtains the information about the problem just solved and upon this information it controls some parameters of the second one. For example, MASTER checks, if the solved problem is linear or not. For linear problem it checks, if it is the covering problem or not. MASTER also computes W_i, $i \in (\overline{1,m})$ and puts the constraints in optimal order, and separates (2.5) into subsystems according to section 4.2.

On the ground of this information we may automatically change: the order of applying the fathoming criteria and the procedure of multiplication of characteristic functions. If for example $0.3 \leqslant W_i \leqslant 0.6$, $i \in (\overline{1,m})$, then we consider (3.2) as the first constraint.

In the H-R program the multiplication of characteristic functions is done in the following way: we determine all families for the first (with the greatest priority) constraint. Then we take the last family with the greatest number of fixed variables and solve $g_2(X) \geqslant 0$ over it and so on up to solving $g_m(X) \geqslant 0$ and computing f_{ak} or improving its value. Next the H-R program checks a list of created families and if this list has run out, then it takes the next family from S_1. Such organization of the multiplication requires small storage capacity to execute it.

6. COMPARISON OF THE COMPUTATIONAL RESULTS

The efficiency of an algorithm should be measured by means of number (or its estimation) of properly defined operations needed to solve suitable chosen typical tasks. But it is very difficult to define such operations in discrete programming, therefore the computer time needed to solve the subjectively chosen examples is considered as the measure of the efficiency of an algorithm. This time obviously depends on these examples, the computer and the language. This creates such a situation as described in [1]. The partial solution would be a statistic approach to comparison of results for example in such a way as in [13, 15].

In order to reduce subjectivism of our results we took all numerical examples from the references [6, 11, 16, 17]. Table 1 presents the comparison of results for linear tasks and Table 2 for nonlinear ones. We can see that the H-R algorithm is the best especially for nonlinear problems. We should observe that the examples [17] are rather specially constructed for Taha's algorithm and in spite of all the H-R algorithm, is faster that the remaining one. The example from the second part of Table 2 cannot be solved either by Taha's or by Watter's algorithm.

Table 3 shows how optimal order of constraints influences the efficiency of computation. We can observe that the majority of these examples has wrong order of constraints and putting the constraints in

TABLE 1. Computational results for linear problems

Problem				Computing time in seconds for method		
Number	Size m	Size n	Source	Lawler and Bell	Mimstep 16	H-R algorithm
1	10	10		1	13-21 (b)	1
2	15	15		2-6 (a)	71-227	4
3	20	20		10-60	481-3377	22
4	30	30		650-1325 (c)	6658	136
5	40	40		-	-	354
6	15	15	Haldi	-	921	27
7 (d)	35	15	"	30	711-1195	47
8 (e)	50	32	Słomiński	-	796	29
9 (e)	87	48	"	-	24960	2383
Computer:				IBM 7090	ODRA 1204	

(a) Results for different examples.
(b) Results for different versions of the algorithm.
(c) Computation non completed.
(d) This problem was solved by Freeman [4] by means of the IBM 7044 computer in 150 seconds.
(e) This problem has a quadratic goal function.

optimal order is especially efficient for large problems (up to 10 times).

TABLE 2. Computational results for nonlinear problems

Problem				Computing time in seconds for method			
Num-ber	Size			Source	Taha	Watters	Lawler and Bell	H-R algorithm
	m	n	r					
1	3	5	5	Taha	0.3	0.5	0.6	< 1
2	3	10	5		0.2	0.6	8.3	< 1
3	3	20	5		0.2	3.6	919.4	1
4	7	5	10		1.9	4.7	0.7	< 1
5	7	10	10		0.6	2.8	3.8	1
6	7	20	10		3.3	21.1	> 3000.0	2
7	7	30	10		1.0	19.0	-	3
8	7	5	10		5.1	3.0	-	< 1
9	6	10	15		2.1	2.9	5.6	2
10	6	10	15		8.3	6.3	-	1
11	6	10	15		2.4	6.5	-	1
12	6	10	15		3.4	8.7	-	1
13	6	10	15		11.2	14.1	-	1
14	7	23	75	Lawler a. Bell	-	-	13	68
15	7	23	105		-	-	24	89
16	10	27	106		-	-	438	187
17	10	27	156		-	-	634	298
18	10	27	316		-	-	716	407
C o m p u t e r:					IBM 7040 For problems 14-18 IBM 7090			ODRA 1204

TABLE 3. Efficiency of solving the equivalent problem

Problem				Computing time in seconds		
	Num-ber	Size		the best order	references order	the worst order
		m	n			
Linear	1	15	15	4	5	22
	2	20	20	22	38	235
	3	30	30	136	3000	5000
	4	15	15	4	6	8
	5	20	20	10	23	39
	6	30	30	57	411	632
	7	50	32	29	88	235
Non-linear	8	7	23	89	273	347
	9	10	27	298	3000	-
	10	7	20	2	3	3

1. CONCLUSIONS

1. The H-R algorithm can be generalized in such a way as it was done in [8] with pseudo-Boolean programming.

2. On the ground of recent results we may guarantee to solve any problem with about 50 variables and for the problem of special structure with considerably more variables. We may observe that the efficiency of the H-R algorithm considerably increases in case of the implementation it for highly parallel fourth generation computers such as ILLIAC IV.

3. Many authors [2, 3, 7] have pointed out the fact that the equivalent problem corresponding to a given one may considerably increase the efficiency of the known algorithms. But this question hasn't been investigated so far. On the ground of our results we may say that the equivalent problem need not be a linear one.

REFERENCES

1. Anonymous, Mathem. Programming 2, 260-262 (1972).
2. Bradley H.G., Manag. Sci. 17, 354-366 (1971).
3. Bradley H.G., Discrete Math. 1, 29-45 (1971).
4. Freeman R.J., Operat. Res. 14, 935-941 (1966).
5. Fortet R., Cahiers du Centre de Rácherche Opérationnelle, 1, 5-36 (1959).
6. Haldi J., Working pap. No. 43. Graduate School of Business, Stanford Univ. 1964.
7. Geoffrion A.M., Marsten R.E., Manag. Sci. 18, 465-491 (1972).
8. Hammer P.L., Rudeanu S., Boolean methods in operation research and related areas. Nerlin 1968.
9. Hammer P.L., Rudeanu S., Operat. Res. 17, 233-261 (1969).
10. Hu T.C., Integer programming and networks flows. New York 1969.
11. Lawler E., Bell M., Operat. Res. 15, 1098-1112 (1967).
12. Mao J.C., Walingford A.B., Manag. Sci. 16, 51-60 (1968).
13. Nugent Ch.E., Vollmann T.E., Ruml J., Operat. Res. 16, 150-173 (1968).
14. Randolph P.H., Swinson G.E., Walker M.E. In: Applications of mathematical programming techniques. London 1970.
15. Ritzman L.P., Manag. Sci. 18, 240-248 (1972).
16. Słomiński L., Proc. of the Conf. on Control in Large Scale Systems of the Resources Distribution and Development, Jabłonna 1972 (in Polish).
17. Taha H., Manag. Sci. 18, B 328-343 (1972).
18. Walukiewicz S., Arch. Autom. i Telemech. 15, 455-483 (1970).
19. Watters L.J., Operat. Res. 15, 1171-1174 (1967).

NUMERICAL ALGORITHMS FOR GLOBAL EXTREMUM SEARCH

J. Evtushenko

In many problems of operation research in which systems containing uncertain parameters are designed, optimum solutions often are based on minimax strategies. Utilization of such an approach for solving practical problems is restricted by lack of numerical methods. To this time, as far as I know, there do not exist any general numerical methods for obtaining minimax strategies in multistage games. Some first results in this direction obtained (the numerical solution of a class of one-step processes).

In present paper a numerical method for determination of minimax estimation is presented. This method is based on numerical method of finding global extremum of a function [1, 2].

1. Suppose that a function $f(x)$ satisfies Lipshitz conditions whith constant C :

$$|f(x^1) - f(x^2)| \leq C \| x^1 - x^2 \|$$

We shall consider the problem

$$\max_{x \in X} f(x) \tag{1}$$

where X is a compact set, $x = (x_1, x_2, \dots, x_n)$.
We shall call the vector x_* a solution of problem (1) if

$$|\max_{x \in X} f(x) - f(x_*)| \le \varepsilon \tag{2}$$

where ε - is the accuracy of a solution.

If for any sequence $x^1, x^2, \dots, x^k$ the value

$$F_k = \max [f(x^1), f(x^2), \dots, f(x^k)]$$

is found then for all x belonged to the spheres

$$\| x^i - x \| \le \frac{1}{C} [\varepsilon + F_k - f(x^i)] \quad (i = 1, 2, \dots, k) \tag{3}$$

the condition (2) holds. If the spheres (3) entierly cover the domain X then the problem (1) is solved and magnitude F_k is an approximate maximum of $f(x)$ on X. In [1] the simplest algorithm of such covering is presented. In the case when $f(x)$ is differentiable function the local methods of maximum searching are used as auxiliary methods which essentially accelerate computation.

Similar approach was used for finding global extremum of a function which gradient satisfies Lipshitz conditions. The programm was made which used A L G O L - 60 for computing.

2. We shall now consider determination of minimax estimation for $r/2$ steps process:

$$J = \min_{x_r} \max_{x_{r-1}} \min_{x_{r-2}} \max_{x_{r-3}} \dots \min_{x_2} \max_{x_1} K(x) \tag{4}$$

where x_i - has dimension $p(i)$ and $x = (x_1, x_2, \dots, x_r)$ - is a $\sum_{i=1}^{r} p(i)$ dimensional vector. The extremum with respect to any vector x_i is searched on a compact domain $X_i \subset E_i$, where E_i is an i -dimensional Evclidian space. The function $K(x)$ satisfies a Lipshitz condition on the domain $\Omega = X_1 \times X_2 \times \dots \times X_r$.

The method of seeking a global extremum is used step by step for solving problem (4). This numerical method permits us to solve problem (4) to an arbitrary fixed accuracy. A number of modifications are developed which use local methods for acceleration of convergence.

In the Computer Center of the Academy of Sciences special programs have been developed for solving (4). As examples of problems solved by this method we can mention the two following problems

$$J_1 = \min_{x_4} \max_{x_3} \min_{x_2} \max_{x_1} \left[(x_3 - x_4)x_1 - x_2\right]^2$$

$$J_2 = \min_{y_m} \dots \min_{y_1} \max_{x_1} \left[\sum_{i=1}^{m} \sin 2\pi(i x_1 + y_i)\right]^2$$

where $0 \le x_i \le 1$, $0 \le y_i \le 1$.

Numerical computations show that these methods work effeciently.

3. Consider the game problem

$$\mathcal{J}(\mu_*, \nu_*) = \min_{\mu} \max_{\nu} \iint_{XY} K(x, y)\mu(dx)\nu(dy) \quad (5)$$

where μ, ν - probability measures defined on sets X, Y respectively. Function K is assumed to be continuous on $X \times Y$. It is easy to show that for any probability measure μ, ν:

$$\mathcal{J}(\mu_*, \nu) \le \mathcal{J}(\mu_*, \nu_*) \le \mathcal{J}(\mu, \nu_*)$$

$$\mathcal{J}(\mu_*, \nu_*) = \min_{\mu} \max_{y \in Y} \int_X K(x, y)\mu(dx) \quad (6)$$

Let $\varphi(\mu)$ denotes the solution of prombes:

$$\varphi(\mu) = \max_{y \in Y} \int K(x, y)\, \mu(dx)$$

Function $\varphi(\mu)$ is convex. Instead of problem (5) we can solve problem (6). For any measure μ we shall find $\varphi(\mu)$ using method for finding global maximum. For minimization $\varphi(\mu)$ we can use the local numerical method, which we shall discribe now.

4. Using approximation, we can our problem put into the following mathematical form: Find a vector $z = (z_1, z_2, \dots, z_s)$ which minimizes convex function

$$\min_z F(z) \quad (7)$$

subject to $z \in Z = \{ z : z \ge 0, \sum_{i=1}^{s} z_i = 1 \}$.

If $F(z)$ is differentiable then for solving problem we consider the system:

$$\frac{dz_i}{dt} = z_i\left[\sum_{j=1}^{s} z_j \frac{\partial F}{\partial z_j} - \frac{\partial F}{\partial z_i}\right], (i = 1, 2, \dots, s), z(0) = z_0 \quad (8)$$

Let $z_0 \in Z_0 = \{ z : z > 0, \sum_{i=1}^{3} z_i = 1 \}$.
We can prove (see [3]), that the limit solution $\lim_{t \to \infty} z(z_0, t)$ of the system (8) is the solution of problem (7) for any $z_0 \in Z_0$.

If $F(z)$ is nondifferentiable, but differentiable in any direction (as in our case), we can use following discrete version:

$$z_i^{r+1} = z_i^r \left[1 + \alpha_r \left(\sum_{j=1}^{3} l_j^r z_j^r - l_i^r \right) \right], (i = 1, 2, \ldots, 3) \quad (9)$$

where the vector l^r belongs to the set S^r of support functionals:

$$S^r = \{ l : l \in E_3, \sum_{j=1}^{r} l_j (z_j - z_j^r) \leq F(z) - F(z^r), \forall z \in Z \}$$

A sequence α_r such that

$$0 < \alpha_r < a, \lim_{r \to +\infty} \alpha_r = 0, \sum_{r=1}^{\infty} \alpha_r = \infty$$

If a is sufficiently small then the limit of sequence z^r is a solution point for any $z_0 \in Z_0$.

References

I Евтушенко Ю.Г., Журнал вычислительной математики и математической физики II, I390-I403 (I97I), Москва.

2 Евтушенко Ю.Г., Журнал вычислительной математики и математической физики I2, 89-I04 (I972), Москва.

3 Евтушенко Ю.Г., Жадан В.Г. Журнал вычислительной математики и математической физики I3, 583-598 (I973), Москва.

GRADIENT TECHNIQUES FOR COMPUTATION OF STATIONARY POINTS

E. K. Blum

Department of Mathematics, U. of Southern California
Los Angeles, California 90007

Let J be a real functional defined on a subset of a Hilbert space H. $u \in H$ is a stationary point of J if some derivative of J is zero at u. In particular, if u is a minimum point of J and the derivative of J exists at u, then the derivative is zero. Thus, extremum points are stationary points, but of course the converse need not be true. We shall present some gradient methods for determining stationary points of a rather general type. We consider sets of non-isolated stationary points and non-convex functionals and give conditions for convergence of the gradient methods. We then give applications to the optimal control problem of Mayer and to the generalized eigenvalue problem $Ax = \lambda Bx$, where A and B are arbitrary bounded linear operators from one Hilbert space to another.

The gradient methods presented here are based on the intuitive idea that convergence can be expected whenever there is a neighborhood of the stationary point u in which the cosine of the angle $\beta(x)$ between the gradient $\nabla J(x)$ and the vector $x-u$ is bounded away from zero. For a problem with equality constraints, the angles $\alpha_i(x)$ between the gradients of the constraints and $x-u$ also enter into consideration. We shall first consider the problem with equality constraints. $\langle u, v\rangle$ denotes the inner product.

Let R be the real line and let D be a subset of H. Let $g_i : D \to R$, $1 \leq i \leq p$, be real functionals. The sets $C(g_i) = \{x \in D : g_i(x) = 0\}$ and $C = \bigcap_{i=1}^{p} C(g_i)$ are called "equality constraints". Let $J : D \to R$ be another functional, called the "objective (or cost) functional". We denote the Fréchet (or strong) gradients of these functionals at x by $\nabla J(x)$ and $\nabla g_i(x)$. (See [2], [5], [6] or [8] for pertinent definitions.) Then the differential of J at x

with increment h is $dJ(x; h) = J'(x)h = \langle \nabla J(x), h \rangle$. The subspace, G_x spanned by the vectors $\{\nabla g_i(x)\}$ is called the "gradient subspace" at x. Its orthogonal complement, T_x, is called the "tangent subspace". Thus, $H = G_x \oplus T_x$ and $\nabla J(x) = \nabla J_G(x) + \nabla J_T(x)$. We call $\nabla J_T(x)$ the "tangential component" of $\nabla J(x)$.

Definition 1. $u \in D$ is called a stationary point of $\{J, g_i\}$ if $\nabla J_T(u) = 0$ and $u \in C$.

If u is a local minimum point of J on the equality constraint C, then under appropriate conditions u is a stationary point of $\{J, g_i\}$. This is a direct consequence of the Lagrange Multiplier Rule. There are many versions of this rule. We state one in the following theorem.

Theorem 1. Let $J(u) \leq J(x)$ for all $x \in N_u \cap C$, where N_u is some neighborhood of u. Let $\nabla J(y)$ and $\nabla g_i(y)$ exist at all points $y = u + t_0 h + \sum_1^p t_i \nabla g_i(u)$, where $h \in T_u$ and $t = (t_0, t_1, \ldots, t_p)$ is in some neighborhood $N(h)$ of $0 \in R^{p+1}$. Let $\nabla g_i(y)$ be continuous in $t \in N(h)$ and $\nabla J(y)$ continuous in t at $t = 0$. If $\nabla J(u) \neq 0$, then there exist real $\lambda_0, \lambda_1, \ldots, \lambda_p$ not all zero such that $\lambda_0 \nabla J(u) + \sum_1^p \lambda_i \nabla g_i(u) = 0$. Furthermore, if $\{\nabla g_i(u)\}$ is a linearly independent set, then $\lambda_0 = 1$ and $\lambda_1, \ldots, \lambda_p$ are unique and not all zero. Thus, $\nabla J_T(u) = 0$. and u is a stationary point of $\{J, g_i\}$.

Proof: See [2].

Notation: $\bar{x} = x / \|x\|$ is the normalized vector.

Definition 2. A stationary point u of $\{J, g_i\}$ is called quasiregular if there exists a neighborhood $N = N(u)$ such that for $x \in N$ the following four conditions are satisfied: (i) $\nabla J(x)$ and $\nabla g_i(x)$ are uniformly continuous in x and $\nabla J_G(x) \neq 0$, $\nabla g_i(x) \neq 0$; (ii) the Gram matrix $(\langle \nabla g_i(x), \nabla g_j(x) \rangle)$ is nonsingular (normality); (iii). For the angle $\theta(x) = \arcsin(\|\nabla J_T(x)\| / \|\nabla J(x)\|)$, when $\nabla J_T(x) \neq 0$ the gradient $\nabla\theta(x)$ exists and $\|\nabla\theta(x)\|$ is bounded away from zero. When $\nabla J_T(x) = 0$, the one-sided differential $d\theta(x; \bar{h}^+)$ exists for all

$h \in H$ and $d\theta(x+h; \bar{h}^+) \to d\theta(x; \bar{h}^+)$ uniformly as $\|h\| \to 0$; (iv) For $x \in N$ not a stationary point, let U_x = {stationary points u such that u is the closest stationary point to x on line segment [x, u] }. For $u \in U_x$, let $\Delta x = x-u$ and $\alpha_i = \arccos < \overline{\nabla g_i(x)}, \overline{\Delta x} >$, $\beta = \arccos < \overline{\nabla \theta(x)}, \overline{\Delta x} >$.

($0 \le \alpha_i, \beta \le \pi$.) Then there is a constant $\gamma > 0$ such that $\sum_1^p \cos^2\alpha_i + \cos^2\beta > \gamma$ if $\nabla J_T(x) \neq 0$ and $\sum_1^p \cos^2 \alpha_i > \gamma$ if $\nabla J_T(x) = 0$.

A gradient procedure for determining quasi-regular stationary points is given by the following formulas.

$$x_{n+1} = x_n + s_n h(x_n). \tag{1}$$

$$h(x) = h_G(x) + h_T(x). \tag{2}$$

$$h_G(x) = - \sum_1^p \frac{g_i(x)}{\|\nabla g_i(x)\|} \overline{\nabla g_i(x)} . \tag{3}$$

$$h_T(x) = \begin{cases} \dfrac{-\tan \theta(x)}{\|\nabla \theta(x)\|} \overline{\nabla \theta(x)} & \text{if } \nabla J_T(x) \neq 0 \\ 0 & \text{if } \nabla J_T(x) = 0. \end{cases} \tag{4}$$

$$d/2 < s_n < d, \text{ where } 0 < d < 1/2\,p. \tag{5}$$

We call this the "angular gradient procedure" or a "mixed strategy procedure".

As an example, consider $J(x) = <Ax, x>$ where $A : H \to H$ is a linear bounded self-adjoint operator. We impose one equality constraint, C, defined by $g(x) = \|x\|^2 - 1$; i.e. C is the unit sphere. It is easily proved that u is a unit eigenvector of A if and only if $\nabla J_T(u) = 0$. (See [2].) In [2] it is also shown that

$$\nabla \theta(x) = \frac{2 <Ax, x>}{\|x\| \; |<Ax, x>| \; \|\nabla J_T(x)} \left(\frac{<Ax, x>}{\|Ax\|^2} A^2 x + \frac{<Ax, x>}{\|x\|^2} x - 2Ax \right)$$

if $\nabla J_T(x) \neq 0$ and $\cos\ \theta(x) \neq 0$. The following theorem can be proved. (See [2], [3], [4].)

Theorem 2. Let A be a bounded self-adjoint operator on H. If $\lambda \neq 0$ is an eigenvalue of A of multiplicity 1 and λ is an isolated point of the spectrum of A, then any unit eigenvector belonging to λ is a quasiregular stationary point of $\{J, g\}$.

(It appears that the angular gradient method can be used to compute such intermediate eigenvalues and eigenvectors even when λ is not isolated. In the latter case, the stationary points are not isolated.) For example, if $Ax = \int_0^1 K(t, s)\, x(s) ds$ is an integral operator with a symmetric kernel K continuous on the unit square, then A is a compact self-adjoint operator on L_2. Its spectrum is at most denumerable and a non-zero eigenvalue must be isolated. By a suitable choice of $x_0(s)$, the angular gradient procedure will converge to intermediate eigenfunctions. The general convergence result is as follows.

Theorem 3. Let u be a quasiregular stationary point of $\{J, g_i\}$ and N a quasiregular neighborhood of u. There exists $r > 0$ and positive $k < 1$ such that for any initial vector in the ball $B(u, r)$ the angular gradient procedure converges to a stationary point u^* and $\|x_n - u^*\| \leq k^n[(1+k)/(1-k)]\, \|x_0 - u\|$.

Proof. See [2].

As a second example, consider the Mayer optimal control problem. We are given the differential equations of control $dx/dt = f(x, u)$, with $x \in R^m$ and $u \in R^q$, the boundary conditions $\psi_i\, (a, x(a), b, x(b)) = 0$, $1 \leq i \leq p$, and the cost function $J(a, x(a), b, x(b))$. It is required to determine a control function $u(t)$ which produces a trajectory $x(t)$ satisfying the boundary conditions and minimizes J. We shall restrict $u(t)$ to be piecewise continuous on $[a, b]$ and have values in some open set of R^q. To simplify the example, suppose $x(a)$ is prescribed. Then $x(b)$ depends only on $u = u(t)$ and ψ_j and J become functionals of u. Take H to be the cartesian product of q copies of $L_2[a, b]$ with inner product

$< u, v > = \int_0^1 \sum_{i=1}^{q} u_i(t)\, v_i(t)\, dt$. As is well-known, the gradients are given by $\nabla J(u) = J_x(t)\,(\partial f/\partial u)_0$ and $\nabla \psi_i(u) = \psi_{ix}(t)\,(\partial f/\partial u)_0$, where $(\partial f/\partial u)_0$ is the matrix of partial derivatives $(\partial f_i/\partial u_j)$ evaluated along a solution $(x(t), u(t))$ of the differential equations, $J_x(t)$ is the solution of the adjoint equations $dz/dt = -(\partial f/\partial x)^T z$ with values $J_x(b) = (\partial J/\partial x(b))_0$, and the $\psi_{ix}(t)$ are solutions of the adjoint equations with $\psi_{ix}(b) = (\partial \psi_i/\partial x(b))_0$. However, the gradient $\nabla \theta(x)$ would be too difficult to compute. Therefore the angular gradient procedure is modified by taking

$$h_T(x) = -\left(1 / \|\nabla J_G(x)\| \; | < \nabla \theta(x), \overline{\nabla J_T(x)} > | \right) \nabla J_T(x),$$

and approximating the differential $< \nabla \theta(x), \overline{\nabla J_T(x)} >$ by a finite difference $[\theta(x + s\,\overline{\nabla J_T(x)}) - \theta(x)]/s$.

To obtain convergence of the modified angular gradient procedure we replace the condition (iv) in Definition 2 by the following: (iv')

$\sum_1^p \cos^2 \alpha_i + \cos \alpha_0 \cos \beta / |\cos \beta_0| > \gamma$, where $\beta_0 = \arccos < \overline{\nabla \theta(x)}, \overline{\Delta J_T(x)} >$ and $\alpha_0 = \arccos < \overline{\nabla J_T(x)}, \overline{\Delta x} >$. This method has been applied successfully to optimum rocket trajectory (minimum fuel) problems. It has also been tested successfully on the classical brachistochrone problem.

Now, we consider a related method for the unconstrained problem [10],[12]. To motivate it, we consider the generalized eigenvalue problem $Ax = \lambda Bx$, where A and B are bounded linear operators on one Hilbert space H to another H'. Let $J(x) = (1/2) \| Ax - (< Ax, Bx > / < Bx, Bx >) Bx \|^2$, when $Bx \neq 0$. Then it is not difficult to show [10] that $J(x) = 0$ if and only if $\nabla J(x) = 0$. Thus, the eigenvectors are precisely the stationary points of J. Since there can be a subspace E_λ of eigenvectors, the set of stationary points need not be isolated.

To compute such stationary points we can use the gradient method, $x_{n+1} = x_n + h(x_n)$, where

$$h(x) = - \frac{2\, J(x)}{\|\nabla J(x)\|^2} \nabla J(x) \qquad (6).$$

A straightforward calculation yields that $\nabla J(x) = (A-R(x)B)^* (A-R(x)B)x$, where $R(x) = \langle Ax, Bx\rangle / \langle Bx, Bx\rangle$ and $*$denotes the adjoint operator. The method has been tried successfully on the finite-dimensional case, where A and B are square matrices. It should be especially effective in case where A and B are band or sparse matrices and when only certain intermediate eigenvalues are being sought.

The step in (6) is a special case of a more general method which can be applied to find certain kinds of non-isolated stationary points.

Definition 3. A set E of stationary points is a C-stationary set for J if for $\epsilon > 0$ there exists a neighborhood N of E and a constant $c > 0$ such that for $x \in N$, $\nabla J(x)$ is continuous in x and there exists a unique nearest point $x^* \in E$ and for $x \in N-E$ the following conditions hold: (i) $\nabla J(x) \neq 0$; (ii) $\cos^2 \alpha(x) \geq c$, where $\alpha = \arccos \langle \overline{\nabla J(x)}, \overline{\Delta x} \rangle$ and $\Delta x = x - x^*$; (iii) $| \cos \alpha(x) - 2(J(x) - J(x^*))/\|\nabla J(x)\| \, \|\Delta x\| \, | < \epsilon$. N is called a C-stationary neighborhood.

The following convergence theorem is proved in [10].

Theorem 3. Let E be a C-stationary set for J and let N be a C-stationary neighborhood of E. For $x \in N$ define

$$h(x) = \begin{cases} \dfrac{-2\,(J(x) - J(x^*))}{\|\nabla J(x)\|^2} \nabla J(x) & \text{if } \nabla J(x) \neq 0, \\ 0 & \text{otherwise.} \end{cases} \qquad (7)$$

and let $x_{n+1} = x_n + h(x_n)$. There exists a neighborhood M of E and a positive constant $k < 1$ such that for any initial vector $x_0 \in M$ the inequality $\|x_n - E\| \leq k^n \|x_0 - E\|$ holds for $n \geq 0$ and the sequence (x_n) converges to a point in the closure of E. Furthermore, for arbitrary $\delta > 0$ the neighborhood

M can be chosen so that

$$\left| k^2 - (1- \inf_{x \in M-E} \cos^2 \alpha(x)) \right| < \delta .$$

Of course, the step h(x) in (7) is only computable if we know $J(x^*)$. (In the generalized eigenvalue problem, $J(x^*) = 0$.) It is possible that a close approximation to the value $J(x^*)$ would be available in practice and this might suffice. However, pending further study, application is limited to those problems in which $J(x^*)$ is known.

The step-sze in (7) is an approximation to the distance in the gradient direction from x to the point nearest to x^*. Thus, the method of theorem 3 could be called a "gradient method of closest approach." In this respect, it differs from the steepest descent method and other gradient methods [2], [5], [6], [7], [9]. Its application to eigenvalue problems in infinitedimensional spaces (e.g. integral equations) would generally involve discretization errors of some kind [1], [11]. This requires further investigation.

References

1. K.E. Atkinson, Numerical Solution of Eigenvalue Problem for Compact Integral Operators, TAMS 1967.
2. E.K. Blum, Numerical Analysis and Computation (Ch. 5, 12), Addison-Wesley, 1972.
3. __________, A Convergent Gradient Procedure in pre-Hilbert Spaces, Pacific J. Math., 18 , 1 (1966).
4. __________, Stationary points of functionals in pre-Hilbert spaces, J. Comp. Syst. Sci Apr. 67.
5. J. W. Daniel, The Approximate Minimization of Functionals, Prentice-Hall, 1971.
6. A. A. Goldstein, Constructive Real Analysis, Harper 67 .
7. E.S. Levitin and B. T. Polyak, Constrained Minimization Methods, Zh. vychisl. Mat. mat. Fig., 1966 (Comp. Math and Math. Phys).

8. M. Z. Nashed, Differentiability and Related Properties of Nonlinear Operators - in Nonlinear Functional Analysis and Applications, ed. L. B. Rall Ac. Press 1971.

9. S. F. McCormick, A General Approach to One-step Iterative Methods with Application to Eigenvalue Problems, J. Comp. and Syst. Sci. Aug. 72.

10. E. K. Blum and G. Rodrigue, Solution of Eigenvalue Problems in Hilbert Spaces by a Gradient Method, USC Math. Dept. Prepring Apr. 72.

11. H. Wielandt, Error bounds for Eigenvalues of Symmetric Integral Equations Proc. AMS Symp. Applied Math., C, 1956.

12. G. Rodrigue, A Gradient Method for the Matrix Eigenvalue Problem $Ax = \lambda Bx$ Kent State U. Math. Dept. Dec. 72.

13. S. McCormick and G. Rodrigue, A Class of Gradient Methods for Least Squares Problems for Operators with Closed and Nonclosed Range, Claremont U. and U.S.C. Report

PARAMETERIZATION AND GRAPHIC AID IN GRADIENT METHODS

Jean-Pierre PELTIER

Office National d'Etudes et de Recherches Aérospatiales (ONERA)

92320 - Châtillon (France)

Abstract

The first part reports an experiment in which a graphic interactive console was used to operate a gradient-type optimization program.

Some indications are provided on the program sturcture and the requirements for the graphic software. Conclusions are drawn both upon advantages and difficulties related to such project.

The second part deals with parameterization of optimal control problems (i.e. solution through non-linear programming). A local measure of the loss of freedom pertaining to such technique is established. Minimization of this loss leads to the concept of optimal parameterization. A first result is given and concerns the metric in parameters space.

PART I : Console and gradient

1. Introduction

In the past, interactive graphic display consoles have been used, in the field of optimization, to select the desired model (i.e. state equations, constraints...) and initiate computations (i.e. provide initial values as in STEPNIEWSKI 1969). The experiment carried on at ONERA is original in that the interaction deals with the optimization procedure itself. A conclusion of previous computational experience had shown that, in general, a rapid solution of large, highly non linear optimal control problems requires sequential use of several numerical techniques although each of these is, on the paper, sufficient. This is why ONERA developed a fairly large optimization program, TOPIC (Trajectory OPtimization for Interception with Constraints), offering a range of options. Options can fall into seven categories :

1) controls (how will they be modelized)
2) constraints (penalization, Lagrange multipliers ...)
3) search direction-local analysis (metric choice, semi direct technique ...)
4) search direction-global analysis (takes past step into account e.g. variable metric)
5) step size (fixed, linear search techniques ...)
6) convergence index (Kelley 1962, Fave 1968 ...)
7) technical options (e.g. integration procedures).

Although some options seem to be non-independant, it is best to program them as if they where, the present trend being to re-introduce in various algorithms possibilities which, at first, seemed not to be compatible.

In order to facilitate comparisons of methods and speed up computations, a graphic display console has been interfaced with the program. The console has a treble action.

(i) monitor computations so that an operator can juge of their worthiness ;
(ii) aid an operator diagnostics and enable direct action ;
(iii) facilitate edition of results.

2. General structure of the program

The presence of a graphic package together with an already voluminous computing program results in a quite large memory requirement (about 400 k octets) so the program structure has to be carefully studied and be compatible with overlay techniques. TOPIC structure is showed in figure 1. It is entirely written in FORTRAN IV language. Each of the block can be divided in a set of overlayed segments except for MAIN, MAIN CONSOLE and MEMORY blocks. The MEMORY block stands for labelled commons which contains all variable values such as : current control, state and gradient histories, algorithm memories and so forth. Thus these values are preserved during overlay operations.

The MAIN program is reduced to a switch and calls for initialization and input routines, to the console and optimization driver (s) in block 1.

Fig. 1 - Program structure

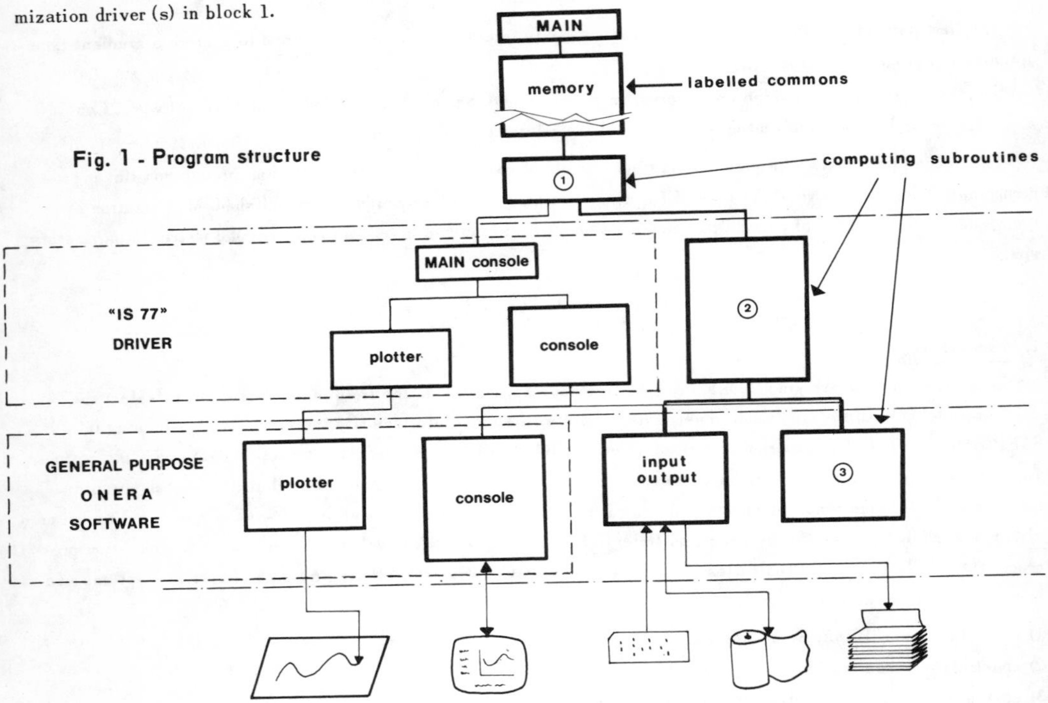

3. Computing programs

A very general gradient program has to perform the following twelve tasks :

1 - build a design vector from actual parameters plus control functions (if required).
2 - Load initial state.
3 - Forward integration of state (load state tables).
4 - Compute performance index from final state.
5 - Algorithm decides wither to go on 6 or 11
CALL TO CONSOLE.
6 - Compute final adjoint.
7 - Backward integration to compute gradient (or direct finite differences).
8 - Projection of functional gradient into design vector space.
9 - Algorithm decides to go on 10 or 11 (or to exit)
CALL TO CONSOLE.
10 - Search direction modification (non local methods).

11 - Step computation.

12 - Control or design vector modification.

Of these tasks 1, 2, 3, 7 and 8 can be deleted in the case of a function extremalization. It has been sound efficient to schedule calls to console after points 5 and 9.

The list of tasks gives a good idea of program(s) in block 1, i.e. driver program(s). Such routines are just big switches and call specialized routines to perform each of the 12 tasks. When the console package returns to the driver, a flag enables to restart computations at any of the 12 points.

Block 2 in figure 1 contains algorithm routines, integration routines, model routines plus least squares, linear system solution and so forth.

Block 3 routines are divided into two classes :

- specific : routines handling control functions and design vector ;
- general purpose : matrices, vectors and functions handling, such as performing + × $\|\cdot\|$ operations on such elements.

Such structure leads to an unusual programming of algorithm routines which cannot call directly subroutines to perform tasks but instead return control to the driver with a proper set of flags to indicate what is needed.

4. The graphic package

It is, in fact, composed of two distinct packages :

(i) The ONERA general purpose graphic package : SYCEC ;

(ii) a special package, IS77, specific to the program.

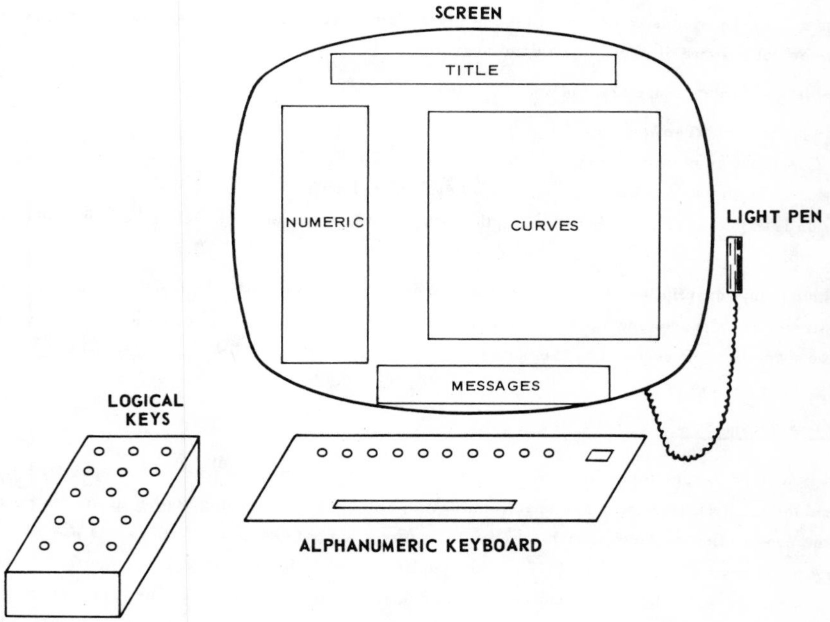

Fig. 2 - Console system

SYCEC divides the console screen into 4 areas, as shown on figure 2. We are going to review the use IS77 makes of the 3 more important areas, with emphasis on the monitor set up which is more delicate :

A) The curve area : 3 curves are plotted against the number of function evaluation :

. the current value of the performance index (thin curve),

. the current value of the unaugmented (i.e. true), performance index (dotted curve),

. for visual aid, the best value reached for the performance index (heavy curve).

Comparing the first two histories, the operator sees how penalty terms are behaving. Comparing the first and the last curves tells him immediately if an improvement is currently achieved : the two curves merge. After each gradient computation sensitivity functions with respect to controls are plotted on the screen. Of course in intervention phase (i.e. when the operator has a manual control) all interesting curves such as controls, state, pseudo-state (constraints) ... should be available for examination and even for manual changes (through alphanumeric keyboard or directly with the light pen). For parametric optimization problems, histograms are plotted instead of functions.

B) The numeric area : it allows 20 numbers, real or integers, to be displayed without mnemonics. It has been found necessary to reserve 8 lines for flag displays showing main options on the algorithms, printout volume, iteration count. Thus the operator knows what is actually performed by the computing program. Remaining 12 lines display the performance index, the unaugmented performance index, values of final state and pseudo state, final constraints and the current step size value for the linear search. To compute a distinct pseudo-state for each of the current-state constraint has been found more informative than to lump them up all in a single variable (which is possible), in spite of the extra dimension requirement in the integration procedure.

Here again, in intervention phase IS77 brings upon request in the numerical area any significant number of the problem for examination and modification.

C) The message area performs an important task of giving to numerical data on the screen their full meaning by keeping the operator aware of where these data come from.

So diverse standard messages let the operator know :

- where the console package was called from in Block 1 ;
- which IS77 subroutine is in action ;
- if an action is expected from him (e.g. depress a key from a given set) ;
- if no action is expected. This is used for lengthy input/output operations. The message is supposed to keep the operator cool.

To conclude this description of IS77, subroutine "Main console" has just a logical function and calls, either on automatic or manual mode, specialized subroutines ; Block 3 routines should be available to the graphic package for simple computations as required by the operator.

5. Difficulties encountered

(i) TOPIC is a program specific in its formulation to one optimization problem, and IS77 is specific both to the problem and the program. Develop a non specific graphic package (i.e. depending neither upon problem formulation nor upon the constitution of the memory block) would require a big programming effort. Values should be passed through arguments instead of commons. IS77 would have its own input file (it has one, but short) to define, through a set of flags priorities for the monitor option. And as a handbook could not indicate (as it is presently done) the signification of numbers, this would have to be done on the screen. A file of mnemonics would have to be loaded and each numerical value should appear with its name. For such a work, the console software would have to be rebuilt from scratch.

(ii) At it is, on the IBM 2250 console which is used, the possibility for manual intervention goes through a program stop demanding an operator action. Thus there is no possibility to have a (normal) monitoring phase interrupted unexpectedly by the operator, although this would be the best use for the system. Actually a simple 2 positions switch would be sufficient for this purpose, but such switch does not exist on the 2250.

(iii) The economy of the system is delicate :

the time t_g necessary to update the screen, in monitor option, is constant (about 3-4 seconds) while the time necessary for computations varies widely with the problem. A good reference is the central unit time per function evaluation, t_c. If t_c is too small compared to t_g, the use of a complex console system is unjustified to monitor (on production) such unexpensive calculations. If t_c is too long ($>$ 1 minut) the screen is static and the operator (a specialist) time is wasted although computations are expensive enough to be closely followed. For the problem delt with through IS77, t_c varies from 10 secs to 1 minut which has been found reasonable.

6. Conclusion

After a year and a half of use, the graphic system has proved it's efficiency for :

- acquiring rapidly an insight into a new problem,
- building a reasonable first guess for the algorithms.
- getting fast results (a night on the console is equivalent to a week of normal procedure !).

Moreover it helped us to reach some interesting conclusions upon constraints handling, global algorithms, integration methods and parameterization of optimal control problems (such conclusions might have been reached without the graphic system but may be less rapidly).

However development of a really general graphic software is an expensive task which now awaits the conclusion of a present phase implementing a general purpose, versatile, multiple option optimization program.

PART II : Parameterization of optimal control problems

1. Introduction

Optimization problems in functional space and, among these, optimal control problems aim to determine, in general, a vector valued function U^* of, let us say, time over a given (finite) interval $\mathcal{E}$, which optimizes some performance index. Any problem of practical importance in the field has to be solved on a digital computer and this eventually transforms the function into a set of many parameters. Already this approximation raised some theoretical examination (KELLEY DENHAM 1968). Then, variable metric algorithms, which have represented a major improvement in the field of parametric optimization, with their convergence properties, tempted several groups into using these techniques to solve optimal control problems (BRUSCH 1970 - JOHNSON-KAMM 1971, SPEYER & al 1971, the author 1971). Of course this technique impose a modeling of controls (figure 3) involving a limited number of parameters (an order of magnitude less than the average number of discrete points in functional programs). Of course such a technique delivers only a suboptimal solution V^*_n which depends upon the way controls are parameterized and the number n of parameters. A first question which arizes is to know wether or not one can find $n(\varepsilon)$ large enough so that $\| U^* - V^*_n \| < \varepsilon$ i.e. if $V^*_n \to U^*$ when $n \to \infty$.

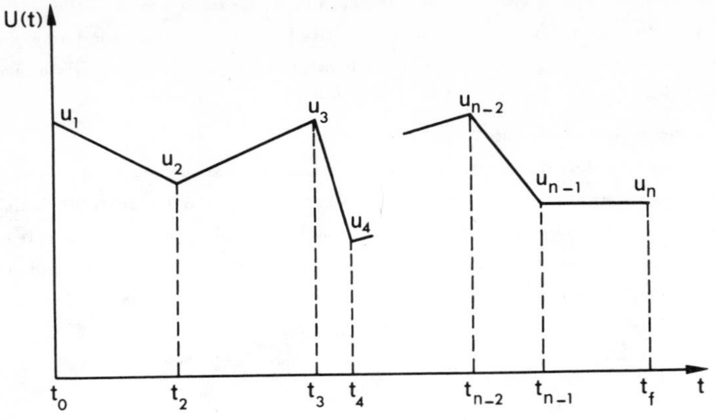

Fig. 3 - Parameterization example

This question of theoretical importance has been solved by CULLUM 1972.

The contribution of this paper is to try to select, for a given number of parameters, the best parameterization for the control.

2. Steepest slope in a hilbert

Let $U \in \mathbb{U}$ and J(.) an application of $\mathbb{U}$ into $\mathbb{R}$. The slope β of J(.) in direction dU and at point $\bar{U}$ is defined as

(1)
$$\beta \triangleq \lim_{s \to 0} \left\{ \frac{| J(\bar{U} + s\, dU) - J(\bar{U}) |}{\| s\, dU \|} \right\}$$

where $s \in \mathbb{R}$ and $\| \cdot \|$ is the selected norm in $\mathbb{U}$. Whenever a Frechet derivative $\frac{\partial J}{\partial U}(\bar{U})[\cdot]$ exists at point $\bar{U}$,

(2)
$$\beta \equiv \frac{\frac{\partial J}{\partial U}(\bar{U})[dU]}{\| dU \|}$$

Let $\mathbb{U}$ be Hilbert and $\mathcal{M}(.)$ be an application of $\mathbb{U}$ onto $\mathbb{U}^*$, it's a dual space. $\mathcal{M}(.)$ is linear, continuous, symmetric and coercive. Let $\langle \cdot , \cdot \rangle$ be the duality product (i.e. the application of $\mathbb{U} \times \mathbb{U}^*$ into $\mathbb{R}$ defined by U*(U)) and define a dot product in $\mathbb{U}$ as

(3)
$$(\cdot , \cdot)_{\mathcal{M}} \triangleq \langle \mathcal{M}(\cdot) , \cdot \rangle$$

The norm in $\mathbb{U}$ will be the associated norm $\| . \|_{\mathcal{M}}$. A classical step in the gradient technique consists into defining a steepest slope direction dU* which maximizes β.

This direction is found to be

(4)
$$dU^* = \mathcal{M}^{-1}\left(\frac{\partial J}{\partial U}(\bar{U}) \right) \qquad \text{where} \quad \frac{\partial J}{\partial U}(\bar{U}) \in \mathbb{U}^*$$

and the corresponding slope is

(5)
$$\beta^* = \left\langle \frac{\partial J}{\partial U}(\bar{U}) , \mathcal{M}^{-1}\left(\frac{\partial J}{\partial U}(\bar{U}) \right) \right\rangle^{\frac{1}{2}}$$

Let us define

(6) $$\eta(\bar{U}, dU) \triangleq \frac{\beta(dU)}{\beta^*} \leqslant 1$$

3. The V(.) application

Similarly, let $\mathbb{A}$ be hilbert with a metric associated to application $\mathcal{N}(.)$ and K(.) an application in $\mathbb{R}$, $\langle .,. \rangle$ the duality product and V(.) an application of $\mathbb{A}$ into $\mathbb{U}$. At point $\bar{a} \in \mathbb{A}$, let $\frac{\partial V}{\partial a}(\bar{a})[\cdot]$ be the Frechet derivative of V.

The steepest slope direction of K(.) in $\mathbb{A}$ at point $\bar{a}$, da*, is transformed through $\frac{\partial V}{\partial a}(\bar{a})$ [.] into direction

(7) $$dU = \frac{\partial V}{\partial a}(\bar{a})\left[\mathcal{N}^{-1}\left(\frac{\partial K}{\partial a}(\bar{a})[\cdot]\right)\right]$$

at point $\bar{U} = V(\bar{a})$.

As example, suppose $\mathbb{U}$ to be $\mathcal{L}^2(\mathcal{E})$, $\mathbb{A}$ to be $\mathbb{R}^n$. V(a, t) is some control model depending upon a n-vector $|a\rangle$. Dot products are respectively defined as $\int_{\mathcal{E}} \langle .[M(t)]. \rangle\, dt$ and $\langle .[N]. \rangle$ where N and M(t) are symmetric, continuous, positive definite matrices. Let $\frac{\partial K}{\partial a}(a)[.] \equiv \langle g_k(\bar{a}) | \cdot \rangle$ and

$$\frac{\partial J}{\partial U}(\bar{U})[\cdot] \equiv \int_{\mathcal{E}} \langle H_U(\bar{U},t) | \cdot \rangle\, dt$$

where $\langle . | . \rangle$, not to be taken for $\langle . , . \rangle$, is the usual euclidian product in $\mathbb{R}^n$.

Notation H_U is used by similarity with optimal control problems but, so far, is just a notation.

In this example,

(8) $$\beta(dU) = \frac{\int_{\mathcal{E}} \langle H_U(\bar{U},t)\left[\frac{\partial V}{\partial a}^T(\bar{a},t)\right][N^{-1}]\, g_K(\bar{a}) \rangle\, dt}{\sqrt{\int_{\mathcal{E}} \langle g_K(\bar{a})[N^{-1}]\left[\frac{\partial V}{\partial a}^T(\bar{a},t)\right][M(t)]\left[\frac{\partial V}{\partial a}(\bar{a},t)\right][N^{-1}]\, g_K(\bar{a}) \rangle\, dt}}$$

where appear terms

(9) $$\langle g_J | \triangleq \int_{\mathcal{E}} \langle H_U\left[\frac{\partial V}{\partial a}^T(\bar{a},t)\right] dt$$

(10) $$[N'] \triangleq \int_{\mathcal{E}} \left[\frac{\partial V}{\partial a}^T(\bar{a},t)\right][M(t)]\left[\frac{\partial V}{\partial a}(\bar{a},t)\right] dt$$

4. Parameterization efficiency

In the case where K(.) is defined as the composed application

(11) $$K(\cdot) \equiv J(V(\cdot))$$

the chain derivation formula gives

(12) $$\frac{\partial K}{\partial a}(\bar{a})[\cdot] = \frac{\partial V}{\partial a}^*\left(\frac{\partial J}{\partial U}(\bar{U})\right)[\cdot]$$

where $\frac{\partial V}{\partial a}^*(\cdot)[\cdot]$ is the adjoint of $\frac{\partial V}{\partial a}(\cdot)[\cdot]$.

In the special case where V(.) is biunique (i.e. $\mathbb{U}$ and $\mathbb{A}$ can be identified), it is possible to select in $\mathbb{A}$ the $\mathcal{N}(.)$ metric so that

$$dU = \frac{\partial V}{\partial a}\left(\mathcal{N}^{-1}\left(\frac{\partial V^*}{\partial a}\left(\frac{\partial J}{\partial U}(\bar{U})\right)\right)\right) \tag{13}$$

be identical to dU^* (see (4)). Then $\eta = 1$.

In the general case (and in the case of parameterization) $\mathbb{U}$ and $\mathbb{A}$ cannot be identified and even for an injective V(.) application $\frac{\partial V^*}{\partial a}$ is not so. Therefore $\frac{\partial V}{\partial a}(\mathcal{N}^{-1}(\frac{\partial V^*}{\partial a}(.)))$ cannot be identified to $\mathcal{M}^{-1}(.)$ and $dU \neq dU^*$ $\forall \mathcal{N}(\cdot) \Rightarrow \eta < 1$.

However, it is sufficient that $\frac{\partial V}{\partial a}(.)$ be injective in order to confer to application

$$\mathcal{N}'(\cdot) \triangleq \frac{\partial V^*}{\partial a}\left(\mathcal{M}\left(\frac{\partial V}{\partial a}(\cdot)\right)\right) \tag{14}$$

all the required properties, starting with coercivity, to be associated with a metric in $\mathbb{A}$. Moreover $\mathcal{N}'(.)$ precisely defines the image-metric in the case of identifiable $\mathbb{A}$-$\mathbb{U}$ spaces. It is non-local (i.e. constant $\forall\ \bar{a} \in \mathbb{A}$) if V(.) is linear.

In any case we shall define the parameterization efficiency at point $\bar{a}$ as the value of η (dU), (where dU is defined in (13)), which is less thane one. η (dU) can be considered as a local measure of the loss of freedom induced by parameterization of the functional problem. In the above defined example of suboptimization in $\mathbb{R}^n$ of a problem defined in $\mathcal{L}^2(\mathcal{C})$, $\langle g_k | \equiv \langle g_J |$ and the value of η is given by

$$\eta = \frac{\langle g(\bar{a})\,[N^{-1}]\,g(\bar{a})\rangle}{\sqrt{\int_{\mathcal{C}} \langle H_u(\bar{v},t)\,[M(t)]\,H_u(\bar{v},t)\rangle\,dt}\;\sqrt{\langle g(\bar{a})\,[N^{-1}N'N^{-1}]\,g(\bar{a})\rangle}} \tag{15}$$

5. Optimal parameterization

Having built a index of quality for a parameterization, it is normal to try to maximize it.

Maximization of η can be carried on on several steps :

1) $\mathcal{M}(.)$ being given, select $\mathcal{N}(.)$ in $\mathbb{A}$ so that η is maximum.
2) For a given $\mathcal{N}(.)$, V(.) can usually be imbedded into a family of transformations depending upon a given set of parameters $|b\rangle$ which are not subject to optimization i.e. do not belong to the $|a\rangle$ set. It is possible to solve an accessory optimization problem, maximizing $\eta(V(|\bar{a}\rangle, |b\rangle, .), dU(|\bar{a}\rangle, |b\rangle, .))$ with respect to $|b\rangle$.
3) Eventually it is possible to compare optimum values of η for various kinds of V(.) transforms and select a preferred parameterization technique.

The problem is that as η is locally defined such accessory optimization problems will lead to local solutions and local conclusions. However, if it can be done rapidly, the parameterization can be modified whenever the algorithm does not use past step informations.

A first conclusion can be reached on optimal parameterization and it is a non-local conclusion :
in order to maximize η with respect to $\mathcal{N}(.)$, one should select $\mathcal{N}(.) \equiv \mathcal{N}'(.)$ as defined in (14) which, incidentally, simplifies (15) as, in our example [N] has to be equal to [N'] of (9).

6. Conclusion

A quality measure of the parameterization of a functional optimization problem has been introduced with the efficiency coefficient η . This notion will enable comparisons of parameterization techniques, hopefully. A first conclusion has been reached on optimal parameterization which leads to the selection of the proper metric (when it exists) in the reduced space of controls. Proofs have not been given and will await applications to practical examples for a more complete development.

Acknowledgment

May Mr C. Aumasson, from ONERA, find here the author's gratitude for his many discussions and criticisms over parameterization. The author is also indepted to Miss G. Mortier, from ONERA Computer Center, for developing special features in SYCEC and for her help in the graphic project.

References

PART I

FAVE, J. - Critère de convergence par approximation de l'optimum pour la méthode du gradient, in computing methods in optimization problems, springer verlag 1969, p. 101-113 (proceedings of the 2nd International Conference on Computational methods and optimization problems, San Remo, sept. 1968).

KELLEY, H.J. - Methods of gradients, in Optimization Techniques, G. Leitmann ed., Ac. Press, 1962, p. 248-251.

STEPNIEWSKI, W.Z. , KALMBACH, C.F. Jr. - Multivariable search and its application to aircraft design optimization. The Boeing Company, Vertol division, 1969.

PART II

BRUSCH, R.G. , SCHAPPELLE, R.H. - Solution of highly constrained optimal control problems using non-linear programming. AIAA paper 70-964 and AIAA Journal, vol. 11 n° 2, p. 135-136.

CULLUM Jane - Finite dimensional approximations of state constrained continuous optimal control problems. SIAM J. Control, vol. 10 n° 4, Nov. 1972, p. 649-670.

JOHNSON, I.L. , KAMM, J.L. - Near optimal shuttle trajectories using accelerated gradient methods AAS/AIAA paper 328. Astrodynamics specialists conference, August 17-19 1971, Fort Landerdale Florida.

KELLEY, H.J. , DENHAM, W.F. - Modeling and adjoint for continuous systems 2nd International Conference on Computing Methods in Optimization Problems, San Remo, Italy, 1968 and JOTA vol. 3, n° 3, p. 174-183.

PIGOTT, B.A.M. - The solution of optimal control problems by function minimization methods. RAE Technical Report 71149, July 1971.

SPEYER, J.L. , KELLEY, H.J. , LEVINE, N. , DENHAM, W.F. - Accelerated gradient projection technique with application to rocket trajectory optimization. Automatica, vol. 7, p. 37-43, 1971.

LES ALGORITHMES DE COORDINATION
DANS LA METHODE MIXTE D'OPTIMISATION A DEUX NIVEAUX

G. GRATELOUP*
Professeur

A. TITLI*
Attaché de Recherches
au C.N.R.S.

T. LEFEVRE*
Ingénieur de
Recherche

I - INTRODUCTION

Pour contourner les difficultés théoriques et de calcul qui se présentent lors de la résolution des problèmes d'optimisation de grande dimension, un moyen efficace est certainement l'introduction de méthodes d'optimisation à deux niveaux, utilisées notamment en commande hiérarchisée dans les structures de commande à deux niveaux.

Pour cette tâche, que l'on supposera être l'optimisation statique d'un ensemble de sous-processus interconnectés, on peut alors utiliser la notion de "division horizontale du travail" faisant apparaître des sous-problèmes résolus de façon locale, les actions locales étant coordonnées par le niveau supérieur de commande, de façon à obtenir l'optimum global.

Or, chacun des problèmes inférieurs étant défini par 2 fonctions (modèle du sous-processus et critère associé), il y a trois modes possibles de décomposition-coordination :

- par l'intermédiaire de la fonction critère
- par l'intermédiaire du modèle
- par action sur les 2 fonctions.

Ce troisième mode qui est étudié ici, utilise comme grandeur de coordination les variables d'interconnexion entre sous-systèmes et les paramètres de Lagrange associés.

Dans cette communication, les sous-problèmes locaux d'optimisation sont définis, et différentes possibilités de coordination sont proposées pour le niveau supérieur de commande. Sont examinés notamment les coordonnateurs type gradient, Newton, à itération directe et gradient-itération directe.

La résolution d'un problème de répartition optimale des énergies, dans un système de production hydroélectrique, permet de mieux comparer certains de ces algorithmes coordonnateurs.

* Laboratoire d'Automatique et d'Analyse des Systèmes du C.N.R.S.
B.P. 4036
31055 TOULOUSE CEDEX - FRANCE

II - DECOMPOSITION DANS LA METHODE MIXTE

II.1 Position du problème (Problème "séparable")

Supposons que le processus complexe à optimiser soit divisé en N sous-systèmes comme celui représenté sur la figure 1.

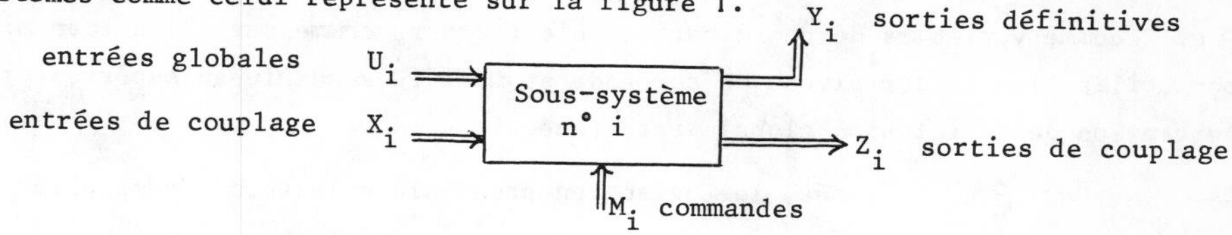

Figure 1

U_i, X_i, M_i, Z_i, Y_i sont des vecteurs à m_{U_i}, m_{X_i}, m_{M_i}, m_{Y_i} composantes respectivement.

Pour un vecteur d'entrée globale U donné, le sous-système est complètement décrit en régime statique par les équations vectorielles :

$$Z_i = T_i(M_i, X_i) \tag{1}$$

$$Y_i = S_i(M_i, X_i) \tag{2}$$

L'interconnexion entre les sous-systèmes est représentée par :

$$X_i = H_i(Z_1 \ldots Z_i \ldots Z_N) \tag{3}$$

La fonction objectif du système est supposée donnée sous forme "séparable, additive".

$$F = \sum_{i=1}^{N} f_i(X_i, M_i) \tag{4}$$

Le problème global est de maximiser (4) en présence des contraintes égalité (1) et (3). A ce problème d'optimisation, on peut associer le Lagrangien :

$$L = \sum_{i=1}^{N} f_i(X_i, M_i) + \sum_{i=1}^{N} \mu_i^T (T_i - Z_i) + \sum_{i=1}^{N} \rho_i^T (X_i - H_i(Z_1 \ldots Z_i \ldots Z_N)) \tag{5}$$

La solution optimale doit nécessairement satisfaire les conditions de stationnarité de ce Lagrangien, à savoir :

$$L_{X_i} = 0 = \frac{\partial fi}{\partial Xi} + \left(\frac{\partial Ti}{\partial Xi}\right)^T \mu i + \rho i \tag{6}$$

$$L_{M_i} = 0 = \frac{\partial fi}{\partial Mi} + \left(\frac{\partial Ti}{\partial Mi}\right)^T \mu i \tag{7}$$

$$L_{Z_i} = 0 = -\mu i - \sum_{j=1}^{N} \left(\frac{\partial Hj}{\partial Zi}\right)^T \rho j \tag{8}$$

$$L_{\mu_i} = 0 = T_i - Z_i \tag{9}$$

$$L_{\rho_i} = 0 = X_i - H_i(Z_1 \ldots Z_i \ldots Z_N) \tag{10}$$

II.2 Décomposition, formulation des sous-problèmes (G.Grateloup, A.Titli)

Cependant, pour simplifier la résolution de problèmes de grande dimension, on répartit le traitement de ces équations entre deux niveaux de commande, non arbitrairement, mais de façon à obtenir une forme "séparable" des équations au niveau inférieur.

Dans la méthode proposée, une telle répartition est obtenue en choisissant ρ et Z comme variables de coordination, c'est-à-dire comme variables transmises pour utilisation au 1er niveau de commande et modifiées au niveau supérieur jusqu'à l'obtention de la solution globale recherchée.

Pour ρ et Z donnés, le Lagrangien prend alors la forme "séparable" suivante

$$L = \sum_{i=1}^{N} L_i = \sum_{i=1}^{N} f_i(X_i,M_i) + \rho_i^T X_i - \rho_i^T H_i(Z_1 \ldots Z_i \ldots Z_N) + \mu_i^T (T_i - Z_i) \tag{11}$$

L'examen de L_i permet de formuler chaque sous-problème en termes de commande optimale ; ainsi, le sous-problème n° i s'écrit :

$$\max \left[f_i(X_i,M_i) + \rho_i^T X_i - \rho_i^T H_i(Z_1, Z_i, Z_N) \right] \tag{12}$$

$$\text{sous } T_i(X_i, M_i) - Z_i = 0 \qquad \text{pour } \rho \text{ et Z donnés.}$$

Il apparaît bien dans (12) qu'à la fois critère et modèle sont utilisés pour la coordination.

Sur le plan analytique, la résolution de chaque sous-problème correspond au traitement des équations (6), (7), (9), les équations restant à résoudre au 2e niveau étant :

$$\begin{aligned} L_\rho(X, Z) &= 0 \\ L_Z(Z, \mu, \rho) &= 0 \end{aligned} \tag{13}$$

L'équation (9), qui doit être compatible pour ρ et Z donnés, impose

$$m_{X_i} + m_{M_i} \geqslant m_{Z_i} \tag{14}$$

Le transfert d'informations nécessaire entre niveaux de commande est représenté fig.2

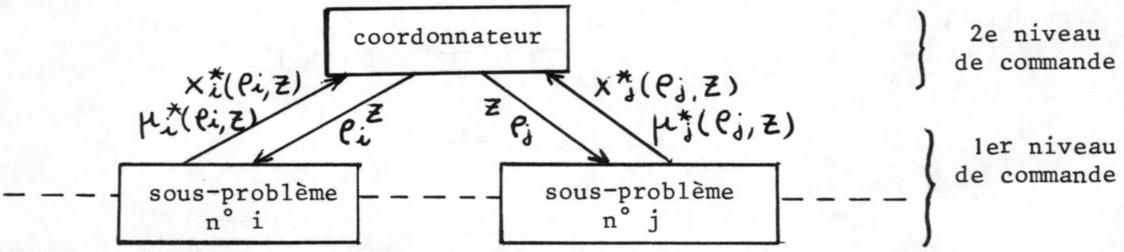

Figure 2 : Transfert des informations dans la méthode mixte

II.3 Décomposition des problèmes non séparables (A. Titli, T. Lefèvre, M. Richetin)

Dans l'hypothèse de "non séparabilité" du problème retenue ici, c'est-à-dire lorsque le couplage entre les sous-systèmes intervient non seulement par les équations d'interconnexions classiques entre les entrées et les sorties, mais aussi, par l'intermédiaire des fonctions critères, on aboutit à la formulation suivante du problème d'optimisation :

$$\max F = \max \sum_{j=1}^{N' \leq N} f_j(X'_j, M'_j, W)$$

$$\left.\begin{array}{l} \text{sous } Z_i = T_i(X''_i, M''_i, W) \\ X''_i = H_{1i}(Z) \qquad i = 1 \text{ à } N \\ W_{X_i} = H_{2i}(Z) \end{array}\right| \qquad (15)$$

en mettant en évidence le vecteur W formé avec les composantes de X_i, M_i qui assurent ce couplage supplémentaire.

Après regroupement, si $N' \leq N$, de certaines équations de modèle et de couplage, il est possible d'écrire (15) sous la forme ci-dessous :

$$\left.\begin{array}{l} \max \sum_{j=1}^{N'} f_j(X'_j, M'_j, W) \\ \text{sous } Z'_j = T'_j(X'_j, M'_j, W) \\ X'_j = H'_j(Z') \qquad j = 1 \text{ à } N' \\ W_{X_j} = H''_j(Z') \end{array}\right| \qquad (16)$$

La décomposition de ce problème global d'optimisation peut alors être obtenue :

- soit en insérant W dans les variables de coordination Z et ρ
- soit en ajoutant au problème initial des contraintes de la forme

$$\alpha_i^j - \begin{bmatrix} X_j \\ M_j \end{bmatrix} = 0, \; j \in j_i$$ ensemble des indices des sous-problèmes en interaction avec le sous-problème i.

Dans ce cas, un terme de la forme $\sum_i \sum_{j \in J_i} \beta_i^T \left[\alpha_i^j - \begin{bmatrix} X_j \\ M_j \end{bmatrix}\right]$ est ajouté au Lagrangien global et les variables de coordination peuvent être Z, α, ρ, β.

III - COORDINATION DANS LA METHODE MIXTE

III.1 Coordonnateur type gradient :

Par analogie avec la méthode de Arrow-Hurwicz pour la recherche d'un point col, on peut utiliser l'algorithme coordonnateur suivant :

$$\left.\begin{array}{l} \frac{d\rho}{dt} = - L_\rho \\ \frac{dZ}{dt} = L_Z \end{array} \Longrightarrow \begin{array}{l} \rho(t+1) = \rho(t) - K L_\rho \\ Z(t+1) = Z(t) + K L_Z \end{array} \quad K>0 \right\} \qquad (17)$$

Différentes études de convergence de ce coordonnateur ont été faites (A. Titli).

III.2 Coordonnateur type Newton

Il est possible également, pour assurer la coordination, d'appliquer un algorithme de Newton-Raphson à la résolution de l'ensemble des 2 équations vectorielles :

$$L_Z = 0 \quad \text{et} \quad L_\rho = 0$$

en écrivant : $\frac{dW}{dt} = - \left[\frac{dL_W}{dW}\right]^{-1} L_W$

ou : $$W(t+1) = W(t) - c \left[\frac{dL_W}{dW}\right]^{-1} L_W \; ; \quad W = \begin{bmatrix} Z \\ \rho \end{bmatrix}, \; 0 < c \leq 1 \qquad (18)$$

On montre (A. Titli) que si cet algorithme est applicable, il est asymptotiquement stable.

III.3 Coordonnateur à itération directe

Dans le cas d'un couplage linéaire ($X_i = \sum_{j=1}^{N} C_{ij} Z_j$), et si certaines conditions sur les composantes sont satisfaites, ($\sum_i m_{X_i} = \sum_i m_{Z_i}$), il est alors possible de calculer Z à partir de $L_\rho = 0$ et ρ à partir de $L_Z = 0$, mettant ainsi en oeuvre une méthode à itération directe.

Une étude générale de la convergence de ce mode de coordination a été effectuée (T. Lefèvre).

III.4 Coordination mixte : itération directe-gradient :

Dans ce genre de coordination, qui ne nécessite pas $\sum_i m_{X_i} = \sum_i m_{Z_i}$ certaines équations du niveau supérieur permettent une détermination directe de certaines variables de coordination (B_1), les autres variables (B_2) étant déterminées par un algorithme de type gradient :

$$B_1^{i+1} = F_1(A^i) \qquad (19)$$

$$B_2^{i+1} = B_2^i + \alpha \cdot K \cdot F_2(A^i) \qquad (20)$$

i : indice d'itération, $K > 0$, $\alpha = \pm 1$ suivant la nature des variables B_2 (variables physiques ou paramètres de Lagrange).

Remarque : Cette méthode mixte peut être généralisée au cas des problèmes d'optimisation statique avec contrainte inégalité et d'optimisation dynamique (A. Titli).

IV - EXEMPLE D'APPLICATION : REPARTITION OPTIMALE DES ENERGIES DANS UN RESEAU DE PRODUCTION HYDROELECTRIQUE (T. Lefèvre)

IV.1 Formulation du problème

Nous traitons ici un problème similaire à celui abordé par MANTERA.

Soit le réseau série-parallèle décrit par la figure 3, et composé de deux rivières, comportant chacune deux centrales ; les quatre centrales fournissent leur énergie à un système de 5 charges dont 2 modulables ("conforming loads") et 3 non modulables ("non conforming loads").

Ces centrales sont couplées au système de charges par un réseau électrique devant fournir une puissance P_D. Chaque centrale est définie par un modèle (Fig. 4) :

$$q_n^{\,L} = g_n^{\,L}(P_n^{\,L}) \quad \begin{cases} L = \{1, 2\} \\ n = \{1, 2\} \end{cases}$$

Les pertes de production P_{TGL} sont données par : $P_{TGL} = P_{TG^1 L} + P_{TG^2 L}$

$$\text{avec : } P_{TGL^i} = \sum_{n=1}^{2} P_{TGL^i\,n}$$

Les pertes en ligne sont de la forme :

$$P_L = \sum_{i=1}^{4+3} \sum_{j=1}^{4+3} (P^i)^T B^{ij} P^j$$

Le problème global est :

$$\min_{P_n^L \; \begin{cases} L = \{1, 2\} \\ n = \{1, 2\} \end{cases}} \left[P_L + P_{TGL}^{\;1} + P_{TGL}^{\;2} \right]$$

sous : $P_L + P_D - P_T = 0$

$$\left.\begin{array}{l} q_2^{\,1} - q_1^{\,1} - Q_{11}^{\;1} - Q_{12}^{\;1} \leqslant 0 \\ q_2^{\,2} - q_1^{\,2} - Q_{11}^{\;2} - Q_{12}^{\;2} \leqslant 0 \end{array}\right\} \text{Contraintes sur les débits.}$$

$$\left.\begin{array}{l} 0 \leqslant P_1^{\,1} \leqslant P_1^{\,1}\max \\ 0 \leqslant P_2^{\,1} \leqslant P_2^{\,1}\max \\ 0 \leqslant P_1^{\,2} \leqslant P_1^{\,2}\max \\ 0 \leqslant P_2^{\,2} \leqslant P_2^{\,2}\max \end{array}\right\} \text{Contraintes sur les puissances.}$$

et a pour lagrangien :

$$\begin{aligned} L = {} & P_L + P_{TGL}^{\;1} + P_{TGL}^{\;2} + \rho(P_L + P_D - P_T) + \mu_2^1 [q_2^1 - q_1^1 - Q_{11}^{\;1} \\ & - Q_{12}^{\;1}] + \mu_2^2 [q_2^2 - q_1^2 - Q_{12}^{\;2} - Q_{11}^{\;2}] + \gamma_{11}^{\;1}(-P_1^1) \\ & + \gamma_{12}^{\;1}(-P_2^1) + \gamma_{11}^{\;2}(-P_1^2) + \gamma_{12}^{\;2}(-P_2^2) + \gamma_{21}^{\;1}(P_1^1 - P_1^1\max) \\ & + \gamma_{22}^{\;1}(P_2^1 - P_2^1\max) + \gamma_{21}^{\;2}(P_1^2 - P_1^2\max) + \gamma_{22}^{\;2}(P_2^2 - P_2^2\max) \end{aligned} \tag{21}$$

Pour arriver à une décomposition de ce problème en deux sous-problèmes relatifs aux deux rivières, il est nécessaire de décomposer P_L, ce qui peut se faire par introduction de deux pseudo-variables $P'^{\,2}_1$ et $P'^{\,2}_2$ et de deux contraintes égalité supplémentaires : $P_1^{\,2} - P'^{\,2}_1 = 0$

$$P_2^{\,2} - P'^{\,2}_2 = 0$$

(λ_1^2, λ_2^2 seront les paramètres de Lagrange associés à ces contraintes). P_L prend alors la forme séparable : $P_L = P_{L1} + P_{L2}$.

Si pour une méthode mixte, nous choisissons : $P'^{\,2}_1$, $P'^{\,2}_2$, ρ , λ_1^2 et λ_2^2 comme variables de coordination, le lagrangien (21) prend alors une forme séparable $L = L_1 + L_2$, avec :

$$L_1 = P_{L1} + P_{TGL}^{1} + \rho\left(P_{L1} + P_D - (P_1^1 + P_2^1) - (P_1^3 + P_2^3 + P_3^3)\right)$$
$$+ \mu_2^1\left[q_2^1 - q_1^1 - Q_{11}^1 - Q_{12}^1\right] + \gamma_{11}^1(-P_1^1) + \gamma_{12}^1(-P_2^1)$$
$$+ \gamma_{21}^1(P_1^1 - P_1^1 max) + \gamma_{22}^1(P_2^1 - P_2^1 max)$$
$$L_2 = P_{L2} + P_{TGL}^{2} + \rho\left(P_L^2 - (P_1^2 + P_2^2)\right) + \mu_2^2\left[q_2^2 - q_1^2 - Q_{11}^2 - Q_{12}^2\right]$$
$$+ \gamma_{11}^2(-P_1^2) + \gamma_{12}^2(-P_2^2) + \gamma_{21}^2(P_1^2 - P_1^2 max) + \gamma_{22}^2(P_2^2 - P_2^2 max)$$
$$+ \lambda_1^2(P_1^2 - P'^2_1) + \lambda_2^2(P_2^2 - P'^2_2)$$

correspondant à deux problèmes d'optimisation au niveau inférieur.

IV.2 Modèle linéaire

La résolution numérique de ce problème de répartition optimale de l'énergie dans un réseau hydroélectrique a été faite sur un ordinateur 1130-IBM travaillant en simple précision.

Niveau inférieur : au niveau inférieur, deux méthodes ont été mises en oeuvre pour résoudre les problèmes d'optimisation relatifs aux deux sous-problèmes :

- une méthode lagrangienne utilisant un algorithme quadratique (Newton-Ralphson) ;
- une méthode de pénalisation utilisant l'algorithme de Davidon.

Niveau coordonnateur : deux algorithmes ont été utilisés pour traiter les variables de coordination :

- un algorithme de type gradient
- un algorithme mixte (gradient-itération directe).

Résultats : 1° Algorithme du gradient

L'étude de la convergence de cet algorithme montre que la valeur optimale de la constante d'itération K est : $K^* = 0.00065$

Au bout d'une centaine d'itérations, l'erreur totale ε_T :

$$\varepsilon_T = \sqrt{(L\rho)^2 + (L\lambda_1^2)^2 + (L\lambda_2^2)^2 + (LP'^2_1)^2 + (LP'^2_2)^2}$$

est de l'ordre de : $\varepsilon_T \simeq 0.85$ et décroit très lentement (figure 5) à cause des pseudo variables P'^2_1, P'^2_2 et des paramètres de Lagrange associés λ_1^2, λ_2^2, qui varient extrêmement peu à chaque itération. Pour remédier à cette faible rapidité de convergence, on a utilisé un algorithme mixte : gradient sur ρ et itération directe sur les variables dont l'évolution est très lente : $\lambda_1^2, \lambda_2^2, P'^2_1, P'^2_2$. Cette méthode s'avère très efficace, comme on peut le constater sur les résultats présentés ci-dessous.

2° Algorithme mixte

Cet algorithme converge en un nombre minimum d'itérations pour $K^* = 0.0004$. L'erreur totale est toujours définie par :

$$\mathcal{E}_T = \sqrt{(Le)^2 + (L\lambda_1^2)^2 + (L\lambda_2^2)^2 + (LP'^2_1)^2 + (LP'^2_2)^2}$$

La figure 6 montre l'évolution au cours de la convergence de l'erreur (pour P_D = 10 MW).

La linéarité supposée du modèle des centrales simplifie beaucoup la résolution des optimisations locales. Mais un modèle plus proche de la réalité (modèle exponentiel ou de classe C1) peut aussi être utilisé.

IV.3 Modèle exponentiel

Au niveau inférieur, les deux méthodes décrites précédemment sont utilisées. Cependant, la méthode de pénalisation est plus performante.

Au niveau supérieur, seul l'algorithme mixte est retenu et les résultats obtenus sont, ici encore, très performants (cf. figure 7 qui donne l'évolution de l'erreur).

L'utilisation de ce modèle non linéaire, de type exponentiel, est plus réaliste et n'introduit aucune difficulté supplémentaire de mise en oeuvre. Elle conduit même à un temps de convergence plus faible.

IV.4 Modèle de classe C1

Au niveau inférieur, seule la méthode de pénalisation utilisant l'algorithme de Davidon a été retenue, car les méthodes procédant à un calcul direct du Hessien, sont oscillantes sur cet exemple.

L'algorithme mixte donne, ici encore, de bons résultats (cf. figure 8).

V - CONCLUSION

Dans cette communication, nous avons présenté une méthode mixte de décomposition-coordination des problèmes d'optimisation de grande dimension et défini les tâches de chaque niveau de commande.

Nous avons montré que le coordonnateur type gradient, toujours applicable, présente des conditions de stabilité et que le coordonnateur type Newton est toujours convergent, s'il est applicable.

Les conditions d'utilisation d'une coordination à itération directe ont été dégagées. Cette coordination apparaît intéressante pour le traitement des problèmes non séparables.

La résolution d'un problème de répartition optimale des énergies dans un système de production hydroélectrique (problème hautement non séparable et délicat à résoudre), nous a permis de mieux comparer, sur le plan des applications, certains de ces différents coordonnateurs. En particulier, l'efficacité de l'algorithme de coordination à itération directe ou mixte (itération directe + gradient) a été mise en évidence.

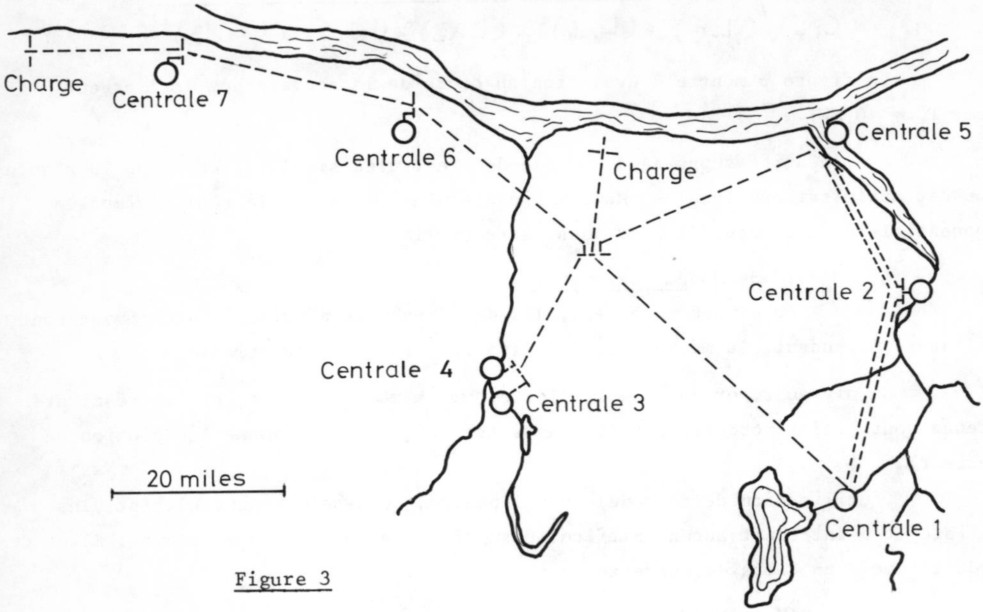

Figure 3

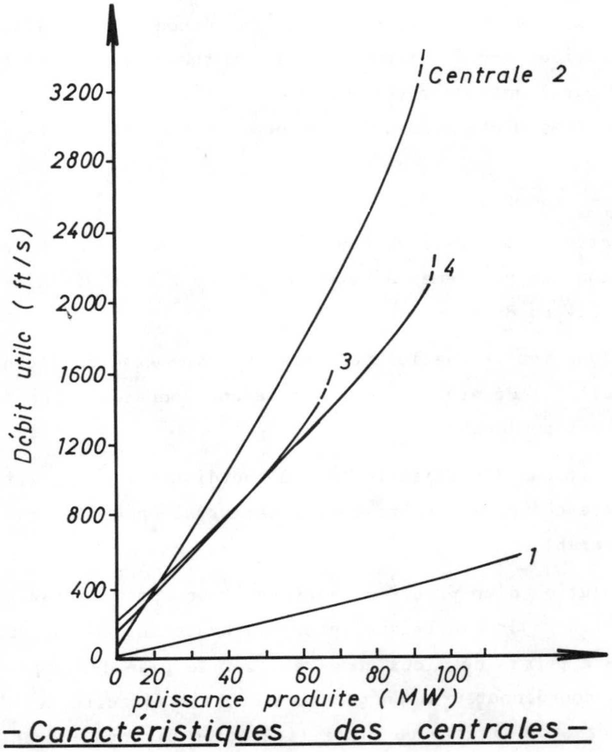

Figure 4

-Caractéristiques des centrales-

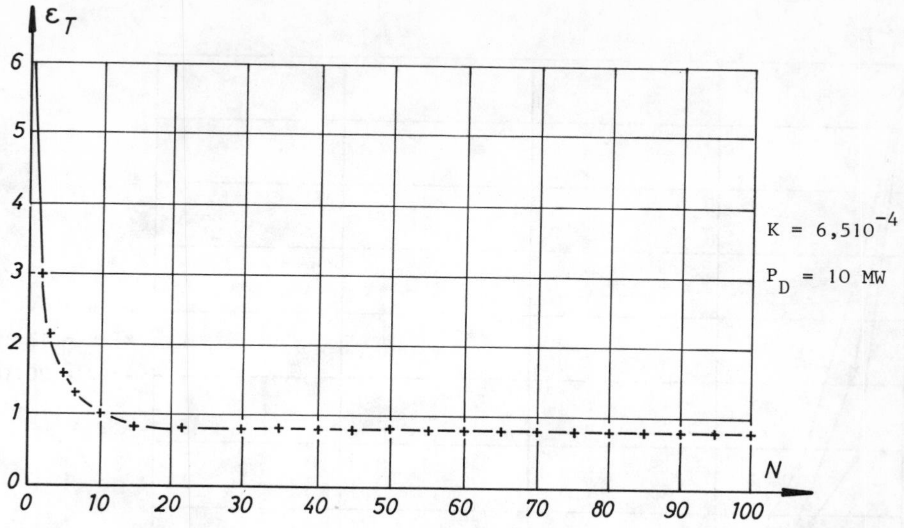

Figure 5 : Modèle linéaire. Algorithme du gradient. Evolution de l'erreur.

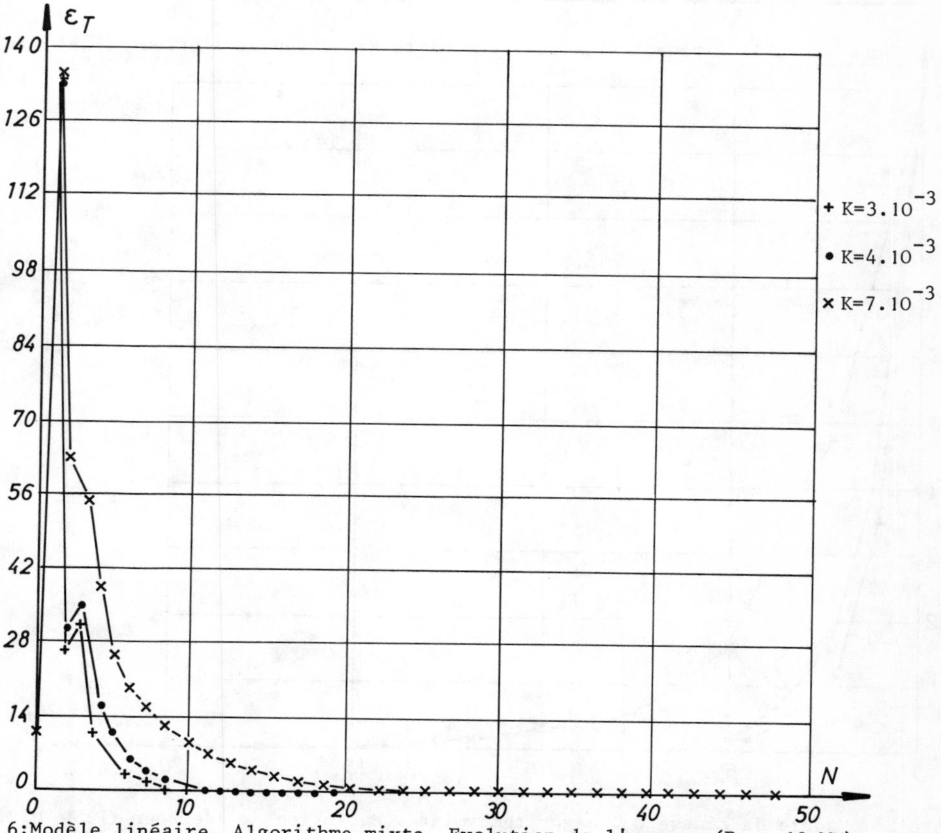

Figure 6:Modèle linéaire. Algorithme mixte. Evolution de l'erreur (P_D = 10 MW)

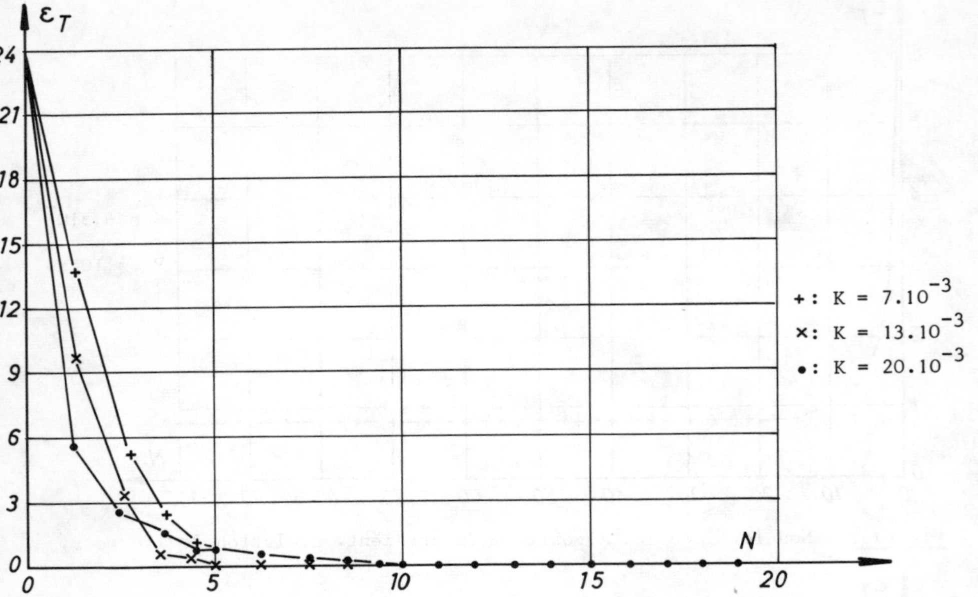

Figure 7:Modèle exponentiel.Algorithme mixte.Evolution de l'erreur (P_D=10 MW)

Figure 8 : Modèle de classe C_1. Algorithme mixte. Evolution de l'erreur (P_D = 10 MW)

BIBLIOGRAPHIE

ARROW K.J., HURWICZ L., UZAWA H. : Studies in linear and non linear programming. Stanford University Press. 1964.

GRATELOUP G., TITLI A. : Combined decomposition and coordination method in large dimension optimization problems. A paraître dans International Journal of Systems Science.

LEFEVRE T. : Etude et mise en oeuvre des algorithmes de coordination dans les structures de commande hiérarchisée. Thèse de Docteur-Ingénieur. Université Paul Sabatier, Toulouse, décembre 1972.

MANTERA I.G.M. : Optimum hydroélectric-power generation scheduling by analog computer. Proc. I.E.E. Vol. 118, n° 1, January 1971.

TITLI A. : Contribution à l'étude des structures de commande hiérarchisées en vue de l'optimisation des processus complexes. Thèse de Doctorat ès-Sciences Physiques. Université Paul Sabatier, Toulouse, juin 1972.

TITLI A., LEFEVRE T., RICHETIN M. : Multilevel optimization methods for non-separable problems and application. A paraître dans International Journal of Systems Science.

APPLICATIONS OF DECOMPOSITION AND MULTI-LEVEL TECHNIQUES TO THE OPTIMIZATION OF DISTRIBUTED PARAMETER SYSTEMS

Ph. CAMBON - L. LE LETTY

CERT/DERA - Complexe Aérospatial
TOULOUSE - FRANCE

ABSTRACT

The resolution of optimal control problems for systems described by partial differential equations leads, after complete (or semi) discretization, to large-scale optimization problems on models described by difference (or differential) equations that are often not easy to solve directly.

The hierarchical multi-level approach seems to be well suited to a large class of synthesis problems for these complex systems. Three applications will be presented here :

- minimum energy problem for the heat equation with distributed input (this is the classical test example)

- minimization of a steel index for the parabolic equation with one-sided heating

- reheating furnace with a non linear boundary condition (radiation)

INTRODUCTION

Solving an optimal control problem (or an identification problem) for systems described by partial differential equations leads, after complete discretization, to a large-scale optimization problem which is often difficult to solve directly in a global way.

The decomposition and multi-level hierarchical techniques with coordination seem well suited to the solution of a large class of these complex synthesis problems. Applications of these techniques will be made here to optimal control problems with the aid of two-level hierarchical structures.

I - GENERAL PRINCIPLES OF MULTI-LEVEL TECHNIQUES

Let us briefly present here the basic concept and the general principles of the hierarchical techniques. Associated first with the names of M.D. MESAROVIC as for the control aspect by hierarchical structures and L.S. LASDON and also some other authors as for the matehmatical programming aspect for large-scale optimization problems, these techniques are now actively studied for complex dynamic systems. In distributed parameter systems, some applications have also been studied by D.A. WISMER and Y.Y. HAIMES.

The basic idea consists for a too complex "system-objective function problem" to be solved directly in a global approach to define a number of subproblems by subsystems and subcriteria sufficiently simple to be efficiently treated by classical methods and algorithms and to coordinate the interconnected set by higher levels (one or more higher levels).

The control structure (or the optimization structure) consists then of units at several levels giving a pyramidal hierarchical structure (figure 1)

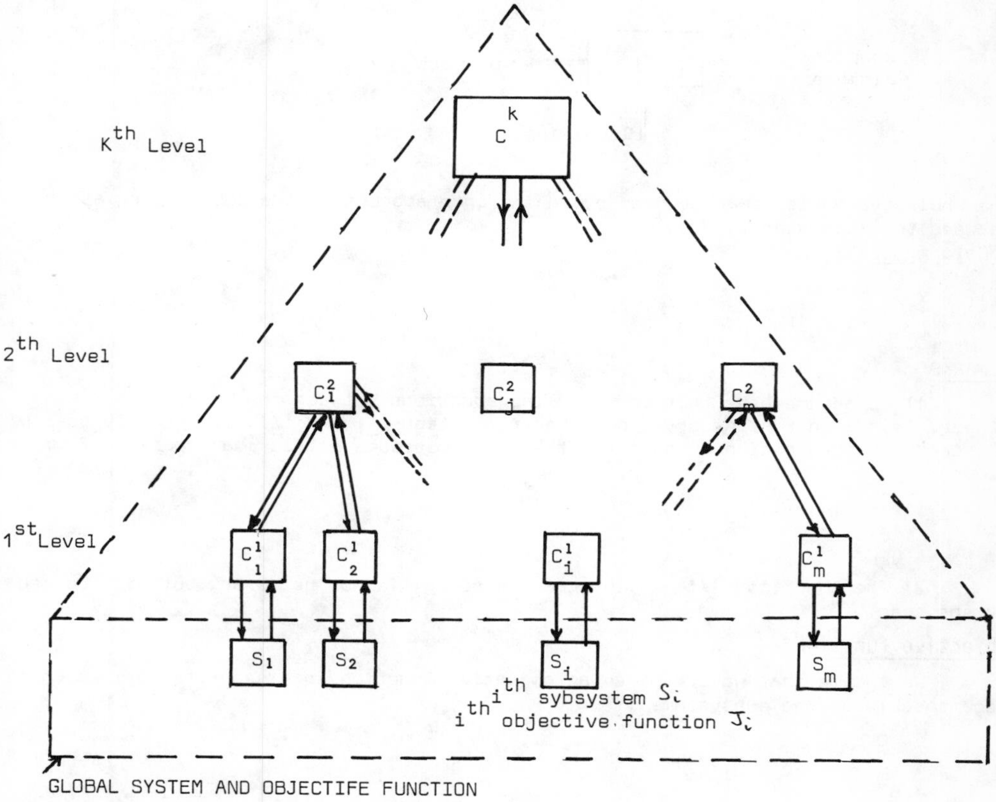

FIGURE 1

This structure is called "several levels-several objective functions" ; several objective functions for the control units at a given level have different objectives which can be moreover in conflict due to the fact that the subproblems are separately and independently solved at a lower level while in fact interconnected. The aim of the higher levels called higher level controllers is then to coordinate the set of operations of lower level in order to achieve the optimal solution of the whole problem at the top of the structure.

The classical structure is a two-level structure with a single unit at the second level. It is of course the simplest one but it is sufficient for most of the problems encountered.

II - TWO-LEVEL HIERARCHICAL STRUCTURE COORDINATION METHODS

Let us consider a system S which is decomposed into subsystems S_i, i = 1 to L.

The system S_i is represented by its model equation :

(1) $\underline{Z}_i = \underline{T}_i\,(\underline{X}_i, \underline{M}_i)$; $\underline{X}_i \in R^{p_i}$, $M_i \in R^{m_i}$, $\underline{Z}_i \in R^{q_i}$

where $\underline{Z}_i$ are the outputs, $\underline{M}_i$ the local inputs and $\underline{X}_i$ the coupling inputs from the other subsystems.

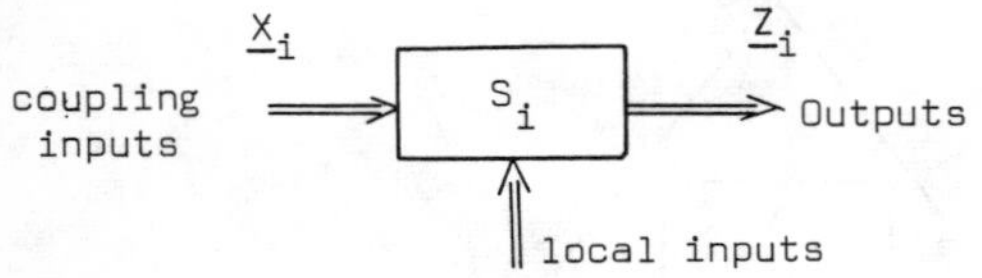

The whole system is reconstructed by taking into account the coupling equations between the subsystems.

Coupling equation

(2) $\underline{X}_i = \sum_{j\neq i} C_{ij}\,\underline{Z}_j$, i,j = 1 to L

REMARKS

1) We assume here that the coupling equations between the subsystems are linear. This is not necessary ; non linear equations $\underline{X}_i = \underline{C}_i(\underline{Z})$ are equally possible but in some coordination methods (as the non admissible method) a separable form is needed :

$$\underline{X}_i = \sum_{j\neq i} C_j\,(\underline{Z}_j)$$

2) The condition j≠i (no internal coupling) is not necessary but will be usually the case.

Objective function

We assume that we are given an objective function in a separate form which is decomposable on the subsystems :

$$J = \sum_{i=1}^{L} J_i\,(\underline{X}_i, \underline{M}_i, \underline{Z}_i)$$

which can be written (by (1))

(3) $$J = \sum_{i=1}^{L} J_i \,(\underline{X}_i, \underline{M}_i)$$

The global problem which is :
"Minimize J under the constraints (1) and (2), i = 1 to L" leads to the determination of the saddle point of the Lagrangian :

$$L = L\,(\underline{X}_i, \underline{M}_i, \underline{Z}_i, \underline{\mu}_i, \underline{\rho}_i)_{i=1,L}$$

(4) $$= \sum_{i=1}^{L} [J_i(\underline{X}_i,M_i) + \underline{\mu}_i^T \,(\underline{Z}_i - \underline{T}_i\,(\underline{X}_i, \underline{M}_i) + \rho_i^T\,(\underline{X}_i - \sum_{j\neq i} C_{ij}\,\underline{Z}_j)]$$

We assume that J_i and T_i are continus and have continuous derivatives with respect to the variables.

Then, the equations are :

$$i=1,L \begin{cases} \underline{L}x_i = \dfrac{\partial J_i}{\partial \underline{X}_i} - \left(\dfrac{\partial T_i}{\partial \underline{X}_i}\right)^T \underline{\mu}_i + \underline{\rho}_i = \underline{0} & (5) \\ \underline{L}M_i = \dfrac{\partial J_i}{\partial \underline{M}_i} - \left(\dfrac{\partial T_i}{\partial \underline{M}_i}\right)^T \underline{\mu}_i = \underline{0} & (6) \\ \underline{L}z_i = \underline{\mu}_i - \sum_{j\neq i} C_{ji}^T\,\underline{\rho}_j = \underline{0} & (7) \\ \underline{L}\mu_i = \underline{Z}_i - \underline{T}_i\,(\underline{X}_i,\underline{M}_i) = \underline{0} & (8) \\ \underline{L}\rho_i = \underline{X}_i - \sum_{j\neq i} C_{ij}\,\underline{Z}_j = \underline{0} & (9) \end{cases}$$

The application of the principles of two-level hierarchical techniques will consist here to split the treatment of these stationarity conditions into two levels in order that the first part correspond at the first level to a set of separated optimization problems and that the second level realize the coordination which is here the resolution of the remaining equations.

The second level controller will be of iterative type using the first level information after solving the sub-problems which will be done either also by an iterative scheme or by a direct method depending on the problem.

Three now well-known methods will be used.

1) Admissible" method

(or coordination by the $\underline{Z}_i$, also called coordination by the model).
The $\underline{Z}_i$, i = 1,L, are given to the first level by the second level controller.

At the first level, we solve :

$$\underline{L}x_i = \underline{L}M_i = \underline{L}\mu_i = \underline{L}\rho_i = \underline{0}$$

giving :

$$\begin{cases} \underline{X}_i = \underline{X}_i(\underline{Z}) \\ \underline{M}_i = \underline{M}_i(\underline{Z}) \\ \underline{\mu}_i = \underline{\mu}_i(\underline{Z}) \\ \underline{\rho}_i = \underline{\rho}_i(\underline{Z}) \end{cases} \qquad i = 1, L$$

This first part of the equations represents the subproblems :

Min $J_i\ (\underline{X}_i, \underline{M}_i)$

with :

$$\begin{cases} \underline{Z}_i = \underline{T}_i\ (\underline{X}_i, \underline{M}_i) \\ \underline{X}_i = \sum\limits_{j \neq i} C_{ij}\ \underline{Z}_j \end{cases} \quad , \ \underline{Z}_i \text{ given}$$

The local model and coupling equations are satisfied justifying the name "admissible".

The optimal $\underline{Z}$ remain to be found (coordination by the model) ; then the combination of the solutions of the sub-problems will achieve the optimal solution of the global problem with : $L = \sum\limits_i L_i$ and $J = \sum\limits_i J_i$.

At the second level, the equations $\underline{L}_{Z_i} = \underline{0}$ are then solved by an iterative scheme, for example, a steepest descent method :

$$\underline{Z}_i^+ = \underline{Z}_i - k\ \underline{L}_{Z_i} \quad , \quad i = 1, L$$

or a Newton-Ralphson method :

$$\underline{Z}_i^+ = \underline{Z}_i - k \left(\frac{\partial \underline{L}_{Z_i}}{\partial \underline{Z}_i} \right)^{-1} \underline{L}_{Z_i}$$

or some other optimization algorithm, with :

$$\underline{L}_{Z_i} = \underline{\mu}_i - \sum_{j \neq i} C_{ji}^T \cdot \underline{\rho}_j$$

which is calculated from the values of the variables given by the first level.

The algorithm is operating until $\| \underline{L}_{Z_i} \| \leq \varepsilon$. This coordination method is applicable when :

Dimension $(M_i) \geq$ Dimension $(\underline{Z}_i)$

condition which limits its applicability.

2) "Non admissible" method

(coordination by the $\underline{\rho}_i$, also called coordination by the objective function).

First level :

The $\underline{\rho}_i$ are fixed

$$\underline{L}_{X_i} = \underline{L}_{M_i} = \underline{L}_{Z_i} = \underline{L}_{\mu_i^\circ} = 0 \quad \text{gives :}$$

$$\begin{cases} \underline{X}_i = \underline{X}_i\ (\underline{\rho}) \\ \underline{M}_i = \underline{M}_i\ (\underline{\rho}) \\ \underline{Z}_i = \underline{Z}_i\ (\underline{\rho}) \\ \underline{\mu}_i = \underline{\mu}_i\ (\underline{\rho}) \end{cases}$$

Second level :

Determination of new values of $\underline{\rho}$ by the algorithm :

$$\underline{\rho}_i^+ = \underline{\rho}_i + \omega(\ )\ \underline{L}_{\rho i} \qquad , \qquad i = 1,L$$

where

$$\underline{L}_{\rho i} = \underline{X}_i - \sum_{j \neq i} C_{ij}\ \underline{Z}_j$$

In this method, when the optimal solution is not yet obtained, we have : $\underline{L}_{\rho i} \neq \underline{0}$: the coupling equations are not satisfied which justify the name "non admissible".

The global lagrangian can be written :

$$L = \sum_i [J_i + \underline{\mu}_i^T\ (\underline{Z}_i - \underline{T}_i) + \underline{\rho}_i^T\ \underline{X}_i - \underline{Z}_i^T \sum_{j \neq i} C_{ji}^T\ \underline{\rho}_j]$$

and is of a separate form for the variables considered at the first level. The corresponding optimization sub-problems are :

$$\begin{cases} \text{Min } [J_i + \underline{\rho}_i^T\ \underline{X}_i - \underline{Z}_i^T \sum_{j \neq i} C_{ij}\ \underline{\rho}_j] \\ \text{with} \\ \qquad \underline{Z}_i = \underline{T}_i\ (\underline{X}_i, \underline{M}_i) \end{cases}$$

This method does not imply any dimensionality condition on the variables in the general case where the model equation is non linear in $\underline{X}_i$ and $\underline{M}_i$ (second order terms are however needed, excluding non linearities of the form $|X_{ij}|$ or $|M_{ij}|$).

In the case where linearity occurs in $\underline{X}_i$ and $\underline{M}_i$, the objective function f_i needs to include non linear terms in $\underline{X}_i$ and $\underline{M}_i$ up to the second order at least. In practice, it is necessary to examine the compatibility of the equations or to reformulate the coordination method.

3) Mixed method

(or coordination by the $\underline{\rho}$ and the $\underline{Z}$).

We have :

At the first level ($\underline{\rho}$ and $\underline{Z}$ fixed)

$$\begin{cases} \underline{L}_{X_i} = \underline{0} \\ \underline{L}_{M_i} = \underline{0} \\ \underline{L}_{\mu_i} = \underline{0} \end{cases} \quad \text{giving} \quad \begin{cases} \underline{X}_i = \underline{X}_i\ (\underline{Z}, \underline{\rho}) \\ \underline{M}_i = \underline{M}_i\ (\underline{Z}, \underline{\rho}) \\ \underline{\mu}_i = \underline{\mu}_i\ (\underline{Z}, \underline{\rho}) \end{cases}$$

At the second level :

$$\begin{cases} \underline{Z}_i^+ = \underline{Z}_i - k_Z \underline{L}_{Zi} \\ \underline{\rho}_i^+ = \underline{\rho}_i + k_\rho \underline{L}_{\rho i} \end{cases} \quad , \quad i = 1,L$$

In the case where $\sum_i$ Dimension $(\underline{X}_i) = \sum_i$ Dimension $(\underline{Z}_i)$ a non iterative direct method can be used to solve :

$$\underline{L}_{\rho_i} = \underline{0} \quad \text{for the } \underline{Z}$$

and

$$\underline{L}_{Z_i} = \underline{0} \quad \text{for the } \underline{\rho}$$

from the values of the variables given by the first level.

III - APPLICATIONS. RELAXATION SCHEMES

We will give applications of the different coordination methods to optimal control problems arising from systems described by partial differential equations.

In order to gain in memory requirements on the computer and to gain also in convergence speed, we have been led to avoid the application of the multi-level approach in its usual conception by using relaxation schemes for the resolution of the first level equations. We use then for these equations the "new" values of the coordination variables $\underline{Z}_i$ and $\underline{\rho}_i$ for the next first level sub-problem (S_{i+1}, J_{i+1}).

IV - FIRST PROBLEM. OPTIMAL DISTRIBUTED INPUT FOR THE HEAT EQUATION WITH MINIMUM ENERGY

This is the classical test problem whose solution is well known and could be obtained by easier and faster ways (either using the adjoint equation or approximating the problem on a truncated basis of the eigenfunctions of the operator $\partial^2/\partial x^2$). It is also the example given by D.A. WISMER in a multi-level approach using the maximum principle after semi-discretisization and decomposition.

1) The problemn and discretization

$$\begin{cases} \dfrac{\partial y}{\partial t} = D \dfrac{\partial^2 y}{\partial x} + u(x,t) & \text{in } Q \\ y(x,0) = y_o(x) & \text{in } \Omega \\ y(0,t) = y(1,t) = 0 & \text{in } \Sigma \end{cases} \qquad \begin{array}{l} x \in [0,1] \\ t \in [0,T] \end{array}$$

Final condition : $y(x,T) = y_o(x)$

Cost function : $J(u) = \|u\|^2_{L^2(Q)}$

After discretization : t_i $(i = 1,N)$; x_j $(j = 1,M)$, we have :

$$\begin{cases} \dfrac{y_{ij} - y_{i-1,j}}{\Delta t} = \sigma\,(y_{i,j+1} - 2\,y_{ij} + y_{i,j-1}) + u_{ij} \qquad i = 2,N \quad j = 2,M-1 \\ \qquad \text{where } \sigma = D/(\Delta x)^2 \\ y_{1j} = y_{\circ j} \quad . \quad \text{initial condition} \\ y_{i1} = y_{iM} = 0 \quad : \quad \text{boundary conditions} \\ y_{Nj} = y_{dj} = \text{final desired condition} \\ J(\underline{u}) = \sum_i \sum_j u_{ij}^2\, \Delta x\, \Delta t \end{cases}$$

2) <u>Decomposition and coordination. First solution</u>

A first decomposition is to consider each node (i,j) of the discretization grid as a subsystem S_{ij}

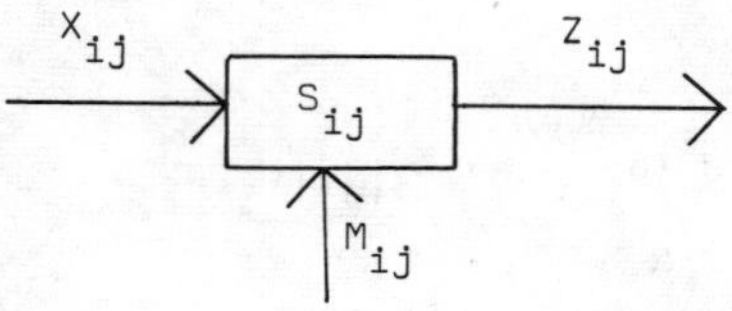

with :

$$\begin{cases} Z_{ij} = y_{ij} \\ X_{ij} = y_{i-1,j}/\Delta t + \sigma\,(y_{i,j+1} + y_{i,j-1}) \\ M_{ij} = u_{ij} \end{cases}$$

Then, we have :

<u>Model equation</u> :

$$Z_{ij} = (X_{ij} + u_{ij})/\alpha\ , \qquad \alpha = \frac{1}{\Delta t} + 2\sigma$$

<u>Coupling equation</u> :

$$X_{ij} = Z_{i-1,j}/\Delta t + \sigma\,(Z_{i,j+1} + Z_{i,j-1})$$
$i = 2,N$
$j = 2,M-1$

The objective function is :

$$J(\underline{u}) = \sum_{i=1}^{N} \sum_{j=1}^{M} u_{ij}^2 \quad \Delta x \quad \Delta t$$

equivalent to :

$$J(u) = \sum_i \sum_j u_{ij}^2$$

The global lagrangian is written :

$$L = \sum_{i=2}^{N} \sum_{j=2}^{M-1} \left\{ u_{ij}^2 + \mu_{ij} [Z_{ij} - (X_{ij} + u_{ij})/\alpha] + \rho_{ij} [X_{ij} - Z_{i-1/\Delta t} - \sigma (Z_{i,j+1} + Z_{i,j-1})] \right\}$$

$$+ \sum_{j=2}^{M-1} \lambda_{Nj} [Z_{Nj} - y_{dj}] + \sum u_{ij}^2 \quad \text{on the boundaries } i=1,\ j=1,\ j=M$$

The initial and boundary conditions are taken into account in the coupling constraints by their particular values, respectively for $i = 1$ and $j = 1$ and M.

The stationarity conditions for $1 < i < N$ are :

$$\begin{cases} L_{X_{ij}} = -\mu_{ij} / \alpha + \rho_{ij} \\ L_{u_{ij}} = 2\, u_{ij} - \mu_{ij} / \alpha \\ L_{Z_{ij}} = \mu_{ij} - \rho_{i+1,j}/\Delta t - \sigma\, (\rho_{i,j-1} + \rho_{i,j+1}) \\ L_{\mu_{ij}} = Z_{ij} - (X_{ij} + u_{ij}) / \alpha \\ L_{\rho_{ij}} = X_{ij} - Z_{i-1,j} / \Delta t - \sigma\, (Z_{i,j+1} + Z_{i,j-1}) \end{cases}$$

The admissible method can be used as Dim. M_{ij} = Dim. Z_{ij}. We have then :

At the first level :

After immediate direct resolution :

$$\begin{cases} X_{ij} = Z_{i-1,j} /\Delta t + \sigma\, (Z_{i,j+1} + Z_{i,j-1}) \\ u_{ij} = \alpha\, Z_{ij} - X_{ij} = (Z_{ij} - Z_{i-1,j})/\Delta t - \sigma\, (Z_{i,j+1} - 2\, Z_{ij} + Z_{i,j-1}) \\ \mu_{ij} = 2\, \alpha\, u_{ij} \\ \rho_{ij} = \mu_{ij/\alpha} = 2\, u_{ij} \end{cases}$$

At the second level

$$Z_{ij}^{+} = Z_{ij} - k\, L_{Z_{ij}}$$

where :

$$L_{Z_{ij}} = \frac{2}{\Delta t} (u_{ij} - u_{i+1,j}) - 2\, \sigma\, (u_{i,j+1} - 2\, u_{ij} + U_{i,j-1})$$

with

$$u_{ij} = (Z_{ij} - Z_{i-1,j})/\Delta t - \sigma\, (Z_{i,j+1} - 2\, Z_{ij} + Z_{i,j-1})$$

There remain to treat :

- for i=N the supplementary condition $L_{\lambda_{Nj}} = Z_{Nj} - y_{dj} = 0$ by $\lambda_{Nj}^{+} = \lambda_{Nj} + \omega(Z_{Nj} - y_{dj})$
- the terms $\Sigma\, u_{ij}^2$ on the boundaries i=1, j=1, j=M. Here, we obviously have : $u_{ij} = 0$

We have then (N-1) (M-2) coupled sub systems. Several examples have been done, with different step sizes : N=31,61 ; M=41,61. For N=31, M=41, we have 1170 sub-systems and 2370 variables (state variables and inputs) where 1170 variables are coordination variables.

In the relaxation scheme for the admissible method, a new Z_{ij} is calculated at each node $(i,j) \in \{ i=2,N\ ,\ j=2,M-1 \}$ and the $\underline{Z}$ already calculated are used in $L_{Z_{ij}}$.

Note here that we can solve for the Z_{ij} from $L_{Z_{ij}} = 0$. The memory requirements are only :

Tableau Z_{ij} $(i=1,N \; ; \; j=1,M)$

and

Vector λ_{Nj} $(j=2,M-1)$

The problem has been solved on an IBM 360-44 computer for different step sizes (cf. tables). For N=31 and M=41, 125 iterations are needed for a 10^{-3} relative precision on :

$$\sup_{i,j} \frac{L_{Z_{ij}}}{Z_{ij}} \quad \text{and} \quad \sup_{j} \frac{\lambda_{Nj}}{\lambda_{Nj}}$$ which is too long.

3) Second solution. New decomposition and mixed second-level controller

A more interesting decomposition consists to consider each column of the discretisation on the time axis as a sub-system with :

$$\underline{Z}_i = \{y_{ij} \quad , \quad j=2,M-1\}$$

$$\underline{M}_i = \{u_{ij} \quad , \quad j=2,M-1\}$$

Then, we have now for the vectors $\underline{Z}_i$, $\underline{u}_i$, $\underline{X}_i$ the same model and coupling equations as in the first solution :

$$\underline{Z}_i = (\underline{X}_i + \underline{u}_i)/\alpha$$

$$\underline{X}_i = \underline{Z}_{i-1}/\Delta t + \sigma\,(Z_{i,j+1} + Z_{i,j-1}) = (A\,\underline{Z}_i)$$

The mixed coordination method gives :

At the first level

$$L_{\underline{X}_i} = L_{\underline{u}_i} = L_{\underline{\mu}_i} = \underline{0}$$

which gives the explicit solution :

$$\begin{cases} \underline{\mu}_i = \alpha\,\underline{\rho}_i \\ \underline{u}_i = \underline{\rho}_i/2 \\ \underline{X}_i = \alpha\,\underline{Z}_i - \underline{u}_i \end{cases}$$

At the second level, we have :

$$i=2,N-1 \begin{cases} L_{Z_{ij}} = \alpha\,\rho_{ij} - \rho_{i+1,j}/\Delta t - \sigma\,(\rho_{i,j+1} + \rho_{i,j-1}) = 0 \\ L_{\rho_{ij}} = \alpha\,Z_{ij} - Z_{i-1,j}/\Delta t - \sigma\,(Z_{i,j-1} + Z_{i,j+1}) - u_{ij} = 0 \end{cases} \quad j=2,M-1$$

$$i=N \quad \left\{ L_{ZN_j} = \alpha\,\rho_{Nj}/2 + \lambda_{Nj} - \frac{\sigma}{2}\,(\rho_{N,j+1} + \rho_{N,j-1}) = 0 \right.$$

Then, a direct solution consisting of a sweep forwards on S_i from S_1 to S_n with $\underline{Z}_i = \underline{y}_0$ giving the $\underline{Z}_i$ from $L_{\underline{\rho}_i} = \underline{0}$ and of a sweep backwards from S_N to S_2 with λ_{Nj} given by $L_{Z_{nj}}$ and giving the $\underline{\rho}_i$ from $L_{\underline{Z}_i} = \underline{0}$ gives the solution.

The iterative scheme is then on λ_{Nj} :

$$\lambda^+_{Nj} = \lambda_{Nj} + \omega\,(Z_{Nj} - y_{dj})$$

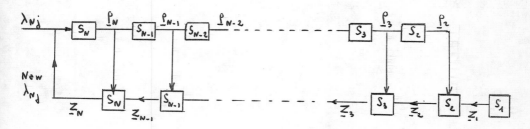

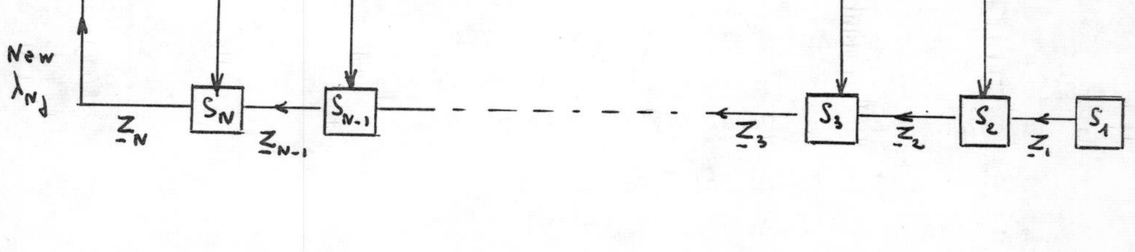

4) Results

Several cases have been solved some of which are shown in the tables. The initial condition $y_o(x)$ is a symetric triangle about $x = 0.5$ and the final condition $y_d(x)$ is a symetric polynomial of order 4 on x. The advantages of the relaxation scheme over the usual conception can be briefly described by the following :

1) Gain in memory requirements :
NM variables instead of 2 NM for the admissible method
NM+ 2M variables instead of 2 NM for the mixed method

2) Gain in computing time by a factor of 2 for both cases in this example (the mixed method seems to be superior in most cases).

V - STEEL INDEX OPTIMIZATION WITH ONE-SIDED HEATING AND INPUT CONSTRAINT

1) The problem

$$\begin{cases} \frac{\partial y}{\partial t} = D \frac{\partial^2 y}{\partial x^2} & x \in [0,L] \\ y(x,0) = y_o = cst & t \in [0,T] \\ \left.\frac{\partial y}{\partial x}\right|_{x=0} = 0 & \\ \left.\frac{\partial y}{\partial x}\right|_{x=L} = u(t) & \end{cases}$$

Cost function

$$J(u) = \int_0^T R(t)\ (u(t) - \gamma)^{\alpha}\ dt$$

Final conditions

$$y(0,T) = TM$$
$$y(L,T) = TS$$

where : $\alpha=2$

$$R(t) = \begin{cases} 0 \text{ if } u(t) \le \gamma \\ 1 \text{ if } u(t) > \gamma \end{cases}$$

Input constraint $\qquad u(t) \le u\ M$

This is the heating of a slab with a decarbonation performance index. An oxydation criterion of the same nature could be used in the same way. The quantity

$$Pd = \lambda \sqrt{\int_o^T R(t)\ (u(t) - \gamma)^2\ dt}$$ represents the decarbonation depth.

The numerical values are :

$\gamma = 850°C$, $uM = 1250°C$, $TM = 1150°\ C$, $TS = 1200°\ C$, $T = 6000$ s

2) Discretization. Decomposition and coordination

The previous second decomposition will only be considered. We have :

$$S_i = \begin{cases} \underline{Z}_i = \{y_{ij}\ ,\ j=1,M\} \\ \underline{X}_i = \{y_{ij}\ ,\ j=1,M\} \end{cases}$$

Model equation : $\underline{Z}_i = \underline{X}_i$

Coupling equation : $A\ \underline{X}_i = \underline{Z}_{i-1}$

with :

$$A : \begin{cases} \alpha\ X_{ij} - \tau\ (X_{i,j+1} + X_{i,j-1}) = Z_{i-1,j} & j=2,M-1 \\ \alpha\ X_{i1} - 2\ \tau\ X_{i2} = Z_{i-1,1} & j=1 \end{cases}$$

$$\tau = \sigma \times \Delta t \qquad , \qquad \alpha = 1 + 2\sigma$$

Lagrangian :

$$L = \sum_{i=2}^{N} \{R_i (X_{iM}-\gamma)^2 \times \Delta t + \frac{1}{\varepsilon}\ \text{Penal}\ (u_i) \times \Delta t \triangleq f_i$$
$$+ \underline{\rho}_i^T\ [A\ \underline{X}_i - \underline{Z}_i] + \underline{\mu}_i^T\ [\underline{Z}_i - \underline{X}_i]\}$$
$$+ \frac{1}{\varepsilon_1}\ [X_{N1} - TM]^2 \times \Delta t + \lambda_M\ [X_{NM} - TS]$$

where $\text{Penal}(u_i) = \begin{cases} 0 & \text{if} \quad u_i \le uM \\ (u_i - u_M) & \text{if} \quad u_i > uM \end{cases}$

We penalized the input constraint and the final condition $X_{N_1} = TM$

The variables are : X_{iM} , $\{X_{ij}, Z_{ij} \; ; \; j=1,M-1\}$, μ_{ij} $(j=1,M)$, ρ_{ij} $(j=2,M-1)$. The mixed method has been applied for : $i < N$, $j=1,M-1$ and $i=N$, $j=2,M-1$ and the non-admissible method has been applied for : $i = N$, $j=1$

First level :

$$\begin{cases} i < N \\ L_{X_{iM}} = -\mu_{iM} - \tau\,\rho_{i,M-1} + \dfrac{\partial f_i}{\partial X_{iM}} \\ L_{x_{i,M-1}} = -\mu_{i,M-1} + \alpha\,\rho_{i,M-1} - \tau\,\rho_{i,M-2} \\ L_{\mu_{ij}} = Z_{ij} - X_{ij} = 0 & j = 1,M \\ L_{X_{ij}} = -\mu_{ij} + \alpha\,\rho_{ij} - \tau\,(\rho_{i,j+1} + \rho_{i,j-1}) & j = 3,M-2 \\ L_{X_{i1}} = -\mu_{i1} + \alpha\,\rho_{i_1} - \tau\,\rho_{i_2} \end{cases}$$

gives $\underline{\mu}$ and $\underline{X}$

$$\begin{cases} i = N \\ L_{XN_1} = \alpha\,\rho_{i_1} - \tau\,\rho_{i_2} + \dfrac{2}{\varepsilon_1}(X_{N_1} - TM) \\ L_{XN_M} = -\mu_{NM} + \dfrac{\partial f}{\partial X_{NM}} + \lambda_M \end{cases}$$

gives X_{N1} and X_{NM}

Second level :

$$\begin{cases} i < N \\ L_{Z_{iM}} = \mu_{iM} \\ L_{Z_{ij}} = \mu_{ij} - \rho_{i+1,j} \; ; & j = 1,M-1 \\ L_{\rho_{ij}} = \alpha\,X_{ij} - \tau\,(X_{i,j+1} + X_{i,j-1}) - Z_{ij} \; ; & j=1,M-1 \\ L_{\rho_{i_1}} = \alpha\,X_{i_1} - 2\,\tau\,X_{i_2} - Z_{i-1,1} \end{cases}$$

$$\begin{cases} i = N \\ L_{Z_{Nj}} = \mu_{Nj} \; ; & j = 1,M \\ L_{\lambda_M} = X_{N_M} - TS \end{cases}$$

The solution (with ε and ε_1 fixed) is then as follows :

1) λ_M and ρ_{N1} fixed, $\underline{\rho}$ and $\underline{Z}$ are given by a direct forward-backward procedure.

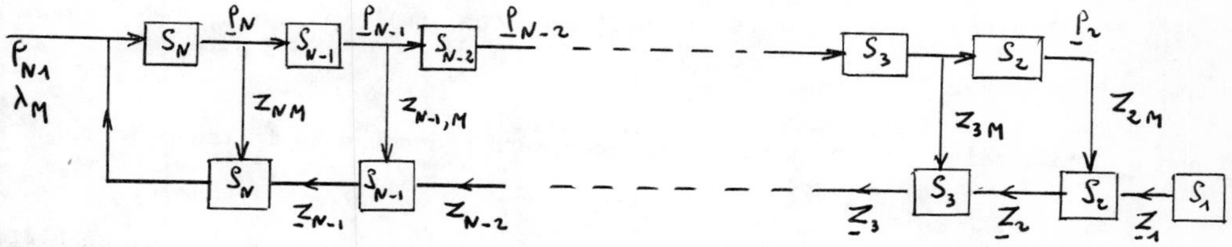

2) Then : $\begin{cases} \rho N_1^+ = \rho N + \omega\, L_{\rho N_1} \\ \lambda_M^+ = \lambda_M + \omega\,(X_{NM} - TS) \end{cases}$

RESULTS

An example is given in Table II (temperature at the surface and at x=0, input u(t)). The results are not sensitive to the step sizes. A comparison is made with a theoritical solution obtained after Laplace transformation and transforming the problem into a minimization of a time integral criterion with an integral constraint.

VI - NON LINEAR CASE. REHEATING FURNACE WITH A RADIATION BOUNDARY CONDITION

1) The problem. Decomposition

$$\begin{cases} \dfrac{\partial y}{\partial t} = D \dfrac{\partial^2 y}{\partial x^2} \\ y(x,0) = y_o(x) \\ \left.\dfrac{\partial y}{\partial x}\right|_{x=0} = 0 \\ \left.\dfrac{\partial y}{\partial x}\right|_{x=L} = \lambda \left[(u(t)+273)^4 - (y(L,t)+273)^4\right] \end{cases}$$

Final conditions $y(0,T) = TM$, $y(L,T) = TS$

Constraint : $u(t) \leq uM = 1400°C$

$$J(u) = \int_0^T R(t) \times [y(L,t) - \gamma]^2 \, dt \quad , \quad R(t) = \begin{cases} 0 & \text{if } y(L,t) \leq \gamma \\ 1 & \text{if } y(L,t) > \gamma \end{cases}$$

The same decomposition gives analogue model and coupling equations. The Lagrangian is :

$$L = \sum_{i=2}^{N} f_i + \pi_i \left[X_{iM} - X_{i,M-1} - \lambda\Delta \times (u_i + 273)^4 - (X_{iM}+273)^4\right]$$
$$+ \sum_{i=2}^{N} \Big\{ \sum_{j=2}^{M-1} \rho_{ij} \left[\alpha X_{ij} - \tau(X_{i,j+1} + X_{i,j-1}) - Z_{i-1,j}\right]$$
$$+ \rho_{i1} \left[\alpha X_{i1} - 2\tau X_{i2} - Z_{i-1,1}\right]$$
$$+ \sum_{i=2}^{N} \Big\{ \sum_{j=2}^{M} \Gamma_{ij} \left[Z_{ij} - X_{ij}\right] + \Gamma_{i,M+1} \left[Z_{i,M+1} - U_i\right]$$
$$+ \frac{1}{\varepsilon_1} \left[X_{N1} - TM\right]^2 + \lambda_M \left[X_{NM} - TS\right]$$

2) Coordination by the Mixed Method

The mixed method is used for $i < N$, $j = 1, M-1$ and $i = N$, $j = 2, M-1$.

The main difference is here we have, after first level resolution , to treat at the second level the terms :

$$\begin{cases} L_{u_i} = \dfrac{2}{\varepsilon} \text{ Penal } (u_i) \times \Delta t - 4_i\lambda\Delta x \times (u_i + 273)^3 \\ L_{X_{iM}} = -\tau \rho_{i,M-1} + \pi_i \left[1 + 4\lambda\Delta \times (Z_{iM} + 273)^3\right] \\ L_{\pi_i} = Z_{iM} - Z_{i,M-1} - \lambda\Delta x \left[(u_i + 273)^4 - (Z_{iM} + 273)^4\right] \end{cases}$$

The following procedure is used :

1) - If $u_i \geqslant uM$, $\varepsilon = 0$ and we have $u_i = uM$

Z_{iM} is obtained from $L_{\pi i} = 0$ (non linear equation)

π_i is obtained fro $L_{X_{iM}} = 0$

$\underline{Z_i}$ and $\underline{\rho_i}$ are obtained by the same forward-backward procedure as in the previous example.

2) If $u_i < uM$

Z_{iM} is obtained from $L_{X_{iM}} = 0$ (non linear equation), $\rho_{i,M-1}$ being known, after obtaining :

$$\pi_i = \frac{\text{Penal } (u_i) \times \Delta t}{2\lambda\Delta x (u_i + 273)^3} = 0 \text{ from } L_{u_i} = 0$$

u_i is obtained from $L_{\pi_i} = 0$ (non linear equation)

3) If u_i thus calculated is $\geq uM$ we go again to 1).
If $u_i < uM$, go to 2.

The stop criterion is on $L_{\rho N_1}$.

The results (shown on the figures III) have been obtained on an IBM 360-44 computer. With N=61, M=11, and a satisfactory initialisation of ρ_{N1} the execution time for the non linear case has been 7mn 24s where it was 3mn 7s for the linear case. Non linearity increases of course the number of iterations and the computing time but it however difficult to say to which accuracy the non linear equations at the second level should be solved during the evolution of the iterative procedure for the whole problem. The results are also little sensitive to step sizes variations.

VII - CONCLUSION

The application of multi-level hierarchical techniques seems to be very interesting to solve optimal control problems for systems described by partial differential equations. The difficulties in optimization problems related to the large dimensionality of the systems obtained after discretization of partial differential equations makes this approach promising (cf. also Sr. WISMER). Moreover this approach allows to attach in the some way problems with constraints, performance indexes different from the most usual case of quadratic functionals, and also non linear problems which will be often the case in pratical situations.

Here the applications have been made for optimization on a digital computer. It will be very interesting to approach these problems in view of a hybrid computation or in wiew of parallel computing. This will introduce different techniques of decomposition and coordination.

BIBLIOGRAPHY

[1] R.KULIKOVSKI *"Optimization of large scale systems"* Congrès IFAC Varsovie 1969

[2] D.A. WISMER *"And efficient computational procedure for the optimization of a class of ditributed parameter systems"* Journal of Basic Engineering June 1969

[3] Ph.CAMBON - JP.CHRETIEN - L.LE LETTY - A. LE POURHIET *"Commande et optimisation des systèmes à paramètres répartis"* Convention D.R.M.E n° 70 34 166, Lot n° 4 - C.E.R.T. - D.E.R.A.

[4] Ph.CAMBON *"Application du calcul hierarchisé à la commande optimale de systèmes régis par des équations aux dérivées partielles"* Thèse de Docteur Ingénieur - Université de TOULOUSE, Juillet 1972

[5] A.TITLI *"Optimisation de processus complexes"* (Cours donné à l'INSA de TOULOUSE 1971)

[6] R.KISSEL - G.GOSSE - M.GAUVRIT - D.VOLPERT - J.F. LE MAITRE *"Identification et optimisation des fours"* Convention D.G.R.S.T. n° 70 7 2507 Société HEURTEY - D.E.R.A.

[7] A.FOSSARD - M.CLIQUE - Mme N.IMBERT *"Aperçus sur la commande hierarchisée"* Revue RAIRO - Automatique n°3 - 1972

[8] Y.Y.HAIMES *"Décomposition and multi level approach in the modeling and management of water resources systèmes"* NATO Advanced study Institute on Decomposition as a tool for solving large scale problems CAMBRIDGE July 1972

I – PLANCHES I : PREMIER EXEMPLE
COMMANDE RÉPARTIE – ÉQUATION DE LA CHALEUR

APPLICATION DU CALCUL HIERARCHISE
METHODE ADMISSIBLE (précision 10^{-3})
Température Θ (x,t) t = 0, $\frac{T}{3}$, $\frac{2T}{3}$, T

t = 0

t = $\frac{T}{3}$

t = $\frac{2T}{3}$

t = T

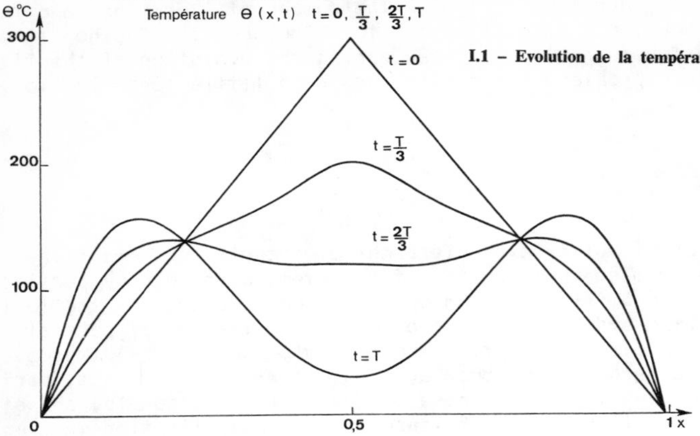

I.1 – Evolution de la température – Méthode admissible

APPLICATION DU CALCUL HIERARCHISE
METHODE ADMISSIBLE
Commande $\mathcal{U}$(x,t) t = 0, $\frac{T}{3}$ $\frac{2T}{3}$, T

t = T
t = $\frac{2T}{3}$
t = $\frac{T}{3}$
t = 0

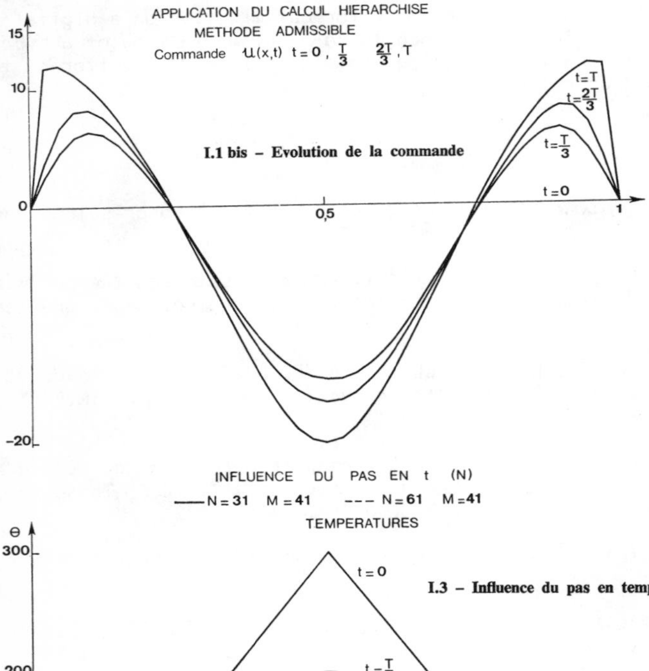

I.1 bis – Evolution de la commande

INFLUENCE DU PAS EN t   (N)
—— N = 31   M = 41   – – – N = 61   M = 41
TEMPERATURES

t = 0

t = $\frac{T}{3}$

t = $\frac{2T}{3}$

t = T

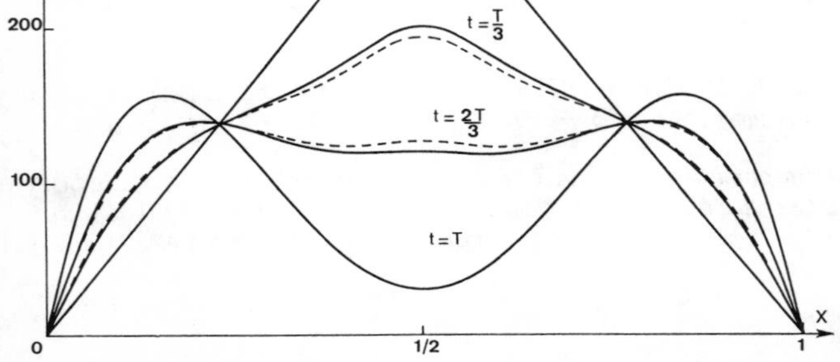

I.3 – Influence du pas en temps (température)

II – PLANCHES II : DEUXIÈME EXEMPLE
CRITÈRE MÉTALLURGIQUE – SIMPLE CONDUCTION

RESOLUTION PAR APPROCHE THEORIQUE

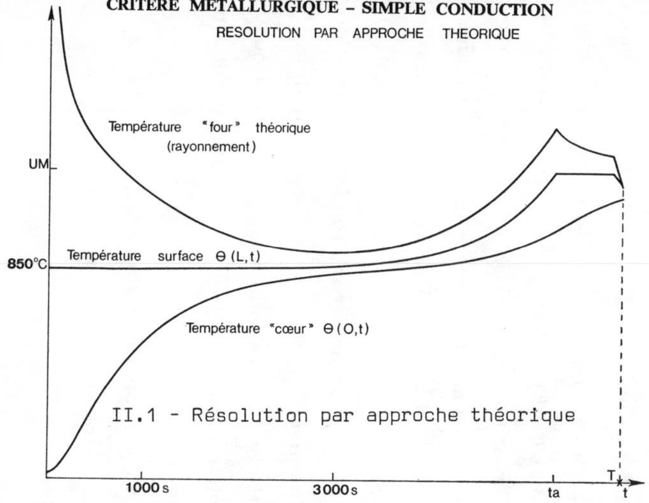

II.1 - Résolution par approche théorique

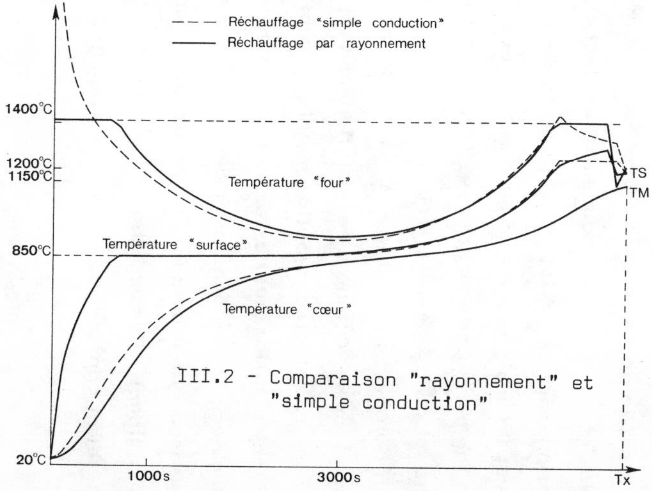

III.2 - Comparaison "rayonnement" et "simple conduction"

III – PLANCHES III : TROISIÈME EXEMPLE
FOUR DE RÉCHAUFFAGE – RAYONNEMENT

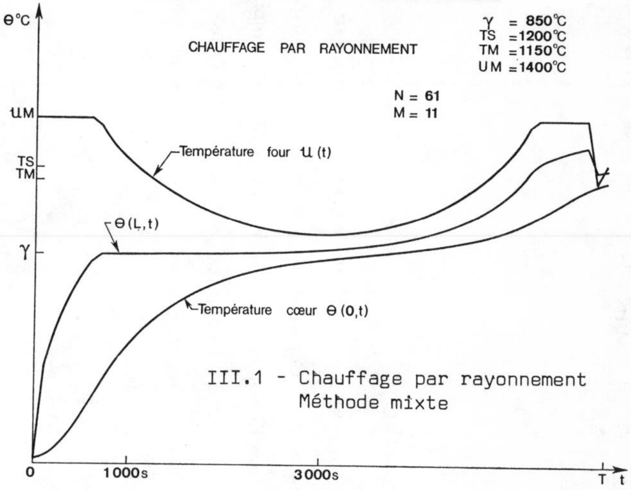

III.1 - Chauffage par rayonnement
Méthode mixte

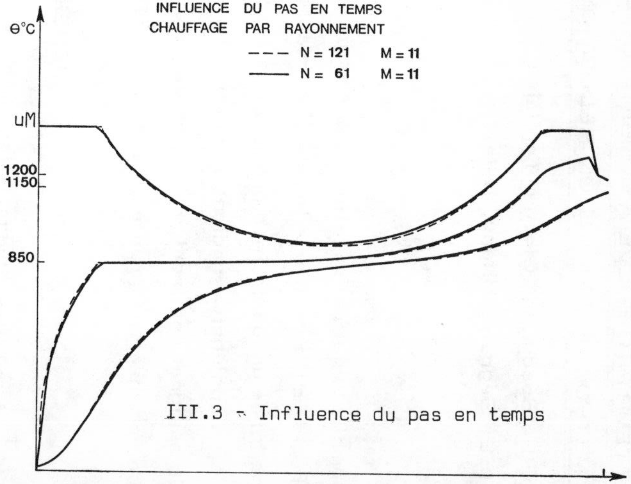

III.3 - Influence du pas en temps

ATTEMPT TO SOLVE A COMBINATORIAL PROBLEM IN THE CONTINUUM BY A METHOD OF EXTENSION-REDUCTION

Emilio Spedicato, Giorgio Tagliabue
CISE, Segrate, Milano, Italy

ABSTRACT

Combinatorial optimization problems in n variables are formulated as nonlinear programming problems in $(n-1)^2$ variables and n^2 constraints. Methods for solving the large unconstrained optimization problem generated are considered, with emphasis on conjugate-gradient algorithms based on the homogeneous model. The quadratic assignement problem is considered as an application example and results from the nonlinear programming approach are discussed.

GENERALITIES

Let P be an $n \times n$ permutation matrix and $f=f(P)$ a functional of P. The functional f is supposed to describe some property of a discrete model consisting of n elements which may be interchanged. The matrix P describes interchanges and a matrix P^* is sought which minimizes f. In principle, P^* can be determined by enumeration; in practice such a procedure is not feasible, even for relatively small values of n. Optimal and suboptimal techniques using only a subset of the n! permutation matrices exist for estimating the minimum of f when f has special structures; the main feature of such techniques is that they operate in the discrete space of the permutation matrices. It is possible, however, to operate in the continuum in the following way: we formally define f in R^{n^2} (extension) replacing P in its definition by a general $n \times n$ matrix X; we introduce a set of constraints (reduction) which make X to be a permutation matrix; we give an arbitrary (or in some fashion sensibly selected) starting matrix X_o and minimize f subject to the constraints. Every solution must be a permutation matrix and a local minimum for the continuous problem; if the local minimum is also a global minimum then it is an optimal solution (generally not unique!) to the original combinatorial problem. If it is only a local non global minimum, then it may be or not be a suboptimal solution.

A set of constraints defining a permutation matrix is the following:

(1) $x_{ij}(x_{ij}-1) = 0$ $\quad i, j = 1, 2, \ldots, n$

(2) $\sum_{i=1}^{n} x_{ij} = 1$ $\quad j=1, 2, \ldots, n-1$

(3) $\sum_{j=1}^{n} x_{ij} = 1$ $\quad i=1, 2, \ldots, n-1$

(4) $\sum_{i,j=1}^{n} x_{ij} = n$

Conditions (1-4) are necessary and sufficient for X to be a permutation matrix; see the Appendix for more about them.

System (1-4) consists of n^2+2n-1 equations, and the 2n-1 linear equations may be used to eliminate 2n-1 components of X. Therefore every combinatorial optimization problem which is a function of a permutation matrix can be expressed in the continuum by a nonlinear programming problem in $(n-1)^2$ variables and n^2 constraints.

In realistic problems of plant layout formulated as quadratic assignement n may be more than one hundred, and the associated nonlinear programming problem is a very large one. In order to deal with it through the penalty function approach, algorithms for unconstrained minimization in very many variables have to be developed.

CONJUGATE-GRADIENT ALGORITHMS FOR UNCONSTRAINED MINIMIZATION

Basic formulas

Algorithms for unconstrained minimization of a function F=F(z) of m variables with continuous first derivatives g=g(z) may use only function values (direct search methods) or also gradient values. Here we do not consider methods of the first kind, because they often have only linear rate of convergence and some of the most efficient of them (say Nelder and Mead [1] and Powell's [2] methods) have an $O(m^2)$ storage requirement. Newton's and Quasi-Newton's methods are very efficient algorithms using gradient values, but they have to be discarded for large problems because both storage requirement and time per iteration are $O(m^2)$. Fortunately a rather fast rate of convergence (often (m+1)-superlinear) and limited storage requirement can be obtained using conjugate gradient algorithms of the Fletcher-Reeves [3] kind. These methods are based on the iteration

(5) $z_{k+1} = z_k - a_k s_k$

where a_k is a scalar such that $F(z_{k+1}) \leq F(z_k)$ and the "search vector" s_k is defined as

$$
(6) \qquad \begin{aligned} s_o &= g_o \\ s_k &= g_k + b_k s_{k-1} \qquad (k=1,2,\ldots.) \end{aligned}
$$

and b_k is a scalar.

The following five choices of b_k are considered here:

$$
(7) \quad b_k = \frac{g_k^T g_k}{g_{k-1}^T g_{k-1}} \qquad \text{(Fletcher-Reeves [3])}
$$

$$
(8) \quad b_k = \frac{(g_k - g_{k-1})^T g_k}{g_{k-1}^T g_{k-1}} \qquad \text{(Polak-Ribiere [4])}
$$

$$
(9) \quad b_k = \frac{(g_k - g_{k-1})^T g_k}{(g_k - g_{k-1})^T s_{k-1}} \qquad \text{(Sorenson [5])}
$$

$$
(10) \quad b_k = \frac{g_k^T g_k}{g_{k-1}^T g_{k-1}} \left(\frac{F_k}{F_{k-1}} \right)^{t_k - 1} \qquad \text{(Fried [6])}
$$

where the scalar t_k is the nonzero solution of equation

$$
(11) \quad \left(\frac{F_k}{F_{k-1}} \right)^{t_k} = 1 - \frac{a_{k-1} s_{k-1}^T g_{k-1} t_k}{2F_{k-1}}
$$

a variation of formula (10) is obtained when the scalar t is the nonzero solution of equation

$$
(12) \quad \left(\frac{F_k}{F_{k-1}} \right)^{t_k} \left(1 + \frac{a_{k-1} s_{k-1}^T g_k t_k}{2F_k} \right) = 1 - \frac{a_{k-1} s_{k-1}^T g_{k-1} t_k}{2F_{k-1}}
$$

Methods using formulas (7), (8), (9) determine the minimum of a positive definite quadratic function in no more than m iterations when a_k satisfies the "exact linear search" condition

$$
(13) \quad g_{k+1}^T s_k = 0
$$

In such a case values of b_k defined by formulas (7), (8), (9) are the same, but they differ on nonquadratic functions or when equation (13) does not hold. Formulas (10-11) and (10-12) are the same if equation (13) is satisfied and coincide with formula (7) on quadratic functions. They allow to determine the minimum of a homogeneous function $F = \frac{1}{2r} (x^T Kx)^r$, where K is a positive definite matrix, in no more than m+1 iterations, if equation (13) holds. Formula (10-12) is a variation that we propose to Fried's method; its rationale is that if equation (12) is solved in stead of equation (11) then the last two search vectors are K-conjugate even if exact linear search is not performed.

The five methods have been incorporated in a polyalgorithm whose strategy is similar to that one adopted in a Quasi-Newton polyalgorithm previously described [7]; in particular a_k is determined by a parabolic search based on Fielding's [8] method; storage requirement on the IBM 1800 is less than 2500+4 m words; details are given elsewhere [9].

Numerical experiments

Extensive comparison of the five conjugate gradient methods described in the above section was made during the development of the polyalgorithm. Whereas a detailed analysis can be found elsewhere [10], the following comments are in order:

- more accurate linear search reduces the number of iterations, keeping the same precision in determining the minimum. This is especially marked on quadratic and homogeneous functions where exact linear search is a condition for termination. Therefore if m function evaluations are equivalent to one gradient evaluation, it is strongly recommended to use high precision in the linear search because the number of function evaluations that the search requires is substantially independent on m.
- the five algorithms behave similarly with a marginal superiority of the Polak-Ribiere and the Sorenson methods over the Fletcher-Reeves and of Fried's variation over the original method. The superiority of the last methods over the others is evident when high precision linear search is used, but is lost when low precision is adopted.
- some cases show that, where theoretically m iterations would be required for termination, in practice this number is larger when m is small but may be substantially less when m is large. This is a promising result for the goodness of the algorithm on functions in very many variables.

- strict termination is strongly sensible to exactness of the linear search.

Table 1 clearly evidences this for Fried's method. The function minimized was a homogeneous function with $r = 3, k_{ij} = d_{ij}$ (d_{ij} the Kronecker delta), starting point $x_o = (1, 2, \ldots, m)$ and m equal four or twenty. The exact value for a_k is given by formula

$$(14) \qquad a_k^* = (2rF_k)^{\frac{1-r}{r}} \frac{g_k^T s_k}{s_k^T K s_k}$$

- Fried's functional model is the function $F = \frac{1}{2r} (x^T kx)^r + q$, with q identically zero. If $q \neq 0$ we can obtain (m+2)-termination making initially two steepest descent searches, and using two equations of the type (11) to solve simultaneously for r and q. However experiments show that even if q is $\neq 0$ and large the efficiency of the method is not radically changed. A theoretical explanation is that if F_k converges to $F^* \neq 0$, then Fried's method tends to behave as Fletcher-Reeves method.

- Fried's method may be interpreted as scaling b_k in Fletcher-Reeves formula. Termination is kept scaling in the same way the Polak-Ribiere parameter and this modification gives marginal improvement to the algorithm.

TABLE 1 - Effect of exactness in the linear search

FUNCTION	METHOD	$a_k/a_k^{(*)}$	F	ITERATIONS
m = 4	Polak Ribiere	1	1E-6	5
		1.01	2E-5	4
		1.1	2E-8	4
m = 4	Fried's variation	1	3E-38	4
		1.01	7E-9	4
		1.1	4E-6	4
m = 20	Polak Ribiere	1	6E-6	13
		1.01	2E-6	13
		1.1	3E-6	12
m = 20	Fried's variation	1	5E-6	11
		1.01	5E-6	11
		1.1	9E-6	11

APPLICATION TO THE QUADRATIC ASSIGNEMENT PROBLEM

The quadratic assignement problem arises when n elements must be assigned to n different locations in order to minimize a cost function which can be written as

$$(15) \qquad f = \frac{1}{2} \sum_{i,j,k,l=1}^{n} x_{ij} x_{lk} a_{il} b_{kj}$$

where $x_{ij} = 1$ if elements i and j are interrelated, $x_{ij} = 0$ otherwise. Clearly the matrix $X \equiv \{x_{ij}\}$ is a permutation matrix whereas $A \equiv \{a_{il}\}$ and $B \equiv \{b_{kj}\}$ can easily be interpreted as "exchange" and "distance" matrices.

A compact writing of the cost function f is for instance

$$(16) \qquad f = \frac{1}{2} \operatorname{Tr}\left[X^T A X B \right]$$

The matrix of derivatives of f, $D \equiv \{d_{ij}\}$, where $d_{ij} = \frac{df}{dx_{ij}}$, can be written in the form

$$(17) \qquad D = \frac{1}{2} (AXB + A^T X B^T)$$

The quadratic function f generally is not positive definite; it can assume negative values, even if it assumes nonnegative values (f = 0 only for a degenerated problem) on the space of permutation matrices.

In the usual combinatorial framework the quadratic assignement problem is dealt with by optimal and suboptimal techniques. The first ones give a global minimum (which generally is not unique!), but they are unfeasible when n is as large as twenty, say. Such are implicit enumeration methods and Lawler's[11] reduction to a (larger) linear assignement. Suboptimal techniques are generally based on heuristics; well known methods have been published by Armour and Buffa [12], Gilmore [13], Hillier and Connors [14], Graves and Whinston [15]. Heuristic algorithms are feasible for problems up to fifty variables, say, and the solution is generally good. More information about these methods can be found elsewhere [16].

Finally, the implicit enumeration algorithm for minimizing a quadratic function in zero-one variables under quadratic constraints due to Hansen [17] might be usefully applied to the quadratic assignement; however experimental results are not known to us. In the solution of the nonlinear continuous programming problem we have to deal with two main points, once a method for unconstrained optimization is available. They are how to treat the constraints and how to choose the initial point.

Conditions (2), (3), (4) are readily eliminated, deleting the last row and column from matrix X and letting an (n-1) by (n-1) unknown matrix $\tilde{X}$. Derivatives $\tilde{d}_{ij}$ with respect to $\tilde{x}_{ij}$ are given by the formula

$$(18) \qquad \tilde{d}_{ij} = d_{ij} + d_{nn} - d_{in} - d_{nj}$$

Conditions (1) are dealt with in our approach by building the penalty function

$$(19) \qquad \tilde{f} = f + \sum_{i,j=1}^{n} k_{ij} \left[x_{ij}(x_{ij}-1)+h_{ij} \right]^2$$

where the "loss" coefficients k_{ij} are positive and both k_{ij} and the "correction" coefficients h_{ij} are constant for every unconstrained minimization. A Lagrangian function approach has been discarded as the number of equality constraints is greater than the number of independent variables. A question is how to change the k_{ij}'s and the h_{ij}'s in order to force convergence to a feasible point. We used two methods. In the first, following Miele et al.,[18] we have the h_{ij}'s identically zero while the k_{ij}'s are all equal to a value k; initially k=100 and then it is modified according to the formula

$$(20) \qquad k \leftarrow 10k \sqrt{(\tilde{f}-f)/ke}$$

where e is a convergence parameter. As Miele's formula gives fast rate of convergence to a feasible point we limited to four the number of cycles of unconstrained minimization. In the second method we followed Powell[19]. In the first cycle the h_{ij}'s are identically zero and the k_{ij}'s are all equal to $k = 100$. Then, according to some tests, the k_{ij}'s and the h_{ij}'s are modified at every cycle by formulas like multiplication by a constant or the mapping

$$(21) \qquad h_{ij} \leftarrow h_{ij}+x_{ij}(x_{ij}-1)$$

Convergence is supposed when $\max \left| x_{ij}(x_{ij}-1) \right| \leq .04$ and no more than 4 cycles are allowed.

Table 2 shows how a matrix X is generated (a case with n=6 and starting point P4). With both strategies the maximum number of computer operations required by the nonlinear programming problem is easily bounded. No more than four cycles are allowed; no more than 100(n−1) function evaluations are allowed in each cycle, corresponding to about 10(n−1) iterations and ten function evaluations per linear search. Now the number of operations required to calculate f grows as $7n^3+30n^2$, if three auxiliary matrices are used to store matrix products, otherwise it is $O(n^4)$; to calculate D this number grows as $7n^3+36n^2$, if two auxiliary matrices are used. We suppose that additions and multiplications take the same time and that shift time is negligible. Time required by conjugate gradient algorithm is $O(n^2)$, therefore for n sufficiently large the number of operations is bounded by $2801\, n^4$. The most efficient of Gilmore's algorithms requires $O(n^5)$ computer operations; CRAFT needs $O(n^3)$ operations per iteration; for small n the number of iterations is generally low; its probabilistic dependence on n is not known to us.

TABLE 2 – Generation of a permutation matrix

CYCLE	ITERATIONS	MATRIX X
1	28	.631, .065, .549, .037, –.183 .228, .376, .024, .401, –.075 .019, .053, .022, .202, .67 .09 , .43 , .43 , .14 , –.08 .05 , .06 , .00 , .12 , .73
2	23	1.16, –.1, –.18, –.03, –.09 –.16, –.17, .17, 1.21, –.44 .11 , .15 , .96, –.03, –.18 –.06, 1.2 , –.16, –.05, –.01 –.06, –.09, .1 , –.07, 1.1
3	20	1.00, 0, 0, 0, 0 0 , 0, 0, 1, –.02 0 , 0, .99, 0, 0 0 , 1, 0, 0, –.01 0 , 0, 0, 0, .99

Our bound is realistic; 2801 is a high coefficient, so only for large n, say $n > 100$, the time required by the continuum approach is estimated to be less than that required by heuristic procedures. An advantage of heuristic methods is that they calculate only variations of f due to exchange of elements which require $O(n)$ computer operations.

The second main point is how to choose the initial matrix X. It is well known that if the object function is not convex, then different initial points may lead to different minima. In our problem every feasible point is a local minimum and also there may be many global minima. For quadratic assignement problems arising from plant lay out multiplicity of global minima is often a consequence of geometrical symmetry. In more general cases a rough estimate of the number of global minima can be done as follows. Let us assume that A and B are integer valued and $\max |a_{ij}| \leq p$, $\max |b_{ij}| \leq p$, p being an integer. Then on the space of permutation matrices it holds

$$(22) \qquad 0 \leq f \leq \frac{1}{2} p^2 n^2$$

Assuming that the integer values of f are distributed uniformly on the space of permutation matrices then the average number of global minima is $2^2.3.4.\ldots(n-2)/p^2$. This implies that even if different starting points may give global minima, generally we can expect only local minima. The choice of the starting points could be made using values given by combinatorial procedures; however we did not explore this possibility(*) and we chose arbitrarily initial unfeasible points. The five choices considered are given in Table 3.

TABLE 3 Starting points

CASE	x_{ij}
P1	10 if $j + i(n-1)$ is even, otherwise −10
P2	0 if $\sin[j+i(n-1)] \leq .5$, otherwise 1
P3	Change inequality sign in case P2
P4	$1/(n-1)^2$
P5	Stochastic sequence 0, 1

Results are given in Table 4 for n=3, 5, 6, 7. They were the same for Miele's and Powell's methods, except a few cases which can be explained by the fact that when there are many minima, the actual one calculated may depend on the sequence of the penalty coefficients. Matrices A and B are found in Nugent et al.[20]. For n=7 (36 variables and 49 constraints) the time of execution is about twenty minutes on the IBM 1800.

Notation n. m. means that the case was not considered; n. c. means that convergence to a permutation matrix was not obtained. In this case the final matrix showed some identical rows.

TABLE 4 Results

n	Nb.	N. w.	P1	P2	P3	P4	P5
3	n. m.	n. m.	15	17	15	25	17
5	25	31	n. c.	34	35	n. c.	30
6	43	46	49	62	52	49	43
7	74	84	n. m.	96	n. c.	84	n. m.

We can observe that under certain conditions a penalty function approach using conjugate gradient algorithms cannot work. For instance, if $a_{ij}=b_{ij}= 1 - d_{ij}$ (d_{ij} the

---

(*) the knowledge of derivatives of f which is a byproduct of the continuum approach could be used in the combinatorial heuristics to suggest which elements should be interchanged

Kronecker symbol) and initially $x_{ij} = r$, then every variable is changed by the sa me amount at every iteration and a permutation matrix is not generated.

Column headed N.b. and N. w. contain the best and the worst results quoted by Nugent for some heuristic procedures.

The following conclusions can be made:

1) the nonlinear programming approach generally gives a permutation matrix in two or three cycles
2) the final matrices do not usually correspond to global minima and values of f are rather scattered
3) the heuristic combinatorial procedures are therefore superior both for quality of solution and time of execution in the range considered for n. For larger n, where the continuum approach might become competitive, the computer time is unfortunately too demanding to make experiments possible.

CONCLUSION

Combinatorial optimization problems have been expressed as nonlinear programming problems and efficient techniques for handling the large unconstrained minimiza tion problem arising have been considered. The choice of the initial points is a cri tical problem which has not been solved satisfactorily. Solutions for the quadratic assignement problem are inferior to those given by combinatorial techniques; in fact, the necessity of exploring a succession of local minima reproduces in a certain sense the original combinatorial problem.

APPENDIX

The definition of a permutation matrix X is equivalent to say that the elements of X must satisfy equations

(1.1) $\quad x_{ij}(x_{ij}-1) = 0 \qquad i, j = 1, 2, \ldots, n$

(1.2) $\quad \sum_{i=1}^{n} x_{ij} = 1 \qquad j = 1, 2, \ldots, n$

(1.3) $\quad \sum_{j=1}^{n} x_{ij} = 1 \qquad i = 1, 2, \ldots, n$

Summing equations (1.2) and (1.3) respectively over j and i we obtain identities. Therefore system (1-4) is readily generated. Also the following Theorem holds:

Theorem: Conditions (1-4) are not redundant.

Proof. Suppose firstly that one of relations (1), say $x_{lm}(x_{lm}-1) = 0$ can be deleted. Put $x_{li}=1$ for $i \neq m$, $x_{jm}=1$ for $j \neq l$, $x_{lm} = -n+2$ and all remaining x_{ij}'s equal to zero. Then all conditions are satisfied but the matrix is not a permutation matrix.

Without loss of generality, suppose now that one of equations (2), say $x_{il}=1$, can be deleted. System (2-4) can be written in this case

(1.4) $\sum_{j=1}^{n} x_{ij} = 1$ $\quad i = 1, 2, \ldots . n$

(1.5) $\sum_{i=1}^{n} x_{ij} = 1$ $\quad j = 1, 2, \ldots . n$, $j \neq l, l+1$ if $l < n$, otherwise $j \neq l, l-1$.

Now put the diagonal elements of X equal to one except $x_{l+1, l+1}$, if $l < n$, otherwise $x_{l-1, l-1}=1$ if $l = n$. Put $x_{l, l+1} = 1$ if $l < n$, otherwise $x_{l, l-1} = 1$. Put all other x_{ij}'s equal to zero.

Then conditions (1.1-1.4-1.5) are satisfied but the resulting matrix is not a permutation matrix.

Finally, observe that a permutation matrix satisfies the unitarity condition $X^T X = XX^T = I$. This equation is easily derived from equations (1-3), its nonlinearity being connected with the nonlinearity of equation (1.1).

REFERENCES

1. Nelder, J. A. and Nead, R.: A simplex method for function minimization, Comput. J., 7, 308-313, 1965
2. Powell, M. J. D.: An efficient method of finding the minimum of a function of several variables without calculating derivatives, Comput. J., 7, 155-162, 1964
3. Fletcher, R. and Reeves, C. M.: Function minimization by conjugate gradients, Computer J., 7, 149-154, 1964
4. Polak, E. and Ribiere, G.: Note sur le convergence de methodes des directions conjugees, University of California, Berkeley, Dept. of Electrical Engineering and Computer Sciences, working paper, 1969
5. Sorenson, H. W.: Conjugate Direction Procedures for Function Minimization, Journal of the Franklin Institute, 288, 421-441, 1969
6. Fried, I.: N-step Conjugate Gradient Minimization Scheme for Nonquadratic Functions, AIAA Journal, 9, 2286-2287, 1971
7. Spedicato, E.: Un polialgoritmo per la minimizzazione di una funzione di più variabili, Atti del Convegno AICA su Tecniche di Simulazione e Algoritmi, Milano, Informatica, Numero speciale, 1972
8. Fielding, K.: Function minimization and linear search, Algorithm 387, Commun. of ACM, 13, 8, 1970

9. Spedicato, E.: Un polialgoritmo a gradiente coniugato per la minimizzazione di funzioni nonlineari in molte variabili, Nota tecnica CISE-73.012, Milano, 1973

10. Spedicato, E.: CISE-Report to appear

11. Lawler, E.L.: The Quadratic Assignement Problem, Management Sci, 9, 586-599, 1963

12. Armour, G.C. and Buffa, E.S.: A Heuristic Algorithm and Simulation Approach to Relative Location of Facilities, Management Sci., 9, 294-309, 1963

13. Gilmore, P.C.: Optimal and Suboptimal Algorithms for the Quadratic Assignement, SIAM J., 10, 305-313, 1962

14. Hillier, F.S. and Connors, M.M.: Quadratic Assignement Problem Algorithms and the Location of Indivisible Facilities, Management Sci., 13, 42-57, 1966

15. Graves, G.W. and Whinston, A.B.: An Algorithm for the Quadratic Assignement Problem, Management Sci., 16, 453-471, 1970

16. Casanova, M. and Tagliabue, G.: CISE-Report to appear

17. Hansen, P.: Quadratic Zero-One Programming by Implicit Enumeration, in Numerical Methods for non-linear Optimization, (F.A. Lootsma, ed.), Academic Press, 1972

18. Miele, A., Coggins, G.M. and Levy, A.V.: Updating rules for the penalty constant used in the penalty function method for mathematic programming problems, Aero-Astronautics Report n. 90, Rice University, Houston, 1972

19. Powell, M.J.D.: A Method for Nonlinear Constraints in Minimization Problems, in Optimization, (R. Fletcher, ed.), Academic Press, 1969

20. Nugent, C.E., Vollmann, T.E. and Ruml J.: An experimental comparison of techniques for the assignement of facilities to locations, Operations Research, 16, 150-173, 1968.

This work was supported by the Consiglio Nazionale delle Ricerche, in the framework of the research contract CISE/CNR n. 71.02207.75- 115.2946